Liquefied Natural Gas Project Management

（Volume Ⅰ）

Xing Yun

Petroleum Industry Press

内容提要

本书全面总结了一般项目管理的九大知识体系：集成管理、范围管理、计划管理、质量管理、费用管理、人力资源管理、信息管理、风险管理、采购/合同管理。除此以外，还特别撰写了具有LNG项目特点的征地及站址比选、可行性研究及报告编写、健康安全环保（HSE）管理、财务管理、LNG远洋运输管理、验收管理、试生产运行管理这七个方面管理内容，构成了LNG项目更完善的管理体系。

本书是从事LNG项目管理和技术人员必不可少的一本参考书和实用手册，也可作为大学项目管理院系相关专业教师、研究生、高年级本科生的一本参考教材。

图书在版编目（CIP）数据

液化天然气项目管理 =Liquefied Natural Gas Project Management：英文/邢云著，—北京：石油工业出版社，2021.5

ISBN 978-7-5183-3434-6

Ⅰ.①液… Ⅱ.①邢… Ⅲ.①液化天然气－天然气输送－项目管理－英文Ⅳ.①TE83

中国版本图书馆CIP数据核字（2020）第136398号

出版发行：石油工业出版社

（北京安定门外安华里2区1号　100011）

网　址：www.petropub.com

编辑部：（010）64523541　图书营销中心：（010）64523633

经　销：全国新华书店

印　刷：北京中石油彩色印刷有限责任公司

2021年5月第1版　2021年5月第1次印刷

787×1092毫米　开本：1/16　印张：75

字数：1870千字

定价：360.00元（全两册）

（如出现印装质量问题，我社图书营销中心负责调换）

Content Summary

China National Offshore Oil Corporation (CNOOC) is Chinese first set foot in liquefied natural gas (LNG) of the national company. At present, Guangdong Dapeng, Fujian and Shanghai and other LNG projects have been put into production, there are a number of LNG projects under construction and preliminary research, creating a Chinese LNG industry.

"Liquefied natural gas project management" is the first English book in the LNG industry, the book seeks to combine theory with practice, from the actual LNG project management to set up the theoretical systems, a comprehensive summary of the general project management 9 knowledge systems, including integrated management, scope management, schedule management, quality management, cost management, human resource management, information management, risk management, procurement and contract management. In addition, the book also write the related contents with special LNG project features, such as land acquisition and site selection management, feasibility study and project check and approval management, HSE management, financial management, LNG marine transportation management, acceptance and post evaluation management, and trial production and operation adjustment management of these 7 aspects, which constitutes the 16 LNG project management systems.

The book is engaged in LNG project management and technical personnel essential reference book and practical manual, also can be used as the faculties of the university project management department related professional teachers, graduate students, undergraduate students a reference material.

Preface

Liquefied Natural Gas Project Management is the previous publication of "*LNG Project Management Manual*" for the second edition in English. Today the publication of this work is the industry once again works worth celebrating !

China National Offshore Oil Corporation (CNOOC) is the first national corporation to set foot in the introduction of liquefied natural gas projects in the country. CNOOC leadership with strategic vision, made a wise decision as early as more than 20 years ago, promotes the development of the LNG industry, the emerging field in our country. Subsequently, Guangdong, Fujian, Shanghai, Hainan LNG projects have been put into production, and there are a number of LNG projects under construction and preliminary research, writing a new chapter in Chinese LNG industry.

At that time, I also wrote a preface for the "LNG Project Management Manual" . I mentioned: the "LNG Project Management Manual" is based on the summary of the LNG projects which have been put into operation and under construction and the preliminary studies, it is a new work of project management combining theory and practice, which fills a gap in the industry. Now, English version of "*LNG Project Management*" is published when a number of LNG projects are completed and put into production in China, in addition to the CNOOC, LNG projects of CNPC, Sinopec and other social units also have been put into operation, showing the Chinese coastal area LNG receiving terminals scattered like stars, LNG tanks stand tall and straight, trunklines stretched ups and downs to formed the huge industrial scene. With the rise of Chinese LNG industry, the international exchange of the relevant LNG industry, international cooperation are in the ascendant, international LNG forum hosted by the Chinese also affects the world. To this end, the publication of the English version of "*LNG Project Management*" , for Chinese technical and management experts and foreign counterparts' exchange and cooperation, will play the bridge and positive role.

Mr. Xing Yun is a veteran of CNOOC. He is in charge of the LNG business research management, standardization management, human resource management and project management, by more than 10 years of hard work and effort in his spare time, he wrote the work. The publication of the work is an important milestone in China LNG industry as well as the National Development and Reform Commission (NDRC), various ministries and commissions of the state, the provincial government where LNG project are located and the CNOOC after more than 20 years of hard work, in addition to the LNG project hardware construction, and in the critical moment by his hand, to publish the theoretical results—software, the work embodies all the Chinese LNG project participants: all researchers, designers, builders, managers and operators' common struggle of the classic works. Here I would like to point out some early participants, now retired veteran of CNOOC, the old experts, first-line management and technical personnel, because their selfless dedication, hard work and valuable experience, make the work more profound and detailed.

The characteristic of the work is: combining theory and practice, from the actual LNG project management to set up the theoretical system, the general project management theory has been sublimated and improved, a comprehensive summary of the general project management 9 knowledge systems, including: integrated management, scope management, schedule management, quality management, cost management, human resources management, information management, risk management, and procurement / contract management. In addition, the work also wrote the related contents with special LNG project features, such as land requisition and site selection management, feasibility study and project check and approval management, HSE management, financial management, LNG marine transportation management, acceptance and post evaluation management, trial production and operation adjustment management of the 7 contents constituted a more complete LNG project management systems. It summarizes the management process from each of the management content, namely the 3 stage process, which are the basis and management system; actual operation procedure management; and control process and objective effect management. This is more convenient for LNG project management and technical personnel to master and apply, but also for other similar engineering projects for reference.

Through the preparation of the work, summary, finishing a large number of valuable engineering information and project management experience and in the concepts of project management to increase the new connotation, and achieved gratifying results. But we also recognize that Chinese LNG introduction of project management is a huge system engineering, there are many aspects waiting for us through other LNG projects and continuous improvement, the project from microcosmic to macroscopic and internal relations need to be summarized and improved, but we finally have the first English version of the LNG project management book, and it will throw out a minnow to catch a whale. We hope that through the work publication, there will be more and better LNG project management theory literature and work will be published.

The work is based on the previous "*LNG Project Management Manual*" on the second edition, with English writing, this will help Chinese LNG industry technical and management experts to technical exchanges with foreign counterparts, through 20 years of project practice, China has fully mastered the LNG project from design, construction, operation of a full set of technology and management methods, the work for Chinese LNG industry going to the world and contracting foreign LNG project construction will also play a role in demonstration and guidance.

Academician of Chinese Academy of Engineering: Zeng hengyi

Foreword

The introduction of liquefied natural gas (LNG) for improving Chinese energy structure and its efficiency, energy conservation and emission reduction, environmental protection has played a decisive role. The National Development and Reform Commission (NDRC) put the development and utilization of natural gas on the agenda 20 years ago, China Offshore Oil Corporation (CNOOC) responded the call of the government, actively implemented the Chinese government's emission reduction targets, in the introduction of LNG, the project construction and operation, made a breakthrough and made a national company contribution that should be done for the country. Currently, CNOOC has been completed and put into production: Guangdong, Fujian, Shanghai, Zhejiang, Tianjin and Hainan LNG project, and a number of projects have entered the construction and early stage of research.

We cannot forget, entrusted by the NDRC, CNOOC began to introduce foreign LNG industry research since middle 1990s. By December 1996, Chinese LNG industry pioneer—CNOOC led through detailed investigation for a whole year, offered the NDRC a weighty "Southeast Coastal Areas Using Liquefied Natural Gas and Project Planning Report" . In October 1998, the State Council approved the first LNG pilot project in Guangdong Province, and CNOOC carried out preparatory work for the project, since then opened the prelude of the introduction of LNG in Chinese. At that time, in the face of the emerging LNG industry, the project team overcame many difficulties, learned warfare through warfare, and consulted the foreign counterparts, gradually explored a way to develop the LNG project in Chinese. Now, when we face the LNG tank stands tall and straight, scattered and orderly gasification facilities, wave rolling pipeline network, and info graphic symbolized electronic control center, we can proudly say that our efforts were not in vain !

We cannot forget the morning of June 28,2006, Premier Wen Jiabao and Australian Prime Minister Howard came to the beautiful Guangdong Shenzhen Dapeng Bay, attended Chinese first import LNG project–Guangdong Dapeng LNG project commissioning ceremony. Two premiers jointly pressed the start button, announced that the Guangdong Dapeng LNG receiving terminal officially was put into operation, the two premiers made speeches respectively, for congratulations of mutually beneficial cooperation between China and Australia in Guangdong Dapeng LNG project operation smoothly. Since then, Chinese LNG industry was born !

We cannot forget that in the middle 1990s, the NDRC stood at the height of the national level, from the perspective of improving Chinese energy structure, protecting the environment and promoting the rapid and healthy development of local economy, all along leading and driving to the LNG projects, in the overall leadership and coordination of the LNG projects, made full use of the development strategy of domestic and foreign two kinds of resources and two markets, played a major role. Under the "By domestic market exchanges resources and technology", and "Domestic goods transported by national company, and domestic ship made in China shipyard" of policy oriented, the NDRC vigorously promoted Chinese enterprise holding LNG project, holding LNG transport, sharing the LNG upstream resources, self-building LNG vessel and gas turbine localization, made a series of fruitful results. In order to solve the energy supply and demand, especially the urgent need of clean energy, the sustained and rapid development of Chinese economy and society, as well as environmental protection and transportation and other prominent bottleneck problems and achieve sustainable development goals, broke out of a road with Chinese characteristics of energy supply and security strategy. Especially in the pilot project area selection, coordination and local government relations, LNG resource countries selection, LNG resource investment negotiations, LNG resources import tariffs, product tax, project Engineering, Procurement and Construction (EPC) and supervision, LNG transport industry policy, transport ship building, downstream users supporting the project implementation, and the cultivation of the gas market in the operation of the project, the NDRC gave the project company policy guidance and strong support, it was precisely because of the introduction of a series of policies and regulations for the industry which played a decisive and protection role in the project smooth implementation and to put into production.

We cannot forget, since the middle 1990s, the coastal provinces, city Development and Reform Commission and the relevant departments, from the project site selection, the project social support conditions, policy support for pipeline routing, the local downstream user project's coordination, to the project safety, health, environmental protection inspection and acceptance, gave the LNG project companies strong support and cooperation, also played a key role in all LNG projects which have put into operation. At present, in addition to CNOOC, China National Petroleum Corporation (CNPC), Sino Petrochemical Corporation (Sinopec) and local enterprises are either constructing or preparing to construct a number of LNG projects, the layout of the LNG projects in Chinese coastal line has been completed.

The LNG project has been rooting, blossoming and bearing fruit in Chinese, this is the result of 20 years of joint efforts of Chinese LNG industry. *LNG Project Management Manual* published 4 years ago, marking LNG project from the introduction of technology to the mature process of LNG project construction and operation in Chinese, the publication of the book was the process of theory promotion and summary. In recent years, Chinese LNG industry in the international technical exchanges have become increasingly frequent, the Chinese voice in international conferences on technology and management also affects the world. To this end, in the past 3 years of preparation, planning and revision, I re-examined the *LNG Project Management Manual* book content, some sections have been rewritten or further improved. For more conform to the international communication and language communication convenience, I decided to write in English, in order to adapt lofty goal of "based on domestic, go to the world". *LNG Project Management* will be available to offer some necessary language and theory for China LNG technical and management personnel contracting international LNG project for construction and operation, this will be another milestone in Chinese LNG industry.

In addition to describing the hard facilities: tanks, jetty, pipeline, the work has obtained the software results-LNG project management knowledge system. It has witnessed CNOOC is in the dominant position of China LNG industry and is of great significance in the world LNG industry forefront. For the further development of CNOOC at home and abroad in the upstream, midstream and downstream market, the work will provide valuable experiences, in order to ensure and improve the engineering construction quality and management level of the future LNG project.

The staff provided source material, also paid their own efforts and sweat, most of them were on-site command and directly involved in project management and technical personnel, the book contents are enriched enough through practice. Although the book was written for a long time, and more staff provided information, writing work, there were many difficulties, but under my organization and coordination, all kinds of difficulties were overcome, especially according to their own experiences, accessed to information and interviewed the personnel who participated in the project construction and management at the time, sorted out and summarized the management work of several LNG projects. The book respects for history, seeks truth from facts, analyzes the problems and sums up the experience, as well as keeps the book's practicality and readability.

Of course, *LNG Project Management* is Chinese first LNG project management book in English, after all is in the "Feeling the stones across the river", the exploration of the summary, although the chapters have been carefully revised or rewritten, but it also has shortcomings and omissions. We also have an understanding of the process of *LNG project management* theory, resources and markets, especially in the formation of project company, shareholders coordination, EPC turnkey, project company personnel involved in the overall management, "Take or Pay" terms of the contract implementation, there are a lot of experience and lessons to be learned. In the process of translation in English, English vocabulary and terms for the LNG industry also have a query, matching, idiomatic usage and lining with international practice also have a deepening process, so these are need to be improved in the future in the book second edition.

From the perspective of project management, LNG project has the characteristics of investment in diversification, operation in socialization and management in marketization. The project had attracted the attention of all sectors of society from the beginning. The summary of the theory and practice of project management in enterprises are very meaningful things. Its advantage is that the enterprises directly engageed in project management activities, could get first-hand practical information. Guangdong Dapeng LNG project was constructed by CNOOC and BP in cooperation, both through work and communication, working language is English and Chinese, it provided a convenient for Chinese personnel learning LNG professional vocabulary in English, this would be conducive to put the international project management knowledge system application to practice directly, combined with the Chinese conditions, but it could be found the lack of the original theory. Enterprise is the core strength to enrich the theory of project management and make it become more mature.

In order to make the LNG industry rise in Chinese, CNOOC had established the specialized organizations from early 1990s, in addition to the research of LNG industrial technology and construction, also conducted a special research on the LNG international standards, compiled country (industry) level LNG standard system and CNOOC enterprise standard system table, and step by step to prepare national, industry and enterprise standards, which laid the foundation for the LNG project management. At the same time, CNOOC also attached great importance to the LNG industry chain research work, a number of research results are also made in the research on the key technologies of LNG, including natural gas liquefaction technology, key equipment localization research and development, the research on the design and construction of large LNG storage tank with full containment, the research on scheme selection of LNG receiving terminal at sea, the research on offshore skid mounted LNG plant, the research on pipeline transportation of natural gas pipeline network and LNG terminal system using digital technology, all of above researches effectively support and deepen the industrial chain for development potential. In order to summarize the experience of LNG project management as soon as possible, CNOOC organized people who involved in the projects to carry out a systematic summary for the Guangdong Dapeng and Fujian LNG projects, I have always been the organizer and participant of the above work, the project management achievements on-site greatly enriched the theories of project management, also formed my idea of the book- "Original Edition" framework, laid the foundation for collection of the necessary information. So the "Original Edition" was based on the absorption of domestic and international project management theory and practice, combined with China's first LNG project in Guangdong, Shanghai, Fujian and the subsequent LNG projects, included pre-feasibility study, feasibility study, the implementation of resource transport, LNG and jetty sites selection, material equipment acquisition, construction contractor selection, project construction, and commissioning etc. of a series of project management practice research. Through the first batch of project participants involved in more than 20 years of hard work and experience, published Chinese first *LNG Project Management Manual*, it filled a gap in the research of LNG project management in our country, it's worth for LNG industry to celebrate.

So, I would like to take this opportunity to thank hundred experts and managers who engaged in LNG project management and construction. I particularly want to thank the CNOOC's old leadership and experts, they are: Wang Yan, Tang Zhenhua, Zhao Xiuguang, Yang Keming, Zhao Xungu, Miu Yuanfei, Qiu Jianyong, etc., they supported the work writing, and put forward a lot of valuable amendments; I would also like to thank the technical and management experts who worked in the front line of the project, they are: Qu Sheng, Zhao Wei, Li Feng, Luo Ziyuan, Yin Hong, Zhou Wei, Huang Qun, Yu Yi, Jia Shidong, Zou Hongyan, Li Yinxi, Tan Yi, Peng Jiali, Hou Jianguo, Zhang Rongwang, Pan Heshun, Cheng Gongyi, He Feng, Lu Xiangdong, Lv Xianlang, Tang Xiaobo, Wei Guanghua, Du Dayong, Zhang Shuansuo, Yu Yuechun, Liu Wanshan, Zhu Hongdong, Zhao Deting, Tong Lianxing, Chen Zhengzhong, Huo Yanning, Xia Fang, Zhang Qingxu, Zhou Lin, Liu Zengwen, Wu Peikui, Zheng Hongtao, Chen Hui, Ma Jingzhu, Bi Xiaoxing, Chen Feng, Wu Yueqing, Liu Ziping, Liang Jie, Han Songqun, Zhang Chao, Zhou Chan, Song Pengfei, Chen Ruiying, etc.. They gave great support for the "Original Edition" work resources to collect and provide raw materials. It was because of their selfless dedication, hardworking and sweat and valuable experience, the content of work "Original Edition" is heavy and detailed. And I would also like to thank my family, without their support and help, this work will not meet with readers in the period.

"LNG Project Management" (Second Edition) English version is based on "Original Edition". In order to make it publish as soon as possible, I organized young technical staff of Technology R&D Center, CNOOC Gas & Power Group to translate "Original Edition" in English, they also paid efforts and hardships, they are: Zhao Sisi, Zhang Dan, Sui Zhaoxia, Ma Zhengyu Hu Xianwen, Tian Meng, Wong Ranran, Tang Deyuan, Zhao Ye, Xiao Li, Wu Jianhong, Zou Mohan, Yan Weiyi, Huang Yu, Duan Pinjia, Feng Jie, Chen Haiping, Wang Xiulin, Pu Hui, Chang Xinjie, Chen Jie, Tian Liang, Li Xinxin, Yu Yixin, Li Lezhong, Tang Ying, Jiang Shixin, Su Zhang, Gu Feng, Zhong Mihong, Cui Jingyun, Hou Jianguo, Zhang Yu, Song Pengfei, Wang Chengshuo, Gao Zhen, Yu Jiaojiao, Dou Xing and Cui Yanfei, etc.

Since I started concepting and writing "Original Edition" in January 2005, it took me 6 years. The English reprint "Second Edition" took another 4 years. It should be said that I devoted a lot of effort and energy, and it can be said that it is a long time, because 10 years exceeds one LNG project from the opportunity study, construction to put it into production time; it can be said that it is short time, because I completed data collection, writing, translation and revision in a hurry, it seems that there are many flash points not being discovered, there are a lot of theoretical knowledge is not sublimation.

The work—— "Second Edition" is required to maintain the basic elements of the theory of project management, but not to take the pure theory discuss. The work emphases on the general project management 9 knowledge systems: integrated management, scope management, schedule management, quality management, cost management, human resource management, information management, risk management, procurement and contract management. The author wrote from the project's social, macro, hierarchical and special consideration, but also wrote the special project with the characteristics of LNG related contents, such as: land requisition and site selection management, project feasibility study and project check and approval management, HSE management, financial management, LNG marine transportation management, acceptance and post evaluation management, and trial production and operation adjustment management of the 7 contents, constitute the LNG project 16 management systems (see: LNG Project

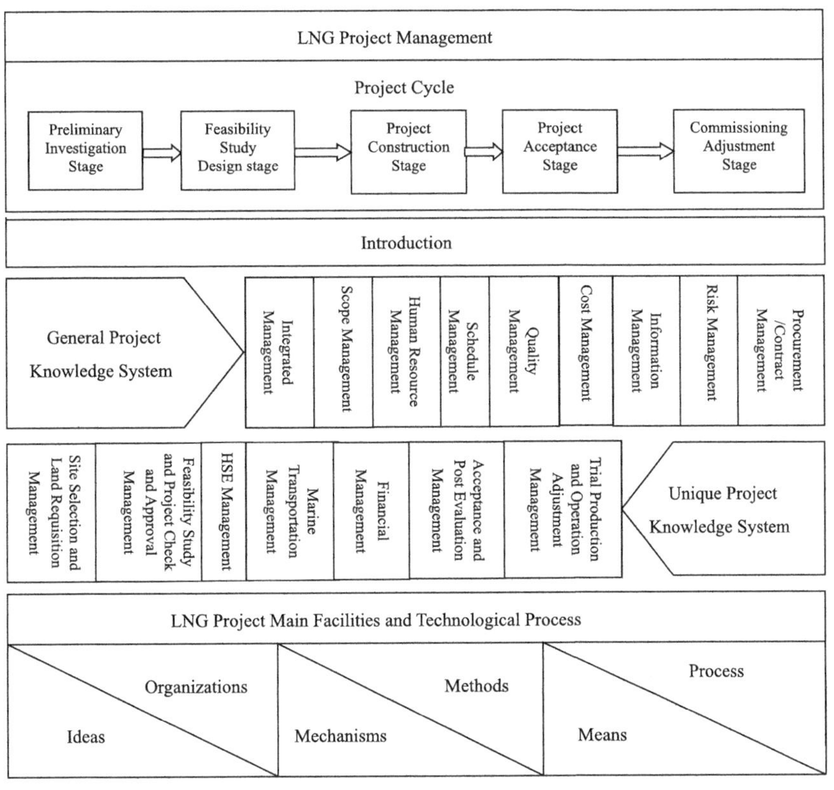

LNG Project Management Framework

Management Framework), and for the LNG project management integrity, also specifically added: the introduction and LNG project major facilities and processes of these 2 chapters, the work is divided into 18 chapters. Due to the author's original intention is to provide a practical manual for the management of Chinese LNG projects, so, the above mentioned some China's LNG project practice, and in recent years, the international exchange and cooperation experience should also be reflected in the work, project management has the characteristics of both scientific and artistic characteristics, the examples in the book are only for references. This work is essential to the management and technical staff engaged in LNG project as a reference book and a practical manual, but also can be used as a reference material for teachers and students of the university project management department.

Despite the limited level of my theory and English, the work is the first English monograph on the management of LNG projects in china. The official publication of the information which can be learned and referred is less, so this work certainly exists many shortcomings, I will make further changes after listening to the readers' advice, in order to make the next edition more perfect.

Author: Xing Yun 邢云

October 2020 in Beijing

CONTENTS

(Volume Ⅰ)

Chapter 1 Introduction

1.1 General Concepts of LNG Project Management

1.1.1 Definition of LNG Project Management

1.1.1.1 Definition of the General Project

In people's production activities, managers broadly define the project as an event in the future to be completed within a certain time, including the construction of a power plant, the construction of an LNG terminal, laying a pipeline, developing an new product or convening a meeting and so on. The project can also be defined as a one-time task to meet the predetermined quality requirements within a certain period of time, personnel and other resources.

American Project Management Institute (PMI) definition: the project is a time limited task to create a specific product or service (including: "Time" refers to every project has a clear beginning and end; "Specific" is a project formed by the products or services on key characteristics different from other similar products and services).

1.1.1.2 Definition of General Project Management

In the case of time limited, cost constraints and quality requirements, by making full use of existing resources and rationally allocating resources to achieve predetermined quality specifications and environmental health and safety (HSE) goals, it is one-time task. Here's management emphasis on the organizational resources to fully utilize and reasonable allocation, to create project stakeholders satisfied with the products and services.

1.1.1.3 Definitions and Concepts of LNG Project Management

In order to achieve the development strategic target of organizational gas industry, LNG project management is to through a special, temporary organization and operating mechanism, to organize their projects involve personnel who apply the theory, skills, tools and techniques to the LNG project activities in the expected time and funding conditions, to make sure quality, quantity, and secure operation for LNG resources supply, shipping, receiving, gasification, pipeline transportation, and use many links to complete a given task.

At present, China LNG project is mainly involved in upstream resource supplier selection, downstream user's development, implementation of LNG ocean transportation company, LNG receiving terminal, jetty and pipeline sub-project design and construction, which is engaged in a major LNG project company content. At the same time, contributed to the downstream supporting projects and extension project is also an important part of the project company. For the operation of the LNG project and the establishment of the project company plays a role as a bridge in order to achieve the completion of the whole industry chain and put into parallel operation, to achieve win-win situation.

1.1.2 Relationship with LNG Project Management and Other Curriculum

1.1.2.1 Relationship with General Project Management Theories

First of all, the LNG project management will make the general project management theory as its basis, as illustrated in Figure 1-1, it also includes the project scope management, schedule management, cost management, quality management, human resource management, information management, risk management, procurement and contract management, integrated management of the 9 knowledge systems. From the social, macro and hierarchical considerations of the LNG project, the book is also wrote especially in particular LNG project land and terminal site selection management, preparation of feasibility study and project check and approval management, HSE management, financial management, LNG marine transportation management, acceptance and post evaluation management, and trial production and operation adjustment management those 7 managing contents which constitutes 16 LNG project management systems. And for managing the integrity of LNG project, this book specifically increase the introduction, and LNG project main facilities and technological process of 2 chapters. This book is divided into 18 chapters.

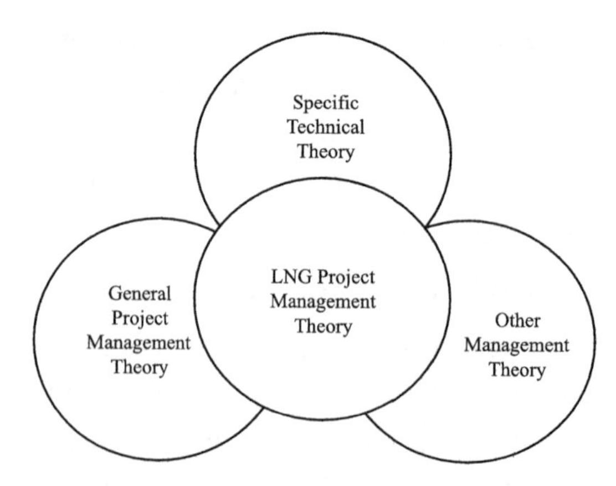

Figure 1-1 Relationship between LNG project management theory and other curriculum

1.1.2.2 Relationship with Specific Technology Theories

From an international perspective, LNG technology is a systematic engineering of high technology, has been relatively mature. From a domestic perspective, the technology of natural gas liquefaction and storage have twenty years of history, the SHANGHAI JIAO TONG UNIVERSITY, SUN YAT-SEN UNIVERSITY, South Chinese Unicersity of Technology, HARBIN

INSTITUTE of TECHNOLOGY, CHINESE ACADEMY of SCIENCES and so on, have studied in different forms of theoretical research, for instance, the SHANGHAI JIAO TONG UNIVERSTY research group led by professor Gu Anzhong, after years of research, published "Liquefied Natural Gas Technology" monograph, which has greater impact in both academia and industry at home and abroad and is also the theoretical basis of my book— "Liquefied Natural Gas Project Management".

1.1.2.3 Relationship with Other Management Theories

Other theories mentioned here refer to the theories of other subjects, include: statistics, operations research, management science, computer science, human resource management, organizational behavior, enterprise culture and other subjects. These theories are more or less reflected in the chapters of "LNG Project Management", which enrich the theory and practice of the book.

1.1.3 LNG Project Management Life Cycle

1.1.3.1 General Concept

The general project life cycle is usually divided into four stages: identifying needs, propose solutions, implementation of the project, the end of the project. Because of the special nature of the LNG project, the project management life cycle is divided into the following five stages (Figure 1-2): Preliminary investigation stage, project feasibility study and project design stage, project construction stage, project acceptance stage, and trial production and operation adjustment stage. As the LNG project management according to the project life cycle, each stage has its specific content.

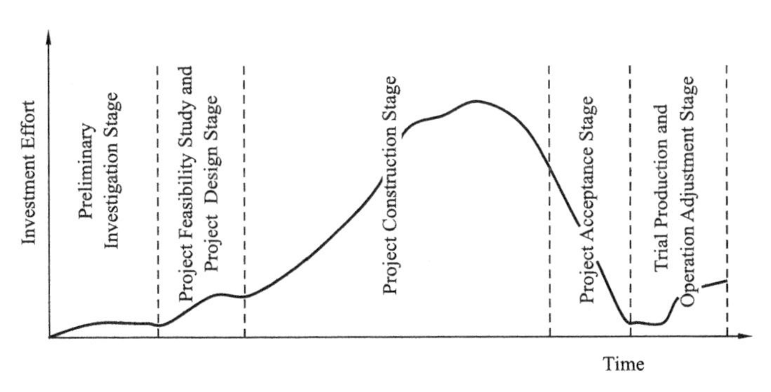

Figure 1-2 LNG project management life cycle

1.1.3.2 LNG Project Preliminary Investigation Stage

Because natural gas is a special commodity and energy strategic materials, demanding for natural gas is often closely associated with local economic construction and development, from the perspective of local government, under the guidance of the economic development strategy, GDP target, industrial layout, pillar industries cultivating, industrial restructuring, the coordinated development of industry, agriculture and services, environmental protection and so on, all those need for the new demand of add energy, as an energy source, natural gas should be an ideal substitute for coal or other primary energy sources. At the same time, the initiator companies who want to develop LNG project must respond the local energy demand, especially the demand for natural gas, which proposed LNG project proposals to obtain local government support. In the future, Project Company provides natural gas to local users, from the user to collect fees and obtain the corresponding profits, also making users to afford the corresponding gas prices to meet local demand for natural gas, to meet local demand for natural gas, bringing economic growth and stimulating the coordinated development of related industries. Only the two requirements are met, the investment and construction of the LNG project is possible to be achieved (Figure 1–3). The main outcome of this stage is the opportunity research or project proposal and pre-feasibility study report. Key personnel in this stage include the early project team members, the pre-feasibility study unit (sometimes are affiliated research institutions of LNG project initiator companies), state and local government Development and Reform Planning Department.

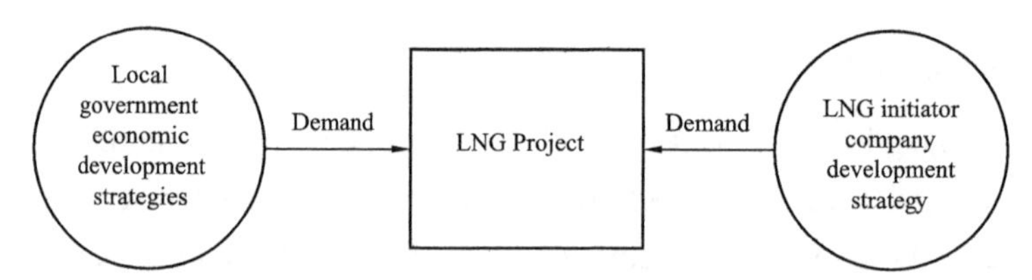

Figure 1–3 LNG project demand formation relationship graph

1.1.3.3 LNG Project Feasibility Study and Design Stage

Under the joint efforts of the local government and LNG investment and initiator company, special team will be set up by LNG project initiator company (described in detail in the fourth chapter) to conduct a pre-feasibility study on LNG projects, which include gas market survey, natural gas supplier selection, natural gas transportation mode selection, LNG terminal site selection, design of terminal, jetty and storage tank, the main route of gas pipelines, matching users determination, the overall project economic evaluation and so on. On the basis of LNG project pre-feasibility study report, after much discussion by experts, to proceed with the project feasibility report, at the end the report approved by national authorities in order to obtain a LNG project license. Front-end engineering design (FEED), project preliminary design, and detailed design is conducted on this basis, to prepare for the comprehensive construction project. The main participants of this stage include project company, the design unit of the sub projects, research units, government departments at all levels, collaborative units, etc..

1.1.3.4 LNG Project Construction Stage

The implementation of the LNG project is a process by the ideas and plans, text and drawings into one entity. At this stage, is invested in manpower, financial and material resources of the most period, the input of human resources include: project company employees, national, local and territorial government people, project designers, project construction units, personnel

involved in the supervision units and foreign, domestic suppliers, each item of special technical consultants, etc.; financial resources include capital of the shareholders, financing, loans, etc.; material resources include the necessary equipment, civil engineering materials, special steel, auxiliary materials and other supporting, etc.. This stage the project company general manager should play a full role of overall planning and coordination, the project company staff of the various departments should act as the go-between and respective duties.

1.1.3.5 LNG Project Acceptance Stage

The main participants in this stage include the above mentioned project participants, also include the state departments, local authorities, such as the Quality Supervision Bureau, Safety Production and Management Bureau, Health Bureau, Environmental Protection bureau, etc.; the LNG project company, and even shareholders will prepare the LNG project acceptance specifications, also include attached to the 3 sub projects, but also with professional characteristics of acceptance, such as fire control acceptance, occupational health acceptance, occupational safety and health acceptance, environmental protection acceptance, lightning protection design acceptance, as well as the unique pressure vessel acceptance. After the acceptance of the above links, entering the final acceptance, generally led by the national authorities, shareholders and projects involving units and personnel participate in the full acceptance work and conduct the project performance evaluation, including funding, specifications, quality, health, safety and environmental protection. After acceptance of the above categories, the parts not meeting the quality requirements should be rework and remedied, the design modification and budget should be multi recognized and settled with the contractors and the supervisor, the LNG project receiving terminal, jetty and pipelines which will be put into use should report to local government authorities, and also recognized by environmental, industry and commerce, taxation, community and other government sectors. Also it needs to do the commissioning and adjustment, prepare for the formal handover to the LNG operations department or company.

1.1.3.6 LNG Project Trial Production and Operation Adjustment Stage

LNG is a special commodity, the investment and the initiator company is usually an integrated oil and gas companies or big downstream companies to ensure the huge capital investment. Enterprises or companies throughout the chain need to sign long-term contractual relationship, with generally upstream and downstream by the "take or pay" contracts to lock, supply and use of natural gas is generally more than 20～30 years. Investment and the initiator company in general is also LNG production and operation of the company, at least, China's current situation is in this way. The LNG production and management company is responsible for the long-term business, so, from the stage of project management cycle, we put the initial stage of project production and operation also included in the cycle. This stage is general from LNG project pre-commissioning acceptance to trial production and operation adjustment, the end of the period is when LNG project achieves the design requirements of maximum annual receiving amount or passes final acceptance, through the official handover ceremony, and complete the work of LNG Project Company.

1.1.4 Concept of LNG Whole Life-cycle Project Management

Currently, the concept of whole life-cycle project management is respected by theoretical circle (Figure 1-4), which is an attempt to put project management theory for subsequent expansion. In terms of Chinese LNG project, the LNG project initiator company is also a long-term of the LNG production and operation company in the future, so the whole life-cycle of the project management has practical significance. The followings will explain the relationship and related issues between LNG project management and LNG project whole life-cycle management.

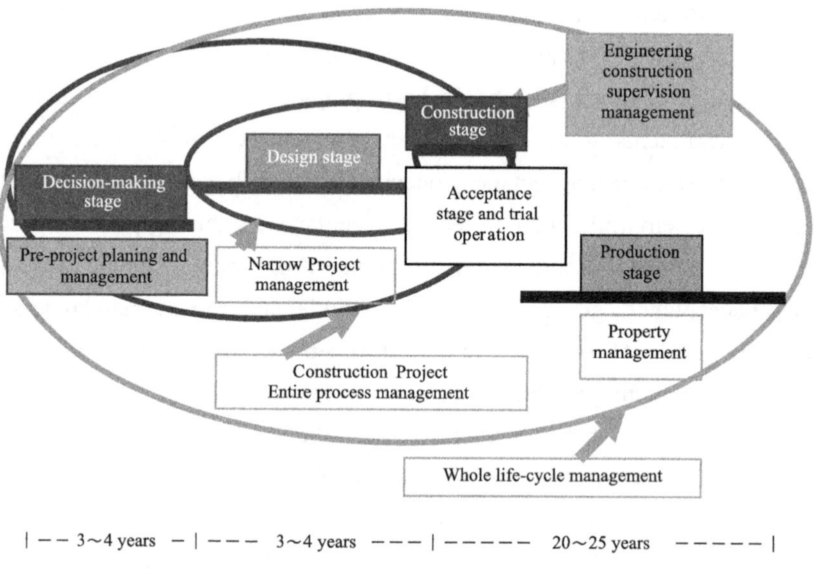

Figure 1-4 LNG project whole life-cycle management

1.1.4.1 Relationship between LNG Project Management and LNG Project Whole Life-cycle Management

The above description of the LNG project can be divided into five stages, the preliminary investigation stage corresponding to the pre-project planning and management, project feasibility design stage, project construction stage, project acceptance stage, project commissioning stage corresponding to project management, and the above ones belong to the whole process of project management; plus production stage would constitute full life-cycle project management. Therefore, project management is a part of the whole life-cycle of the project management.

1.1.4.2 Whole Life-Cycle Project Management is in line with the foundation of LNG Operating Company Management

Since Chinese LNG's project investment, construction and operations are usually the same company, so the whole life-cycle of the project management fits well with the management philosophy of company's operations. This would be in accordance with the whole life-cycle ideas to manage from the pre-project to project operation.

(1) Forward-thinking in early stage.

It is precisely because a company to carry out the project has always been, which requires the early project team in the early stage of the market research stage to do in-depth and meticulous work, and carry out a careful analysis to the investment region and the economic development of surrounding area, which include the current industrial energy demand, industry type, industrial layout, environmental requirements, etc., but also predict the next 5 to 10 years or even longer-term economic trends, which include the amount of clean energy growth, industrial restructuring trends, the coordinated development of various industries, environmental objectives set, etc. Only the above situation is very clear, in order to accurately estimate the energy demand trend, we should not only consider the ability to receive the phase 1 of the jetty, but also consider the size and scalability of the phase 2, and lay a solid foundation for future production and operation of the project.

(2) Project final acceptance and production and operation to achieve seamless connection.

General project final acceptance and production and operation are corresponding to different organizations, it will inevitably bring about problems before and after the handover and it needs time to gradually familiar with the process. Due to the particularity of Chinese LNG project, it creates a seamless connection conditions for the final acceptance and production and operation, but the project company must also be in the late stages of the construction to prepare for production operations, which include: the establishment of operations organization, personnel transfer and recruitment, training, data transfer and digestion, especially the use of whole life-cycle project management thinking, appropriate arrangements for the production staff to participate in the pre-feasibility study, design, construction, the seamless connection is possible.

(3) Post production operation is the inspection to preliminary work of the project.

Under the guidance of the whole life-cycle project management theory, the end of project management is not the end of the whole work, but the change of management stage. Due to previous staff participation, so making the personnel as soon as possible to grasp the equipment performance, process conditions and the operation, to avoid the conflict between the two groups of people in the transitional period. At the same time, through the production and operation, in turn, can test the rationality of the preliminary work, including equipment procurement, equipment installation, standard use, process design, personnel and institutional settings, etc., facilitate the transformation of the project and provide reference for other projects.

1.1.5 LNG Project Chain

LNG project chain is defined as the link between the related units or organizations in the project. The following describes the upstream, midstream and downstream industry chains, the industry chain, business chain, and economic chain involved in LNG project. Through the introduction of each chain to get people to understand logical relationships between industries involved in LNG project.

1.1.5.1 Industry Chain

(1) Chain impact concept.

LNG project management is not just the LNG terminal and end-users, the two aspects, it is actually an industrial chain (Figure 1-5) management, which involves the upstream parts: gas production, gas impurity separation process; midstream: gas liquefaction, LNG ocean transportation, LNG receiving terminal, LNG storage, LNG regasification, land pipelines; downstream: gas power plants, heat and power cogeneration, chemical gas, city gas, industrial gas, vehicle fuel, etc.. Any link encountering problem in this chain will affect the whole industry chain to ruptured and disjointed, for example: gas production emerging problems, no gas supply, liquefaction plant discontinued, the following aspects paralyzed; LNG ship due to the failure of the ship itself or weather problems can't be transported, the upstream gas field will have to stop, so that downstream users have no gas supply also had to stop production; user problems, no users, upstream each link will be interrupted. Therefore, the above all links must be a chain effect, indispensable, pulling one hair and the whole body is affected.

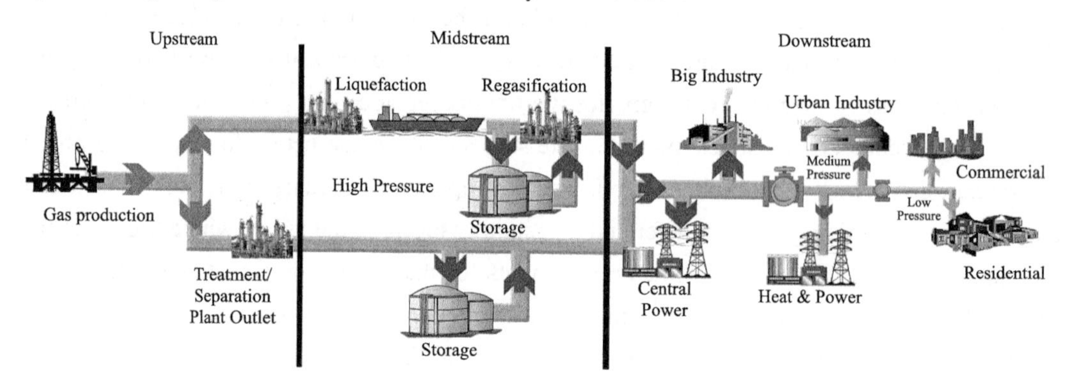

Figure 1-5 LNG project chain

(2) Gas field production.

Gas field production is the first link of LNG project industrial chain. In fact, the natural gas field production belonging to the upstream industry in oil and gas industry, like finding oil, firstly, it must be under the guidance of the oil and gas geological theory to carry out geological surveys, geophysical exploration, including seismic, magnetic prospecting, geophysical data interpretation, the target layer structure (trap) implementation, exploratory drilling, geophysical logging, oil and gas testing, reserves parameter selection and reserve calculation, the overall field development plan preparation, approval, development/production well drilling, wellhead production equipment installation, etc., Finally, natural gas can be produced. If the natural gas field is located in the onshore, reserves reached a certain scale, the users are also there, it is needs to lay the pipeline on land and transport the natural gas to consumers station. If the gas field is located offshore, firstly, offshore production platforms should be constructed, also it needs to lay submarine pipelines and make it connect with land pipeline, finally, supply to the users. The above two cases do not belong to the scope of our discussion.

(3) Gas liquefaction.

When the gas production country gas field production is high, in addition to meet the domestic users, but also rich in natural gas, it can be considered to negotiate with foreign users to seek foreign users. If the users are located in other countries across the sea, transporting the natural gas with the laying of submarine pipelines will not economical, so using the method of liquefied natural gas, turning gas into liquid and using special boats to deliver the gas to the destination, this method has proven to be feasible. Thus, according to the size of foreign users, firstly, gas producer must use a set or several sets of liquefaction devices to shift the gas to liquid by physical changes, but the gas must first go through the separation of non-combustible gases, such as the removal of carbon dioxide, sulfur dioxide and other impurities. The main component is methane gas, up to 75%, a small amount of ethane, propane, nitrogen content of less than 5%, the gas after purification, with higher heat than the general natural gas, the gas volume reduction is typically to 1/600, the LNG temperature is −162℃, ready to ship.

(4) LNG Marine Transport.

The ship for marine transportation of the above mentioned cryogenic LNG, it is conceivable that such fleets are made of special materials, specially designed hull structure. After half a century of research, testing and practice, it has been built a special ship borne transport tasks and several ship designs, such as MOSS type and membrane type (described in Chapter 11). In addition to the special hull, there are issues related to transport modes, such as LNG buyers bearing the transportation (FOB), or LNG sellers bearing the transport (DES). It also relates to shipbuilding, ship owners, ship operating companies and the transport company establishment and the LNG operating company's contractual relationship issues.

(5) LNG receiving.

When the LNG vessel transport liquefied natural gas and berth to the jetty of the receiving terminal, the liquefied natural gas is transported to the storage tanks on the shore of the receiving terminal by the special unloading arms, the storage tanks play a role in receiving and temporarily storing liquefied natural gas.

(6) LNG regasification.

In order to achieve the purpose of users to use conventional natural gas, re-gasification of LNG is a must be carried out firstly, it is a process of physical change from liquid to gas by heating. Generally, based on geographic location of the receiving terminal, the number of supply, the different methods of re-gasification should be taken, such as air gasification, sea water gasification, other heat medium gasification.

(7) Natural gas utilization.

After re-gasification of LNG as normal gas, which is transported to all types of users through pipelines, such as gas power plants, city gas, chemical gas, industrial gas and vehicle fuel. From Guangdong, Fujian LNG user group, gas consumption of gas power plants are account for more than 70%. But some experts believe that the use of liquefied natural gas for gas power generation will cause some of the waste of energy, they advocate more for city gas, chemical gas, industrial gas and vehicle fuel. If the user is far away from the LNG receiving terminals, it needs to construct LNG satellite station in the center of the remote users, LNG is transported to satellite station by LNG tanker, tank ship, and then through re-gasification to deliver to the end-users.

1.1.5.2 Physical Chain, Business Chain and Economic Chain

The above discussion of the various industries is involved in LNG project, what is the logical relationship between the various industries involved in the LNG project? We will elaborate them from the physical chain, business chain and economic chain.

(1) Physical chain.

It involves the exchange of goods on the whole industry chain, that is, liquefied natural gas or natural gas, LNG suppliers deliver the LNG product, if the liquefied natural gas transported by LNG vessels from foreign LNG resource country to China's coastal LNG receiving terminals, LNG is stored in storage tanks, according to the user needs to vaporize LNG to a gaseous state, and then land pipelines (sometimes land pipeline operators and receiving terminal are not the same company) will deliver natural gas to each user stations, or tankers and tank ships transport the LNG to LNG satellite station or users. Physical chain flow is unidirectional — from the LNG supplier to each user (Figure 1-6).

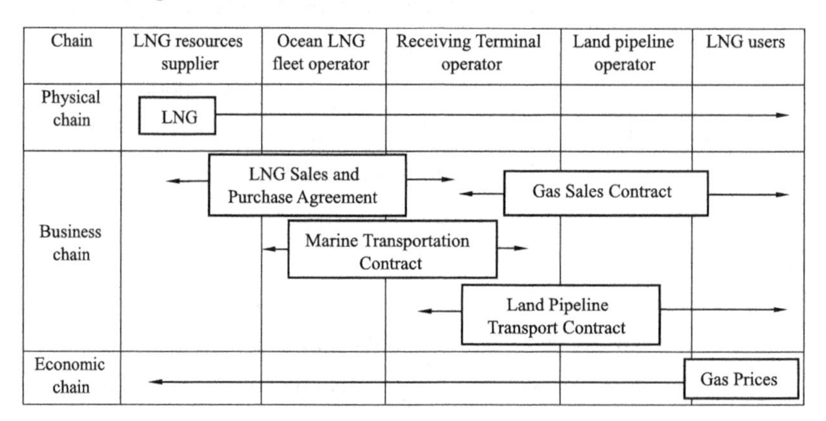

Figure 1-6 LNG project physical chain, business chain and economic chain

(2) Business chain.

A series of business contracts or agreements lock each pairwise relation in the LNG industrial chain. According to Chinese existing LNG project mode of operation, LNG resource suppliers and the receiving terminal operators sign a LNG Sales and Purchase Agreement, to reach relationship of LNG supply and demand. LNG receiving terminal operators and gas users sign Gas Sales Contract, to establish gas supply and gas use relations. In order to ensure LNG transportation, ocean LNG transportation operators and LNG receiving terminal operators sign Marine Transportation Contract, land pipeline operators and LNG receiving terminal operators or uses sign the Land Pipeline Transport Contract. The above relationship is pairwise contractual relationship. If one problem occurs in the process of the implementation of the business contract, the whole industry chain will be broken.

(3) Economic chain.

It is to ensure that various economic benefits can be realized in the form of capital return, by the end user to bear the financial input of its upstream to achieve. Final gas price is determined by the national authorities and the local price department. In general, the price should be considered by the resource supply interface to the end user level increments, the price:

P(final price)= W(wellhead price)+ L(liquefied fees)+ M(marine freight)+ R(receiving, storage, gasification fees)+ D (land pipeline fees)+ O (other costs, such as various tax)

Entire capital flow is from end user to each upstream, contrary to the direction of the physical chain.

1.2 Chinese LNG Projects

1.2.1 Characteristics of Chinese LNG Projects

In addition to the general project such as unique, multi-objective, periodicity, constraints and risk, LNG project involves a wide range, long industrial chain, technical complexity, social influence and other factors, especially the Chinese LNG project which reflects the following characteristics.

1.2.1.1 Chinese LNG Projects are the Unprecedented

Guangdong Province and other coastal areas as the forefront of reform and opening up, economic growth rate is significantly higher than other provinces, also demand for energy is growing. But the original coal-dominated energy consumption structure in southern coastal area stay away from energy production, which generating enormous pressure in environmental and transport, these areas urgently require the use of clean, efficient energy in order to improve the environment and ease transport pressure . As a clean, efficient and inexpensive energy, LNG has focus on the development and utilization target in the beginning of this century. Although several large gas fields have been discovered and developed in Chinese onshore and offshore, it is still difficult to meet the demand for natural gas in the economic development of southeast coast. Britain is Europe's first use of LNG in the world, Japan is the first country in Asia to use LNG and is the largest importer of LNG in the world currently. In Asia, in addition to Japan, South Korea, India and Taiwan province of China has been ahead of our use of the introduction of LNG, while Chinese main land mentioned to import and use of LNG is from the beginning of '90s. Since then, Chinese began a comprehensive work to import liquefied natural gas project, which is a new industry. The introduction of foreign LNG, to improve the structure of energy consumption in Chinese, to protect the domestic energy supply and security, it will have far-reaching strategic significance, also it has a significant impact to country's politics, economy and society.

1.2.1.2 Chinese LNG Project to Get the Strong Support from the National Government

By the end of 1995, the NDRC entrusted CNOOC to study the introduction of LNG project technology. On the basis of investigation and research on the domestic market, CNOOC puts forward the selection of LNG resources and the configuration plan of the receiving base. In October 1998, the State Council made a decision that Chinese importing LNG project "the first pilot in Guangdong". In April 1999, the Guangdong Provincial Development and Reform Commission, and CNOOC submitted the Guangdong LNG pilot project overall project proposal, by the end of 1999, the project proposal was approved by the State Council. Currently, except Guangdong, Fujian, Shanghai, Zhejiang, Dalian, Jiangsu, Tianjin, Shenzhen, Yuedong and Hainan projects have been put into operation. Since the introduction of LNG involves the major issues of national energy security, so, Chinese LNG projects got strong support from the NDRC.

1.2.1.3 Chinese LNG Project is a System Engineering

LNG importing is a systematic project, in addition to LNG project itself, involving several users to invest in downstream projects, including gas power generation projects, city gas projects, industrial fuel projects, natural gas chemical projects, vehicle fuels, the satellite station projects, gas filling stations projects, LNG filling station projects, cold energy utilization projects, etc., and these projects must be constructed with LNG project in the same period and synchronous operation. LNG projects and their supporting projects investment with long construction period, contract negotiations are complex, involving resource development, transport, downstream use, foreign exchange balance and other issues. Many large enterprises at home and abroad will take LNG projects as their strategic investment.

1.2.1.4 LNG is a Special Commodity

The final product of most projects is a commodity, such as the development of new mobile phones, home cars, or the new computer application software, or the opening of new tourist attractions and so on. But for liquefied natural gas has the general merchandise or the use value of the product, and has the social attribute of the special commodity.

(1) Important energy resources.

LNG is not only an important energy resource, but also an important strategic resource, which introduction is beneficial to the safety of national energy supply and to improve our country's position in international trade.

(2) Opportunities for Chinese shipbuilding industry.

The introduction of LNG provides an opportunity for the development of Chinese shipbuilding industry, at the same time, to fill in the gaps in the ocean LNG transportation.

(3) Favorable for environmental protection.

For a region, it is favorable for environmental protection, promoting the development of gas power generation, city gas, natural gas chemical industry and other related industries with a huge social effect.

(4) Industry chain.

Project involves upstream: gas field development, natural gas liquefaction, marine transport, receiving storage, land pipeline transportation, downstream: gas power plants, city gas and other industrial users, to provide the supply and demand chain for various industries and millions of households.

(5) Long contract period.

Each link of the chain locks its supply, transportation, reception and use with long-term contracts, the contract period is generally up to 20 ~ 30 years.

Therefore, LNG plays a pivotal role in the economic development of the country, it not only has a multiplier effect on macroeconomics of the local economy, but also has a micro-economic benefit of specific LNG projects, meanwhile, it will play a positive role in environmental protection, social progress, economic prosperity and people's living standards throughout LNG using in the region.

1.2.2 Chinese LNG Project Involving Unites

LNG project is a new industry for Chinese, and different participants in the project (Table 1-1) play different roles and manage the project from different angles. According to the current investment and management model of LNG projects in Chinese, standing on LNG project sponsors and project management company investor's point of view, the followings are a brief description that the role played by key players in project management, the key players inquire: project company, government departments, investors, LNG suppliers, equipment and materials suppliers, designers, construction contractors, supervision, users. In addition to the general principles of project management, their the specific responsibilities of their management, focusing on the use of management technology and even in the project life cycle needs to be managed in the content of the project are different.

Table 1-1 Related Units Involved in the LNG Project Investment Management

Pre-project	Feasibility Study and Design	Project Construction	Project Acceptance	Project Commissioning
Project Company	- - - - - - -	- - - - - - -	- - - - - - -	- - - - - - - ►
Shareholders	- - - - - - -	- - - - - - -	- - - - - - -	- - - - - - - ►
NDRC	- - - - - - -	- - - - - - -	- - - - - - -	- - - - - - - ►
Provincial Government	- - - - - - -	- - - - - - -	- - - - - - -	- - - - - - - ►
Local Government	- - - - - - -	- - - - - - -	- - - - - - -	- - - - - - - ►
Media	- - - - - - -	- - - - - - -	- - - - - - -	- - - - - - - ►
General Public	- - - - - - -	- - - - - - -	- - - - - - -	- - - - - - - ►
Pre-feasibility Study Unit	- - - - - - -	- - - - - - -	- - - - - - -	- - - - - - - ►
Special Report Study Unit	- - - - - - -	- - - - - - -	- - - - - - -	- - - - - - - ►
	Finance Bank	- - - - - - -	- - - - - - -	- - - - - - - ►
	State/Local Environmental Protection Department	- - - - - - -	- - - - - - -	- - - - - - - ►
	LNG Marine Transport Company	- - - - - - -	- - - - - - -	- - - - - - - ►
	LNG Resources Supplier	- - - - - - -	- - - - - - -	- - - - - - - ►

Continued

Pre-project	Feasibility Study and Design	Project Construction	Project Acceptance	Project Commissioning
	Feasibility Study Unit	------------	---------→	
	Local land and Resources Department	------------	---------→	
	Local Government Land Acquisition Department	------------	---------→	
	Local Resident's Land Requisitioned	------------	---------→	
	Foreign Equipment Suppliers	------------	---------→	
	Domestic Equipment Suppliers	------------	---------→	City Gas
	Material Suppliers	------------	---------→	Other Industrial Users
	Construction Supervision Unit	------------	---------→	Residents Gas
	Tank Construction Contractor	------------	---------→	Gas Power Plant
	Jetty Construction Contractor	------------	---------→	Tax Department
	Pipeline Construction Contractors	------------	---------→	Regulatory Agencies

1.2.2.1 Project Company (Operator) to the Project Management

LNG project company is the core for the project management. Currently, the project initiator company carries out Chinese LNG pre-project works, in advance the initiator company authorizes the pre-project group and the Project Office carries out the market, resources, opportunities research, terminal site selection and pre-feasibility studies. When the LNG project organization get national authorities " Pass" and feasibility study is approved by the initiator company, after the project application report is approved by national authorities, the joint venture company should be established, which is invested by each shareholder, and each shareholder dispatches directors, supervisors formed the board of directors and the board of supervisors. Since then, the general manager of the project company operates the LNG project under the authorization of the board of directors and the general manager is undoubtedly the key to the success of the project. Of course, the project company functional departments, such as engineering department, control department, business department, human resources department, HSE department and other functions also play a gatekeeper role. The project company, in the entire project design, construction, commissioning stages, will focus on FEED, preliminary design and detailed design, selecting qualified construction contractors, supervision companies. Especially in the construction stage of the project, the project company pays close attention to schedule, cost, quality and HSE management.In the late, the project company makes great efforts to each stage acceptance and trial production and operation management. Before the project put into operation, the project engineering department and other staff will be transferred or changed to other posts. For the normal production, the entire project company will adjust organization for adapting to the production operations management. Project Company is the subject of liabilities in project construction and operation.

1.2.2.2 Government Departments to the Project Management

From narrow sense, broad concept of government departments can be divided into the central government, local governments (the provinces and municipalities) and the grass-roots government (local government). Because natural gas is a kind of strategic resource, increasing use of natural gas is an important measure in improving the energy structure of China, therefore, LNG project has been highly regarded by the government. Under Chinese current investment environment, three levels of government for the success of the project play a vital role.

In the preliminary investigation stage, the national authority NDRC will conduct project pre-project audits and issue project "Pass", which base on preliminary proposals submitted by the early project team. Local governments in accordance with regional economic development and social needs, the LNG project will be included in the overall economic planning, by the local government to recommend local enterprises to participate in LNG project with project initiator and to establish Project Office. The government department such as Land Resources, Environmental Protection, Planning, Construction, Forestry, Agriculture, Water conservancy, Transportation, Public security, Electricity, Telecommunications and other sectors will participate in the selection of the receiving terminal, jetty, coastal line, regional pipeline engineering land acquisition and put above content included in their planning. In the project feasibility study stage, Environmental Protection Departments should review and approval the the project environmental impact assessment report. The NDRC will reply to the project sponsor by approval document, and national or local business sector will conduct company registration to project company. The above three levels of government departments will also participate in the project construction stage, project acceptance stage, the trial operation stage of the review, supervision and approval.

1.2.2.3 The Investors to the Project Management

In China, currently as the LNG project shareholders, usually participate in the project company's major decision-making process. From the investment, the shareholders of the project company can invest directly through the capital input, but also can get loans and subscribe for shares and other ways to invest in the project. Thus, the management of the finance bank to loan project is doing every loan as a project to manage, the project life cycle is divided into 6 phases, such as project selection, project preparation, project evaluation, project negotiations (including the signing of the loan agreement), project implementation (mainly supervise and control the use of loans) and project evaluation. No matter what kind of investors should not abandon or neglect their investment in the project management. They naturally care about the success of the project, whether it is profitable, or whether it can get back the principal and interest. Therefore, Investors, in particular project sponsors, must have the appropriate management of the project. Although their primary responsibility is investment decisions and their management focus on the construction stage of the project, and the primary means used in the project are assessment, quality inspection and acceptance, the investors who want to really get the expected return, should pay more attention to the project operation in addition to the entire life cycle of the project monitoring and management. The general operation of the project is 20 to 30 years, due to the huge investment in LNG project, the investment recovery is more than 10 years. The operating period which is good or bad is the key to the outcome of the project.

1.2.2.4 The Design Unit to the Project Management

Design results of the project can be provided by internal members of the investor, or by external units. In either case, the designer should accept and cooperate with the management of the project, but also manage the design task itself.

The design of the project is often more innovative and uncertain than the other work of the project, therefore, the management methods and technology also have its characteristics that can't be ignored.

At present, the international LNG projects should be technically mature, it is not like other ordinary products, such as the new family car, new mobile phones and the final product of LNG project is LNG or natural gas in accordance with the contract. The design results are mainly embodied in the gasification of LNG into natural gas and through the design of jetty, storage tanks, land pipelines and other functions to transport, therefore, design unit must firstly grasp the international and domestic more mature latest technological achievements or patents used in above sub-projects, and then intensively communicate with the LNG project company, under the premise of not violating the budget of the project funds, to actively introduce the new technologies or patents. If the cost of new technologies and patents exceed the budget, but it does give the project put into operation to bring greater economic benefits, designer should convince the project company to make appropriate adjustments for investment. In short, the designer should be responsible for investors, extensive and in-depth communication with Project Company throughout the design process, to make the project sub projects design can withstand the test of the history. At the same time, designer should give a more accurate estimate on the design task, the time and cost required for the completion. The design work is often a process of repeated comparison and modification, the progressive rule of conventional network planning techniques, for example: critical path method (CPM), project evaluation and review technique (PERT) are often not fully applicable, it needs to adopt special planning techniques. Design work is a creative work, which should pay more attention to the designer with self-realization and self-achievement in the management of human resources personnel. Evaluation of the design results is difficult to have a unified scale, so it often used experts with Delphi method.

1.2.2.5 LNG Suppliers to the Project Management

LNG supply is the foundation for project establishment. Regardless whether LNG is in buyer or seller market environment, an honest LNG supplier should conduct closely communication with LNG project owners, and sufficiently understand the local need for LNG or natural gas and the project scale. When the local government agrees to build the LNG project and the downstream user market is optimistic, LNG supplier should make efforts to make the project successful. The LNG supplier management on project is mainly reflected in four aspects. Firstly, in the establishment stage of LNG project, LNG supplier conduct resources allocation and supply frequency planning on the basis of project scale.Secondly, sufficient communication is conducted with LNG carrier, LNG project owner on LNG transportation, rational transportation routes are designed and measurement system is delivered. Thirdly, LNG supplier signs with the owners "LNG supply and sales contract" .Fourthly, before official production, normal LNG supply period and supply rate are explored and established, and at the same time, rational suggestion is proposed on project safe operation, to ensure LNG project safety and steady production.

1.2.2.6 The Equipment and Materials Supplier to the Project Management

Equipment and material supply are the premise for LNG project construction. Equipment and materials suppliers communicate and negotiate with LNG project owner in advance, sufficiently get to understand the project scale, equipment, and material bidding requirements, in order to obtain bidding permit, finalists bidding range and procedures. The management is mainly reflected in four aspects. Firstly, in bidding document release period, equipment and materials suppliers need to actively respond to owner release, and offer qualified bidding tender. Secondly, in bidding clarification period, honestly communicate with owners about the technical and price bottom line, highlight the strengths and weaknesses of equipment and materials and credibility is earned with honesty. Thirdly, fairly compete with other equipment and materials suppliers, with their equipment and materials technical performance, safety and reliability, high performance price ratio and other advantages to be selected to short list, and finally sign a supply contract. Fourthly, supply period and after goods supplied, cooperate closely with the owner, from inspection, commissioning and after the formal operation after-sale service, let the owners feel good faith from a series of services and get the next order.

1.2.2.7 The Contractors to the Project Management

Project construction and implementation must meet the requirements of the investment and operators to achieve the goals of the project. Through the planning and design of the project, these goals are usually more specific and clear. The project contractor management responsibilities of the project are mainly based on the project objectives of the implementation process of the schedule, cost, quality and safety of a comprehensive plan and process control, and other relevant management.

The project contractors may be an internal organization of investors and operators, or from external parties. Early, Chinese LNG receiving terminal, especially the tank was generally used for turnkey project (EPC) tender. If the successful tenderer is foreign company, according to relevant laws in Chinese, it is requested that foreign company set up joint venture with Chinese similar construction company. At present, the general contracting method for LNG storage tank construction by the Chinese company has been appeared. The jetty and pipeline project are generally used in the domestic tenders, and the successful bidders are the domestic companies.

Builders must accept the supervision and management of investment and operators, and maintain close communication and coordination with the owners. In order to achieve the project implementation, the contractors may also participate in the procurement process of the owner (such as bidding, negotiation, etc.).

After the completion of the project, the contractors shall accept the acceptance of the government and Project Company, and do a good job of ending and handing over the project. In some cases, the project contractor is also the project designer, accepts the full commission of Project Company.

1.2.2.8 The Supervisors to Project Management

Project supervisors are generally selected through official bidding procedures by LNG Project Company on the basis of project requirements. Supervisors are primarily responsible for quality supervision on various links of engineering construction during project construction period. Firstly, Supervision Company must possess qualification for supervising construction projects of the same type. Meanwhile, supervisors must have clear idea of the objectives of LNG Project Company, and fully understand the key points of all sub-project construction. Especially the supervisors have to be very familiar with international and national quality standards. And it is requested that they had supervision experience and acceptance of same type construction projects.

1.2.2.9 Users to Project Management

In most cases, the users are the recipient and user of the final outcome of the project. For the LNG project, the final product - natural gas or LNG, users are mentioned above, such as gas power plants, city gas, industrial gas, natural gas chemicals, fuel vehicles and other users.

In some cases, these users are LNG project downstream supporting projects, so users are more concerned on the upstream project. They will track LNG project upstream jetty, receiving terminal, pipeline project feasibility study and concern signing long-term gas supply agreement with upstream (containing "Take or Pay" clause), in order to ensure their long-term supply demand; at the same time, users pay more attention to themselves projects construction and production preparation. If users' supporting projects are part of LNG overall project, it should be set up, submitted and approved, constructed and put into production simultaneous with LNG project.

1.2.3 Introduction of LNG project in China

1.2.3.1 Guangdong Dapeng LNG Project

By the end of 1996, the state formally approved the first phase project of Guangdong Dapeng LNG pilot project. The LNG pilot project is the system engineering, Including jetty, LNG terminal and trunk line project, gas power plant projects, city gas pipe network projects and LNG transportation, and other legal person projects. Among them the core projects are jetty, receiving terminal and pipeline projects.

Guangdong Dapeng LNG receiving terminal is located in the Chentoujiao, east of Dapeng

Bay, Shenzhen. At the first phase of the project, the design scale is 3.85×10^6t/a, there are 4 $16\times10^4m^3$ storage tanks and jetty could dock (14.7~21.7)$\times10^4m^3$ LNG cargo ship were built. The first phase of gas transmission pipeline project includes branch, trunk length is about 400 km, the project was completed and put into production by the end of June 2006, and the total investment is about 7 billion Yuan. The LNG provider for the first phase is Australian ALNG Group. At the second phase of the project, the terminal project scale will reach 10×10^6t/a.

For the operation of the Guangdong LNG project, Sino foreign joint venture company was set up, with Chinese side holding. The partners and shares: CNOOC's Gas and Power Group holds 33%, Pearl River Delta Investment Company (BP) holds 15%, Guangdong investment Company (BP) holds 15%, Shenzhen Gas Group holds 10%, Guangdong Yuedian Group holds 6%, Guangzhou Gas Company holds 6%, Shenzhen Energy Group holds 4%, Hong Kong Electric (natural gas) Company holds 3%, Ganghua Investment Company holds 3%, Dongguan Fuel Industrial Company holds 3% and Foshan Gas Company holds 2.5%.

On October 11,2003, the NDRC approved the "Guangdong Dapeng LNG Overall Pilot Project Feasibility Report".

On December 28,2003, Guangdong LNG receiving terminal project officially started. French (Italian) STTS Group as winning bidder of receiving terminal designer, procurement and construction contract (EPC), responsible for the construction of Guangdong Dapeng LNG importing terminal and jetty facilities.

On February 23,2004, Guangdong Dapeng LNG Ltd officially registered and the registered capital is 2.1 billion Yuan.

On April 30,2004, Guangdong Dapeng LNG Ltd., the BP Group and Guangdong Provincial Government held a business contract signing ceremony in Beijing Great Hall, and signed the first domestic LNG project — the Guangdong Dapeng LNG project financing agreement, natural gas sales contract (10 users), the EPC contract and time charter contract memos and other 19 contracts.

On November 6,2004, the key projects of Guangdong Dapeng LNG pilot project - gas transmission trunk line and Guangzhou gas periphery high pressure pipeline project (Pearl River shield crossing engineering and directional drilling crossing river engineering) were started in Panyu, Guangzhou.

On December 12,2004, Guangdong Dapeng LNG Ltd. and ALNG Group held "LNG Resource supply and Sales Contract" in force ceremony in Australia.

On September 28,2006, the project began its commercial operation, and supplied 5.1 billion cubic meters of high quality natural gas to Chinese Pearl River Delta Region and Hong Kong every year.

By the end of 2016, Guangdong Dapeng LNG terminal ushered in the Q-Flex "Onaiza" - ship, which marked the LNG jetty operation, has been a full 10 years, safely unloading LNG 800 ships, receiving a total of more than 52 million tons of LNG.

1.2.3.2 Fujian LNG project

Fujian LNG project was completely invested and operated by the Chinese themselves – CNOOC's Gas and Power Group and Fujian Government's Investment and Development Corporation respectively invested 60% and 40%, the project was totally invested 5.5 billion Yuan. The total investment in the first phase of the overall LNG projects is more than 20 billion Yuan.

The first phase of the Overall Projects consists of a jetty, a receiving terminal, and gas transmission trunk line projects, 3 new gas-fired power plant projects and 5 city gas distribution projects. LNG project includes jetty, receiving terminal, and the trunk and branch transmission pipeline connected to the power plants and urban gas gate station.

The receiving terminal is located in Xiuyu port, Putian City, Fujian Province, on the north shore of Meizhou Bay. The first phase is able to receive 2.6×10^6t/a LNG; 2 16×10^4 m^3 storage tanks (Currently has reached 6) were built; in the receiving terminal, a special type of T butterfly wing layout jetty was built, which can dock 20.69×10^4 m^3 LNG Q-Flex transport ship. The total length of the trunk line is about 369 km and length of the branch line is about 53 km.

The first phase of project was put into operation in April 2008, and began to supply gas to Putian, Eastern Xiamen and Jinjiang power plants and the city gas users in Fuzhou, Putian, Quanzhou, Xiamen, Zhangzhou. The total receiving unloading scale of LNG will come up to 5×10^6t/a. At the same time, the gas supply to projects of the second phase of Putian power plant, Eastern Xiamen power plant, Quanzhou power plant and the third phase of Fuzhou Huaneng Power Plant will be increased. Indonesia Tangguh Co. is the first phase of the project the LNG supplier. Transport ship factory for construction is Hudong Zhonghua Shipbuilding (Group) Co., Ltd..

By the end of 2016, Fujian LNG project had received a total of more than 21×10^6t LNG.

1.2.3.3 Shanghai LNG Project

Shanghai LNG project consists of a special LNG jetty, LNG receiving terminal and submarine gas transmission trunk line. According to the plan, it should receive LNG 6.0×10^6t/a, and the project would be implemented in 2 phases. The design of the construction scale in the first phase is LNG 3×10^6t/a, mainly the constructions include one $(16.5\sim20)\times10^4m^3$ LNG jetty and heavy cargo wharf facilities, three $16\times10^4m^3$ storage tanks, and 40 km submarine gas pipeline through Lingang New City Gas distribution station, entering the Shanghai city natural gas high pressure trunk network system. The second phase of the expansion of the site is reserved (the overall project also includes LNG maritime transport, ancillary gas power plants and urban gas, etc.). The site is selected in the Shanghai international shipping center Yangshan Port. A total investment of the first phase of the project budget is about 4.6 billion Yuan, and it was put into operation by the end of 2009.

On December 31, 2004, CNOOC Gas and Power Group and Shanghai Shenergy (Group) Limited Company jointly set up the Shanghai LNG Co.. The partners and shares: Shenergy

(Group) Limited Company accounts for 55% of the share ratio, while CNOOC Gas and Power Group accounts for 45%. After that, the Co. started full operation of Shanghai LNG project.

LNG project located in Shanghai indicates that the CNOOC has taken an important step in the practice of its gas strategy project. As an effective supplement to the other energy, this project will help to improve the people's living standards, and to benefit the local environmental protection.

The project has been completed in September 2009, On October 25,2009. Shanghai LNG project held a ceremony to celebrate the first LNG transport ship—Malaysia International Airlines' "Diamond Princess I" transferred $13.7\times10^4m^3$ LNG and berthed at Shanghai special LNG jetty, which marked that the LNG project in Shanghai had changed from project construction into production stage.

By the end of 2016, Shanghai LNG project had received a total of more than 19×10^6t LNG.

1.2.3.4 Zhejiang LNG Project

Zhejiang LNG project is CNOOC's fourth LNG projects in the coastal. The project is to build a scale of 3×10^6t/a in the first phase and the scale increased to 6×10^6t/a in the second phase. The project consists of jetty, terminals, supporting pipeline section. The constructions include 3 $16\times10^4m^3$ storage tanks; one jetty with berthing (8.0~26.6) $\times10^4m^3$ LNG ship and water intake and drainage project. Supporting natural gas pipeline length is 38.7 km, and the design pressure is 7.0MPa. The first phase of the project design capacity is $4.2\times10^9m^3/a$, and the second phase of the project is $8.2\times10^9m^3/a$. Zhejiang LNG project is a total investment of about 6.97×10^9Yuan. Supporting natural gas pipeline project total investment is about 640×10^6 Yuan. And through the supporting pipeline connected to Zhejiang province trunk line, the gas is transported to the city network.

In the LNG project, CNOOC's Gas and Power Group holds 51%, Zhejiang Energy Group Co., Ltd holds 29% and Ningbo Electric Power Development Company holds 20% in project, respectively. CNOOC Zhejiang Ningbo LNG Co., Ltd. was set up to operate the project. Trunk line project by Zhejiang province natural gas development Co., Ltd. is responsible for the construction. LNG was bought from Qatar.

On March 10,2004, Zhejiang Province and CNOOC signed the "the Cooperation Agreement of Zhejiang Province's Introduction of liquefied natural gas and the Application Project" in Beijing.

On December 3,2004, the NDRC formally issued (development and Reform Office of energy 〔2004〕No. 2279) the document, approved to promote the work of the LNG project in Zhejiang.

On June 29,2009, the NDRC approved the project and marking the Zhejiang LNG receiving terminal project has made substantial progress.

On September 19,2012, the first LNG ship arrived to the jetty, which means the LNG project came into operation stage. By the end of 2016, the LNG project had received a total of more than 6.4×10^6t LNG.

1.2.3.5 Dalian LNG Project

Dalian LNG project is the first LNG project constructed by CNPC in Chinaese northeast area and is the first batch of large-scale projects for CNPC importing LNG resources from overseas. The project is located in Da Gushan Peninsula Catfish Bay, Dalian City bonded zone, Liaoning Province. Dalian LNG project includes three parts: jetty, receiving terminal and gas pipeline, and the project is built in two phases.

The first phase of construction scale is 3×10^6t/a, gas supply capacity of $4.2\times10^9m^3$/a, began to production in 2011. The second phase of construction scale is 6×10^6t/a, gas supply capacity of $8.4\times10^9m^3$/a. The total investment is 10×10^9Yuan.

On April 18, 2008, civil engineering began, and by the end of 2009, the roofs of tank No. 1 and No. 2 had risen. Gas transmission pipeline is composed of Dalian to Shenyang trunk line, Dalian branch line and Fushun branch line. The full-length of trunk line is 389 km, and the length of branch is 86 km.

On November 16, 2011, the first ship loaded with LNG docked at the jetty, to carry out the unloading operations.

On December 29, 2011, the natural gas pipeline for commercial transmission was begun from Dalian to Shenyang, marking the official input of production and operation.

Dalian LNG project mainly received LNG resource from Australia, Qatar and other countries. The project mainly supplies natural gas to the users in Liaoning province. The trunk line and planning line in the north-east gas pipeline network have been connected, formed a multi gas sources supply.

1.2.3.6 Jiangsu LNG Project

Jiangsu LNG project is invested by CNPC, Pacific Oil and Gas Co. Ltd. (which is a subsidiary company of Singapore Golden Eagle International Group Co.) and Jiangsu Guoxin Asset Management Co., Ltd. CNPC holds 55%, Pacific Oil and Gas Co. Ltd. holds 35%, and Jiangsu Guoxin Asset Management Co., Ltd. holds 10% stakes respectively in the joint venture company.

Jiangsu LNG project consists of the artificial island, the receiving terminal, the jetty, the submarine pipeline and the gas transmission trunk line. The receiving terminal is located in the Yellow Sea coastal shoal radiation Xitaiyangsha artificial island, Rudong County, Jiangsu covers an area of $0.3km^2$. The special LNG jetty was build for the $26.7\times10^4m^3$ of large LNG transport ship to berth. The receiving terminal is constructed in two phases. The first phase of the project is designed to receive 3.5×10^6t/a LNG, and the second phase is supposed to be 6.5×10^6t/a, so that the average annual supply of LNG can reach $8.7\times10^9m^3$. At the moment, the first phase 3 16×10^4 m^3 storage tanks and supporting facilities were built, and it could supply $4.8\times10^9m^3$ natural gas every year. The long-term goal of the project is to supply $13.5\times10^9m^3$ natural gas every year, and the total investment is over 10×10^9Yuan.

The LNG of Jiangsu LNG project is from Australia and Qatar, etc.. In order to meet the market demand for clean energy in the Yangtze River Delta, through the gas trunkline connected with the West-to-East Gas Trunkline and Ji-Ning liaison pipeline to form a multi gas source complementary and mutually standby of safety gas supply.

On February 28,2007, Jiangsu LNG project had formally received approbation from the NDRC. In April 2008, the pile foundation of the LNG storage tanks of Jiangsu LNG project has been under construction, which marked that the project started to be constructed formally. On May 24,2011, this receiving terminal received and unloaded $14.5\times10^4 m^3$ LNG from Qatar transmitted by ships, which marked the project was put into operation formally. It began to enter commercial operation on November 8,2011.

1.2.3.7 Qingdao LNG project

Sinopec is also aspiring to the LNG project. Sinopec's Qingdao LNG project total investment is 4.5×10^9Yuan, the design scale of the first phase of the project is 3×10^6t/a, long term for 5×10^6t/a. According to the market demands, the project could gradually develop to the scale of receiving 10×10^6t/a (equivalent to $12\times10^9 m^3$ natural gas), so it could meet the needs of the industrial production in Shandong Peninsula and all sectors of society as much as possible. On October 28,2010, it was reported that in accordance with the information from Qingdao City Environmental Protection Department, the environmental impact assessment of Shandong LNG project has begun its public summons. The project includes LNG jetty, LNG receiving terminal and natural gas pipelines. The jetty and receiving terminal are located in Dongjiakouzui in the southeast of Jiaonan City and two $16\times10^4 m^3$ storage tanks would be constructed. When the project was completed, the LNG imported from overseas can be transmitted to Dongjiakou jetty. Natural gas transmission pipeline includes two parts: the north pipeline starts from the LNG first distribution station in Jiaonan City to Laixi distribution stations in Laixi City, and it covers four county-level municipalities including Jiaonan City, Jiaozhou City, Jimo City and Laixi City; the southwest pipeline starts from Rizhao distribution station (the enlarged one) to Linyi distribution station, via Jiaonan city, Rizhao City, Linyi City under the jurisdiction of Qingdao.

On November 4,2009, Sinopec announced that its subsidiary company and Exxon Mobil's subsidiary company have signed a framework agreement about the regular supply of 2×10^6t/a LNG in order to ensure the need of Shandong Qingdao LNG receiving terminal. The investment of the first phase project has been raised to 10.5×10^9 Yuan and has been put into operation in 2014.

1.2.3.8 Other LNG projects

In recent years, Chinese energy demand especially clean energy demand is increasing, which provides a huge space for the three major oil companies in Chinese to develop their own natural gas strategy, and the three major oil companies have teamed up with local governments and enterprises in succession. At present, according to report, the national authorities have approved nearly 60 LNG projects (Table 1-2). It is necessary to point out that CNOOC continues to lead the LNG industry, CNPC, and Sinopec followed. Chinese power groups, such as CHINA HUADIAN CORPORATION (HUADIAN), CHINA GUODIAN CORPORATION (GUODIAN), GUANGDONG YUDEAN GROUP CO., Ltd (YUDEAN), CHINA HUANENG GROUP

(HUANENG) also meaningfully enter the LNG industry, while private enterprises are also reluctant to show weakness, such as ENN, Xinjiang Guanghui Industry Investment Group (Guanghui), Hanas, Shandong Dongming Petrochemical Group (Dongming), JOVO, Baota Petrochemical Group (Baota), POLY-GCL Petroleum Group (GCL) also catch up from behind, the Golden Eagle Group with the international background is also the early bird catches, all of above form the LNG projects sites distribution in the coastal of Chinese with the potential of feudal lords vying for the throne, great things may be done by mass effort, which will be conducive to the development of Chinese clean energy.

Table 1-2 LNG Receiving Terminal Projects have been put into production, under construction and planning in China

No.	Project name	Construction Site	Scale	Operator	Production Time
Have been put into production					
1	Guangdong Dapeng LNG Project	Dapeng Bay, Shenzhen	385+470*	CNOOC	2006.8
2	Fujian LNG Project	Putian, Fujian	260+240	CNOOC	2008.4
3	Shanghai No. 5 channel LNG project	Shanghai	50	Shenergy Group	2008
4	Shanghai LNG Project	Yangshan Port, Shanghai	300+300	Shenergy Group	2009.10
5	Dalian LNG Project	Dalian Bay, Liaoning	300+300	CNPC	2011.7
6	Jiangsu LNG Project	Rudong East Sun Island, Jiangsu	350+300	CNPC	2011.6
7	Zhejiang LNG Project	Linbo Beilun, Zhejiang	300+600	CNOOC	2012
8	Dongguan Jiufeng LNG Project	Humen Port Island, Guangdong	150	JOVO	2012.12
9	Tianjin FLNG Project	Nanjiang Harbor Districtin, Tianjin	220+380	CNOOC	2013
10	Tangshan LNG Project	Caofeidian, Tangshan	350+300	CNPC	2013.12
11	Zhuhai LNG Project	Gaolan Island Pingpai Mountain, Zhuhai	350+350	CNOOC	2014
12	Hainan LNG project	Yangpu Heiyan Port, Hainan	300	CNOOC	2014
13	Yuedong LNG project	Jieyang, Guangdong	200+200	CNOOC	2014
14	Qingdao LNG Project	Jiaonan Dongjialou, Qingdao	300+300	SINOPEC	2014
15	Shenzhen LNG project	Dapeng Bay, Shenzhen	400	CNOOC	2015
16	Beihai LNG project	Beihai, Guangxi	300+300	SINOPEC	2016
	Subtotal		4415+4040		

Continued

No.	Project name	Construction Site	Scale	Operator	Production Time
Under construction and Planning					
17	Macao LNG Project	Huangmao Island, Macao	300	SINOPEC	2019
18	Zhangzhou LNG project	Zhangzhou Fujian	200	CNOOC	
19	Jiangsu LNG project	Binhai, Jiangsu	260+340	CNOOC	
20	Tianjin LNG Project	Binhaixinqu Langang Industry Zong, Tianjin	300+500	SINOPEC	2015
21	Guangxi LNG Project	Fangchenggang, Guangxi	60	CNOOC	
22	Qinzhou LNG Project	Qinzhou, Guangxi	400	CNPC	2020
23	Fuqing LNG Project	Fuqing, Fujian	600	CNPC	
24	Weihai LNG Project	Weihai, Shandong	300	CNPC	
25	Wenzhou LNG Project	Wenzhou, Zhejiang	300	SINOPEC	
26	Tianjin LNG Project	Tianjin Harbor South Industrial Zone	300	SINOPEC	
27	Lianyungang LNG Project	Lianyungang, Jiangsu	300	SINOPEC	
28	Sailo Shantou LNG Project	Shantou, Guangdong	300	Guodian Guangdong	
29	Huadian Huizhou LNG Project	Huidong Contry, Huizhou, Guangdong	400	HUADIAN	
30	Jiangmen LNG Project	Guanghai bay, Taishan City, Guangdong	300	HUADIAN	2017
31	Ganyu LNG Project	Lianyungang Ganyu port area, Jiangsu	300	HUADIAN	2017
32	Haojiang LNG Project	Guangdong Haojiang Guang'ao Port Area	150	HUADIAN	
33	Chengmai LNG Project	Qiaotou Town, Chengmai County, Hainan	600	HUADIAN	
34	Hunan inland River LNG strategic reserve center (Yangtze River Project)	Linxiang Industrial Park, Binjiang Industrial Demonstration Zone	50	HUADIAN	
35	Qidong LNG Project	Qidong City, Jiangsu	60	Guanghui	
36	Zhoushan LNG Project	Zhoushan Economic Development Zone, Zhejiang	300	ENN	
37	Huang Mao Island LNG Project	Huang Mao Island, Zhuhai	300	Hanas	
38	Putian LNG Project	Putian, Fujian	300	Hanas	

Continued

No.	Project name	Construction Site	Scale	Operator	Production Time
39	Rizhao LNG Project	Rizhao, Shandong	300	Hanas	
40	Yangtze River LNG Project (Fransfer station)	Jiangyin Lingang Development Zone Shi Zhuang District, Jiangsu	100	Hanas	
41	Cangzhou Huanghua LNG Project	Cangzhou, Huanghua port, Hebei	260	Golden Eagle	
42	The Pacific oil and gas (Yangjiang) LNG peak shaving Gas storage	Jishu Operation Area, Hailing Bay, Yangjiang Port, Guangdong	200	Golden Eagle	
43	Qinzhou LNG Project	Qinzhou, Guangxi	300	Dongming	
44	Chaozhou LNG Storage and Distribution Station	Chaozhou Minyue Economic Cooperation Zone, Guangdong	100	Baota	
45	Penglai LNG Project	Penglai City, Shandong	280	Baota	
46	LNG Transfer Station, Nantong Coastal Industrial Park	Industrial Park of Qidong Economic Development Zone	60	GCL	
47	GCL Caofeidian LNG Project	Caofeidian Development Zone, Tanghai County, Tangshan City	700	GCL	
48	GCL Linhai LNG Project	Toumen Port New District, Linhai City, Zhejiang	300	GCL	
49	GCLTaizhou LNG Project	Taizhou, Zhejiang	300	GCL	
50	GCL Caofeidian LNG Transfer Station	Caofeidian Development Zone, Tanghai County Tangshan City	700	GCL	
51	Changlian LNG Project	Caofeidian Development Zone, Tanghai County, Tangshan City	650	Changlian Oil/Beijing Energy	
52	Guangzhou LNG emergency peaking gas station	Nansha Xiaohu Island Chemical Zone, Guangzhou	200	Guangzhou Gas Group	
53	YUDEAN Shantou LNG Project	Shantou, Guangdong	300	YUDEAN	
54	Zhoushan Zhejiang LNG Project	Zhoushan, Zhejiang	300	ENN	2017
55	Huaneng Huizhou LNG Project	Huizhou, Guangdong	---	HUANENG	
	Subtotal		11430+840		
	Total		15845+4880		

* Note: 385+470 represents in phase 1, 3.85×10^6t/a, in phase 2 to increase 4.7×10^6t/a, the rest is similar.

Chapter 2 LNG Project Integrated Management

2.1 The Concept of LNG Project Integrated Management

2.1.1 The Definition, Characteristics and Principles of LNG Project Integrated Management

2.1.1.1 The Definition of LNG Project Integrated Management

LNG project management is a comprehensive and all-dimensional project management and process, in order to ensure the coordination and cooperation of various issues in LNG project. It includes the implementation of project integration, the coordination of different relationships in the process of project operation and the overall control of the project change. As "integrated management", it is featured as unity, consolidation and combination, means which includes the project owner's requirements on the users and other stakeholders, and the important actions necessarily should be taken during the whole project for managing their expectations.

LNG project integrated management includes three aspects: the initiator's LNG development strategies, policies, guidelines and program management; the project mainplan including LNG transportation plan making, the project pre-market, the feasibility study, design and construction plan making and implementation; and LNG user group plan making. Its internal relationship and coordination are the main content of the above three aspects.

The meaning of integrated management is quite significant for the LNG project, it involves a long industrial chain of the upstream, midstream and downstream enterprises cooperation. The prominent social impacts involves concerns of national, local, enterprise interests coordination; the large investment is coming up to billions, and the high technology content includes ship building, complex storage tank and cold energy technique.

2.1.1.2 The Characteristics of LNG Project Integrated Management

(1) Wholeness.

Since the LNG project involves the national, local, enterprise interests, in order to keep the project running well, we must take overall consideration, seize every key point and boost the project step by step at the very beginning.

① Considering the overall interests of the country firstly, according to the relationships between national macro-economy and energy needs, we need to select the region with most needing clean energy as a breakthrough of the LNG project. For instance, CNOOC chose Guangdong

Province, the first region beginning reforming and opening and the most active and developed province in China to carrying out LNG pilot projects, which indicates that the project initiator accurately grasped the national macro policy, and actively responded to the demand for clean energy, achieved a specific project and the national energy structure of the overall planning of the anastomosis, these were the key to the success of the LNG pilot project.

② Local economic development planning is the basis of project promotion, however, from the viewpoint of regional economy, there is also a relatively overall development issue. The initiator has to work closely with local government in order to formulate a practical LNG project planning and make the specific LNG project coordinate with local economic development plan, for example, in the process of promoting Guangdong Dapeng, Fujian, Zhejiang Ningbo and Shanghai LNG projects, CNOOC played a quite active role to take the gas power plants and city gas as a breakthrough, just focusing on local economic development plan as a guideline, to increase the proportion of clean energy in the total energy, to solve the power shortage situation and preserve the ecological environment which played a very active role and received strong support from the local government.

③ Initiators' development strategy is the driving force of the project, which involves the overall issues in enterprise's development strategy. For example, according to its upstream development status, CNOOC put forward the development strategy to the downstream market early and especially took the foreign LNG imported as one of its own direction of development trends. And it made continuous breakthroughs in coastal LNG projects so that it led the development of Chinese LNG industry, which should be seen as the fruitful results brought out by the accurate positioning strategy made years ago. But as a company, CNOOC should also consider the competition with other domestic energy competent companies and overall deployment from the government department in order to avoid resulting in unnecessary waste because of the similar enterprises' duplication of the upfront investment and works.

(2) Coordination.

LNG project integrated management should give full consideration to the coordination between upstream, midstream and downstream projects, which includes the coordination between upstream LNG resources and downstream LNG needs, the coordination between upstream LNG resources and midstream LNG transportation, the coordination between transportation and the main LNG projects, and the coordination between the LNG projects and downstream users. However, the accomplishment of all the coordination are relied on the operation of the LNG project entity.

① Upstream resources are the basis of the LNG project, so the initiator of the LNG project or the actual operation team of the project should pay special attention to the implement of resources. At the same time, the annual supply scale of LNG should be based on market research and overall balances about the LNG downstream users. It is necessary to take into account both the annual needs in initial periods and in increasing periods and both the needs in stable periods and in the periods after expanding. Besides that, the resources should base on sales prospects and

keep them synchronous and coordination.

② LNG midstream transportation is an important link that is to guarantee the LNG resources delivered to the LNG project operator. When the annual supply scale of LNG is determined, the LNG project company will work with suppliers to develop transportation plans and arrangements, also inform the information of the shipping company, LNG vessel and transportation plans to the upstream LNG suppliers, so that according to the ship types and transport arrangements, the LNG suppliers could make plans for LNG loading docks and the supply cycle, in order to guarantee the transportation supply. Besides selecting experienced transport companies, LNG project operating companies also need to communicate and coordinate with them about not only the selection of ships with appropriate carrying capacity but also some superfluous carrying capacity, to make plans and arrangements for periodic annual transportation supplies with them and to announce the turnover capacity and the residual quantity of storage tanks, in order to ensure the gas supplies for downstream users in every day, month and year. Pipeline transportation is also an important part. At the moment, when LNG receiving terminal projects constructed, most companies also operate pipeline projects that should be taken into consideration with storage tanks, the market gas demands and so on.

③ LNG project operating company coordinates with downstream users mainly in the early stage of the project market research. Each of the users should supply annual gas consumption and gas consumption change cycle according to its own characteristics and rules of gas consumption, so that LNG project company can take full consideration of all kinds of factors to think over the amount of resources of the whole project and the gas load center, so as to achieve the matching of the LNG resources and the terminal, the receiving station and the route of the gas transmission line.

2.1.1.3 The Principles of LNG Project Integrated Management

(1) Systematic principle.

From above characteristics of LNG project integrated management, what we've said, the overall coordination of the project requires the project integrated management and to be systematic. Arbitrary and artificial Management will inevitably lead to project failure, and the systematic management in single aspect can't ensure the effective operation of the project; the systematic management in every sub-item but with mutual independence also can't completely ensure the ability of the whole system. Only based on the systematic management of every sub-item and effective integration and coordination, it can form the systematisation of the whole project management and make the LNG project ultimately successful.

(2) Combination principle.

If taking LNG project operation entity as an internal work environment, then, in addition to be mentioned, all will be the external environments. The LNG project has to adopt the principle of combining the internal with external environment because of its prominent social impact, long industrial chain and high technology content. First of all, in the LNG project examination and approval, it needs to be approved by State Department, in the LNG resources, transport vessels, related materials and equipment, storage tanks and receiving terminal, it is necessary to work closely with domestic and foreign industries. Also, in sub-projects, like the site selection of LNG project jetty, receiving terminals, the related user's selection of new construction or renovation should get strong supports and active cooperation from local governments.

(3) Transparency principle.

During the project operation process, the LNG project company must play the role of bridge between the upstream and downstream. For the downstream, it needs to give the upstream and midstream information including LNG sources supply origin, supply volume, LNG price. The information of transportation management includes the capacity of the ship, the mode of transportation and the distribution of the supply volume and so on. For the upstream, it is necessary to give the downstream information including user types, the gas peak valley difference, LNG demand, LNG operating company gasification cost, transportation cost and management fees and other information, as well as the natural gas price and so on. Only if making information transparent as far as possible, the project stakeholders can take risks in a harmonious environment and maintain the project operation in order to maximize the interests of all parties.

(4) Comprehensive balance principle.

From the viewpoint of the national, local and enterprises' interests, only if all the interests of these three parties are taken into account and are balanced, the state can approve the project, the local government can support the project and companies can promote the project. From the viewpoint of the industrial chain, the success of LNG project must have the upstream resources as support. At the same time, only to find the downstream users and reach the scale of the project, the economic chain can be formed and the balance point can be eventually found on the basis of the value chain projects, so that a tripartite win-win situation can be realized. Tripartite interests and industry chain links must meet the requirements at the same time.

2.1.2 LNG Project Integrated Management Contents and Their Mutual Relations

2.1.2.1 Main Contents

Given the actual operation of Chinese LNG projects, the promotion of projects must be composed of three parts to form the complete contents of LNG project integrated management, which are the initiator part of the LNG projects, the main part of the LNG project and the LNG user group (Figure 2–1).

(1) The initiator's integrated management.

The initiator's integrated management means that according to the LNG development strategy of the company, project initiator should analyze the domestic LNG industry environment and probe into its advantages, limitations, opportunities and challenges of the LNG investments. The process assets of the initiators will have a great impact on the LNG project, the initiators and local governments sign a cooperation agreement to carry out the LNG project is the first step in the LNG project.

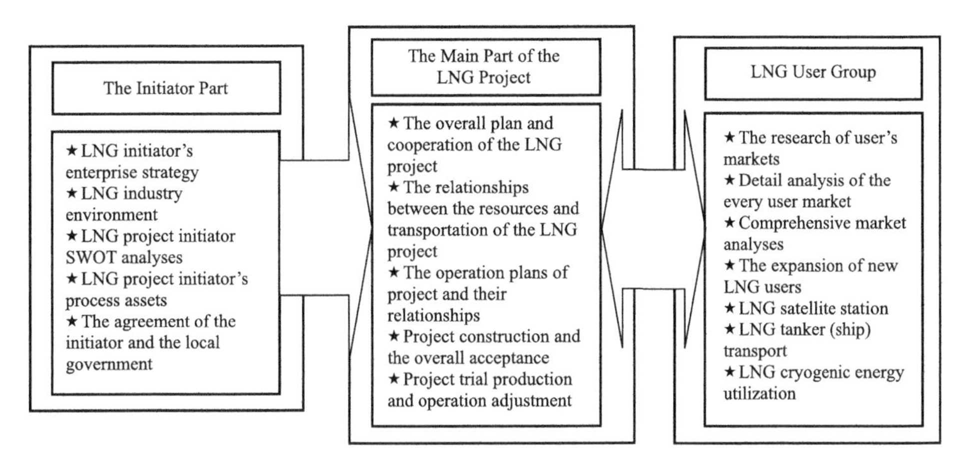

Figure 2–1 Contents of LNG project integrated management

(2) The main part of LNG project integrated management.

The LNG project integrated management mainly includes LNG overall project planning and coordination, the relationship between LNG project resources and transportation, as well as LNG project company to participate in several cases of resources and transport. LNG project company management strategy and planning; LNG project operation plan and interrelation, including site selection and land acquisition and relationship with project feasibility study; the relationship between construction schedule and quality, the relationship between budget and quality, the relationship between schedule and budget, the relationship between health, safety, environmental protection and other investment resources; LNG project company personnel structure and their relationship, the internal communication between the various departments, the coordination between the sub projects; the project owner, contractor and supervisor relationship coordination; the overall project acceptance plan and trial production and operation adjustment of the project and so on.

(3) The integrated management of LNG user's group.

The integrated management of LNG user's group includes gas market research to determine the size of LNG resource, which is an early work of the LNG project. Different users' markets and their relationships and comprehensive gas market analysis are the major parts. The development of the new LNG users is the basis of the subsequent expansion of LNG project so that the recirculation of LNG project management can be formed.

2.1.2.2 Interrelation

The relationship among the above three is: the initiator's integrated management is the driving force to promote the LNG project, the cooperation agreement between the initiator and the local government forms an interest community; the project company composed by the project team, project office or initiator and local companies is the operators of the LNG project and it plays an important role in LNG project management; LNG project users are a prerequisite for the success of the project. Judging from the current situation, the promotion of Chinese LNG project, it needs to consider the problem of both ends, one is LNG resources, it needs to look for LNG resources from foreign oil and gas upstream companies; the other is gas users, it is necessary to develop local existing gas users or the new gas users. LNG project company takes above two aspects to form a community of interests through the project, to form the each other is a prerequisite of supply and demand conditions (Figure 2-2). It embodies that include national macroeconomic development plan and regional economic development plan, initiator's strategic development goals of LNG project investment and local government supporting policies, also include the resource structure of the investment initiator, the layout of local industrial structure and so on. The initiators or shareholders, operate the project by companies authorized by the Board of Directors, the project company and downstream projects through the contract, to lock their respective responsibilities, obligations and rights.

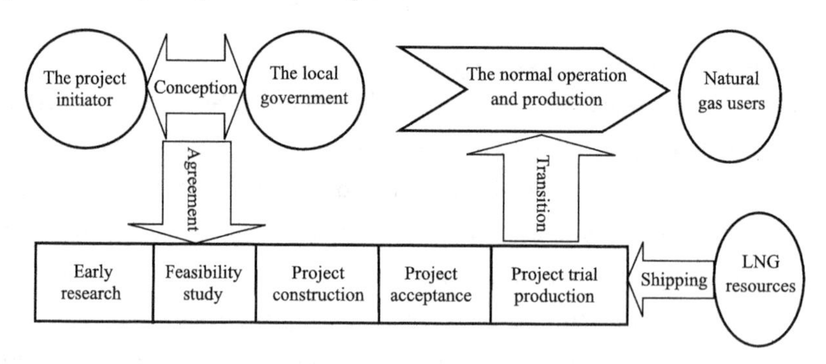

Figure 2-2 LNG project complete operating system diagram

2.2 The Initiator's Integrated Management of LNG Project

2.2.1 The Initiator's Enterprise Development Strategy of the LNG Project

The following is taking the pioneer of Chinese LNG industry — CNOOC as example, describes its relationship between the development strategy of LNG and the formation of LNG project.

2.2.1.1 The Coordinated Development of Upstream and Downstream

CNOOC relying on upstream oil and gas resources exploration, development, production and sales, and gradually grow up to form its current core industry, while driving the development of other complementary, interlocking business segments. As we can see, CNOOC's upstream industries are "Predominant", and then the development of its downstream business segments including natural gas industry, in order to make CNOOC become "Big". Paying too much attention on the "Predominant" would limit CNOOC's development speed; paying too much attention on the "Big" would impact CNOOC's development quality. However, realizing the unity of the "Predominant" and the "Big", which means the synchronous development and growth of the up-

stream and downstream, is one of the important principles of CNOOC's development strategy.

2.2.1.2 The Strategic Navigation of the LNG Industry

Regarding "Rely on upstream businesses and expand downstream business" as its guideline, during the "Ninth Five -Year Plan" period, CNOOC has already worked hard to enlarge the exploration and development of offshore natural gas, and begun to study imported liquefied natural gas actively. Besides that, under the strong support from competent government departments, it led to complete a report called "Southeast Coastal Areas Using Liquefied Natural Gas and Project Planning Report" . It should be said that this is a strategic report on the introduction of LNG projects in Chinese which has played a pioneering role. On this basis, by the end of 1999, the government approved the proposal of the first phase of overall Guangdong LNG pilot project, which marked that China's LNG project officially entered the implementation phase.

2.2.1.3 The Formation of Coastal LNG Industry Trend

In the "Tenth Five -Year Plan" of development strategy, CNOOC began to enter the midstream and downstream industry from the upstream industry, and clearly announced the grand goal— "the construction of the international first-class integrated energy company", set up a wholly-owned subsidiary—CNOOC Gas & Power Group company to manage LNG industry, so that the strategic management of LNG industry became more systematic and standardized. Up to now, LNG strategic development studies from Guangdong to Shanghai, from Fujian to Hainan, and until to other coastal areas have shown broad prospects. Undoubtedly, CNOOC's LNG development strategy is the basis for the development of the industry. The "Eleventh and Twelfth Five-Year Plans" period is the best period for CNOOC to turn the blueprints described in development strategy into LNG jetties, receiving terminals and pipelines.

2.2.2 The Factor's Analysis of LNG Industry Environment

The industry environment here refers to the internal and external factors affecting LNG project operation. It includes national macroeconomic policies and related laws, LNG standards, norms, the industrial policies of corporate sponsors, local development planning and so on.

2.2.2.1 National Macroeconomic Policies and Related Laws

From the viewpoints of national energy policies, clean energy including LNG has always been in the list of the industry encouraged to be invested and developed. At the end of the twentieth century, the state successively approved Guangdong LNG project, Fujian LNG project, Zhejiang LNG project, Shanghai LNG project, Shandong LNG project, Liaoning LNG project, which forms a complete layout of Chinese coastal LNG project development. There is no doubt that these initiatives promote the development of the cause of LNG.

But from national legal level, there is not the supporting legal system, natural gas regulatory system has not yet formed. For natural gas with the high thermal efficiency, environmental protection and clean energy, there is also a lack of economic incentives at the national level, which also has a negative impact on the advance of the LNG project. Thankfully, the competent government departments have already paid special attention to and encouraged the development and use of clean energy, and it is organizing relevant departments to enact natural gas regulations. We have enough reasons to believe that with the growth of this industry and the improvement of its status in the national economy, national laws and regulations also will be gradually perfected in order to help this industry developed as soon as possible.

According to the legal system of our country, enacting organ and effectiveness, it can be divided into following several levels (Table 2-1):

Table 2-1　Chinese Legal and Regulatory Levels

Level	Compilation unit	Illustration
Laws	Standing Committee of the National People's Congress	Normative documents formulated by the state's highest authority and its permanent body, which are to adjust the legal norms in China's economic construction, are the highest guiding principle of Engineering construction
Administrative regulations	The State Council	Normative documents formulated by the highest state administrative organs. They usually appear in the name of regulations, rules, ordinances, etc. Their position are the second only to the law, and they are the legal norm to adjust the relation of some aspects in the economic construction of our country
Local decrees	The local People's Congress and Its Standing Committee	All provinces, autonomous regions and municipalities directly under the central government and city specifically designated in the state plan and special economic zones and the National People's Congress and its Standing Committee, normative documents formulated in the scope of the provisions of laws and administrative regulations within, belong to local regulations, laws and regulations is to adjust the relationship between the local economic construction in some aspects of the relationship
Regulations and standards	Ministries and Commissions under the State Council and the Local Government	The normative documents are formulated and released in accordance with the law, which are used to adjust some aspects of the project construction, and their effectiveness is lower than the laws and administrative regulations

2.2.2.2　The LNG Standards and Specifications

Standards occapy an important position and play a very important role in the development of LNG industry. CNOOC entered the LNG industry for 20 years, and it is also to carry out an in-depth investigation of the foreign LNG industry standards and regulations system, draw on foreign LNG standards for 20 years, and gradually formed Chinese LNG industry regulations and standard system framework. The Figure 2-3 shows that Chinese LNG industry regulation standardization system includes: national laws and regulations, international LNG standards, national LNG standards, industry LNG standards, engineering construction standards and enterprise LNG standards. The six parts form a whole, together play a role in Chinese LNG industry.

(1) Standardization construction of LNG industry in foreign countries.

International LNG industry has matured, and the relative standard system goes to the inte-

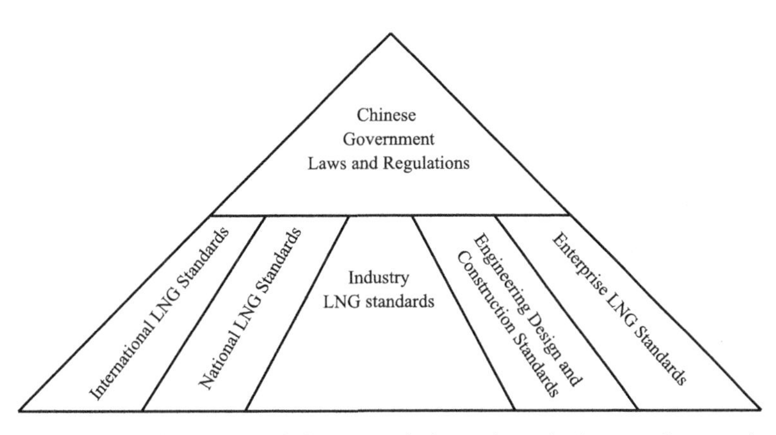

Figure 2-3 Chinese LNG industry regulation and standard system framework

gration. The United States, Europe and Japan, respectively set up their own LNG standard system. International Organization for Standardization (ISO) takes WG10 as a platform for the coordination, planning, preparation of ISO standards. Currently the main work of WG10 working group is: evaluation of existing national LNG industrial standards and specification requirements; evaluation the applicability of the existing LNG international standards and the system in the LNG industry; coordination between each country standard work and the WG10 work combination; the preparation of the ISO standard.

(2) National and industry LNG standards system table.

Our country has done a systemic research on foreign LNG professional standard system by LNG Standardization Technical Sub-Committee, with total of the three stages of work: investigation, collection and summary of foreign standards; recognition and translation of core standards; research, screening and comparison to form the reports. On the basis of the work above, "LNG Professional National and Industry Standard System Table" is compiled.

(3) Enterprise LNG standards system.

Enterprise LNG standards are involved in LNG enterprises must do the preliminary work. CNOOC entered Chinese first several LNG project's construction, accumulated a wealth of practical experience, which provided the possibility for the establishment of enterprise standards. Enterprise standard is an important mean to strengthen internal management of the enterprise, improve efficiency, reduce cost, avoid risk and enhance market competitiveness. It plays a role of coordination and unity in the enterprise internal management and technical work and it is a concrete and refined tool for the project management and technical guidelines. It can be made according to the characteristics and advantages of the project's own, and it is the reflection of project technology, knowledge, wealth, and the essence of the management. It is also an important part of the summary of project success experience and management system. It supplements each other with LNG major national and industry standards, on different occasions, in the service of different objects, cannot replace each other.

2.2.2.3 The Industrial Policy of Initiator Enterprise

In addition to the initiators' industrial strategy of enterprise in the above introduction, enterprise policy of LNG industry also affects the progress of the project. Mainly includes:

(1) LNG investment policy.

Clean energy LNG is one of the countries to encourage investment industry, it is also one of CNOOC's development strategy. In recent decades, CNOOC has invested a lot of manpower and material and financial resources in the early stage of the LNG market research, a complete LNG industry development strategy and planning is formed, and gradually implemented. Investment policy is associated with LNG initiator's enterprise capital strength. CNOOC's capital is abundant, so the LNG project investment has taken a positive incentive. During the period of "11th Five-Year Plan", its own funds will invest 100×10^6 Yuan, the××project total investment will reach×××100×10^6 Yuan.

(2) HR policy.

LNG industry is a new industry in China, lacking professional and technical personnel. But CNOOC actively faces the talent shortage, combines "send-out" training with "invite-in" training. CNOOC basically master the LNG technology and theory by communicating with foreign counterparts and field study. At the same time, by the practical operation of 2 LNG project in Guangdong, Fujian, it also trains up a number of senior management and technical personnel, including project managers, the personnel who in charge with market research, LNG resources procurement, engineering project management, HSE management, risk management, finance and financial management, material procurement, business and contract management, etc. Currently, the work has been carried out, which recruits new college students, graduate students and brings in social talents according to the HR plan.

(3) The project organization structure management policy.

From the structure of organization, the initiator should give full play to the initiative of its enterprises and local enterprises, and let the local enterprises participate in the project. In accordance with the requirements of the company law, to establish the LNG limited liability company, the highest authority is the general meeting of shareholders, determining the company's future and destiny. The company sets up the board of directors and the board of supervisors, making decisions on major issues of the company's business, supervision and authorization. The project company has independent legal person status and shall not be managed by the parent company directly, but to be managed by the board of directors of the project company, authorized by the board of directors, the general manager of project company takes charge of comprehensive management on the daily work.

How much the initiators invest in LNG project company's shares, will determine the size of its project company management and control rights, there will be 2 cases:

① Holding company, and there are 2 kinds of situations: the shares in the project company is more than 51%, we call it absolute holding company; The shares is larger than all investment companies but not reaches 51%, we call it relative holding company. In first case, the initiator acts usually as a president, and the number of directors is the most. According to the rules of order, the board of directors has a certain influence. In second case is that although the shares are at most, the most seats in the board of directors, but its influence is not big than the first one.

② Participating company, namely in the company, its investment and directors are at least less than the other and it has less influence in the board of directors.

(4) Risk control policy.

From the organization, adopts the multistage risk control mechanisms, including the control by the project sponsors investment committee, by the company's shareholders' meeting, by the board of directors, by the board of supervisors, by the project company and by the supervision, etc.

2.2.3 LNG Project Initiator SWOT Analysis

To any initiators who want to devote LNG project, they must analyze that before and after they enter this industry, the existing favorable factors and unfavorable factors, according to their enterprise's actual situation. It's namely the SWOT analysis. The following is an example of CNOOC.

2.2.3.1 Strengths

(1) Ahead in domestic.

For the introduction of LNG industry and technology, CNOOC is in the leading position in domestic, mainly in the following three aspects.

① The project operation. Through nearly 20 years of the early stage of the project research and project operation, CNOOC's LNG project officially started in 1995. The NDRC entrusted CNOOC to compile the liquefied natural gas (LNG) planning, CNOOC reported the southeast coastal area utilized LNG and project report and planning, which laid the foundation of a framework for the future development of LNG in China. After 10 years of effort, on June 28, 2006, the first phase of the Guangdong LNG project was formally put into production, which marked the arrival of the era of large-scale import of LNG in Chinese. Currently the coastal import LNG projects are implemented, totally nine. They locate in Guangdong, Fujian, Zhejiang, Shanghai, Hainan and other places. It is the company with the largest number of LNG projects in Chinese. Through the project operation, we have mastered the core and key of the resources negotiations, business contract and project management. At the same time, it also has the understanding of the main technology, equipment process and practical operation ability. This book "LNG Project Management" is a comprehensive summary on the LNG project management theory and practice. At the same time, external along the industry technology and business are studied, such as small LNG satellite stations, tanker, tank ship transport business entity business operations research. Currently, CNOOC makes the LNG as a leading project, also forms a complete set to participate in the construction and operation of pipe network, LNG satellite stations, CNG stations, LNG filling stations, city gas, which forms a complete set of construction and development of the small liquefaction plants and cold energy utilization projects. The good news is that the industry chain projects have been implemented and achieved good economic benefits.

② Standard formulation. At the beginning of the development of LNG industry in China,

CNOOC has done extensive research and comparison on the international LNG industry standards. As the lead unit of the LNG Standardization Technical Sub-Committee, CNOOC actively pushed forward the construction of the standardization of LNG, chosen foreign standards as a blueprint for Chinese national conditions and established country (industry) LNG professional standard system. Starting from the actual LNG project management, on the basis of a large number of technical and management experience, we also formulated the LNG enterprise standards. It made up the shortage that the national standard in the short term cannot form a complete set and overcome the long process that the approval from project establishment to release cycle and complex procedures, timely supplement LNG enterprise standards. It has been proved to be effective method in LNG project management. All in all, CNOOC's LNG standardization work walks in the front of the domestic industry.

③ Scientific research. CNOOC focuses on key technology of LNG from the start. It sets up the topics such as the natural gas liquefaction technology and key equipment research and development, the large natural gas liquefaction plant research, small LNG plant process optimization and equipment localization research. Based on the core technology to master, it sets up the topics that a large-scale LNG full containment tanks research and the intermediate fluid vaporizer research. In consideration on forward-looking of industrial chain and technical reserves, it sets up offshore LNG receiving terminal project scheme selection study, small mobile LNG module receiving terminal research and offshore skid mounted natural gas liquefaction device research. Closely around midstream and downstream industry as main business, it sets up the LNG non-pipeline transport research and the application of the natural gas pipe network and LNG receiving terminal system digital technology research topics. According to the development of the company's new energy and the exploration to feasibility of the enterprises to develop new industries, it sets up the comprehensive utilization of LNG cold energy research topic and so on. It has received a batch of scientific research achievements, some are already used in production practice.

(2) Resources superiority.

CNOOC has the right to operate Chinese offshore natural gas in China. By the end of 2015, $1105.6\times10^9m^3$ natural gas reserves have been found. According to the second national resource survey, offshore natural gas resources in China are 8.4 resources and exploration potential is tremendous. It is believed that new gas fields will be discovered in the next few years. CNOOC has obtained the LNG with the lower price in Australia and Indonesia, and shared in the upstream gas field. At the same time, it has acquired the LNG resources in Malaysia and Qatar. The several kinds of LNG and natural gas resources complementary is a guarantee on the gas supply.

(3) Market superiority.

Through more than 20 years of market research and development, it is very clear on the coastal natural gas demand, and we have developed the markets in Liaoning, Shandong, Jiangsu, Shanghai, Zhejiang, Fujian, Guangdong and Hainan, which laies a solid foundation for the new projects.

2.2.3.2 Weakness

(1) Large investment.

An annual supply of 3×10^6t of LNG project, the investment of jetty, terminal and pipeline reaches to 7×10^9 Yuan and plus the supporting gas project: power plant, city gas projects with investment amounted to about 20×10^9 Yuan, so, the total investment is large, and the return time is long, means that the failure of the project will face huge economic losses.

(2) Industry experience is insufficient.

As the LNG project is a new industry in our country, whether the project management and related technologies, there is a process of introduction, digestion and absorption, and then to talk about improvement, integration, innovation, in the exploratory stage there must be some risk.

2.2.3.3 Opportunity

(1) The huge demand potential.

With the rapid development of Chinese economy and the growing demand for clean energy natural gas, the LNG industry is rising in China, and there is still much room for development.

(2) Extension of industrial development opportunities.

The LNG project as a leader project, it has led to the development of gas power generation, land gas pipeline and city gas industry, and its extension industry also includes a CNG, LNG filling station, tank car, tank ship transportation, LNG reserve center, LNG satellite station, LNG cold energy utilization, etc.

2.2.3.4 Threats

(1) Existing talent shortage.

Because that LNG in China is an emerging industry, the lack of existing talent is obvious. We should cultivate the technology and management talents in LNG industry by ourselves. As gas and pipeline talents coming into society, it is necessary to have a set of attractive policies. Due to the industry expand rapidly, the shortage of talents is more obvious.

(2) Resource risk.

At present, LNG projects are developing rapidly worldwide. The resources demand in Asia LNG is rising, in addition to Japan, South Korea and Taiwan, the three LNG importing countries and regions which mainly for the continuation of the resources of the old projects. It also increases the demand for LNG in India, Philippines, Indonesia and Singapore. The United States in the next 5 years will transfer a number of LNG receiving terminals to output terminals. The world is in from the LNG buyer's market to a seller's market transition stage, to obtain the competitive advantage in the price of LNG resources is still a risk.

2.2.4 LNG Initiator's Organizational Process Assets

2.2.4.1 Project Management Knowledge Database

(1) Definition of organizational process assets.

According to the definition of "A guide to the project management body of knowledge" by

USA Project Management Association, the organizational process assets refer to accumulated project management processes and procedures, information storage and retrieval of knowledge base in the past production management and practice of LNG project. It directly affects the success of the project. In addition to corporate strategy and industrial policy, the specific contents are just as follows.

(2) Organization's standard management.

If the company has just entered the LNG industry, LNG industry standards can be reference on the original industry standard management system, standard making procedures and guidelines, etc. These processes and procedures can provide reference and experience for the LNG industry standardization work.

(3) Project management experience.

Also, the original industry experience in project management, such as the project management experience of CNOOC in the upstream of offshore engineering, can be provided as reference for LNG project management.

(4) Corporate governance.

To the new LNG project company, it also needs a set of methods and procedures etc. Corporate governance, such as the initiator's corporate governance experience can be used as the reference.

(5) Financial control procedures.

It includes financial management system, management regulation, fund allocated, review, approval, authorization, etc.

(6) Corporate culture.

The survival of an enterprise has created the enterprise culture, and the development of a business depends on the influence and promotion of the enterprise culture. There is no doubt that the new company needs to innovate on the basis of the original corporate culture.

2.2.4.2 Organization Integrated Information Database

LNG project initiator has been formed in the process of asset size and information accumulation, which can be applied to LNG project company, or by modifying and improving, to be used in LNG project management.

(1) Project archives.

It refers to the project file management processes and management measures laid down by the initiator's organization.

(2) Engineering database.

It refers to the initiator's supplier library, individual items cost, overall cost, unit price, material price and equipment price, etc.

(3) Risk database.

It includes the lesson database, problems and defects of the database, etc.

(4) Expert database.

It refers to the various professional categories of experts, including the internal enterprise,

but also including enterprise external.

(5) The configuration management database.

It includes the organization's formal standards, guidelines, work procedures, file formats, templates, etc.

(6) Financial database.

It includes man-hour, cost, budget, plan, etc.

2.2.4.3 LNG Project Decision-Making Process and Procedures

(1) Change of national approval procedures.

At present, the competent authorities for the major projects involving national economy and people's livelihood has changed from the examination and approval system to the check and approval system. The purpose is to increase the enterprise subject status of the project investment. The state authorities are mainly from the maintenance of economic security, rational development and utilization of resources, protection of the ecological environment, optimization of major layout, protection of public interest, prevention of monopolies emergence and other aspects to review the project. By Figure 2-4, the LNG project sponsor and the project company established later will promote the LNG projects in accordance with the following procedures.

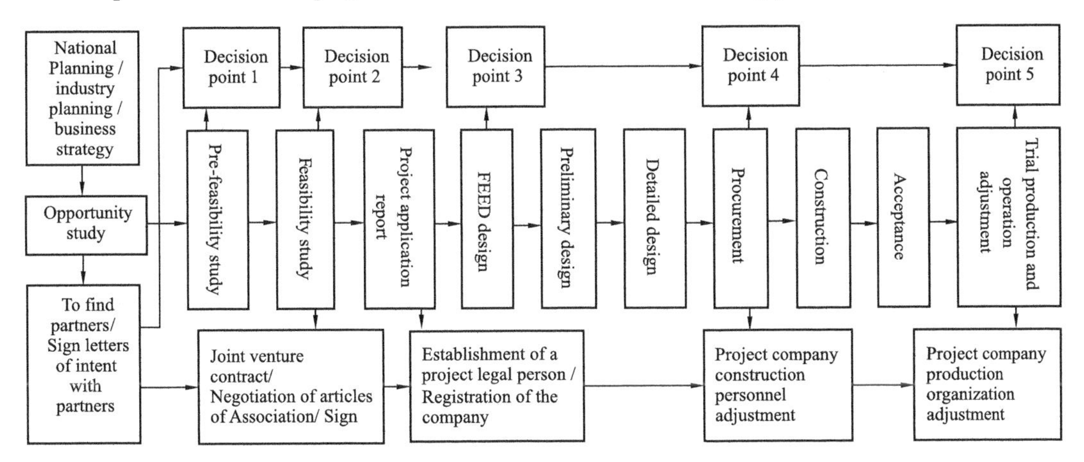

Figure 2-4 LNG project decision-making process and procedures

(2) Project operation procedures.

If the investment initiators want to launch the LNG project, according to the national planning, industry planning and the enterprise strategy. Firstly, it is necessary to carry on the opportunity study of LNG resource and market research, on the basis, to look for the project settled, fully cooperate with the local government and sign a cooperation letter of intent; then, carry on pre-feasibility studies and feasibility studies, offer the LNG project application report on the base, carry on FEED design, preliminary design and detailed design under the approval of the state sector; followed by contractors selection, including construction contractors, supervisors, material and equipment suppliers, and then, engineering construction, acceptance, and commissioning work. The above is the complete path to a successful project, when one of the link in question, the project running will stop.

(3) Corresponding organizational structure.

By Figure 2-4, corresponding organizations and agencies must be set up in order to promote the LNG project. In the early stage of the project, generally the project office and the project team composed of investment initiators carry on the opportunity study, pre-feasibility study, feasibility study, offer LNG project application report at the same time. After the application report approved by state, LNG project company will be set up. After the detailed design is completed, for adapting to the hard work of the construction of the project, it will takes complements and adjustments to the staff in the project company. When entering the related production phase, there will be a major personnel adjustment in the project company, increasing the operation personnel and mastering operation skills as soon as possible, so that the project could smoothly enter the normal production period.

(4) Important decision-making.

There will be a series of major decisions in LNG project operation process, according to its influence degree. We list the five major decision stages for readers as reference. One decision is after the preliminary feasibility study, It is to make a preliminary estimate and advice according to the construction project of the market, the scale of production, construction conditions, production conditions, technical level and economic benefit, environmental protection, funding sources and so on, mainly discussing the necessity and possibility of project set up from the macroscopic view. After completing the above work, the investors can decide whether to set up the project, whether to continue to do the preliminary work or not. Second, after the completion of the feasibility study report, make further conformation or negation to the feasibility of the LNG project. The third is, understanding LNG project investment scale after FEED design, the investors will make investment decisions. The fourth is, after accomplishing the detailed design, the company will conduct the bidding work of contractors and suppliers, investing various resources for concrete construction and acceptance. The fifth is put into operation after commissioning, Project Company will make major decisions in production organization and preparation for personnel, preparing for comprehensive formally production.

2.2.5 LNG Initiators and Local Government Cooperation Framework Agreement

After initiator completing pre-market research and negotiation with local government, the final result is to achieve the construction of LNG project cooperation framework agreement. Its important content is as follows.

2.2.5.1 The Cooperation Intention of Both Sides

The local government economic development and energy structure plan is the foothold of the LNG project, LNG initiator's gas utilization development strategy is a catalyst for LNG project, LNG initiators own LNG development strategy and local economic strategy coordination

and synchronization are the basis for the cooperation of specific LNG projects. Initiators have the resources, technology, management and financial advantages, which are the key to obtain large LNG project investment opportunities in the place.

2.2.5.2 The LNG Project to Achieve a Win-Win Result

By LNG initiator's early market research and discussions with local government departments, 2 sides need to reach a consensus, namely the development of the construction of the jetty, terminal and gas pipeline project for the introduced LNG which play a role as incubator. It will drive the gas power plant, city gas, chemicals, industrial fuel, LNG and gas filling projects propulsion and running, so as to accelerate the local economic development and social progress. At the same time, the implementation of the project initiator's LNG strategy also finds a support point, promotes the expansion of the economic strength and social influence of the initiators. As a result, a LNG project will often reach the local and the initiator of a win-win effect, the 2 sides should seize the favorable opportunity to start and actively promote the work of LNG projects and supporting projects.

2.2.5.3 The Responsibilities and Obligations of Both Sides in the LNG Project

(1) Initiators provide solutions.

The initiators submit to the local government "The construction of LNG project and supporting project plan". Both sides are willing to cooperate in good faith, coordinate closely and seriously in accordance with the "program" to develop and build LNG projects and supporting projects.

(2) The role of the both sides in the project.

Initiators should have the technology, business, management experience and strength for the exploitation of LNG project. The local government needs to agree to by the initiators to take charge of LNG project, in order to promote the progress of the project, initiators hope that the local government assigns one or several effective local enterprises to join research and construction. The two sides agree to the construction of LNG project receiving terminal and other supporting facilities by the parties through consultation.

(3) Both sides of the project operation ideas.

Generally, both sides set up the LNG project development and construction leading (coordination) team, and are responsible for the implementation of major issues in the project implementation process. The head of the leading group shall be led by the local government, and the deputy leader shall be appointed by the initiator, and the members are negotiated by both parties. Under the leading group, Executive Office should be set up, and its members are come from the local competent departments and initiators, responsible for coordination of daily work.

(4) Both sides on the future operation of the company framework assumption.

According to the existing energy management entities in the region, the leading group put forward principles of the future LNG project operation framework, including the range of LNG project company management, such as the initiators and local powerful company or several companies appointed by local government to form LNG project company (including natural gas pipeline network), or a single natural gas pipeline network company and so on, to reach the target of unified planning, construction, management, operation business in the future.

(5) To set up project team and the joint office.

According to the actual situation, by both parties negotiation, one or a number of project teams and the joint office consist of the initiators are set up, responsible for the LNG project, natural gas pipeline project and its supporting projects to start work together to carry out preliminary research, together with the pre-feasibility study when conditions are ripe, and lay a solid foundation for the development and construction of the project.

(6) To establish the project company.

It needs to draw up the structure of the LNG project company after its approval, form a project company led by corporate initiators or local enterprises, determine the company shares proportion of the project, and determine other investors in addition to the local enterprises to participate in the project company, such as foreign participants in the project and to do standard operation according to modern enterprise system.

(7) Project scale.

On the basis of the early market research, combined with the local government's economic development plan, project organization will initially identify the LNG project size for the LNG supply and the late development plan, and seriously study the feasibility of the project.

2.3 LNG Project Main Body Integrated Management

The LNG project main body integrated management is the LNG project in the establishment of a project team, the project office and the project company in three stages directly faces the specific LNG project to conduct a variety of plans, program formulations and variety of coordination management.

2.3.1 The Overall Project Planning and Coordination

2.3.1.1 The Main Components of the Project

In terms of LNG project work links, include four contents (Figure 2–5), namely LNG resource acquisition, LNG marine transportation, LNG users and LNG project construction. The LNG project initiators formulate the LNG development strategy of their own enterprise according to the national energy development plan. Considering the LNG industry environmental factors and on the basis of SWOT analysis from the LNG initiators, they finally sign LNG project cooperation framework agreement with a local government, which marks the prelude to LNG project. Each aspect from the four parts has their own work contents and procedures (Figure 2–5). From a vertical perspective is a progressive relationship and from the horizontal point of view is interdependent relationship, the four parts lock the rights, obligations and responsibility between

the various stakeholders by contracts. LNG Project Company signs sales and purchase agreement with foreign resources suppliers and signs a contract of carriage with the transportation company at the same time. What's more, they sign natural gas supply and receive contract with downstream users. All these make up the whole industry chain and the community of the interests.

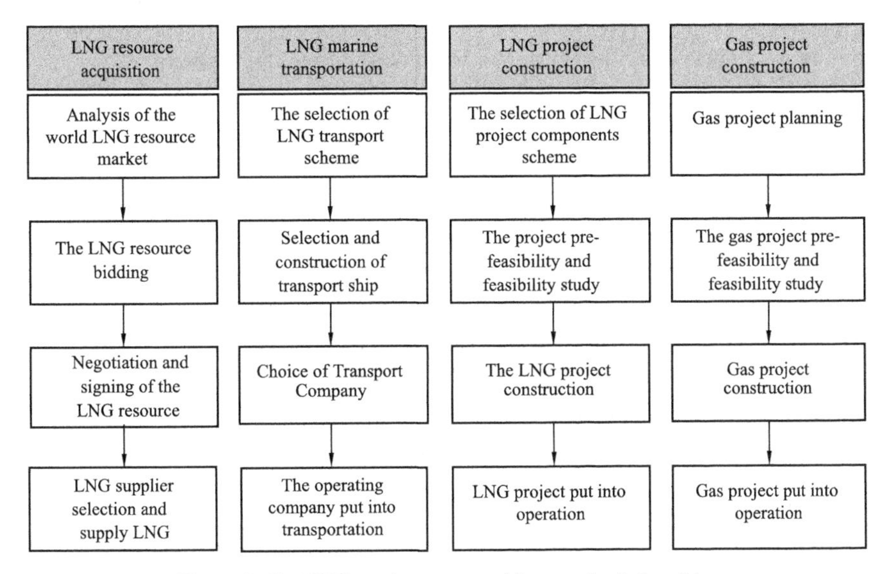

Figure 2–5 LNG project compositions and relationships

(1) LNG resources procurement plan.

LNG resources procurement is the source of LNG project. From the procurement process (Figure 2–5), includes the world's natural gas resources survey, the analysis for world's LNG production in market, selection of LNG resource countries, LNG resources bidding, LNG resources negotiations, the approval of state departments, final selection of suppliers and LNG resources supply. The LNG resource suppliers selected and negotiation are 2 key points.

(2) LNG marine transportation.

LNG marine transportation is an important part of the LNG project. Someone call LNG marine transport is the pipeline in mobile, and the LNG is delivered to the users. It is impossible to achieve the completion of LNG industrial chain without LNG marine transportation. In terms of transportation, includes several aspects of content (Figure 2–6). The first is to solve the transportation problem, also is to choose what type of LNG ships, membrane type or moss spherical. The second is to choose what mode of transportation, the buyer bearing the shipping (FOB) or the suppliers bearing the shipping (DES). The third is the composition of ship-owner company, LNG project company is one of carrier owners or completely rent by the company, building or renting LNG carrier. The fourth is the composition of the transportation company, generally speaking, LNG project company involved in transportation in order to control transport. The last is the composition of the shipping operation company, who is responsible for the marine shipping operations, whether the LNG project company participate in the shipping operation company or not. To complete the transportation task, these problems above must be answered and solved.

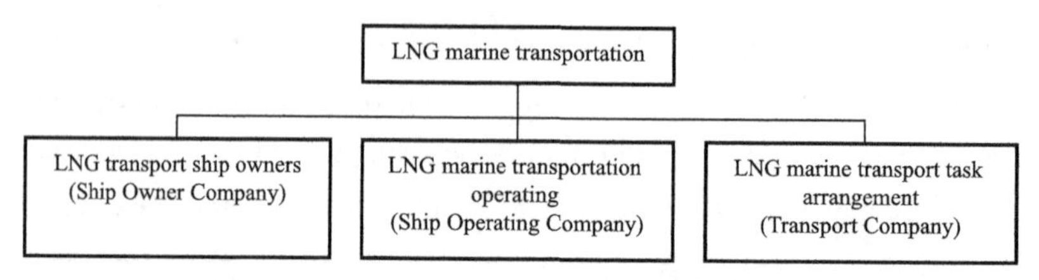

Figure 2-6 LNG marine transport task composition diagram

(3) LNG project management.

If LNG resources acquisition and LNG marine transportation are soft conditions or the necessary conditions of LNG project, then the LNG project construction is the hard condition or the sufficient condition for it. It must provide three hardware parts, including jetty, terminal facilities and onshore trunk line for receiving the LNG. In detail, that include the scope of the project, organization and personnel, the alternative site, project feasibility study, time, budget, quality, HSE and business contracts, accounting, finance, taxation, communication, informatization, completion inspection and acceptance, commissioning, formally putting into production and a series of the formulation, control and adjustment of plans and procedures.

(4) LNG users.

LNG users are the starting point and destination of the project, only to find the users can realize the final link of the LNG project. From the LNG project management scope, the LNG project company is responsible for the market development of gas users and determination of gas use units and enterprises. LNG users give days, months, years volume size of the gas consumption and gas peak valley difference, provide data for LNG project in supply, at the same time, track and coordinate the gas use project's progress, as far as possible to make the gas use projects and the LNG project to achieve the "three simultaneities", namely the simultaneous design, construction and put into production. As far user's project design, construction and production are the own things to various gas users. These gas users include city gas, gas power plants, industrial fuel, chemical raw materials, such as LNG and gas filling.

2.3.1.2 The LNG Project Management Strategy

(1) The definition of project management strategies.

Project management strategy refers to the countermeasures and methods that pointed to the work undertook by the project company, according to their own professional staffing, business policy and project company organization. In terms of the project company, project management can be divided into direct and indirect management. From the degree of control in management, we introduce the concepts: "Management Range" and "Management Strength".

Direct management refers that the project company sets up specialized agencies and arrange specific personnel. Indirect management refers that the project company manages a specific job through outside forces such as consultants, contractors and supervision companies. Management range refers to the scope and content of management. Management strength refers to depth of the management and the degree of control.

(2) The basis of formulating project management strategies.

① Organization. Once the LNG project implementation organization determined, it means to determine the range and strength of LNG project management. If the organization sets up completely, it can undertake effective and comprehensive direct management of the project. For example, the company sets up complete type of work and project design departments, it also sets up a strong international procurement department. What's more, it sets up not only experienced contract business department, but also a control and HSE department that controls and deals with the complicated situation, then the project management range is wide and the strength is big; if institutions are not complete or are equipped with the weak level of technology and management, then the range of project management will be affected. According to the situation, the project general manager can outsource some scope management and adopt the method of indirect management.

② Personnel allocation. The personnel allocation situation in the project team also affects the range and strength of management. As the project team members in the past were proved to be competent in project management and productive, we can use our own power to direct management. If the staff's technology and management level is general, it needs to consider hiring experts or consultants to manage. For example, in LNG tank design, procurement and construction, if the lack of talents in this area, it can only be used in the form of total outsourcing. Through the practice of Chinese current LNG project management experience, equipped with complete professional team members is the first step to improve project management level directly.

③ Time factor. Time factor is also a decisive aspect for the project team for the project management range and strength. When time is enough, we can learn through practice and methods to improve the management level of the team; if the time is tight, we have to consider using the domestic and foreign technology and management personnel to help us achieve its management objectives.

(3) Diagram of project management strategies.

① Totipotent type. Totipotent type is that, within the scope of the project management, the work is done by the project team basically, and the organization sets up to achieve complete and equipped strong personnel. The technical team from design to construction is complete and the management involved in similar project construction and management (Figure 2-7). In fact, this type of team is rare.

② Self dominated type. It can't compare with totipotent type team setup, but basically it is able to carry out a direct project scope management, only in some engineering design and construction, specific technology needs to borrow outside brainpower. At present, Chinese LNG project management team basically belongs to this type.

③ Ancillary type. The main technical and management needs to rely on external forces, and the project team participates in the assistive technology and management.

④ Dependent type. Basically it needs to rely on external technology and management power for project management.

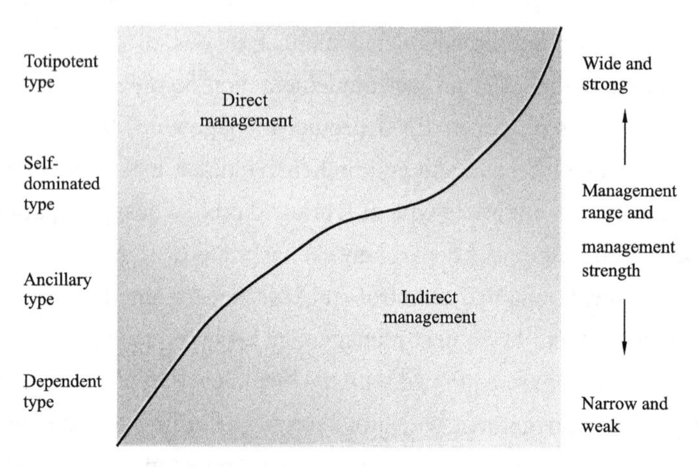

Figure 2–7 Schematic diagram of project company management strategy

2.3.1.3 The Plan Formulation at all Levels

(1) The project integrated planning.

Project integrated plan is generated results through the project and other special planning process, using the method of integrated and balanced developed, used to guide the project implementation and management of integrated, comprehensive, overall and coordinated integrated planning document. To achieve the guidance of the implementation of the project, encourage and inspire the project team morale, measure the project performance and control the project progress, and balance the interests of the project stakeholders.

(2) LNG project scope management planning.

LNG project scope management plan is a guide for the LNG project company to identify record and verify, manage, and control the LNG project scope. It includes the LNG project scope statement, the major work breakdown structure according to the scope of the description, each sub-project deliverable achievements, plan change control process, etc.

(3) Schedule management planning.

It includes project milestones plan and project master schedule control plan, and it can be subdivided into sub-project plans. LNG project must be completed on the implementation of each schedule according to time point, including activity definition, activity sequencing, activity resource estimating, activity duration estimating, schedule, progress control and so on.

(4) Cost management planning.

It refers to the cost management and cost control which is must be invest to complete the LNG project, including cost estimation, cost budget and cost control, etc.

(5) Quality management planning.

Quality management plan is part of the project management plan or affiliate programming. It should cover the quality standards of the early stage of the project selection, preliminary design, FEED design and detailed design in the process of quality control, the quality testing in the implementation, process quality improvement, quality inspection and acceptance of completion.

(6) Human resources planning.

According to project progress to arrange project personnel, including early project team personnel, project office personnel, project company personnel, transition personnel and training, compensation and benefits, recognition and reward, personnel severance arrangements, etc.

(7) Site selection programming.

Project team or the project company carries on the site selecting work on the basis of LNG resources, market conditions, selects jetties, receiving terminals and pipelines route respectively. On the basis of a comprehensive site comparison, select the preferred and alternative sites.

(8) Project feasibility study programming.

Feasibility study is the need of investment sponsors for investment decisions, is also the basic document for the approval of the competent authorities of the state. In the preferred and alternative site selections, it needs to discuss in some detail, such as project resources, market, transportation, construction, economic evaluation and so on. It is the foundation of project investment operation.

(9) Information management planning.

It includes LNG project stakeholders' communication, information release, personal and organizational information received, information transmission technology, methods and channels, etc.

(10) Risk management planning.

It Includes risk management methods, tools and data sources, risk management roles and responsibilities, budget risk control, risk classification, risk probability and impact, risk control, etc.

(11) Procurement management planning.

It includes contract types, expense budget, procurement methods, standard contract forms, supplier management, ordering and delivery management, performance bond, procurement contract management and procurement performance evaluation, etc.

(12) HSE management planning.

It includes HSE management ideas, policies, hierarchy management, management procedures, rules and regulations management, health, safety and environment protection management, and the site HSE management is the top priority.

(13) Financial management plan.

It includes all the following of the LNG project company, such as financial system, project budget, reimbursement procedures for examination and approval authority, capital, capital allocation, project investment, financial analysis, project investment, financing, budget control audit. Financing tax management plan includes financing options to formulate, choice of bank financing, financing amount and duration of interest, financing management, project import tax, tax declaration, tax policy.

2.3.1.4 All kinds of Planning and Programming and Approval System

In practice, the planning above are formulated and audited by project company functional departments, approved by the general manager or deputy general managers. And the implementation departments are responsible for the execution. See Table 2-2.

Table 2–2 The List of LNG Project Planning and programming, Audit, Responsibility, Approval

No.	Level	Name	Preparation and audit department	Responsibility department	Approval
1	1	Project integrated planning	Control/Engineering/Other	Control	General manager
2		The scope of the project control planning	Control/Engineering/Other	Control	General manager
3		Project milestone planning	Control/Engineering/Other	Control	General manager
4		The overall project schedule control planning	Control/Engineering/Other	Control	General manager
5		Project HR planning	HR	HR	General manager
6	2	Project cost control planing	Financial/Control	Financial/Control	General manager
7		Project quality control planning	Control/Engineering	Control/Engineering	Deputy Director
8		Project schedule control planning	Control/Engineering	Control/Engineering	Deputy Director
9		Project HSE control planning	HSE/Engineering/project team	HSE	Deputy Director
10		Project financial control planning	financial	financial	General manager
11		Project risk control planning	Control/Engineering/Other	Control/Engineering	General manager
12		Commissioning schedule control planning	operation/Engineering	operation	General manager
13	3	Design programming	Engineering/Control	Control	Deputy Director
14		Purchase programming	Purchase/Control	Control	Deputy Director
15		Construction programming	Engineering/Control	Control	Deputy Director
16		Commissioning programming	operation/Engineering	operation	Deputy Director
17		Site selection programming	Engineering/market/Government coordination	Engineering	Deputy Director
18		Project feasibility study programming	Engineering/Other	Engineering	Deputy Director
19		HSE programming in stages	HSE/Engineering/project team	HSE	Deputy Director
20		HR programming in stages	HR/Administration	HR/Administration	Deputy Director
21		Risk management programming in stages	Control/Engineering/Other	Control	Deputy Director
22		Cost programming in stages	Financial/Control	Financial	Deputy Director
23		Quality management programming in stages	Control/Engineering	Control	Deputy Director
24		Financial programming	Financial	Financial	Deputy Director
25		Transport programming	Resource transportation	Resource transportation	Deputy Director

2.3.1.5 The Relationship of Four Components of LNG Project

(1) The relationship between resource and transportation

It should be said that the resource and transportation of LNG can be treated as two dependent parts of the LNG project components. There is no direct relationship between them. However, judging from the current situation of international LNG project, there is a link between them more or less in the project. We cannot make conclusion which one is better easily, because it depends on the specific situation of practical LNG project and the final economic evaluation. Some common relationships will be introduced as follow.

① Delivery Ex-ship (DES). This kind of LNG project transportation is adopted by Japan and South Korea. In addition to hold resources, general resource suppliers also have its own ocean transportation capacity, such as BP, Shell, ExxonMobil, who choosing this way also need the capability of ocean shipping, expanding the scale of business by LNG project transportation, having the right of resource transportation in their own hands, not controlled by independent shippers and better to perform the contract.

② Free on board (FOB). This kind of LNG project transportation is adopted by America, Japan and South Korea as well as our top two earliest LNG projects. By this way, it can expand the capacity of LNG marine shipping of the LNG consume country, drive the development of the construction industry of the LNG transport ship and control the right of transport to ensure the supply of the resources of the buyers.

③ The third party to undertake transport task. As the name implies, independent shippers, such as Japan's NYK, MOL and Nippon Yusen, take the mission of transportation. The main reason why some projects choose this way is more economical compared with using their own transportation.

(2) The relationships of resource, transportation and LNG project company.

According to the specific situation of the LNG project, to see whether to participate in resources, transportation, and determine the relationship among the resource, transportation and the LNG project company Figure 2-8.

① LNG project company (or its investment initiators) participates in investment of LNG resource. For instance, CNOOC purchased a certain percentage of A LNG resource in Australia, to some extent, controlling the resource. It is benefit to the resource supply of the purchaser and keep away from the resource risk.

② LNG project company participates in LNG shipping, controlling the transport resource in some degree, is conducive to the project company's transport control, to guard against the transport risk. Such as CNOOC involved in the Shanghai LNG project transport ship owners.

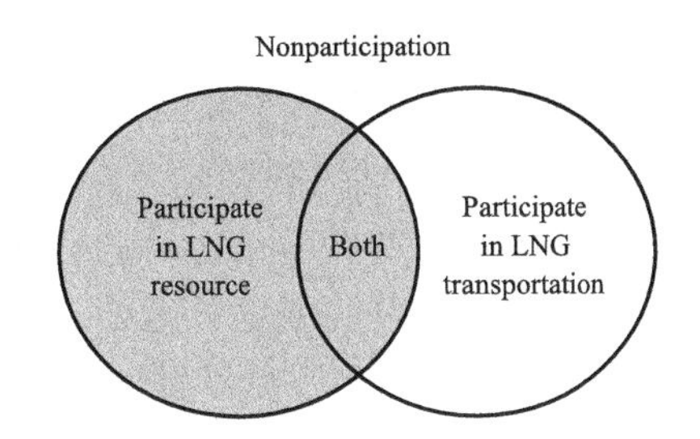

Figure 2-8 The relationship of LNG resource transportation and LNG project company

③ LNG project company (or its investment initiators) not only participates in the purchase of resource shares, but also participates in the transport industry, combined with the advantages mentioned above, but the investment will be increased, also exist the risk of poor management.

④ LNG project company neither participate in resource share purchase, does not participate in the transport industry. Only through a contract to control the resources and transportation risk, to some extent, there is a greater risk of both.

(3) The relationship of LNG resource, transportation and LNG project.

In addition to the LNG project company, LNG resource and transportation are associated with LNG project closely. If the implementation of resources provide LNG products for the project, and the solution of the transport mode provides a transport tool, only both are achieved can make project established.

(4) The relationship between investment sponsors of LNG project company and customers.

Similarly, users of the LNG project are also a key link, LNG cannot be used unless the users are settled, the whole industrial chain can connect finally, so as to realize a physical chain, business chain, economic chain, and ultimately to form the LNG value chain. According to the parent company of LNG project whether to participate in users' project. It can be divided into two cases as follows.

① The investment initiators participate in users' project. For instance, CNOOC cooperated with local power company to establish joint venture, involving Huizhou and Putian power plants. Undertaking the risk also can obtain the interests, so it prevents the risk of users in some degree.

② The investment sponsors do not participates in users' project. It can decrease the investment on this part, but it also increases the risk of the users.

2.3.2 LNG Project Management and Relationship Coordination

2.3.2.1 Permission or Approval Document Related to LNG Project

The workload of transacting the permission and approval documents of a LNG project is huge, according to the different stages of the project, involving a number of governments, external units and internal departments. The cooperation and coordination of these departments and units is the key to make a LNG project succeed.

(1) Departments related to LNG project.

① Government departments at all levels. It involves relevant departments of government at all levels, including industry and commerce, taxation, land and resources, planning, fire control, health and environment department, etc.

② External units. Include feasibility unit, safety pre-assessment units, labor, health assessment, contractors, supervisors, design units, etc.

③ Relevant departments of LNG project company. Include engineering, plan, HSE, finance, public relations, administration, human resource, marketing, business departments, etc.

(2) The list of permission and approval.

The following Table 2-3 is the required permission and approval documents of a LNG project, the situation of other LNG projects may different according to the practical condition, however, the workload is so heavy that ought to cause the attention of LNG project companies.

Table 2-3 Summary of Project Permissions and Approval Documents

No.	Content
1	Project site selection and approval documents
2	Pre-feasibility study, feasibility study report and its evaluation, and approval documents
3	Project evaluation (including lending commitments assessment), demonstration file
4	Environment prediction, survey report, environmental impact report and approval file
5	Labor safety and health assessment report and approval documents
6	Route supplement safety pre-evaluation report and record
7	Geological hazard risk assessment report and approval file
8	Earthquake safety evaluation report and approval documents
9	The preliminary design safety special section and approval documents
10	The design task book, plan and approval documents
11	FEED design, preliminary design, detailed design and approval documents
12	The technical secret material, patent documents
13	Rural migrants demolition, resettlement, agreement, compensation documents of approval
14	Land application, approval documents, the red line diagram, coordinate diagram, administrative area map, land use certificate
15	Temporary land application, approval documents, agreements
16	Routing protocol and its files
17	Construction land use permit
18	Special area construction permit
19	Special application, approval documents
20	Environmental protection, labor, safety, health, fire protection, civil air defense, planning approval documents and acceptance documents
21	Opening report, the task book and ratification
22	Project appraised award application materials, approval documents and certificates

(3) Flow chart.

The following is an example of the LNG project approval flow chart, and the whole process is shown in Figure 2-9.

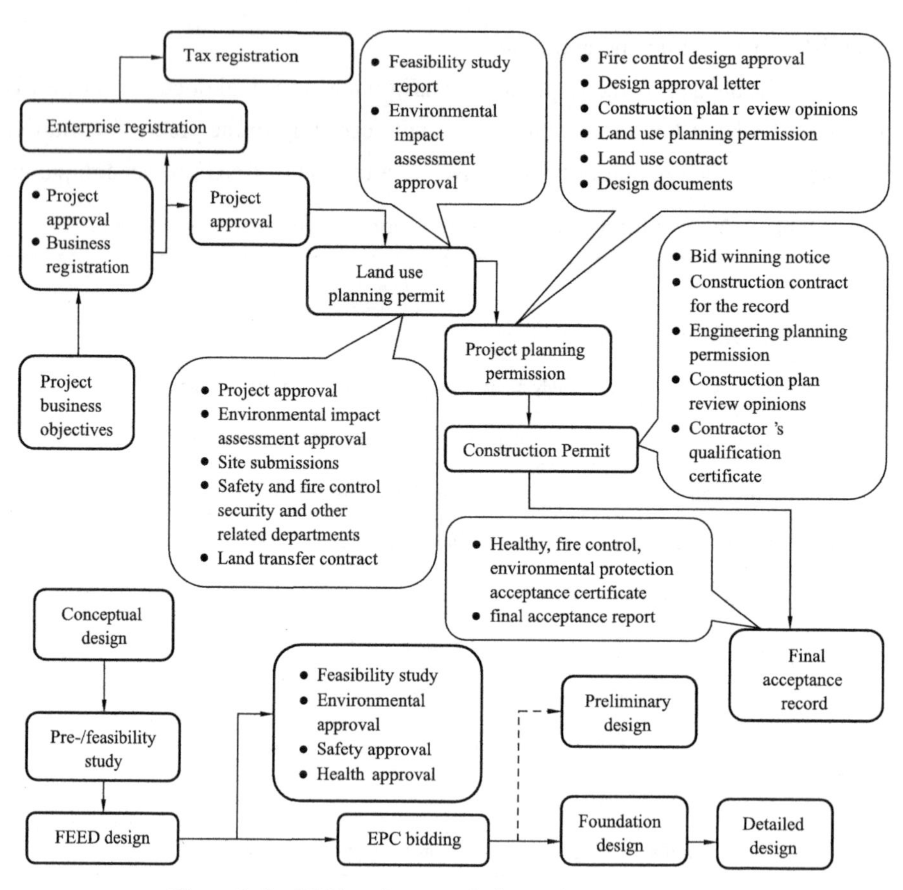

Figure 2-9 LNG project permission and approval process

2.3.2.2 The Relationship among Site Selection, Land Acquisition and Feasibility Research of the Project

(1) Site selection is the premise of feasibility study.

Other than toy manufacture and software development project, LNG project also needs a large quantity source of land and coastline, besides plant and office place, such as Guangdong Dapeng and Fujian LNG project, permanently land acquisition of 132 acre, mainly used in LNG receiving terminal, pipeline block valve chambers, gas transmission stations, moreover the temporary land acquisition for pipeline construction is much more. These land and coastline resources must be through the multisite selection to the final determination. And the feasibility study of the LNG project only in the selection of the preferred site and standby site conditions, can we have a definite object in view to estimate the engineering quantities for jetties, terminals and pipelines and for plane and vertical arrangement. Therefore, site selection is the premise of feasibility study.

(2) The feasibility study can deepen the site selection of content.

The LNG project feasibility study is deepening on the basis of preferred site confirmed. Site selection focuses on the comparison of natural, climate and construction condition, feasibility study is will further work to combine with the layout plan of entire LNG project and gas load

center, study on environment protection and water and soil conservation, forecast the long-short term of LNG market total demand, select process of supporting gas project, study the investment estimation and fund collection and national economic evaluation, and eventually complete the feasibility study report of the project. On the contrast, these works could make effect to site selection on two sides: firstly, the selected site condition is fully demonstrated, or revised in the original conditions, making it more conform to the requirements of the project construction; the second is to make larger adjustment, even overthrow the original site. In either case, it will push up the project.

2.3.2.3 The Integrated Management of Schedule, Budget, HSE, Resource and Quality

(1) The relationship between schedule and budget.

LNG project schedule and budget management are two basis and relevant elements, generally speaking, "Time is money" is also practical in project management. But in advance or delay of time, budget complete or not will have different consequences in project. The different combination of the two elements will cause several consequences as follows:

① The completion of the project on time, and at the same time budget paid, this is the best result, due to the LNG project management, upstream LNG resource supply and purchase agreement to lock the LNG supply schedule, the completion of the project on schedule and trial production is expected to carry out and end, it lays the foundation for formal production. Moreover, according to the plan to complete the budget, indicates no backlog or under pay, it proves the rationality of budgeting. Investment period can turn to production period seamlessly, enter capital recovery and make interests.

② The project is completed and put into operation on time, but the funds are not paid, and the funds are not planned to be paid on time. Also cause conflicts with equipment materials suppliers or contractors.

③ Projects completed in advance, the budget is completed, but because of the inability to provide natural gas supply, can't be transferred to the production period. It explains that is inappropriate schedule management, resulting in an increase in financial costs.

④ The worst situation is neither the project can put into production on time nor the budget is under control. Besides taking the responsibility of "take-or-pay" for upstream and downstream, the company also needs to face the situation of overlarge investment. Plus project could not turn to payback period, the consequence is extremely serious. We need to pay attention to avoid this situation happened.

(2) The relationship between project schedule and quality management.

Quality is defined here is: sub projects in accordance with the prior selection of technical and management standards for design, construction and monomer commissioning and joint commissioning qualified, fully meets the formal production conditions of the sum of the whole project. Schedule and quality also have certain relations, generally speaking, reasonable arrangement of schedule can ensure the quality requirements. Regardless of the project quality and rush blindly will inevitably affect the quality of the project. However, spending too much time on project cannot make sure the quality, either. Long time is not necessarily associated with high quality. Only careful consideration of all aspects of the project, taking into account the time that ensuring the quality, can we make the project quality matching expected requirements.

(3) The relationship between project quality and budget.

The project quality is related to budget closely. Generally speaking, the project quality and budget are directly proportional to the relationship, that is, the higher the budget, the better the quality of the project, but the two are not necessarily linked, not to say that the bigger the budget, the better the quality. The project budget cost determination and control are the premise to guarantee project quality, because the quality determines the value of the project. Therefore, only consider the high quality does not consider the high cost, or regardless of the project quality, just to keep the project cost of management methods are both not scientific.

If we consider the project quality as a fixed value, budget has several ways to reduce as follows:

① To speed up the localization process. Materials and equipment's localization is always the industry policy promoted by the state. Generally speaking, the materials and equipment imported from aboard are more expensive than the domestic ones. The mainly reason is the high labor costs aboard, freight and import tax. Therefore, we should adopt the domestic materials and equipment which can satisfy the project requirement in order to reduce the budget. Certainly, it also could appear counter-example, at the beginning of 21st century, due to the increase of domestic demand for steel tubes, the price of domestic tubes is higher than aboard. Considering the project budget in this situation, it is wise to choose the foreign steel tubes.

② To apply bulk procurement. Using bulk procurement to decrease the unit price of materials and equipment is a way to reduce the budget. We should classify the materials and equipment of the project at the beginning, if the investment sponsors has several LNG projects at the same time, they can make a centralize procurement of one project and several projects joint acquisition, using bulk procurement to reduce the budget. This mode is adopted in domestic and aboard.

③ To use time difference procurement. It could emerge the phenomenon that materials and equipment purchase concentrated lead to demand greater than supply and high price. We can start from market and project plans, choose low price on the materials and equipment procurement, in order to reduce the project budget.

④ To adopt appropriate technical standard. Engineering design and construction of any project will involve the issue of choosing technical standards. There is no precedent of LNG project in domestic, while foreign existing standards with the United States, Britain, Europe, Japan and South Korea, etc. We should take the basis of fully understand the domestic jetties, terminals and the pipeline engineering geology, environmental protection and quality requirements, choose or establish the standards which suitable for our national conditions, to satisfy the quality standards, and reasonable financial costs.

(4) The relationship between project quality and other resources.

The other resources include human resource, information resource, management resource, etc., except financial investment. The success of any project must be guaranteed to match these resources, here the most prominent is the human resources, on the LNG project, the project general manager and other technical personnel, it is best to be equipped with those who have done the LNG project and proved to be competent to do the work and accumulated rich experience, especially the project general manager. But the staff should comply with the ladder and the level of principle, in order to form the overall strength of the team, to avoid friction, good project management only to the best combination of the project team, to achieve the desired quality requirements.

(5) The relationship between project HSE and budget.

To strengthen the relationship between project HSE management and budget should be said that is the positive proportion, in another word, HSE management is based on financial costs, and the strengthening HSE management will finally reflect in personnel increases with the increase of labor costs, the strengthening of HSE is a compulsive requirement according to laws and self-requirements to ensure the quality of the project. However, we also need to consider the proper proportion of HSE management and whole project budget. Increasing investment in HSE blindly could cause unnecessary waste.

(6) The integrated management of project schedule, budget, HSE, other resources and quality.

The above involve the relationship between the 2 of 5 factors. These 5 factors are not simple two-two relations in fact, but extremely important relations of integration. Shown in Figure 2-10, it can vividly express project schedule, budget, HSE, other resources and quality as the apex angles of tetrahedron respectively, and the quality is among these four factors mutually formed a plane. The inclination of quality plane is determined by each apex angle factors coordinated or compromised. We intensify the HSE management, increased the HSE budget of the investment, and the quality plane moving from quality plan 2 to 1. From this point to the other point, the LNG project budget, schedule, HSE, other resources and quality is five direct relevant correlation and

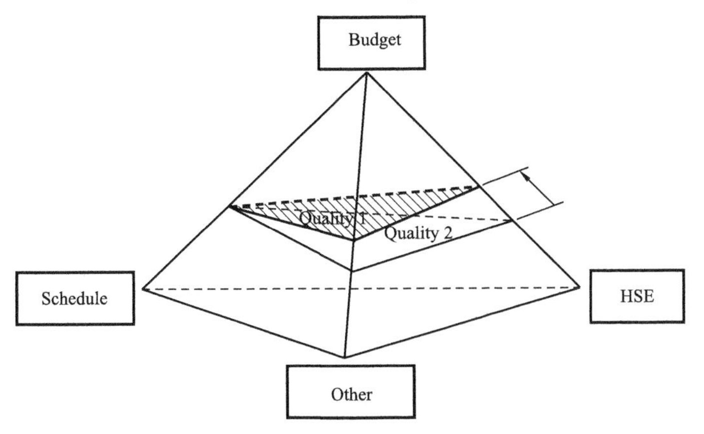

Figure 2-10 Diagram of relationship of Schedule, Budget, HSE, other resources and quality of LNG project

interaction factors, the project construction period in advance or delay will directly influence on the project budget. The project quality can cause directly to the project budget fluctuations. Too much human resource and HSE investment can increase the budget. HSE investment devoting too little will not only cause accidents and loss, but also can increase the clean-up and restoration project expenses. The project budget will directly affect the project schedule and quality. So the project management must be integrated management for the five elements. We need to strive for the coordination of various factors to achieve reasonable quality requirements.

2.3.2.4 Integrated Management of LNG Project Work and Project Goals

Project work and project goal is the relationship between the local and the whole. The goal of the LNG project is to ensure the smooth delivery of qualified natural gas to the users. To do this, the following links needed to be clearly clarified.

(1) Project work and project sub goals.

The overall project is formed by several subprojects. Meanwhile, each subproject consists of every detailed routine work. It can be classified into a number of different management types, such as administration management, public relationship and government coordination management, contract management, equipment procurement management, engineering design and construction management, human resource management, project contract supervision management, HSE management, financial management, project supervision management, team building management, information management and project acceptance management. Only to do every job well, can we ensure the realization of the project goal.

(2) Sub goals and general goal of project.

It is not only the tangible assets, such as LNG jetties, receiving terminals, natural gas transmission pipelines and related supporting facilities included in the sub goals of the overall projects, but also other related management of intangible assets, like implementation of LNG resources, execution of natural gas sale contracts with downstream users, the organization and personnel in place, establishment of all the regulatory regime, etc. It takes all the remarkable sub goals to be accomplished on time that the overall goal can be achieved.

2.3.2.5 LNG Project Integrated Management of All Specialties and Departments

Starting from the establishment of the LNG project team to project office then to project company, problems may occur because of different specialties and departments coordinating. Problems may occur during the project integrated management is going to be expounded in the following chapter in terms of the aspect of the correlation of different specialties and departments.

(1) Coordination of different specialties.

As jobs within a project company are multifaceted from different aspects, project team members are required to be equipped with different professional background to accomplish all

the work defined within the LNG project. From the view of structure of present LNG project, there are as many as 50 majors are involved. Usually different specialties need to work together to get one thing done, such as natural gas specialty, legal staff, commercial contract dealer and economist need to work together on the LNG resource negotiation. Meanwhile, the same major professional staff to do the same thing, they need to learn from each other mutual cooperation and rational division of labor. Except the internal professionals, external staff that involved in the project such as contractor, Supervision Company, external experts need to be reasonable organized and managed. Only by the closely working with different specialties can the goal of accomplishing the overall project be achieved.

(2) Coordination of Different Departments.

There are almost ten departments that involved in the LNG Project Company (see Chapter 4), which are public relationship and government coordination, human resource and administration, financial, market development, department dealing with business affairs, engineering, control division, HSE management, shipping and production arrangements. As it was mentioned before, though specialties in different areas are relatively independents, they should coordinate and cooperate together under the overall project objective. First of all, each department shall accomplish the target within the various stages involved. If any individual department happens to affect the overall progress, then personnel reassignment and enhancing management should be put into use, aiming at the consistency of project. Without the coordination of departments, the project won't be completed as scheduled with high quality and reliable quantity.

2.3.2.6 The Integrated Management of Project Company, Supervision Company and Contractors

(1) The responsibility of the three parties.

① Project company as the main building investor, responsible for the overall management of the project construction, in particular, the general manager of the LNG project company and the company's various departments manager are vital with great responsibilities.

② The supervision company is confirmed by written contract of project company to supervise construction process. Referring to national laws and relevant engineering technology standards, the supervision company takes control and management of quality, cost and schedule in project construction.

③ Contractors is the third-party which participates in the project construction by contracting with project company, carrying out the work according to the work scope and content of contract, based on the national laws and regulations and the stipulations of both sides of the contract requirements to the project company to deliver results, independent responsible for the delivery of the results. Project company or supervision units through the contract manage the contractors on the construction projects.

(2) Unity and opposite of the goal of the project construction.

Project company, supervision company and contractors are the project construction participants undertaking various assignments depending on its own subject position. They cooperate with each other in project implementation, through effective project management to achieve the project goals and interests. In the process of project implementation due to the role and position are different, there are conflicts of interest and the contradiction. Therefore, in the process of project implementation, only increase the area of the common interests of three parties, reduce conflicts of interests, the three cooperate and complement each other closely, in order to ensure the realization of the goal of LNG project construction.

(3) The relationship between the project company and contractor.

As the major part of LNG project construction, Project company is responsible for entire management of project construction. In the beginning of project construction, it should organize the company to analyze the scope of project construction and disposable management resource of market, make business strategy of the project.

Considering LNG project belongs to the new field of natural gas industry, the construction characteristics and difficulties understanding is still in the stage of perceptual knowledge, lack of project practice experience. Usually, under the following circumstances, the project company entrusts the contractor: the first, the project company is lack of some or all of the engineering technology, human resources and management ability in the scope of their work; the second, it has the ability, but lacks labor force, in order to catch the construction period and to borrow outside force. Project company and the contractor are a kind of contractual relationship, through the bidding process, technology, commercial clarification, finally confirming respective duties, rights, obligations stipulated by the contract. The contractor organizes various resources to realize the goal of project quality, progress, cost in accordance with the technical standards prescribed by the contract. Project Company pays the relevant expenses according to the contract. There are bonding points and conflicting points of the interests between project company and contractor.

Bonding point: the goals of project company and contractor finishing the work quality are the same.

Conflicting point: the contractor has its own project profit target. Therefore, in the process of project implementation, contractors will work in their own scope, applying various business strategies for their own interests. The contractor may try to save the expenditure on purchasing material and equipment if the contract allows them to buy in order to obtain more profits space. In the procurement of materials, equipment, if the contract stipulated by the contractor to buy, the contractor may save as much as possible in order to obtain greater profit margins. The contractor may not sent skilled worker according to the requirement on the human resource investment and influence the project quality and time limit for a project. The contractor may not investment enough on HSE management or management does not reach the designated position. On the management of components and the overall goal of the project schedule, the contractor may not be in conformity with the project company, the contractor considers his own more component rather than the goal of the whole project, etc.

(4) The relationship between project company and supervision company.

According to the requirement of "Construction Engineering Quality Management Regulations" and "Provisions of the Construction Project Supervision Scope and Scale Standards", where the state regulations must be implemented in the construction supervision of the projects, supervision commission must be conducted prior to begin construction, and that is entrusted with the qualification conditions of supervision units to monitor project implementation stages.

For LNG project company, hiring supervision company is not only the requirements stipulated by the state laws and regulations, but is also supporting the project company itself technology and project management. Hiring qualified, experienced supervision unit, can effectively assist project company to implement the process of project construction engineering quality, safety, schedule and cost control.

From the legal relationship between the project company and supervision company, it is also a kind of contractual relationship, according to the contract specification of their respective responsibilities, rights and obligations. The Supervision Company in accordance with national laws and regulations and technical standards, as well as the management scope and requirements of the project company, to carry out supervision on construction process of LNG project, in order to achieve the construction of the project in accordance with laws and regulations and standards. According to the contract, the project company pays the cost of supervision service.

Similarly, because of their different status and role, they also exist between the interests of the bonding point and conflicting points.

Bonding point: project company and supervision company in project construction quality standards on the goal is the same, must complete the project tasks in accordance with the agreed international and domestic quality standard.

Conflicting point: as the supervision unit and project company in the contract of the principal and subordinate positions, and independent supervision of legal requirements, when the contradiction between project quality and safety, schedule, cost management, supervision unit is more concerned about the quality and safety issues, and the schedule and cost control are in the secondary position. Supervision service in the essence is technical consulting and management services, in the fierce market competition and labor management are not standardized, so that part of the supervision company in accordance with the requirements are not equipped with similar engineering supervision personnel, it is difficult for engineering supervision and project management communication, which will affect the construction of the project.

(5) The relationship between supervision company and contractors.

Project supervision company is authorized by the project company according to national laws and regulations to undertake supervision for the contractors in the construction process. In the process of engineering construction, the relationship between supervision and being supervised there is no business economic exchanges. The supervision company shall be responsible for the supervision of the construction of the contractor to meet the quality requirements specified in the state regulations and standards. There is the contractual relationship between supervision company and project company.

Because of the different legal status and independent contract role, they have common goals as well as different interests.

Bonding point: supervision company and contractors achieve the goal which is the requirement of project company in the contract by working respectively.

Conflicting point: In the process of project construction, supervision company undertakes in independent supervision position by laws and contractors is in the being supervised position. In the meanwhile, at the same time, the supervision unit should not only undertake the contractual obligations, but also deal with the legal responsibility of the quality of the project. Contractors are mainly liable for project construction and contract responsibility.

(6) The relationship of project company, supervision company and contractors.

Project company, supervision company and contractors come together in the process of project construction, forming division of labor and cooperation under the same goal, guaranteeing their interests respectively and also having contradiction and conflict (Figure 2-11). There are both contract and supervision relationship among them, they treat the contract as a link, each plays his own role in project. Supervision company is responsible for the project company, supervises the contractors for the engineering quality. Contractors shall subject to dual supervision by Supervision Company and Project Company. Project company, the contractor and the supervision company form the triangular relationship, in order to ensure the stability of the triangle, communication and cooperation among them are the key to the success of LNG project.

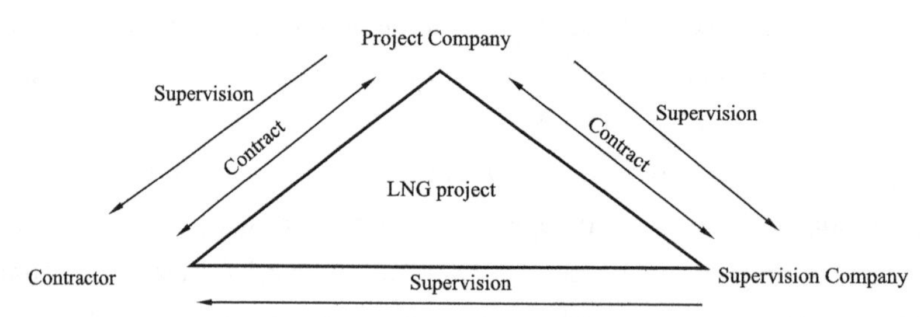

Figure 2-11 Relationship of project company, supervision company and contractor

2.3.2.7 Integrated Management of Project Overall Change Control

(1) General concept of overall change.

The overall change of the LNG project is due to the fundamental change of the original project plan and the major elements of the project, which includes the aspects from the project company, the investor's unit, national authorities and downstream users, and these changes affect the project propulsion seriously. The process of LNG project overall change control run through the project from beginning to the end actually. LNG project involves the domestic and international up stream, middle-stream and downstream industry, newly built and transformed enterprises, each citizen and social community, plus it will spend about 5 years for investment company on project from producing the concept to put into operation, the overall project plan set a few years

ago will inevitably change, even some major changes will happen. So, the management and control of LNG project overall change becomes vital content of LNG project management.

(2) The basis of overall change.

The so-called change is based on the original foundation change. The basis involving: cooperation agreement of each unit of original project company, project feasibility study report, national approval document, project management plan and procedure, client requirement and agreement of LNG purchasing, etc. (Figure 2-12).

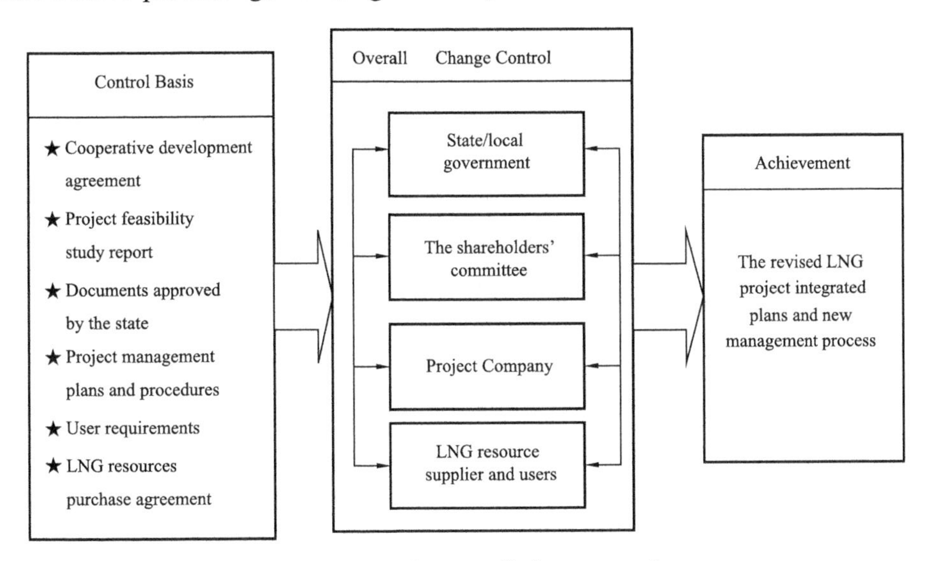

Figure 2-12 Project overall change procedures

(3) Analysis of project overall change.

① Changes in the terms of the project approved by the government

Energy has been the most concerned issue of the national government departments, because it is related to the national economic construction and development of major issues. For the natural gas industry, the national authorities will proceed from the overall interests of the state, the imported gas and domestic gas have an overall consideration, reasonable arrangement for the national oil company, in order to avoid redundant construction and vicious competition. For instance, there are three oil companies tending to develop LNG industry in our country, the government macroscopic guidance and the overall deployment are indispensable, when the state energy layout and some investment changes, Oil companies must act according to the major policies of the state, so that when the project companies for the original project overall plan and the arrangement of the national government have contradiction, should obey the national interests, changing or terminating the project plan.

② The clauses of cooperation contract signed with local government have changed. In the early stage of development of LNG project, the cooperation agreement between initiator's company and local government could change as time goes by. Such as local enterprises to join the project company caused the change of structure of the local economy and ownership, plus the change of the overall layout, etc., are likely to change the project overall plan.

③ The overall plan and plans made in the early stage of project has changed. During the project designing and construction, it may occur the changes of engineering quantity, engineering site ownership management, channel adjustment, pipeline route, compensation of land acquisition, investment of HSE management, large accident happend, and government administration, etc. All the issues mentioned above will increase construction cost and extend the time of the project.

④ The type and quantity of downstream gas using has changed. For the present condition, downstream clients consist of gas power plant, urban gas, industry fuel and chemical raw feed, etc. The change of gas market involves several conditions as follows:

The first is the change of gas using type. All the four enterprises mentioned above, changing for various reasons like gas power plant does not get the project approval, the scale of urban gas market expands. The overall gas load center changes will cause the alteration of LNG project.

The second is the change of the quantity of gas using like the demand of nature gas for local economy development increasing, as the number of new clients increasing or the demand of gas of original enterprises increasing. For instance, the annual supply of Guangdong Dapeng LNG project has increased from 3×10^6t/a to 3.85×10^6t/a. Certainly, the circumstance of decline of LNG demand also could happen, so that we should develop the gas market as soon as possible or change the sales method in order to keep the original gas supply constant, like changing the transport way from pipeline to tanker, etc.

⑤ The element of LNG resource has changed. The head of project industry chain is LNG resource, the project cannot establish unless got LNG resource. Therefore, LNG resource is very essential to sponsors and project company in the early stage of project. The change is the original price, quality, date of delivery. Deliveries of LNG resource contract do not match the request of sellers due to the entry-into-force condition, making the original contract not carrying out. This circumstance is related to the LNG market condition at that time. In the condition of buyer's market, buyer occupies the dominate position in the market, taking advantage of LNG price, even the seller could give additional privilege to buyers for signing the contract. However, in the condition of seller's market, seller occupies the dominate position in the market, selling the gas at a high price. Because not satisfying the requirement of seller, in another word, during the process of turning buyer's market to seller's market, the price increases in Fujian LNG resource purchasing. The change of natural gas price seriously affect the midstream and downstream projects. It also will affect the benefit of project company.

(4) Control flow and procedure of overall change.

It is necessary to make adjustment with the change of circumstances and elements of project. Generally speaking, overall change is proposed by project company. Significant changes in LNG resources, downstream users of fundamental changes, changes in the guiding principles of the government, a major deviation from the investment, project company shareholders or shares significant changes, etc., those should be decided by shareholders'committee. When certain changes involve the general items of the project itself, the board of directors, and even the project company can make decisions, such as the change of the site of project receiving terminal, route of the pipelines or land cost, users to buy gas volume changing, equipment, material supplier selection and price changing, etc. Its investment has not changed significantly and is within the controllable range, the decision can be made by project company. As shown in Figure 2–12, the project company in accordance with the change size, shall report to the shareholders'meeting or the board of directors for approval, important matters shall also be reported to the competent authorities of the state and local governments for approval or filing, some matters should be negotiated with users. The results of the overall change of the project are: LNG project after the revised integrated planning and management procedures, include changes in sub project deliverables quality, time and budget indicators, etc.

(5) The tools and technology of overall change control.

① Project management information system. There are several project management software systems, depending on roles and functions, Project company can select the appropriate software, in addition to timely preparation of project management plan, the software also can quickly carry out the overall change of LNG project, provide company top level about the impact of factor changes on the overall project size.

② Expert judgment. For experts who have participated in the LNG project or similar projects, can visually determine what the impact of a factor change on the project, if the majority of experts'opinions tend to be consistent, we can consider their opinions, through appropriate statistical information and data support, reflected in the project change control.

(6) Results and data archiving of overall change control.

Overall change control documents include the submission of the request, the shareholders and the board of directors approve or reject the documents, the alteration of the LNG project management plan, update the LNG project scope statement, approve the change plan, preventive measures and remedy defects, change the project delivery results of quality, time and budget indicators. The above data archive is an important management work, according to the requirements reported and archived for future reference.

2.3.2.8 The Relationship between Project Acceptance and Approval Nodes

Project acceptance departments consist of relevant government departments at all levels, shareholders and the project company. Due to their different social responsibility, the obligation, contents and focus of the project acceptance is not totally the same.

(1) Acceptance departments at all levels and their relations with the responsibility of the project acceptance.

① The responsibilities of government, shareholders and project company in project acceptance. For large energy LNG projects, the competent departments of government at various levels shall, in accordance with the powers conferred by the state administrative authority, proceed from their respective jurisdiction, and examine them at different stages of the project and from different angles. The shareholders and companies of the project accept the government's inspection based on the LNG project platform and rectify to fulfill the national special project acceptance standards according to acceptance conclusion. Government is the acceptance unit, while the project company and shareholders are the departments to be inspected.

② The responsibilities of the shareholders and the project company in the project acceptance. The shareholders served as investors, performing project acceptance that they have invested is to implement the shareholders' authority, and keep or increase the value of their funds. Generally, the proceeding of the project acceptance is based on enterprise standards of the shareholders. For example, CNOOC and BP invested the first LNG project in China—Guangdong Dapeng LNG project, they paid great attention to project completion acceptance. They selected the specialists as soon as possible to compile administrative regulations and procedures about the acceptance, then reexamined and inspected each period to make sure the project put into production successfully. The shareholders focus on the whole operation of the project. Shareholders are the acceptance departments, and the project company is the unit to be inspected.

(2) The emphasis of the acceptance department in different stages.

① The emphasis of the government in the project acceptance. It includes quality standards supervision, production safety supervision, environmental protection supervision, occupational safety and health supervision, anti-thunder, pressure vessel, and fire control. This relates to the social responsibility of the government departments with law-enforcement inspection and mandatory are the general characteristics.

② The project acceptance of the shareholder often focus on the crucial stage of the project, such as project completion acceptance of the jetty, commissioning, preliminary acceptance, completion acceptance, mechanical completion recheck of the terminal and pipeline, pre-commissioning acceptance, commissioning acceptance and final completion.

③ The project company because of its dominant position in the project, not only focuses on the acceptance of hard tasks, but also focuses on the acceptance of soft tasks. The acceptance of the hardware includes the acceptance of the jetty, receiving terminal and the entity of the pipe, covers inspection lot acceptance, concealed works acceptance, itemized works acceptance, divisional works acceptance, unit works acceptance, sub-project acceptance, project final acceptance (including joint commissioning). In order to achieve the expected results and objectives, the soft task acceptance should be based on the requirements of the soft project quality management of LNG project, to formulate specifications, requirements and evaluation criteria in advance, the acceptance of the software includes the project feasibility study report, safety pre-evaluation report, LNG sales and purchase agreement, gas sales contract, various technical review meetings, consultation, contact and coordination of the internal and external relationships in projects, etc.

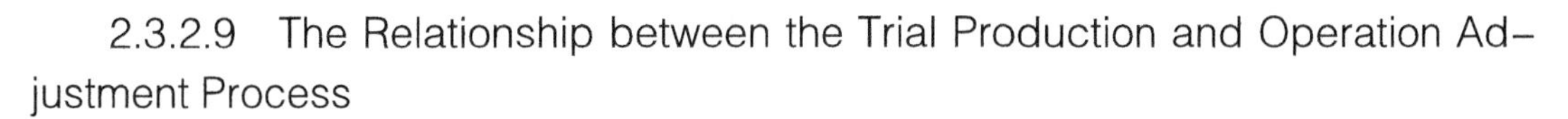

2.3.2.9 The Relationship between the Trial Production and Operation Adjustment Process

Trial production and operation adjustment contains four stages: pre-commissioning, commissioning, trial production and operation adjustment, which is a coherent and whole control process.

(1) The relationship between pre-commissioning and commissioning in the pilot production adjustment.

Pre-commissioning and commissioning are the first two stages in the adjustment process of LNG pilot production. The pre-commissioning means the test of equipment, process system, instruments, apparatus and control, and also includes the blow and dry for the system and equipment as well as the preparation for files. The commissioning includes the system inert and cooling, the stability test of overall system and so on. Only adequately the pre-commissioning does, can the commissioning be started. The sequence of the two stages cannot be reversed and logical errors are not allowed.

(2) The relationship between trial production and operation adjustment.

Trial production operation will carry out the actual LNG unloading, storage, pumping, gasification and natural gas transmission process operation, and it will be on the basis of debugging, really enter the trial production stage. The operation adjustment refers to tests and examinations of equipment, instrument and process system in the different conditions, such as the plan change, unloading speed variation and pumping quantity change, gasification rate changes, gasification volume changes, gas transmission volume changes, BOG safety emissions, flare combustion and emissions, explore the seamless connection between the various interfaces and safety of personnel operating experience, which has the significance of the joint commissioning of the whole project.

2.4 Integrated Management of LNG User's Group

2.4.1 Analysis of Local Economy and Environmental Protection, Industrial Structure and User's Group

2.4.1.1 Local Economy Development and Environment Protection Requirements

(1) Relationship between local economy development and environment protection requirements.

If a LNG project can be established, firstly, it should analyze the necessity of project construction. This is the question also needs to be answered in the feasibility study report and LNG project application. Based on the scope of the LNG users in local administrative divisions, it is necessary to analyze the relationship between the economic development and the GDP growth with energy demand in the divisions, and then quantitative indicators will be given. Economic growth is direct proportion with energy demand at the present stage in our country, so natural gas (NG) will be used widely in city gas, gas power plants, industrial fuel and chemical industry due to effective utilization of NG.

(2) Environment protection requirements.

Analysis from the perspective of environmental protection, economic development should not be at the expense of the environment, we should carefully analyze the local energy structure, including coal, petroleum, natural gas, hydropower, nuclear power and other energy proportion respectively, qualitative and quantitative analysis of the impact of coal and other primary energy on local environmental protection, and the positive role of clean energy in environmental protection, including the analysis of existing urban and industrial boiler fuel coal, fuel oil structure and distribution, the amount of polluted gas emitted from coal fired power plants to the atmosphere every year, Environmental Protection Department of the state environmental protection requirements of the construction of power plants and the future of new power plant fuel planning. The State Environmental Protection Departments requirements on the construction of power plants and the future of new power plant fuel selection planning etc. Natural gas gets more and more favor due to the advantages of clean, high thermal efficiency and quick start.

2.4.1.2 The Analysis of Local Industrial Structure and User's Group Model

(1) The exploiting principles of LNG target market.

① To be a kind of economical alternative energy in market positioning, and based on the low price in equal calorific value, it can substitute LPG, oil products, electricity and artificial coal gas as well as the coal in coal forbidden region;

② To be used in city residents, industrial and commercial user's group (including central air conditioning and the users of heat, electricity and cold cogeneration). The LNG users contain specific large consumers, LNG vehicles, Compressed Natural Gas (CNG) vehicles and so on. LNG may be used as gas power plant for electricity peak shaving;

③ To perform the prudent market development strategy. A new market development needs to form a procedure from general survey, classification, primary selection, optimization, letter of intent and framework agreement;

④ To be used in the region without gas pipe. The tanker transportation market and satellite station gasification should be actively explored and expanded vigorously;

⑤ To be used in the users with affordability and economy. It needs to depend on users with bear ability and economy size to determine the LNG project size, procurement and construction time.

(2) The analysis of local industrial structure.

The industrial structure will be different with the division of the local administrative region. Taking LNG project for an example, the selection of users is based on the current regional industrial structure. From China's coastal provinces and cities, with the reform and opening up early, the economy is active, and local industry or private enterprises is developing rapidly in small

towns, so that some pillar industry gradually formed the distinctive industry gradually in the specific region. After analyzed by the expert, the structure models of local industry are listed as following.

① Building materials industry, including the relative centralization factories producing plat glass, ceramics and cement. For example, in Foshan city, ceramics is well developed and has formed a distinctive industrial group.

② C_1 chemical industry, such as hydrogen, acetylene, methyl chloride, methylene chloride, carbon tetrachloride, carbon black, hydrocyanic acid, carbon disulfide, nitromethane and so on, which are produced by natural gas as raw material.

③ City tourism, including urban residents, hotels, guest house, shopping malls, airports, schools, staff canteens, inns, kindergartens, government agencies, office buildings, natural gas vehicle, central air conditioning, etc.

④ Gas power generation industry, which is a new industry. Currently, with LNG projects in Guangdong, Fujian, Zhejiang and Shanghai, have been built or plan to construct a group of LNG power plant.

(3) The analysis of LNG user's group model.

Generally speaking, LNG users should be based on the existing industrial structure model. In order to achieve economies scale of the LNG project, LNG user's group will break through the industrial structure in some region and extend to surrounding region, which forms the different LNG user's group models as the following items.

① The user group model of gas power generation plus city gas, that is, gas power generation as the main user, the formation of one or more gas load center, coupled with the trunk line along the city gas users together. Guangdong Dapeng and Fujian LNG project belongs to this kind of user's group mode.

② The user group model of C_1 chemical industry plus gas power generation, that is C_1 chemical industry concentrated are formed and city gas users are auxiliary.

③ The user group model of city gas plus gas power generation, that is tourism city with the population of more than five million, the gas consumers include residents hotels, offices, schools and office buildings, together with gas-fueled vehicles and small gas consuming industry formed gas consuming loading centers, and the gas power plant is auxiliary.

④ The user group model of industrial fuel plus city gas, that is for the well-developed small towns in Guangdong, in the need to use natural gas as fuel in many industrial concentration areas, can consider direct supply and combining with the city residents to use gas.

⑤ Mixed user group model, that is the several user group mentioned above mix together and with no priority to any kind of gas consuming users.

2.4.2 The Relationship between LNG Users and LNG Project

2.4.2.1 LNG User Group to Determine the Quantity of LNG Resource

The ultimate result of LNG overall user implementation determines the LNG yearly procurement amount from foreign countries, so as to decide the scale of the LNG supply, gasification and gas distribution. While from the view of development, there are two aspects should be considered.

(1) Recent LNG annual supply.

Mainly taking the existing natural gas market demand as the basis, it refers to determine the initial scale, also it refers to consider the growth of natural gas during the project construction period, to meet user requirements, local users on daily, monthly, seasonal and annual gas consumption should be given with careful analysis, in order to obtain the size of gas used in peak and valley time, furthermore, the location of gas use load centers also needs to be determined, in order to take fully account into the LNG project jetty, receiving terminal, trunk lines of engineering plans and design and construction of process flow.

(2) Long term LNG annual supply.

China has a number of LNG projects in the first phase of construction, immediately followed by the expansion of the scale, into the second phase of construction, and even there is not enough NG to satisfy the consumers in the first construction phase, such as Guangdong Dapeng LNG project. This reflects local economy development accelerating the speed of energy demand, especially for the clean energy. Therefore the land for terminals should be reserved enough when conducting land acquisition at the beginning, including the place and capacity for tanks and vaporizers. It is necessary to predict the relation between local economic development and clean energy requirement preciously and try to prepare early, invest gradually, ensure supply, and promote the development.

2.4.2.2 Relation among LNG users

(1) Relation between major user and small user.

It is rarely if there is only one user in the LNG project. Usually, there are several gas users to form user group with one or more major users. Taking Guangdong Dapeng LNG for example, in the first phase, gas power plants were the major users and formed a major gas consuming loading center including new built power plants such as Qianwan and Meishi and oil to gas power plants. Two newly-built power plants in Huizhou and Eastern formed a minor gas consuming loading center, which support and promote the LNG project quickly. City gas users in Shenzhen, Dongguan, Guangzhou and Foshan play an important complementary role to add the weight of LNG project, and even to play a key role for later expansion and market cultivation.

(2) Relation between old users and new users.

With first Chinese LNG project put into production and the users consuming natural gas, great economic benefit and wide social influence was caused rapidly, which will play a leading and radiating role. Old users considered the expansion of modification to enlarge the business scale. The plan of new enterprises will be considered as agenda by local government, together with the increased proportion of the individual economy in the local economy, new private gas users will appear. What mentioned above will promote a new larger requirement for NG. As a

LNG project operating company, it is necessary to consider the market changing, conduct further market research, and come up with their own development plan as soon as possible, and prepare for the second extension stage or consider new LNG projects.

2.4.3 LNG Industry Chain Extension

The most immediate symbol of a successful LNG project is that downstream user market growing up, new users springing up and requirement increasing every year. If the LNG project company will reach these standards, besides locking the large users of LNG project within the scope of gas transmission trunk line network, it should also pay full attention to developing the small users far off the main line. From the view of LNG industry chain extension, some methods about cultivating the marginal gas market will be introduced as following.

2.4.3.1 LNG Satellite Station

(1) Relation between LNG satellite stations and LNG receiving terminals.

If the above mentioned LNG receiving terminal is compared to the sun, the LNG satellite station is its satellite, which is less than the former in terms of storage and gasification capacity. Actually LNG satellite station is the small LNG terminals and gasifying station. It receives LNG from receiving terminal or liquefaction plants by special tanker (or tanker ship), after gasifying and adding smelly and then feed into city pipe network for civil, commercial and industrial use. There are some reasons for building the LNG satellite stations.

① To reach the minimum scale of the newly built LNG terminals, satellite stations have been adopted to enlarge the sales volume and guarantee the economic effectiveness of the principal LNG receiving terminals.

② A number of towns or districts with a certain distance away from the LNG receiving terminal or the liquefaction plant and the gas transmission trunk line, these areas have a certain number of natural gas users, but fail to extend to the area of the pipeline, or the pipeline extends to the economy of this area is not good, can take this way.

③ Even in the gas transportation pipeline and pipe network coverage area, LNG satellite stations also play a great role as the peak load regulating device and standby gas source.

(2) LNG satellite stations classification.

① Directly to the residents of gas supply, direct canned LNG fuel vehicles or canned small gas tank group. This kind is a tinny LNG station with storing, gasifying and gas distributing. The LNG tankers and containers have been filled with the LNG which is from large LNG receiving terminals and delivered to the satellite stations through road and rail transportation. Then, the LNG will be unloaded into the storage tanks in the satellite stations. After being gasified, the LNG are delivered to residents through the urban pipe network or filled into the bottle group(volume less than 500 L) as liquid supply for residential gasification and canned automobile LNG fuel tank. LNG satellite stations in Japan belong to this kind.

② For pipe gas peak shaving and emergency. The difference between this kind and mentioned above is that the latter kind has liquefying equipment. LNG in the latter kind is from pipeline, but not from large LNG terminals, when pipe gas is redundant, it can be liquefied and stored, and when pipe gas is lacking, it can be gasified and output back with effect of peak shaving. LNG satellite stations in America belong to this kind, as well as LNG accidental gas standby station in Shanghai Pudong NO.5 Groove.

(3) The position and role of LNG satellite stations in the LNG receiving terminal project.

① LNG satellite station is important supplement for the LNG receiving terminal project. Most of NG in our country has been supplied to users in pipe gas, and LNG satellite station is supplement, transition and extension for pipe gas. While LNG satellite station as a main or transition gas source has been widely adopted in the places where natural gas is lack or pipeline cannot arrive, and plays a great role in promoting the acceleration development of NG industry in our country. In the development of natural gas downstream enterprises, satellite station at least can be thought of as a means to open up the market.

② Due to the advantages of the construction in little investment, short construction period, and cooperating with gas transmission pipeline and complementary as well as hour peak shaving in operation, LNG satellite stations can make as peak-shaving facility for urban network and save more than one-third of the construction investment. Especially in highly urbanized areas, the investment of paving pipeline may be very high, also will be restricted by land availability. Because the fluid in LNG satellite station is driven by the pressure difference, air temperature gasification and pressurization, there is no power equipment in technology, so it has the advantage of little energy consumption, less operation cost, and low maintenance cost.

③ LNG in transportation, storage and usage is more secure than liquefied petroleum gas (LPG). LNG is much cheaper than LPG at the same calorific value in the international market. So LNG can replace LPG market, and expand the scale of LNG.

④ The LNG in satellite station conducts the purchase and sale according to the spot and futures trading, completely broke the take-or- pay contract model of the LNG trade in the past 20 years, which makes the sale and purchase of LNG as simple as crude oil and refined products, without delay for a long-drawn-out of LNG purchase and sale contract negotiations.

⑤ LNG satellite station can not only supply gas from many LNG terminals but also from the LNG plants built in any gas field onshore, which not only makes the urban gas supplying more reliable, but also provides a broad market for domestic small gas field development.

⑥ Because LNG satellite stations and land LNG transport technology of is very mature, the safety of LNG storage, transport and usage can be fully guaranteed and the production cost is much lower than that of foreign cost, which has guaranteed the safety and economy of LNG satellite stations construction.

⑦ Natural gas is a kind of high quality fuel, which is the most suitable for urban gas and public transport vehicles to reduce the air pollution and improve the urban air quality, so high quality fuel should be used optimally. We should take the strategic height into consideration to look at the development trend that natural gas market in the future may be mainly supported by the urban gas and public transportation vehicles, so that the liquid carriers, satellite stations supplying gas will inevitably become the main direction of development of natural gas business.

2.4.3.2 Tanker (ship) transport

Tanker (ship) should be used as transport tools supporting LNG terminals and LNG satellite stations mentioned above. Due to the fact that the tanker (ship) solves the transportation problem of LNG, coupled with its flexibility, and without restriction of the route of pipelines, so it has played a positive role in LNG project market extension and new user cultivation.

(1) LNG tanker (ship) transport function.

It is very important to adopt tanker (ship) transport to enlarge the market scale of main LNG terminals, solve the LNG transition of satellite stations in towns, suburbs and outer suburban districts, and guarantee city gas and industry gas in these places, especially for newly built LNG project. Take Fujian LNG terminal for example, based on the LNG consuming scale in the first phase, project company considered LNG transportation by tanker in the beginning.

(2) Combined with the main body of LNG project to study its positioning.

Based on specific LNG consuming market, under the conditions to meet the main users and gas load, it is necessary to focus analysis on LNG tanker (ship) operation mode and business structure of transportation project, includes: whether to set up a separate LNG tanker (ship) transport company, or in cooperation with the LNG project company operation? The transport company whether configures their own tankers, or transport tasks to be taken by social franchise company? What kind of organization and human resource allocation are adopted? Also it is necessary to analyze the safety and accident, investment, competition, resource supply risk and its countermeasures and measures, transportation costs, and the price that the user can bear, and finally financial analysis and investment estimation in order to gain maximum return, at the same time to reduce and control the risk.

2.4.3.3 LNG Vehicle Filling Industry

(1) LNG fuel vehicles are the need to improve environmental quality.

Developing traditional automobile industry has brought the civilization and progress of the society, but also brought pollution to the ecological environment. Developing LNG fuel automobile is the basic way to govern vehicle emissions, improve air quality and the living condition. LNG fuel automobile is a kind of low emission of green one, and the comprehensive emissions will be reduced by more than 82% when compared with gasoline cars.

(2) Developing LNG fuel automobile is the important measures to adjust the energy structure.

With the development of economy and society, our country has become a great nation in energy consumption and energy shortage has been a threat to the country's strategic security. Using the introduced LNG fuel for vehicles will be in favor of energy consuming in our country, and increasing the ratio of natural gas in the energy structure will achieve energy strategic security.

(3) Developing LNG fuel automobile is in line with national industrial policy.

The developed countries paid general attention to the technology development of LNG fuel vehicle industry. Although gas vehicles developed later in our country since 20st century, great attention has been paid to gas vehicles' development, the national clean Automobile Action Coordination Leading Group has been founded, twelve demonstration cities including Beijing, Chongqing, etc. has been determined. What's more, "Clean Vehicles Action Implementation Method" and "The Project Guide of Key Technology Breakthrough and Industrialization about Clean Vehicles Action" have been proposed, the action planning of clean vehicles has been enacted, so the clean vehicles action will promote to breed a new industry.

(4) The development of LNG fuel automobile can improve the economic benefit of transportation industry.

As vehicles fuel, LNG can save the cost about 25%~35% than fuel oil and the economic benefit of gas replacing oil is more reasonable. Estimating conservatively, taxi, public bus, environmental sanitation vehicles, and container truck can save 17000, 31000, 24000 and 42000 RMB each year respectively.

(5) The advantage of CNOOC to promote the LNG vehicles.

CNOOC used its own LNG resources, vigorously developed the LNG automotive filling industry. In 2008, CNOOC built the first batch of LNG vehicles filling stations that could supply clean energy for automobiles in Shenzhen City. In 2009, New Energy Division was set up by CNOOC Gas & Power Group to operate and manage LNG vehicles filling business.

2.4.3.4 Cold Energy Utilization

(1) General concept.

Energy conservation is an important content of the construction of harmonious society. Because LNG has the ultra-low temperature (−162℃) characteristic, while gas for users is normal temperature. In the LNG terminal, generally LNG will be gasified by vaporizer when used, and this process produces plenty of cold energy. Utilization of this kind cold energy has always been an issue to be discussed. How to utilize the cold energy efficiently is also the content considered by LNG project company. The following is the introduction of the cascade utilization of cold energy and the possible ways.

(2) The scale of cold energy utilization.

LNG is natural gas with wtra low temperature, which contains about 840 kJ/kg cold energy and 270 kW·h/t for LNG, so LNG cold energy has huge potential in use. In the 1960s, much LNG began to be used, at the same time, cold energy was developed. Japan utilizes much more LNG cold energy in the world. Cold energy utilization of LNG can be classified 2 aspects: direct and indirect utilization of low cold energy.

① Direct utilization: air conditioning, freezing storage, CO_2 liquefied, dry ice making, air separating, LNG liquefied again and cold energy generating power.

② Indirect utilization: further utilization of cold energy products, such as further utilization of liquefied nitrogen and liquefied oxygen, and freezing food, low temperature beauty, cryogenic comminution, civil freezing construction and low temperature biological engineering, and so on.

③ Cold energy cascade utilization: based on gradually increase of the LNG temperature, taking full advantage of cold energy is the direction of cold utilization in the future. The utilization of cold energy below 0℃ includes LNG liquefied again, air separating (-150℃), petroleum and chemical industry (-100℃), CO_2 liquefied (about -50℃), freezing storage in low temperature. The cold energy above 0℃ can be used in air conditioning and gas turbine inlet air cooling, as shown in Figure 2-13.

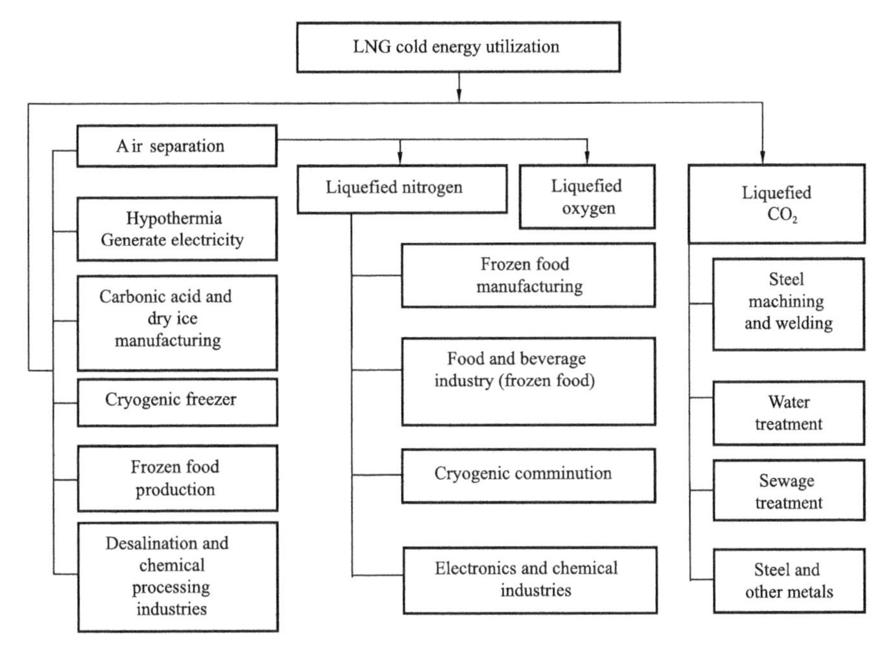

Figure 2-13 Schematic diagram of LNG cold energy utilization

(3) Statements about direct utilization of cold energy.

① Utilizing LNG cold energy to generate power. LNG cold energy power generation is a new non-polluted power generation mode, relying on the power cycle to generate electricity is one of the directions of LNG cold energy recovery, relatively speaking, the technology is relatively mature. The main methods of LNG cold energy power generation include direct expansion method, 2 media method, combined method, mixed media method, and Brandon circulation method and so on.

② Air separation. The traditional way of producing $1m^3$ liquefied air requires about 650 kcal cooling energy. The recovery and utilization of LNG cold energy and 2 stage compression refrigerating machine cooling air to produce liquid oxygen, liquid nitrogen, refrigeration machine is easy to realize miniaturization, consumption can be reduced from 30% to 50% electricity, water consumption can be reduced to 30%, this will greatly reduce the production cost of liquid oxygen, liquid nitrogen, and has considerable economic benefits. So the use of LNG cold energy for air separation, cold energy can be fully applied.

③ Food freezing (keeping fresh). If the LNG base and huge freezing storage are built nearby the terminals, recycle LNG cold energy can be provided to the storage. In the storage using LNG cold energy, containing refrigerant will be chilled to specified temperature, enter into the frozen storage and cooling storage through pipe, and then freezing food by releasing cold energy in chilling disc and tube. Otherwise, according to LNG in different temperature cascade, different cold media are used to exchange heat, and then sent to freezing storage or low temperature freezing device, which enhance greatly the efficiency of cold energy utilization and the whole cost will reduce 37.5% than mechanical freezing.

④ Making liquefied CO_2 and dry ice. Dry ice and liquefied CO_2 are widely used in food industry and beverage industry, which also can enhance the quality of work pieces in the metal welding and casting industry, in the beverage and food industry, can greatly improve the beverage flavor. It can also be used in tobacco industry for tobacco puffing, frozen food preservation and food additive, pharmaceutical, sugar industry, printing and dyeing, wine making, forestry and horticulture, supercritical fluid extraction, scientific research and so on.

⑤ Vehicle refrigerator and vehicle air conditioning. There are two ways to use LNG in the car: power use and cooling use. Power use refers to gasify LNG by pressurizing and heating, promote the impeller to make external work, and heat source is from air or internal combustion engine waste heat. Cooling use refers to liquefy some gas with low temperature by LNG cold energy, or provide cold energy to vehicle compartment. In the hot summer, goods has been freighted into the frozen vehicle after full pre-freezing in the cold storage, cold energy from LNG has been stored in cold storage board. With the transport time increasing and load increasing due to an increase the number times of open door, the cold energy produced by LNG gasifying directly enters the compartment and supplies cold energy together with the cold storage system to maintain the temperature in compartment. If refrigerated vehicle consuming 12 to 15 kg/h LNG, the refrigerating capacity is 28 kw, which is enough to supply after pre-freezing goods with cold energy in short refrigerated transport.

⑥ Manufacturing ice and snow. Analyzing the present date, there is no precedent using the cold energy of LNG to making the skiing and skating rink. A lot of cold energy is required to make artificial skiing and skating rink, which can be provided from a lot of LNG cold energy. Similar to the cold storage and refrigeration, artificial skiing and skating rink requires higher temperature range, so that it is inappropriate to use the entire LNG cold energy to produce ice. Meanwhile manufacturing ice and snow needs a lot of fresh water and land. Otherwise, remote location of LNG terminal, special terrain, and the market of ice world should be also considered comprehensively.

⑦ Recovery storage tank BOG. In order to recycle part of the BOG evaporating from the storage tank, BOG in the selected process will be mixed with cold LNG after pressurizing, and then BOG is condensed into liquid by using the LNG cold energy and returns the process, which can eliminate the waste of evaporated gas from the torch burning.

⑧ Utilizing cold energy to recycle the C_2 and C_{2+} in LNG. Natural gas liquid (NGL) re-

fers to any kind of hydrocarbons such as ethane, propane, butane and condensate (C_{5+} above) extracted from outlet of the well. LPG mainly contains propane and butane. Generally speaking, LNG mainly contains methane, ethane, propane and butane (*n*-butane and *i*-butane) and a small amount of pentane (*n*-pentane and *i*-pentane). After analyzing Guangdong LNG component, C_2 and C_{2+} above take about 8% in volume (volume ratio). Extracting the maximum NGL from LNG needs deepen cryogenic extraction, and most of LPG (above 95%) can be also obtained in design, together with above 70% ethane.

⑨ Cool storage. LNG has been mainly used to generate electricity and city gas, and the load of LNG gasifying varies with time and seasons. The fluctuation of LNG cold energy will have a bad effect on the operation of cold energy utilization equipment, which must be paid more attention to. LNG cool storage devices studied by Japan Osaka Gas Company, using latent heat of phase change materials to store LNG cold energy. When the LNG cold energy is abundant during the day, phase change materials can store the cold energy; when the LNG cold energy is lacking during the night, phase change materials can dissolve and release cold energy to cold energy utilization equipment.

(4) The indirect utilization of the LNG cold energy.

① The cryogenic comminution. The LNG cold energy can be used to crush the materials under low temperature, which could smash the materials that can't be crushed under normal temperature and the powder is smaller. Compared with cryogenic comminution, it can smash the materials into tiny particles and the particles can be separated. There is no particle explosion and smell contamination in cryogenic comminution. The mixture with complex components can be selectively crushed by choosing different low temperature.

② Sewage treatment. High purified ozone can be obtained with Liquefied oxygen. The sewage treated has a high absorptivity to ozone, the treatment can reduce one third of the power consumption than the traditional process and the treatment efficiency is superb.

(5) The issues should be considered when utilizing the LNG cold energy.

① Cold energy cascade utilization. Since the power to produce low temperature increases with the decreasing of the temperature. The lower the temperature is, the more the power is required, the lower the mechanical efficiency is, and the higher the factory's cost is. Consequently, take an overall consideration, the craft process to utilize the LNG cold energy should make full use of the −162℃ low temperature. To realize the cascade utilization is to improve the utilization efficiency of LNG cold energy maximally.

② The amount used and receiving terminal location. As LNG is mainly used by fuel gas and power plant in the city, the change in load between day and night is huge, so, the load of the minimum hours at certain time must be noted. In order to make full use of the LNG cold energy, an easily regulating system must be employed that the load of system could be regulated with the change of the LNG load. Hence, sufficient consideration should be taken between the utilization of LNG cold energy and the supply of the fuel gas. LNG is usually transported by low temperature pipes. Due to the high quality materials of the pipes, the cost is very high. Delivery pressure loses a lot and absorption of heat may lead to the loss of cold energy. Therefore, the cold energy utilization factory should be close to the LNG receiving terminals as much as possible. But, to make the products in circulation and purchase raw material conveniently, cold energy utilization factory prefers to build the location where the transportation is convenient. However, in a lot of case, it is difficult to give considerations to the benefits of the both sides.

③ Safety restrictions. Directly utilizing the LNG cold energy is the best way to use cold energy. But, it is flammable after gasification and heat exchange with materials directly should be avoided in many cases. Take air as an example, explosive gas may generate when the air mix with the leakage of LNG. As a result, intermediate medium are often employed to the utilization of the cold energy.

④ In order to expand the utilization of LNG cold energy, it is necessary to require a larger, continuous gasification. This depends on the further extension and expansion of downstream users. In order to improve the efficiency of cold energy utilization, it is necessary to integrate the system with large scale. Japan has already started research in this area.

⑤ LNG cold energy has been transported to the chemical factory and refinery nearby, to separate the light hydrocarbon products from waste gas in the production. This kind of the cold energy utilization could only be achieved in the industrial cluster, the refinery and chemical factory should been built adjacently to the construction. There is no successfully project case until now, only the Japanese enterprises are conducting some research but much technological difficulties have to be overcome. The realization of LNG cooling refrigerants such as propane, butane and etc., which could be used as cold source for other industries and cold storage, also it is required to be realized in the industrial clusters.

Chapter 3 LNG Project Scope Management

3.1 General Concepts of LNG Project Scope Management

3.1.1 Definitions for LNG Project Scope and Scope Management

3.1.1.1 LNG Project Scope Definition

There are two aspects in LNG project scope, on the one hand, it refers to the "Product Scope" of project that through the negotiation with foreign suppliers, the LNG project team finally signs the LNG sales and purchase agreement which stipulated composition and calorific value of LNG. On the other hand, it means the "Work Scope" or sometimes referred to as the scope of the project, that is the LNG project scale determined by pre-market research, a series of work undertook by project team, project office or project company, such as research, design, construction of each sub-project with good quality and quantity, and conducts all kinds of management and coordination work for the project. It primarily concerns with determining and controlling the processes which should or shouldn't be included the project scope, and finally delivers the qualified gas to each user's gate station or users through the built projects.

3.1.1.2 Definition of LNG Project Scope Management

LNG project scope management refers to hand over the LNG product to users, based on the content of project scope definition, through the planning, organizing, coordinating and controlling the various input resources, providing different deliverables in each project life cycle, and finally achieves the targets of project acceptance and project operation effectively.

This chapter will deeply analyze the work contents involved and the available resources, including the project initiation, scope definition, scope planning, scope verification, scope controlling, the identification and work breakdown of almost all tasks, doing more accurate estimates for project cost, time, human resources, equipment and materials, information under the existing conditions, developing a baseline plan to assess project performance and project control.

3.1.1.3 The Role of LNG Project Scope Management

(1) Project scope is the foundation for making project plans.

Project scope is the foundation for making project plans. For general LNG project, the main scope of the project is almost the same, mainly including LNG resources acquisition, the mode of LNG transportation, the construction of jetty, receiving terminal, trunkline and auxiliary projects, LNG supply contract execution for downstream users. However, among the specific LNG project scopes, the details are largely different. Take the land formation engineering for example, the LNG project in Guangdong Dapeng is greatly different from the LNG project in Fujian, the former is cut mountain and the latter is the reclamation. Most LNG projects consist of three parts including receiving terminal, jetty and trunk line, but the Shanghai LNG project doesn't have sub-projects for pipelines. Some projects have good supporting natural gas market conditions, while others need spend a lot of effort to develop the market. Therefore, different project scope has different project plans. We must make a practical plan on basis of project scope. According to the plans, the important project products can be further divided into smaller, more manageable tasks and work packages.

(2) Project scope is the basis of project resource input.

Defining the project scope can provide foundation for project investor to the resources inputs for the LNG project. The resources input include the manpower, money, materials, time and information. The investor should consider setting up a basic team, especially select a general manager who already managed similar projects before and could bring some management experience and procedures to the project, and recruit shareholders and social workers into team according to the project progress. In terms of investment, currently, 3×10^6t/a LNG project, including jetty, receiving terminal and trunkline sub-projects, needs to invest approximately $(6\sim7)\times10^9$ RMB, and the capital investment is about $(1.8\sim2)\times10^9$ RMB. The strong financial resources is a necessary requirement for investors, plus materials, equipment, data, time period selection and support information, etc. All these above conditions constitute the overall investment scale.

(3) Project scope is the precondition for project change.

The project usually does not run exactly under the initial ideal conditions, during the 4~5 years of the construction project, many initial conditions could be changed, some of which are fundamental. Accordingly, the invest budget will change as the LNG resources, material, equipment, materials market, land acquisition and downstream user's market changed, we must change the original project plan according to such changes. If the upstream LNG resource price makes a breakthrough, we have to make adjustments to the project plan. In short, if the project scope changes, we should do corresponding adjustments to the new conditions, make decision to postpone, suspend or even cancel the project.

(4) Project scope is the criteria for project acceptance evaluation.

Project scope ensures the final results — to provide good quality and quantity of natural gas or LNG to downstream users. At the same time, each project completion to production needs acceptance evaluation, accordingly, there is no doubt that the project scope is the criteria for acceptance evaluation. Through the acceptance evaluation, it can examine whether the early scope we determined is reasonable or not, and discover what aspects are still not perfect. It helps us to summarize and improve the LNG project budget, schedule and resource requirement estimation accuracy, offer the criteria for performance measurement and control of project, which could be used for project owners and participants to measure the degree of success. Simultaneously, the project investors could learn a lesson and accumulate experience, enrich the management process

asset, and do better in the new similar investment projects.

3.1.2 LNG Project Scope Management Processes

3.1.2.1 Three Components of Scope Management Processes

The general flow diagram of LNG project scope management can be seen in Figure 3–1. From this figure we can see that the first is the scope management basis of LNG project, the contents include the basis of project startup, project preliminary scope manual, and definition basis of project scope. The middle part is project scope preparation, which is the core of scope management. From the beginning of LNG product scope to the LNG project work scope preparation, the work breakdown is the basic. LNG project scope verification is an in-depth examination of the tasks. LNG product scope has some relative independence, from the beginning of LNG project work scope, each upper level work is the prerequisite for the next level work, and can feedback to above all levels. The middle part of the work depth and confirm each other directly influence the effect of scope management, and scope control is throughout the project life cycle activities.

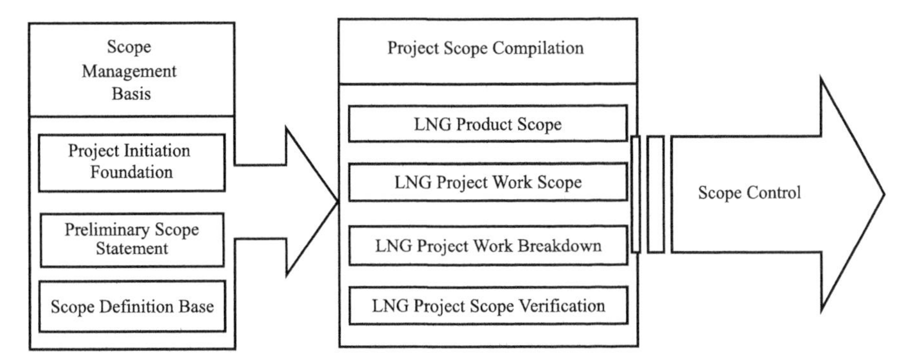

Figure 3–1 LNG project scope management flow diagram

3.1.2.2 The Relationship between the Three Components of the Process

(1) Project scope management basis and scope compilation.

From the general project scope management, project launch is based on its prerequisites, which is stimulated by the LNG project in the market, competition, consumption, scientific and technological progress and legal conditions, contributing to the start of the LNG project. Based on the preliminary scope instruction from the similar projects, the requirement resources should be analyzed, including project products, targets, limitations, assumptions, resources distribution, expense assessment, project organization and management agency. Scope definition is based on the LNG users' market, and LNG resources and LNG project size are described. From the above three aspects of the discussion, it is necessary to provide basic information for the preparation of specific LNG project scope. Firstly, the LNG project scope management should describe the product with physical chemistry properties and danger of LNG or natural gas to their users, having relatively independent. Start with the definition of the LNG project scope of work, it needs to provide the contents of engineering construction must be completed in LNG project, by

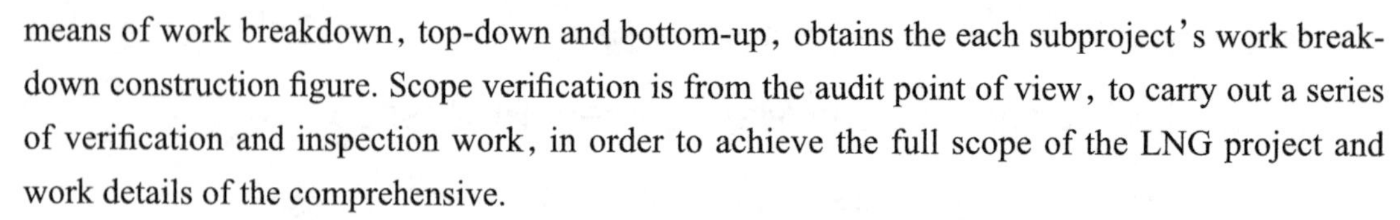

means of work breakdown, top-down and bottom-up, obtains the each subproject's work breakdown construction figure. Scope verification is from the audit point of view, to carry out a series of verification and inspection work, in order to achieve the full scope of the LNG project and work details of the comprehensive.

(2) The relationship between scope compilation and scope control.

The scope compilation decides the amount of work in concrete project. The work breakdown further clears tasks, the content, the target, the payment and the milestone, especially the resources input that needed for undertaking the project management, such as the management team, technology standards, controlling factors, the optional schemes, etc. On the basis of scope compilation, the scope control proceeds hierarchy control and procedure control, and its results are to get the detailed project scope instruction, work breakdown contexture and its directory, as well as project management schedules.

(3) The relationship between scope control and the former two.

From the Figure 3–1, it demonstrates that the scope control is the sublimation and development of scope compilation and scope management basis. Only the scope control is to be done in place, can it guarantee scope compilation not missing and leakage, improve and deepen the understanding of the scope compilation and further understanding of the scope management basis. Conversely, if scope management basis is fully prepared, then can spend less energy making the scope control, and completed scope compilation also lays a solid foundation for scope control. Scope control is the dynamic management of the content, it should be based on the dynamic development of internal and external projects, timely adjustment and coordination, so that the LNG project has always been operated under effective scope control.

Similar to the project schedule, cost, quality management, only the 3 components of scope management tightly linked, confirmed each other, the combination of static and dynamic, timely adjustment, can ensure the project scope management in place.

3.2 LNG Project Initiating and Scope Definition

3.2.1 LNG Project Initiating

3.2.1.1 The Definition of LNG Project Initiating

Aim at one specific gas market, doing the work of LNG market demand and LNG resources investigation by the project initiator or the early project team, means formally starting a new LNG project or confirming whether continue the next stage of an existing project or not. All work belongs to the initiating progresses of LNG project.

Project initiation usually occurs when the LNG project initiator or user encounters one or some "stimulus" and must respond to or respond to such stimulus.

(1) Market demands.

Market demands are the most direct stimulus to start a LNG project. As the 40 years of the reform and opening-up, our country has experienced rapid economic development, along with the increasing energy demands, especially for the clean energy. Natural gas, as one kind of clean energy, is more and more acceptable. The major domestic oil corporations, which focus on the hydrocarbon exploration and development, especially CNOOC, has already targeted the natural gas market at the end of twentieth century, and made large amount of strategic research. At the same time, the economic development of the domestic coastal provinces and cities is in the forefront, and they have increased demand for natural gas. The two aspects accelerate the development of LNG.

(2) Competition demands.

Competition demands means that the LNG users make the natural gas as the fuels or raw materials, to gain the competition advantage in the price of the product. The "Stimulus" comes from the user's preference for the natural gas. For example, the pottery industry in the southern China, which adopted liquefied petroleum gas (LPG) as the fuels before, however, the high price of LPG makes more pressure for the ceramic products in the market, so if using LNG as the fuels which could largely decrease their cost and gain more profit space.

(3) Consuming demands.

Using natural gas as car fuels, it can not only save fuel cost, but also decrease the city pollution. It is a popular industry for both users and the government, the "Stimulus" comes from the new consuming demands—the new consuming market such as LNG filling station appears. The LNG project initiator should make good use of this opportunity, cooperate with local government closely, and actively promote the construction of the LNG filling station, for the benefit of the people.

(4) Technology progress.

Due to technology progress, LNG emerges, as a new natural gas application technology. Because of the development of low-temperature materials, it is possible for gas-liquid conversion, transportation and preservation under the ultra-low temperature. CNOOC has made research on the critical technology in the main part of the industry chain, and some products have in practice. The following step is to enlarge and deepen the research results, apply them to the LNG and natural gas market as soon as possible, and let both LNG project operators and users enjoy the profits and benefits.

(5) Legal requirements.

Environmental protection voice is rising globally, as the responsible countries and local governments have established corresponding laws and regulations, natural gas is just coincided with the requirements of environmental protection. Some local governments have annual reduction of carbon dioxide emissions, sulfur dioxide, dusts as one of the indicators of performance evaluation. Environmental protection provides a good opportunity to develop the use of LNG or natural gas.

3.1.2.2 Basis and Methods for LNG Project Definition

(1) Basis for LNG Project Definition.

① The corporation's development strategy of LNG industry. The corporation here refers to the LNG project initiator. Generally speaking, the project's initiator in the process of the LNG project before, must have made the strategy with a lot of research work, including the relation between LNG industry and corporation's original core industry, the influence of LNG industry to the competitiveness of enterprises, LNG industry risk, and LNG industry could brought new opportunities for development, LNG industry development plan and so on.

② LNG project deliverables description. Deliverables include jetty, receiving terminal (including storage tanks), gas trunk line and other auxiliary projects.

③ Standard selection of LNG projects. Currently, LNG project is a new industry in China, and relevant institutions are working on core standards. In order to push projects, if the domestic is lack of standards, it is necessary to borrow from abroad. Taking Guangdong Dapeng LNG Project as an example, during the design and construction phase, more than 324 domestic and foreign standards were adopted.

④ Related historical information. In order to enter into this industry as soon as possible, foreign LNG information, such as natural gas resource, LNG resource, relative technique, management, LNG transportation, commercial contract, HSE of LNG, must be widely collected.

(2) LNG project decision.

For a LNG project, besides it is a new thing in China, it also has characteristics such as a large investment, complex technology, long industrial chain, wide social influence and so on. So we must start a LNG project very seriously. As LNG project initiator, based on the foundation of extensive research and strategic planning, scientific and democratic decision-making process must be carried out, which usually adopts method of combining quantitative and qualitative methods, such as analytic hierarchy process or quantitative analysis methods such as decision tree or decision table.

3.1.2.3 The Preparation of the Preliminary Project Scope Statement

Preliminary project scope explanation uses the data of same scale of LNG project to describe various tasks and events. It includes products, targets, project constraints, assumptions, resources allocation, cost estimation and project structure of management organization to complete all tasks, etc.

(1) Project product and project target.

Some of the final results of project development are the product produced by the project, such as the new family car development project, the project eventually provides the family car for customers. However, the implementation of the LNG project is not directly related to the product produced. The LNG project provides the user with natural gas, but it is not the LNG project company out of production directly, and it can be the operation of the project entity or independently of a project of other institutions and LNG suppliers through negotiations, signed a LNG supply agreement or contract, to provide natural gas products in accordance with contract quality requirements. In order to deliver the natural gas or LNG to the users, the LNG project company constructs LNG jetty, receiving terminal and pipeline, so the scope of the LNG project is to guarantee the quality and quantity of natural gas or LNG delivery and the scope of the construction project, through the project to transport the natural gas to a user station. To this end, the project needs to determine the supply volume of LNG, the sub goals and the total target of the project, LNG project deliverables, LNG project engineering milestones.

(2) Project restraints and assumptions.

LNG project in the implementation process will encounter some restrictions, such as national laws and regulations, industry standard limits in the design and construction process, HSE restrictions, etc. It assumes that the implementation of the LNG project will rely on a certain future environment. A number of factors encountered in the project are considered to be true and reliable according to project plan, such as LNG resources not confirmed at the beginning of the project, but we think it will be confirmed or implemented in the next few years.

(3) Resource allocation and cost estimation.

The implementation of the LNG project cannot do without the input of resources, such as human, financial and material resources, time and information, in the early stages of a project, must be on the resource input has a basic calculations, in order for the LNG project initiator to have a basic understanding of the resources and facilities available for the project to be implemented, it is necessary to estimate the duration of the project, the project cost estimate, and so on, so as to list in the initiator's investment budget plan.

(4) Project general manager and project management organization.

Selection of project general manager is the first step in project organization construction. As the LNG project initiators should firstly consider the general project manager candidates. The candidates for this position must have the working experience in similar projects as manager or his deputy and the ability to manage the whole project. Besides, related team members on technique and management should be deployed which can be conducted step by step according to the project status.

3.2.2 Basis for LNG Project Scope Definition

3.2.2.1 General Concepts

LNG project scope should be defined on a large number of market investigations and researches domestically and internationally. In order to build LNG project successfully, we must seize "One center and two basic points". The one center refers to all the tasks must be finished during the project process; the two basic points, the first one means the downstream user's type, gas load centers and the annual gas demand volume, the second is foreign LNG resource availability, namely in the foreseeable period, how much can be provided. Above two are complementary to each other. The matching degree between the supplier and the demander on the quantity of LNG is a problem that we are particularly concerned at the beginning of the project. Finally, it is determined by the initiator or LNG project company for the amount of LNG, on the basis of which the scale of the LNG project is determined, and the working range of the LNG project is given according to the scale and field conditions.

3.2.2.2 LNG Users

Types of LNG user group, which is based on current industry structure in the area, has discussed in Chapter 2. Specific users will be discussed in this chapter, which is an important content of LNG scope management.

(1) Gas power plant.

It includes numbers of new plants and oil to gas power plant and annual gas consumption. From the current point of view, in the power peaking big gap city and the electricity load centers, to build peak load power plant of gas turbine or gas steam combined cycle generating units, which can alleviate the contradiction of power load, also can achieve the nearest power supply, optimizing the power grid operation. For LNG project, gas power plant is still a big customer, which occupies 75% and 55% in Fujian and Guangdong Dapeng LNG project respectively. Gas power plants have the advantages of peak shaving, quick start, high thermal efficiency and minimum pollution, etc.

(2) Urban gas.

It is mainly used for urban residents (kitchen, hot water), public welfare (restaurants, hotels, shopping malls, airports, schools, kindergartens, staff canteens, government agencies, offices and other public welfare units), heating (centralized type, unit type), central air conditioning, etc.. Therefore, we should consider the existing city gas network to replace gas and the new city gas network construction, especially to ensure the gas demand of big city, in order to improve the atmospheric environment, improve people's quality of life. From the price of gas bearing capacity, energy efficiency and environmental protection perspective, the city gas is vigorously developing direction in the future.

(3) Industrial fuel gas.

It is used in smelting furnace, heating furnace, heat treatment furnace, baking furnace, drying oven (flat glass, ceramics and cement) as fuel in mechanical and electrical, textile and other industries, because natural gas can improve the product quality and added value.

(4) Industrial raw materials.

It is mainly used for carbon 1 chemicals, such as industrial to produce hydrogen, acetylene, methane chloride, two methyl chloride, carbon tetrachloride, carbon black, nitromethane hydrocyanic acid, carbon disulfide, nitro methane etc..

(5) LNG or gas filling station.

With the strengthening of urban environmental protection, natural gas vehicle has entered the urban transportation industry. For this reason, some domestic cities have allocated special funds for vehicle transformation and related encouragement policy. As the LNG project company should seize the favorable opportunity, to expand the natural gas, the use of LNG industry, actively participate in and cooperate with the local government, develop LNG and gas filling station.

(6) Current LNG project users.

From the Chinese existing LNG users, gas user's is still dominated by gas power generation. It is necessary to introduce the LNG project in the initial stage. Gas is mainly used in gas power plant with large gas consumption for LNG project started and the formation of demonstration and radiation effect, at the same time, city gas is currently the second users. Table 3-1 is a user's list of LNG projects in China. Some experts think that at present, to develop urban and industrial fuel gas user is more appropriate.

Table 3-1 Part of Chinese LNG Project User Statistics

LNG Project	Gas power plant	Urban Users	Others
Guangdong Dapeng	Dongbu Power Plant, Qianwan Power Plant, Huizhou Power Plant, Meishi Power Plant, Zhujiang Power Plant	Shenzhen Urban Gas, Dongguan Urban Gas, Guangzhou Urban Gas, Foshan Urban Gas	Hongkong: Hong Kong and China Gas Other industrial users
Fujian	Putian Power Plant, Jinjiang Power Plant, Xiamen Eastern Power Plant	Fuzhou Urban Gas, Putian Urban Gas, Quanzhou Urban Gas, Xiamen Urban Gas, Zhangzhou Urban Gas	Other industrial users
Shanghai			Shanghai Pipeline Network Company

3.2.2.3 LNG Resource Availability

Top priority of LNG project is looking for opportunity to acquire LNG resources. From the availability of resources, it is divided into two ways.

(1) Import.

At present, the upsurge of economic development and environmental protection of China's high-speed voice, make clean energy — natural gas resource increasingly strong demand, so the import of LNG is an important way for our country at present. From the purchase way according to the contract, including long-term, short-term futures and spot buy etc.

(2) Domestic purchase.

At present, the possibility of purchasing large amounts of LNG in domestic market is less. Although development of the natural gas resourles in China makes great progress, the discovery and exploitation of natural gas resources is not need to keep up with the pace of economic development; although small LNG liquid factories emerg in Chinese, still can't meet the demand of bulk domestic market, just fill some small user market. With the increasing exploration efforts and the discovery of a large gas field in Chinese, the possibility is there in the domestic procurement.

3.2.2.4 LNG Project Size Determination

Take the long-term LNG supply contract as an example, with the LNG user's annual gas consumption and the size of the supply of foreign LNG resources, there can be the following three cases.

(1) Demand and supply to achieve balance.

Demand and supply balance is the most ideal situation, but generally relatively rare. This with the natural gas supply and demand condition and the alternative energy market both at home and abroad, a balance point in time is short. If that happens, LNG project company should seize the opportunity to sign long-term LNG supply-marketing contracts.

(2) Demand is less than supply.

When the world natural gas is greatly discovered, the natural gas supply will be ample, there will be a buyer's market. LNG resource buyers should seize this favorable opportunity to compare suppliers, look for gas suppliers whose price is low but quality is high to sign long-term supply and marketing contracts. At the same time, the buyer should actively develop downstream user market, and strive to expand the scale of the project timely to achieve greater scale effect. For example, during the end of twentieth century when world gas supply was ample, CNOOC seized this chance and procured Australia's northwest shelf LNG, which laid a good foundation for Guangdong Dapeng LNG project.

(3) Demand is greater than supply.

When opposite case as above situation happened, that is, in seller's market, it will bring adverse effect to the LNG project, which means that suppliers' number will decline and gas price will keep a high position. In order to buy relatively cheap LNG, the buyers must consider a combination purchase, that is, joint procurement of multiple LNG projects to reduce the price, or participate in the gas field development to get priority purchase right for favorable price, or sign short-term contracts and spot contract to have a price advantage.

3.3 LNG Project Scope

3.3.1 Product Scope of LNG Project

To purchase the LNG resource for each LNG project is necessary, and the quality of the gas product has to meet the national quality standard that has been regulated by the "General Characteristics of Liquefied Natural Gas" in Europe. Our country adopts the same standard and negotiates with the foreign merchants. The quality and physical chemistry properties of LNG that included in the national standard will be illustrated in the following section.

3.3.1.1 General Concept

LNG is not the original gas produced from the gas field. It has been drying and removing CO_2 and sulfide at 101.325 Pa. After fractionation, decreasing condensation temperature to −162℃, gaseous natural gas transfers to liquefied natural gas with volume shrinking to 1/600 of the original, and the weight is about 45% of the identical volume of water. LNG is colorless, odorless, non-poisonous and low temperature storage is required. It could be heated and gasified according to the specific requirement. LNG is similar to other gaseous hydrocarbons, and one of the characteristics is flammable. The mixture of LNG and air is flammable at the range of volume fraction from 5% to 15%.

3.3.1.2 LNG Properties

(1) Composition.

LNG is a mixture of hydrocarbons composed predominantly of methane and contains minor amounts of ethane, propane, nitrogen or other components normally found in natural gas. The component proportion of different LNG is not the same, while methane should be higher than 75% and the nitrogen should be lower than 5%. Although the major constituent of LNG is methane, it should not be assumed that LNG is pure methane for the purpose of estimating its behavior. In the analysis of the composition of liquefied natural gas, special care should be taken to obtain representative samples not causing false analysis results due to distillation effects. The most common method is to analyze a small stream of continuously evaporated product using a specific device that is designed to provide a representative gas sample of liquid without fractionation. Another method is to take a sample from the outlet of the main product vaporizers. This sample can then be analyzed by normal gas chromatographic methods.

(2) Density.

The density of LNG depends on the composition and usually ranges from 430 kg/m^3 to 470 kg/m^3, but in some cases can be as high as 520 kg/m^3. Density is also a function of the liquid temperature with a gradient of about 1.35 $kg/(m^3 \cdot ℃)$. Density can be measured directly but is generally calculated from composition determined by gas chromatographic analysis. The Klosek Mckinley method as defined in ISO 6578 is recommended.

(3) Temperature.

LNG has a boiling temperature depending on composition and usually ranges from −166℃ to −157℃ at atmospheric pressure. The variation of the boiling temperature with the vapor pressure is about 1.25×10^{-4} ℃/Pa. The temperature of LNG is commonly measured using copper or copper nickel thermocouples or using platinum resistance thermometers.

(4) Examples of LNG.

Table 3-2 shows a typical example of the three LNG recommended by the European Commission for Standardization in general liquefied natural gas (LNG), and it shows the different properties with the components changed.

Table 3-2 LNG Examples

The properties of the bubble point at normal pressure	LNG example 1	LNG example 2	LNG example 3
Molar content (%)			
N_2	0.50	1.79	0.36
CH_4	97.50	93.90	87.20
C_2H_6	1.80	3.26	8.61

Continued

The properties of the bubble point at normal pressure	LNG example 1	LNG example 2	LNG example 3
Molar content (%)			
C_3H_8	0.20	0.69	2.74
$i\ C_4H_{10}$	—	0.12	0.42
$n\ C_4H_{10}$	—	0.15	0.65
C_5H_{12}	—	0.09	0.02
Molecular weight (kg/kmol)	16.41	17.07	18.52
Bubble point temperature (℃)	-162.60	-165.30	-161.30
Density (kg/m^3)	431.60	448.80	468.70
Volume of gas measured at 0 ℃ and 101. 325 Pa/volume of liquid (m^3/m^3)	590.00	590.00	568.00
Volume of gas measured at 0 ℃ and 101.325 Pa/mass of liquid ($m^3/10^3$ kg)	1367.00	1314.00	1211.00

Table 3-3 lists the LNG properties of the supply and marketing contracts signed by Guangdong Dapeng and Fujian LNG projects, which are the top two earliest LNG projects in China. The methane in the components of Guangdong and Fujian LNG project is similar to the example 1 and example 3 in Table 3-2 respectively, but the heavy hydrocarbon in the components and calorific value of Guangdong LNG is relatively higher than Fujian LNG.

Table 3-3 The Quality Indicator Tableof Guangdong and Fujian LNG under the Condition of Converting into Gaseous

Item	Unit	Guangdong LNG	Fujian LNG
Density (Max)	kg/m^3 (☆)	0.805	0.708
Range of high calorific value	MJ/m^3 (☆)	39.890～43.630	38.400～38.900
Range of Wobbe index	MJ/m^3 (☆)	51.600～53.800	50.900～51.100
CH_4 (Min)	%mole content	84.000	96.100
C_4H_{10} and above C_4H_{10} (Max)	%mole content	2.000	0.237
C_5H_{12} and above C_5H_{12} (Max)	%mole content	0.100	0.006
N_2 (Max)	%mole content	1.000	0.451
H_2S (Max)	mg/m^3 (☆)	5.000	4.886
Total sulfur (Max)	mg/m^3 (☆)	30.000	23.700
(☆): Under the standard conditions			

3.1.1.3 LNG Physical Properties

(1) Physical properties of BOG.

LNG is stored in bulk as a boiling liquid in large thermally insulated tanks. Any heat leaking into the tank will cause some of the liquid to evaporate as gas. This gas is known as BOG. The composition of the BOG will depend on the composition of the liquid. As a general example, BOG can contain 20% nitrogen, 80% methane and traces of ethane. The nitrogen content of the BOG can be about 20 times than that in the LNG.

As LNG evaporating, the nitrogen and methane are preferentially lost leaving a liquid with a larger fraction of the higher hydrocarbons. BOG below about −113℃ for pure methane and −85℃ for methane with 20% nitrogen are heavier than ambient air. At normal conditions, the density of BOG will be approximately 0.6 of air.

(2) Flash.

As with any fluid, if pressurized LNG is lowered in pressure to below its boiling pressure, for example, by passing through a valve, then some of the liquid will evaporate and the liquid temperature will drop to its new boiling point at that pressure. This is known as flash. Since LNG is a multicomponent mixture, the composition of the flash gas and the remaining liquid will differ for similar reasons to those discussed above.

As a guide, a 10^3 Pa flash of 1 m^3 liquid at its boiling point corresponding to a pressure ranging from 1×10^5 Pa to 2×10^5 Pa produces approximately 0.4 kg gas. More accurate calculation of both the quantity and composition of the liquid and gas products of flashing multicomponent fluids such as LNG is complex. Validated thermodynamic or plant simulation packages for use on computers incorporating an appropriate database should be used for such flash calculations.

3.1.1.4 Spillage of LNG

(1) Characteristics of LNG spills.

When LNG is poured on the ground (as an accidental spillage), there is an initial period of intense boiling, after which the rate of evaporation decays rapidly to a constant value that is determined by the thermal characteristics of the ground and heat gained from surrounding air. This rate can be significantly reduced by the use of thermally insulated surfaces where spillages are likely to occur as shown in Table 3-4. These figures have been determined from experimental data.

Table 3-4 Evaporation Rate

Material	the rate per unit area after 60 seconds [kg/ ($m^2 \cdot h$)]
Aggregate	480
Wet sand	240
Dry sad	195
Water	190
Standard concrete	130
Light colloid concrete	65

Small amounts of liquid can be converted into large volumes of gas when spillage occurs. One volume of liquid will produce approximately 600 volumes of gas at ambient conditions(table 3-2). When spillage occurs on water, the convection in the water is so intense that the rate of evaporation related to the area remains constant. The size of the LNG spillage will extend until the evaporating amount of gas equals the amount of liquid gas produced by the leak.

(2) Expansion and diffusion of gas clouds.

Initially, the gas produced by evaporation is nearly at the same temperature as the LNG and is denser than ambient air. Such gas will at first flow in a layer along the ground until it is warmed by absorbing heat from the atmosphere. When the temperature has risen to about −113℃ for pure methane or about −80℃ for LNG (depending on its composition), it is less dense than ambient air. However, the gas air mixture will only rise when its temperature has increased so that the whole mixture is less dense than ambient air. Spillage, expansion and dispersion of vapor clouds are complex subjects and are usually predicted by computer models. Such predictions should only be undertaken by a body competent in the subject. Following a spillage, "fog" clouds are formed by condensation of water vapor in the atmosphere. When the fog can be seen (by day and without natural fog), this visible "fog" cloud can be used to show the motion of the vaporized gas and to provide a conservative indication of the flammability range of the gas and air mixture.

In the case of a leak in pressure vessels or in piping, LNG will spray as a jet stream into the atmosphere under simultaneous throttling (expansion) and vaporization. This process coincides with intense mixing with air. A large part of the LNG will be contained in the gas cloud initially as an aerosol. This will eventually vaporize by further mixing with air.

3.1.1.5 LNG Ignition and Explosion

A natural gas air cloud can be ignited where the natural gas concentration is in the range from 5% to 15% by volume.

(1) Pool fires.

With a diameter of more than 10m of ignited LNG pool, the surface radiant power (SEP) of the flame is very high, and it can be calculated by the measured actual forward radiation flux and the determined flame area. The SEP depends on pool size, smoke emission and methods of measurement. SEP will decrease with the increase of the dust carbon black.

(2) Development and consequences of pressure waves.

In a free cloud, natural gas burning at low velocity results in low overpressures of less than 5×10^3Pa within the cloud. Higher pressure can occur in areas of high congestion or confinement such as densely installed equipment or buildings.

3.1.1.6 Containment

Natural gas cannot be liquefied by applying pressure at ambient temperature. In fact, it has to be reduced in temperature below about −80℃ before it liquefies at any pressure. This means that any quantity of LNG that is contained, for example between two valves or in a vessel with no vent, and is then allowed to warm up will increase in pressure until failure of the containment system occurs. Plant and equipment shall therefore be designed with adequately sized vents and (or) relief valves.

3.1.1.7 Other Physical Phenomena of LNG

(1) Rollover.

The term rollover refers to a process whereby large quantities of gas can be emitted from an LNG tank over a short period. This could cause overpressurization of the tank unless prevented or designed for. It is possible in LNG storage tanks for 2 stably stratified layers or cells to be established, usually as a result of inadequate mixing of fresh LNG with a heel of different density. Within cells the liquid density is uniform but the bottom cell is composed of liquid that is denser than the liquid in the cell above.

Subsequently, due to the heat leak into the tank, heat and mass transfer between cells and evaporation at the liquid surface, then the cells equilibrate in density and eventually mix. This spontaneous mixing is called rollover. As is often the case, the liquid at the bottom cell has become superheated with respect to the pressure in the tank vapor space, and the rollover is accompanied by an increase in vapor evolution. Sometimes, the increase is rapid and large. In a few instances the pressure rising in the tank has been sufficient to cause pressure relief valves to open.

Rollover can be prevented by good stock management. LNG from different sources and having different compositions should preferably be stored in separate tanks. If this is not practical, good mixing should be ensured during tank filling. High nitrogen content in peak shaving LNG can also cause a rollover soon after the cessation of tank filling. Experience shows that this type of rollover can best be prevented by keeping the nitrogen content of LNG below 1% and by closely monitoring the boil-off rate.

(2) Rapid phase transition (RPT).

When two liquids at two different temperatures come into contact, explosive forces can occur under specific circumstances. This phenomenon, called RPT, can occur when LNG and water come into contact. Although no combustion occurs, this phenomenon has all the other characteristics of an explosion.

RPT resulting from an LNG spill on water have been both rare and with limited consequences. The universally applicable theory that agrees with the experimental results can be summarized as follows. When 2 liquids with very different temperatures come into direct contact, if the temperature (expressed in kelvin) of the warmer of the two is greater than 1.1 times the boiling point of the cooler one, the rise in temperature of the latter is so rapid that the temperature of the surface layer can exceed the spontaneous nucleation temperature (when bubbles appear in the liquid). In some circumstances this superheated liquid vaporizes within a short time via a complex chain reaction mechanism and thus produces vapor at an explosive rate.

For example, liquids can be brought into intimate contact by mechanical impact and this has been known to initiate RPT in experiments with LNG or liquid nitrogen on water.

(3) Boiling liquid expanding vapor explosion (BLEVE).

Any liquid at or near its boiling point and above a certain pressure will extremely rapidly vaporize if suddenly released due to failure of the pressure system. This violent expansion process has been known to propel whole sections of failed vessels several hundred meters. This is known as a BLEVE.

BLEVE is highly unlikely to occur on an LNG installation because either the LNG is stored in a vessel which will fail at a low pressure and where the rate of vapor evolution is small, or it is stored and transferred in insulated pressure vessels and pipes which are inherently protected from fire damage.

Above the natural characteristics of LNG, one hand is as to determine the range of quality standards in the purchase of LNG, the other hand as and user negotiations as LNG calorific value and component reference, and tell the user security knowledge about LNG.

3.1.1.8 Quality of LNG

LNG quality is determined by LNG resources negotiations, it is directly related to LNG components, according to the negotiation of Guangdong Dapeng LNG resources, determined the following properties:

(1) Qualified natural gas attributes.

① Gross calorific value (by volume): minimum 39.89 MJ/m^3, the highest 43.63 MJ/m^3.

② Density: a maximum of 0.805 kg/Nm^3.

③ Wobbe index (minimum/maximum): 51.6 to 53.8 MJ/Nm^3.

④ Mole content of methane: at least 84.00%; mole content of butane and heavy hydrocarbon: up to 2.00%; mole content of pentane and heavy hydrocarbon: up to 0.10%; mole concent of N_2: up to 1.00%; and impurities, hydrogen sulfide: a maximum of 5 mg/Nm^3; and total sulfur content: a maximum of 30 mg/Nm^3.

Liquefied natural gas must not exist in sufficient quantity of other impurities or solid to affect the receiving LNG or conventional city gas supply or power generation.

(2) Unqualified LNG.

If any LNG does not conform to the above specification, the seller shall compensate the buyer for the defective LNG in accordance with the relevant terms and conditions of the contract.

3.3.2 LNG Project Scope of Work Plan

According to the research of American scholars Clifford F. Gray and Eric W. Larson, project work scope connects all the elements of the project plan. In order to make the LNG project scope complete, the project scope should reflect the common understanding of project stakeholders, and work out project goal to implement detailed planning, and direct the whole team's work during execution, and constitute the criteria for evaluating the change request or to increase the work beyond the scope of the project.

3.3.2.1 The Concept of Plan making for the Project Work Scope

Working out the scope of the project plan refers to the process and process of writing a writ-

ten project scope specification. This project scope specification document will serve as a basis for future LNG project scope decisions.

The project scope plan is given a project scope arrangement, is the basis of the assessment work and performance of the project owner and the project team. Some experts put it defined as key project or a necessary content of project scope statement, including the project goal, LNG project deliverables, LNG project milestones, project management organization structure plan, the technical standards of the LNG project, project constraints, project assumptions premise condition, etc.

3.3.2.2 The Method to Make the Plan for the Project Work Scope

(1) Analysis method of project output structure.

In the formulation of the scope of the project work plan, the project output structure analysis is one of the commonly used method, according to the project cycle to analyze project output structure in each stage, get the task completion process logic relation. For example, in the process of LNG resources implementing, the first, we have to pass the international natural gas resource investigation and to look for LNG producer, scale, how much the volume of LNG can be used by us, and then through the LNG resource bidding. Following determine the LNG resource suppliers short list, after several rounds of business and technical negotiation, finally determine the LNG supplier. These procedures can't be reversed and dislocated.

(2) Project cost-benefit analysis method.

Project cost - benefit analysis methods are borrowed from the financial and economic analysis method, carries on the quantitative analysis of input and output. When the former is smaller than the latter is considered to be the basic feasible, it is not feasible conversely. We have to in the value analysis on components included in the project, ensure that the overall project economically feasible.

(3) Expert judgment method.

Expert judgment method is a frequently used method in LNG project management. That is, to use the expert in some specific technical and management expertise, follow the expert advice on some specific issues. This is particularly important in the emerging LNG industry project management.

3.3.3 LNG Project Work Scope

3.3.3.1 The LNG Project Goals

The first step in defining the scope of the LNG project is to set the project objectives. Project company should pass the efforts of the LNG project initiators, LNG users and all project stakeholders, through various efforts and mode of operation, established in the capital budget, schedule, quality and technology within the scope of the completion of the project mission objectives. The following is our country existing LNG project objectives.

(1) Guangdong Dapeng LNG project goals.

At the end of 1999, the state formally approved the first phase of Guangdong Dapeng overall LNG pilot project, and it has become the first approval imported LNG pilot project in our country. Guangdong Dapeng LNG pilot project as overall project is systems engineering consisting of LNG receiving terminal jetty and trunkline (hereinafter referred to as terminal-trunkline project), supporting gas-fired power plant projects, city gas pipeline projects and LNG transportation, and other corporate projects, its core project is receiving terminal and trunkline. In September 2003, four connections and one leveling engineering of receiving terminal was starting, LNG receiving terminal project has a design scale of 3.85×10^6t/a, three set of 160×10^4m^3 storage tank (at present, has increased to four), receiving terminal jetty can dock $(8.0\sim21.7)\times10^4$m^3 LNG ship, trunkline project in one phase includes trunk and branch line with total length of 387 km. The first phase of the project plan to be completed and put into operation in 2006 by the end of June, the total investment is expected to more than 7 billion Yuan. After the completion of the project, it can supply 5.1×10^9m^3 high quality natural gas every year to Hong Kong and the Pearl River mouth delta region.

(2) Fujian LNG project goals.

First phase of the overall project of Fujian LNG consists of terminal, jetty, trunkline and branch lines. Terminal reclamation engineering construction in August 2003, the first phase of LNG receiving terminal has areceiv capacity of 2.6×10^6t/a, four 16×10^4m^3 storage tank (at present, has increased to 6). In receiving terminal site, special receiving jetty—T butterfly wing type was built, it could be docked 20.69×10^4m^3 LNG Q-Flex transport ship, the trunk line is about 369km, the branch line is about 53 km. The first phase of the project was put into trial production with the productivity 2.6×10^6t on October 1, 2007, the total investment is about 6.2 billion Yuan.

(3) Shanghai LNG project goals.

Shanghai LNG project consists of receiving terminal, jetty and submarine trunkline. At the end of 2005, the four connections and one leveling project was started. First phase construction scale is receiving 3×10^6 LNG, one jetty with mainly $(16.5\sim20)\times10^4$m^3 capacity, 3 LNG storage tanks with 16×10^4m^{3}capacity, an submarine gas transmission trunk line, and other relevant supporting facilities, the first phase of the budget with a total investment of about 6.7 billion Yuan. Project was completed in September 2009. On October 25, 2009, Shanghai LNG project held first ship LNG receiving ceremony, it was the first time for Yangshan deep-water port berthing mainstream world LNG carrier, marked the Shanghai LNG project from construction to commissioning stage.

(4) Zhejiang LNG project goals.

The project consists of the receiving terminal, jetty, 2 of 16×10^4m^3 LNG storage tanks. At the end of 2005, the 4 connections and one leveling project was started, the design capacity of one phase is 3×10^6t/a, and through the project pipeline connected to the planning and construction of gas pipeline network in Zhejiang province trunkline, one jetty with berthing $(8.0\sim26.6)\times10^4$m^3 LNG ship, and transport gas to the Ningbo city gas network and power plant. On June 29, 2009, the project got the approval from the National Development and Reform Commission, marked a substantial progress made in Zhejiang LNG receiving terminal project with total investment of the first phase is about 5 billion Yuan.

Judging from the current situation, the majority of Chinese LNG projects are made by these four architectural engineering projects like jetty, terminal, trunk line and supporting public projects. But some projects make the tank independence from the receiving terminal, and manage it as a single component, such as Fujian LNG project. On the other hand, Shanghai and Zhejiang LNG projects do not have gas transmission trunk line, the reason is that the provinces and cities have been ready-made network before. How many or which LNG project contains sub projects or how to manage them, it should according to specific situation.

3.3.3.2 LNG Project Deliverables

Through the project construction and management, it will eventually reach quality requirements of gas delivered to the users. The project to achieve the ultimate goal is not accomplished at one stroke, but by deliverables accumulation of the project life cycle phase. It includes the preparation of early stage opportunity study report, pre-feasibility report, project proposal, feasibility study report, FEED design, preliminary design, detailed design, the construction phase of the LNG receiving terminal (including LNG storage tank and process facilities, utilities and auxiliary facilities), jetty (including a parking LNG berths, the pier and the work boat docks, etc.), trunk line and outside engineering (including monitoring and dispatching center, 110KV outside the project), the so-called four subproject's deliverables of LNG project. Table 3-5 gives CNOOC involved in the LNG project deliverables.

Table 3-5 CNOOC's LNG Project Deliverables

Engineering	Project			
	Guangdong Dapeng LNG	Fujian LNG	Shanghai LNG	Zhejiang LNG
LNG jetty (can unload LNG ship) ($10^4 m^3$)	1 (8~21.7)	1 (8~16.5)	1 (8~21.7)	1 (8~26.6)
Working Boat Quay ($5\times10^3 t$)	1	1	1	1
Storage tank ($16\times10^4 m^3$)	2+1	2+2	3	3
BOG compressor	2	2	2	3/1
Low pressure pump	9	5+1	4×3	6+1
High-pressure pump	6+2	3+1	5	4+1
Vaporizer: ORV/SCV	7/2	3	4/2	4/2
Trunk/branch (km)	371.6/28.1	301.1/55	51.9/	#
Gas station/block valve chamber	11/24	12/20	1/	
The first phase of the design scale ($10^6 t$)	3.85	2.60	3.00	3.00

Note: ORV-Open Rack Vaporizer, SCV-Submerged Combustion Vaporizer, #Access to the province network.

3.3.3.3 The LNG Project Milestones

The so-called milestone refers to the prominent events at one point on the project cycle, it is sometimes said several core subprojects composed of the start or completion time, or the beginning or end of the deliverables, and also is the time marker for project progress control. How and when, the LNG project application report approved by the competent department of state, How and when, the determination of the LNG project general contractor. How and when, each construction sub project completion, and so on. Generally in the LNG project, engineering construction of milestone point are: Jetty or receiving terminal field leveling construction starting, storage tank engineering starting and completion, jetty or supporting engineering completion, the entire LNG project fully completion, trunk line completion, the entire LNG project fully completion, the first LNG ship unloading, commissioning starting and completion data. Milestone plan is the frame for the entire project from the aspects of large section implementation and completion deadline. Milestone plan is generally prepared by the project company, once set, it will become goal for the whole team of the LNG project company to strive for and cannot be changed arbitrarily.

3.3.3.4 The Organizational Structure Plan of Project Management

According to the progress of the project and the project company's strategy, it should be carried on the organization and staffing, to manage, control and how to change the scope of the project for integrated management, organizations play a role in organizational guarantee (see Chapter 4).

3.3.3.5 LNG Project Technical Standards

In order to ensure that end-users can use qualified natural gas timely, we must act in accordance with international and domestic technical standards in the design and construction of the project, including the domestic standards for jetty design, channel safety distance, water depth for port selection, the international standards in special steel used and construction for storage tank design, and domestic construction safety and environmental protection requirements for receiving terminal, land pipeline, and so on.

3.3.3.6 Alternatives

For a specific LNG project, when the government authorities approve the project, we must organize the site selection and feasibility study, at the same time in the preparation of the preferred program, and consider alternatives program, such as LNG resources, shipping, terminal and jetty site, pipeline routing, network planning, user group programs etc.. Only in the beginning to consider alternatives, when the first program is subverted, there is the second program as an alternative. Small schemes selected always use a common method like brains torming and lateral thinking method, and large schemes are usually selected by experts and collective decision-making.

3.3.3.7 LNG Project Constraints

LNG project team in the project run by the objective and subjective conditions of all the constraints are called constraints, including the budget factors, milestones, quality, HSE and contract factors etc.

(1) Budget factor.

The budget should be made in any project early, so that the project team during project can execute as a cost and the afterwards evaluation standard. In general, the project team engineering and budget departments begin to budget estimate in project design phase, to determine the scope of items such as jetty, receiving terminal, storage tank, and trunkline, etc. The budget will be used as the basis of the capital investment and financing arrangements of the shareholders, also served as the financial and control departments to provide the basis for the allocation of funds. Budget control is one of the key to LNG project managements.

(2) Milestone factor.

The Chapter 7 will introduce milepost list of Guangdong Dapeng and Fujian LNG project, it is every stage of delivery schedule control schedule. In project schedule management, it is particularly concerned by the owner and the contractor, subcontractor, supervision company, sometimes, in order to achieve a milestone point, the owner will sign rewards and punishments protocols with contractors and subcontractors to ensure the realization of important milestones.

(3) Quality factor.

Quality control is an essential factor for LNG projects to transport safe and quality natural gas to users in the future. Unqualitied requirements will cause the users to downtime or can't carry out the production according to plan, also will bring the LNG project own safety risks, or due to quality and rework, Unnormal supply and bring economic loss. A LNG project will form of local supply and demand of production and life circle, if due to quality problems will cause huge social influence. So, total quality management is the most important part of LNG project management.

(4) HSE factor.

HSE factor is another important part of LNG project management, governments at all levels currently always put safety and environmental issues of major projects as the main object of supervision. First of all, from the project pre-feasibility study to feasibility study stage, HSE pre-assessment whether or not be proved, as the key factor for the project established. In the design and construction stages, it will be more HSE supervision as the core of the government management, reflecting in the overall project HSE acceptance and HSE inspection in commissioning and adjustment. The project company shall, from the beginning to the end, pay close attention to HSE management.

(5) Contract terms factor.

In the process of LNG project design and construction, nearly a thousand copies of the contract will be signed, including engineering contract, material supply, equipment procurement,

LNG sale and purchase agreement, gas sales contract. Any contract has the responsibility and obligation clauses, for project company, its liability clause is also restricting factors, must 100% be performed.

3.4 LNG Project Work Breakdown

3.4.1 Definitions and General Practice of Work Breakdown

3.4.1.1 Definition of Work Breakdown

Work breakdown is to decompose project into several major components until each part becomes relatively independent, and easy to management, control and check work unit, also called as a task or work package which visually shows the position and composition in the project. WBS (Work Breakdown Structure) is a structure from the project decomposed as their internal frameworks in order or implementation process, in order to refine and prepare the project plan, implement responsibilities and monitoring. In some application fields, WBS and work breakdown structure list are compiled at the same time. LNG project work breakdown is the basis of time management.

3.4.1.2 General Practice of Work Breakdown

(1) Principles.

① To consider the implementation stages of the project, such as project is generally divided into several parts like preparation, design, procurement, construction, inspection delivery;

② To consider the tender document package, such as the scope of the work of the contractor and the scope of the project business strategy;

③ To consider the actual construction area partition, such as portioned works, unit works of the project of decomposition in construction organization design;

④ To consider the industry division, such as the management of the scope of the division, such as jetty, receiving terminal, gas transmission pipelines;

⑤ To consider the industry enterprise standards, such as oil/petrochemical/electric power industry internal standard requirements;

⑥ To consider progress data collection, such as the level of clarity, step by step to ensure the authenticity of the data;

⑦ To consider the management requirements, such as the decomposition of management, design professionals, procurement and other packages;

⑧ To consider the national laws and regulations, such as regulatory authorities in different industries.

(2) To identify the main links of LNG project.

Chapter 2 shows that the LNG project is mainly composed of LNG resource acquisition, LNG user's market research, LNG marine transport, LNG receiving and gasification, natural

gas pipeline and natural gas utilization of the 6 links. Therefore, the project team should consider those six parts to carry on the organization design and work arrangements. Table 3-6 illustrates the main works of each part.

Table 3-6 The Main Tasks in Each Link of the LNG Project

Working Link	Department/Organization	The Main Task	Goal
LNG resource acquisition	Investors/project team	To sign LNG supply contract	To provide quality and quantity LNG timely
LNG user's market research	Investors/project team	To sign "take or pay" contract	To get reliable market
LNG marine transport	Project team/project company	To sign shipping agreement	To ensure LNG marine transportation in place
LNG receiving and gasification	Project company	To design and construct the jetty, terminal, and tank	To carry out acceptance, safety running
Gas pipeline	Project company	To design and construct Pipeline	To ensure land gas transportation
Gas utilization	Production preparation/ Operation	To carry out commissioning/ normal production	To receive LNG and execute contracts

(3) Decomposition of project elements.

① Sub-projects, generally refers to the sub-project structure, such as Fujian LNG project, which is divided into 5 sub-projects (some project is divided into 4 sub-projects): LNG storage tanks (including four 16×10^4 m^3 full containment concrete ground storage tanks and ancillary facilities); receiving terminal (including LNG gasification device, utilities and auxiliary facilities); jetty (including a anchor accommodating 8 to $1.65\times10^5 m^3$ LNG vessels and work boats berth, 345m trestle bridge, etc.); gas pipeline (including 301 km trunk, three branches of 55 km, starting from the main trunk Putian Xiuyu first station, connecting Fuzhou, Putian, Quanzhou, Xiamen and Zhangzhou in five cities, a total length of 356 km pipeline project) and other sub-projects (including monitoring and dispatching center in Fuzhou, 110kV outside engineering, auxiliary buildings, outside terminal service center in four parts).

② Engineering classification. During the implementation process of Guangdong Dapeng LNG project, according to the construction deploy ment, design and easy for quality management, the entire project is decomposed into several sub-projects, and then divided into several unit works and sub-unit works, further according to professional engineering or construction site, divides into several divisional works and sub- divisional works, and then based on the types of work, equipment groups or system section, divides into itemized works.. Site construction work is all one by one itemized works to complete. The receiving terminal is composed of a total of 25 unit works, in marine engineering including 6 unit works, in storage tank including 4 unit works, in process zone foundation and installation including 4 unit works, in buildings and outside environments engineering including 10 unit works, in slope protection 1 unit works. The

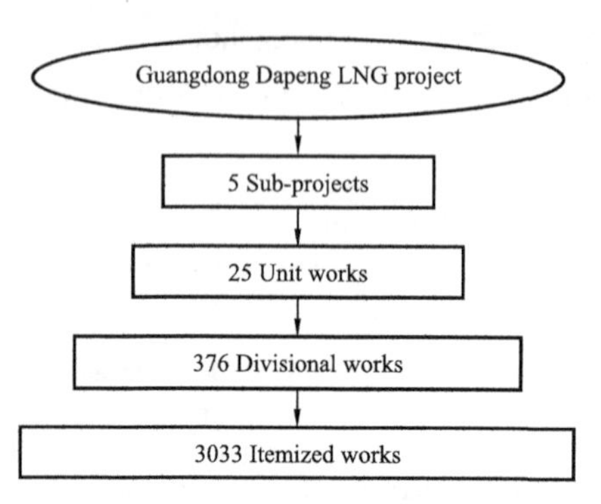

Figure 3-2 Decomposition structure diagram of guangdong dapeng LNG project

unit works are divided into 376 divisional works, 3033 itemized works. The engineering division of statistics is shown in Figure 3-2.

④ The sub project level decomposition.

The general practice of the work is to dissect the contents of the project layer by layer, starting with project milestones, then preparing the overall schedule. For some of the major sub-projects, but also is divided into several tender's plan, finally must be divided into execution plan. Each of the sub-project is decomposed to facilitate the specific operation, not all branches must be decomposed to the same level, may be the last decomposition level is not consistent, that some may be decomposed into the fourth layer, some may be decomposed into fifth, sixth or seventh layer, there is no uniform requirements, decomposition to the execution plan standard is to look at the last level whether to achieve the operational level. Actually, It is not the more detailed the better, too small decomposition may cause management ineffective and energy consumption, low resource utilization and even reduce the efficiency of work. The project team needs to balance the relationship between the work breakdown structure to match available resources, neither too thick nor too thin.

3.4.1.3 Work Breakdown Styles

(1) Decomposition according to the time sequence of the operation of the project.

Decomposition according to the time order of the project cycle, those are the four stages of layer down decomposition: the project start, project plan, project implementation and project closeout. The characteristics of this method are that the time period are clear, schedules are easy to be controlled.

(2) Decomposition according to composition of the project.

We can follow the composition of project work to breakdown. As the LNG project is concerned, there are four main sub-projects: jetty, terminal, gas lines and utilities. The characteristics of this method are clear boundaries of project, much easier to management and control.

(3) Decomposition according to project milestones.

Using the LNG project milestones as the basis for decomposition work, such as project feasibility study, FEED design, preliminary design, detail design, the terminal site formation start, tanks start and completion, jetty and supporting project completion, the entire terminal completion, the pipeline completion, the entire LNG project fully completion, the first LNG ship into the jetty, date of commissioning, the end of trial operation. The characteristics of this method are to seize a major phase of the project deliverables, making all stages management and schedules control much easier.

(4) Breakdown by mixture method.

Using all the methods mentioned above. Generally speaking, work breakdown is not using only one method from top to bottom in a form of decomposition, but the mixed application of two or more methods mentioned above. For example, in the first level, decomposition forms according to the project compositions; in the second level, decomposition forms by time sequence. Or in the first level, by milestones, and in the second level, by the method of the project compositions. Theoretically, what methods adopted, there are no distinction between good and bad, it's just the customary practice of project company plan control department and the contractor related.

3.4.2 Work Breakdown of a LNG Project

The work breakdown is based on the work contents defined by project scope. Because the LNG project is a new industry in domestic, with industry wide, complex technology, various links, it's impossible for the project company to undertake all the jobs of the project. It is necessary to learn from foreign experience of management, design, construction, the work breakdown of LNG project is done by the project company and contractors together.

3.4.2.1 Work Breakdown Procedure

Beginning with the project scope, the work breakdown of the sub-project is usually done in two steps: the first, preliminary or structure breakdown, assumed by the planning control department of the project company; the second, detailed breakdown by the contractors, and finally, after discussions between the company and the contractors, till the work scope and work breakdown should be satisfied by project company and the stakeholders. The procedure is shown by Figure 3-3. From the sub-projects to level two, level three, level four or lower levels, the contractor should break the work down to practically operating work packages. If it can be divided into more detailed jobs, it should be broken down more. Until the work packages are in the lowest level, the work breakdown structure should be drafted. It's likely that the planning control department doesn't agree with the result of work packages defined in the beginning, finally, the both sides through consultations can reach a unified opinion.

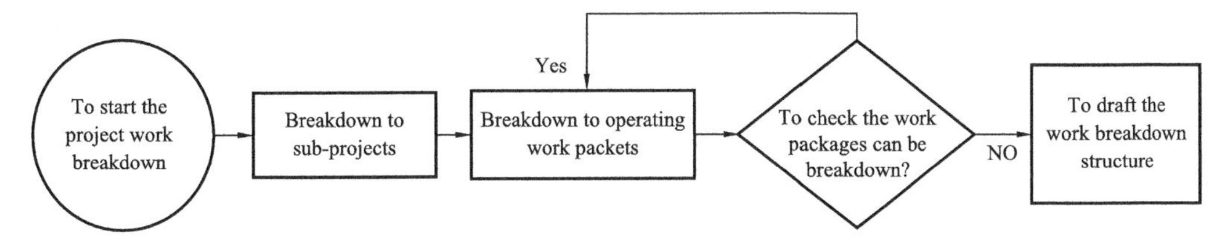

Figure 3-3 Work breakdown process from top to bottom

3.4.2.2 Work Breakdown of Project Company

(1) Two levels of work breakdown of Fujian LNG project

Figure 3-4 is a WBS made by the Fujian LNG project control department, and the sub-projects of Fujian LNG WBS are divided into five parts: LNG storage tanks, receiving terminal, jetty engineering, gas transmission trunk line and off site engineering. Each part is mainly divided into

the basic design, project bidding, detailed design, materials procurement, construction, pre-commissioning based on their characteristics in chronological order progressing to conduct work breakdown of the project from top to bottom. Each sub-level work breakdown is not necessarily uniform, but according to the project company to facilitate the control and management.

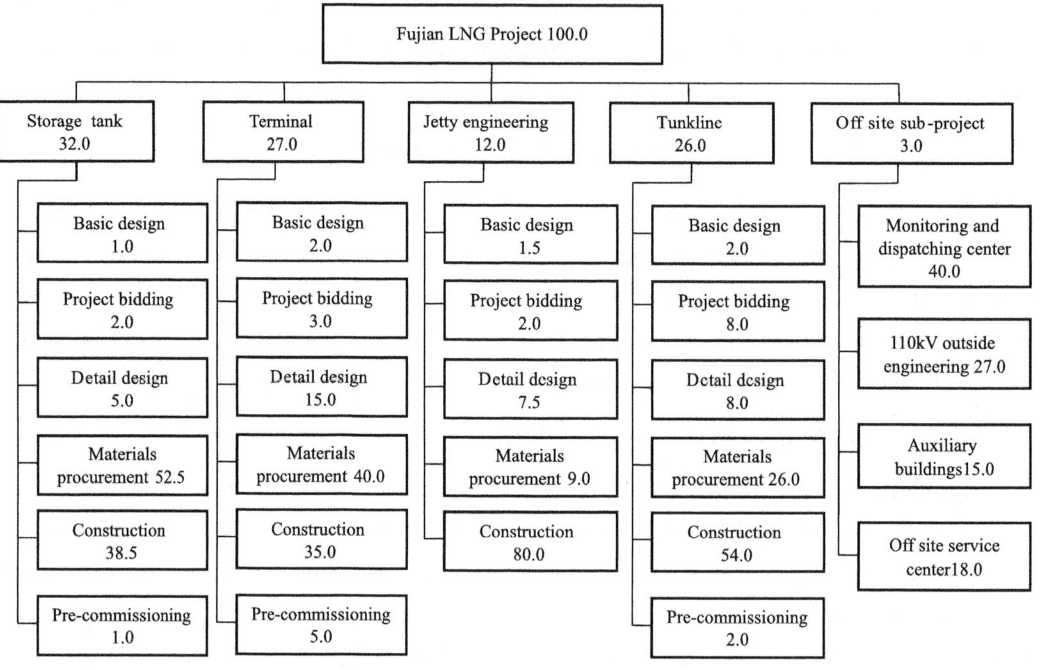

Figure 3-4 The WBS of fujian LNG project

(2) Guangdong Dapeng LNG project WBS.

The project WBS takes the whole project to conduct decomposition according to the single sub-projects, expand-unit works, unit works, divisional works, and itemized works. It takes the construction of the project as a family tree, layer by layer decomposition, and finally arrive the bottom of the branches end, and it becomes a work or a few work in network planning (or activity). So, the numbers of level of detail will directly affect the decomposition of activities in network planning. Too detailed, will increase network map, it is not conducive to read and management; decomposition too coarse, will lack of necessary guidance for schedule control. The degree of WBS of Guangdong Dapeng LNG project is considered the following factors.

① The level of detail of the WBS decomposition should be coordinated with the function of the schedule. Owner's schedule is mainly used for schedule control, which should be thicker, usually decomposed into the level of divisional works (up to itemized works). For the contractor WBS schedule to be implemented, the contractor shall, on the basis of the owner's WBS framework, according to the scope of the project needs to conduct self-subdivision.

② The frame structure of WBS should take into account the specific requirements of engineering subcontract, in order to avoid the occurrence of a decomposition unit across two contracts.

③ Because the entire pipeline more than 380 km across Shenzhen, Huizhou, Dongguan, Guangzhou and Foshan, so WBS should also be considered in land acquisition, regional geography, cross and other factors.

④ With the degree of decomposition, each part of the project to do basic consistent, so as to be able to correctly determine the logical relationship between activities in network planning.

⑤ The level of detail of the WBS decomposition should correspond to the activities, it can be easily identified in the construction site and it is conducive of the progress inspection and progress control.

3.4.2.3 Detailed Work Breakdown Structure

(1) From top to bottom of the work breakdown structure.

Figure 3–5 is a WBS Figureafter the work breakdown based on the actual situation of a LNG project. The first stage of work is generally be done by the previous project team starting from the early opportunity study, pre-feasibility study, to project feasibility study report work; The second stage of work could be completed by the project office, or also by the Project Office commissioned a design or research unit to complete, but also by the two together to accomplish; the third stage of work is usually commissioned by the project design or research units to complete. The real decomposition work is after completion of the feasibility study report, before entering the LNG jetty, receiving terminal and pipeline design bidding of three sub projects, done by the control department, Project Company to make the main work breakdown structure. Contractors could make more detailed work breakdown structure based on project company work, and mainly list the contents of each sub-project and below the bottom level of work activities. Also detailed work breakdown can be down jointly by control department, project company and the contractors, including design, procurement, construction division and its breakdown, but not including a lot of coordination, consultation, and liaison work.

(2) Bottom-up work breakdown.

Bottom-up work breakdown is generally completed by the contractor on the basis of the project company work breakdown structure and the description. Starting from the smallest unit of work, namely starting from the bottom layer of the activity, and gradually the back stepping to the upper level of task, if the bottom is divided into seventh level, pushed up into sixth level, and so on, this work also can be controlled by the contractor and the project company control departments work together. The following is based on a LNG project jetty, receiving terminal and trunkline three sub projects as an example from the bottom up work decomposition process. In order to better identify the next step of work, uniform numbers are generally used, which correspond to a work package, but also provide the basis for future project cost management.

① Jetty work breakdown. In Figure 3–6, Jetty engineering ID is No.1 sub-project (level 1); It consists of 2 secondary sub-projects (level 2): hydraulic engineering (1.2) and the land formation (1.1); then down for the design, procurement and construction project (level 3).

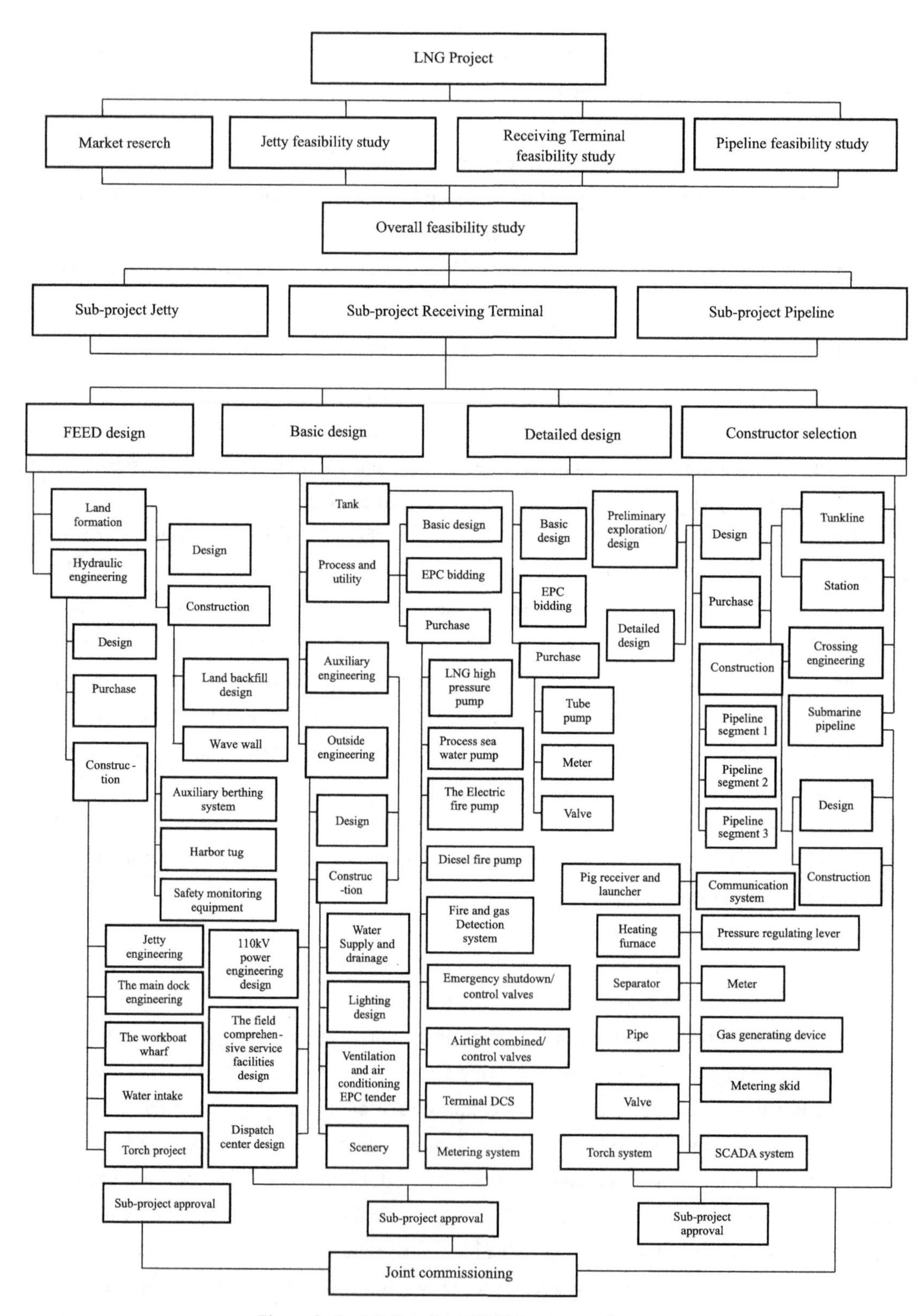

Figure 3–5 LNG project WBS from top to bottom

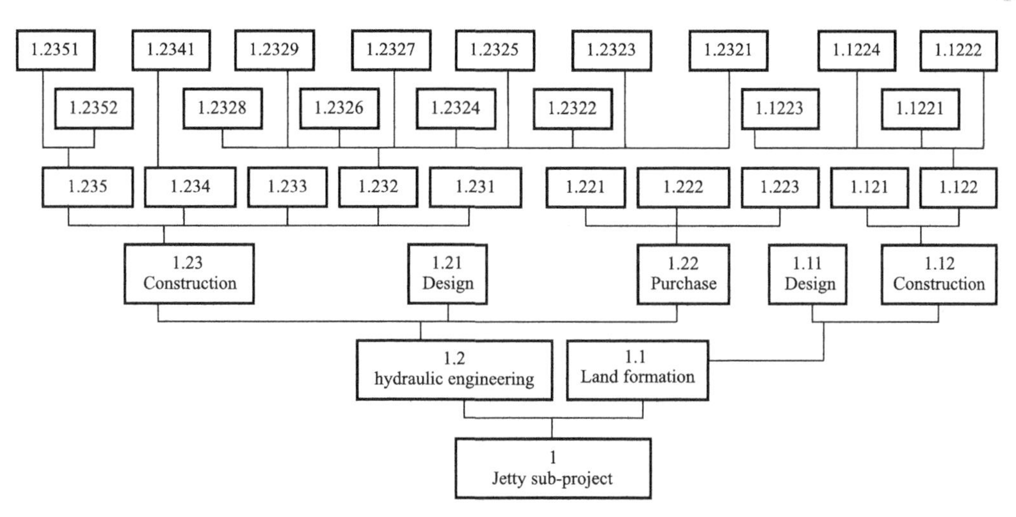

Figure 3–6 Jetty WBS from down to up

Note: The relationship between the number in diagram and the corresponding work:
1.2351—Torch tower; 1.2352—Torch trestle; 1.2341—Intake of hydraulic structures; 1.2321—Work platform; 1.2322—Berthing pier; 1.2323—Mooring pier; 1.2324—The pedestrian bridge; 1.2325—Trestle bridge; 1.2326—Boarding ladder; 1.2327—Joint Inspection Facilities; 1.2328—Drainage of hydraulic Engineering; 1.2329—Large temporary Engineering; 1.1221—Reef blasting; 1.1222—Dredger; 1.1223—Sweeping sea; 1.1224—Navigation system; 1.235—Torch engineering; 1.234—Water intake; 1.233—Working boat quay; 1.232—Main dock engineering; 1.231—Auxiliary berthing system; 1.221—harbor vehicle and vessel; 1.222—Safety supervision equipment; 1.223—Harbor engineering; 1.121—Land reclamation; 1.122—wave wall

The contractor on the basis of Project Company on the decomposition, takes over the work from the bottom up, such as hydraulic engineering construction, divided into fifth level of torch tower (No. 1.2351)and the torch trestle (1.2352), they make up fourth level of torch engineering (No. 1.235). According to each contractor of division of work and practices and set for the minimum level of division, no uniform requirements. As the Figurebelow the lowest level of the hierarchy is divided to fifth level, a total of 16 work packages, fourth level consists of 10 secondary tasks, and not all the branches are divided into the same level, which is determined by the nature of the work, such as the design is break down to third levels.

② The work breakdown of the receiving terminal can be seen in Figure 3–7.

③ The work breakdown of the pipeline can be seen in Figure 3–8.

3.4.2.4 WBS Directory

(1) Explanation.

The work breakdown structure directory is matched with the work breakdown structure use. It must include all tasks to be performed (also called work packages) in the project, and work breakdown structure directory can refine the WBS. Tasks described in a directory should be completed, which include any activities in the scope of the project.

(2) Example.

The following is based on an LNG project including the lowest layer of the work breakdown structure list (Table 3–7).

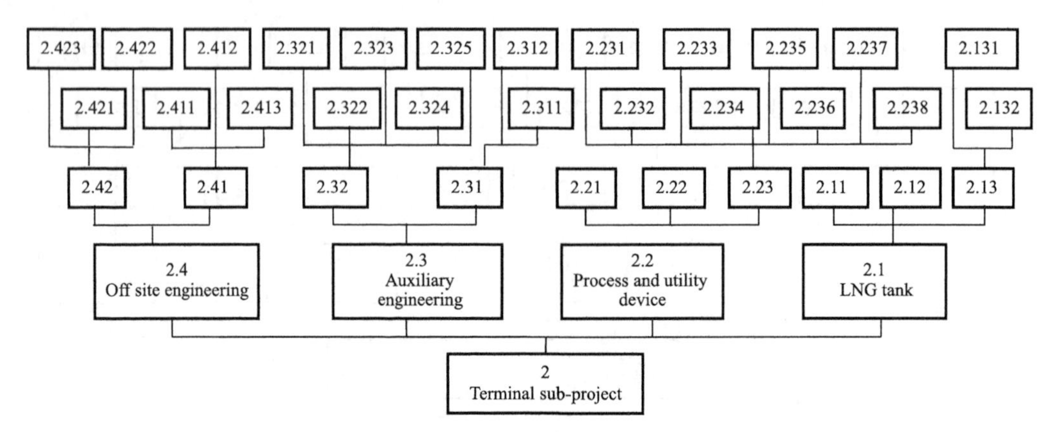

Figure 3-7 Receiving terminal WBS from down to up

Note: The relationship between the number in diagram and the corresponding work:
2.421—110KV Outside power supply engineering; 2.422—Field integrated service facilities; 2.423—Company dispatching control center; 2.411—110KV Outside power supply design; 2.412—On-site comprehensive service facilities design; 2.413—Company dispatching control center design; 2.321—Construction; 2.322—Water supply and drainage; 2.323—Lighting; 2.324—Ventilation and air conditioning; 2.325—Scenery; 2.311—Planning and design; 2.312—Architectural design; 2.231—LNG high-pressure pump; 2.232—Sea water pump; 2.233—Electric fire pump/diesel fire pump; 2.234—Fire and gas detection system with; 2.235 Emergency shutdown valve and control valve; 2.236—Terminal DCS; 2.237—110KV Air tight combination switch; 2.238—Metering system; 2.131—Tank pump; 2.132—Online instrument and valve 2.41—Design; 2.42—Construction 2.32—Construction; 2.31—Architectural design; 2.21—Preliminary design; 2.22—EPC contract package (process) ; 2.23—Owners purchase (process) ; 2.11—Preliminary design (process) ; 2.12—EPC contract package (tank) ; 2.13—Owners purchase (tank)

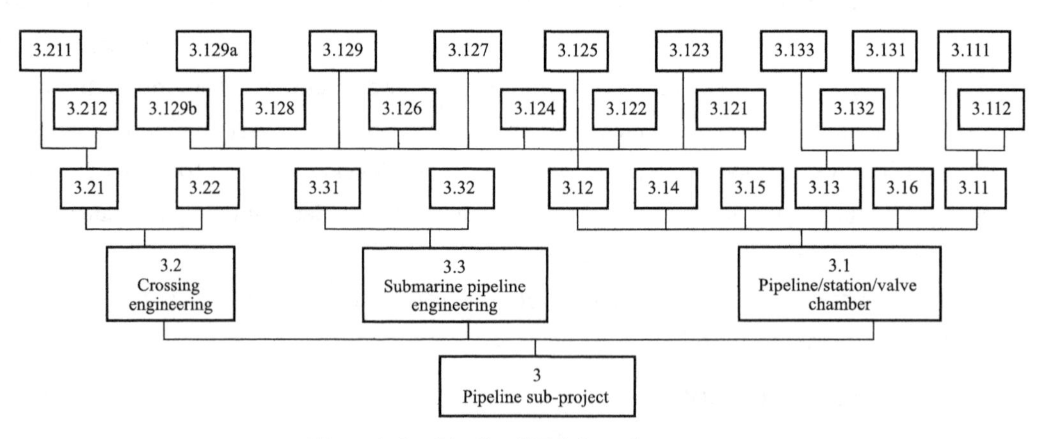

Figure 3-8 Pipeline WBS from down to up

Note: The relationship between the number in diagram and the corresponding work:
3.211—Preliminary survey and design (crossing) ; 3.212—Detailed survey and design (crossing) ; 3.129a—Heating furnace/Separator/Pig receiver & launcher; 3.129b—Torch system; 3.129—Flowmeter metering skid; 3.128—Long distance power unit valve chamber; 3.127—Gas power plant; 3.126—SCADA system; 3.125—Valve; 3.124—Pipe; 3.123—Meter; 3.122—Pressure regulatory pry; 3.121—Communication system; 3.133—Trunk(Branch)line 3 ; 3.132—Trunk(branch) line 2 ; 3.131—Trunk line/Branch line 1 ; 3.111—Preliminary survey and design (trunk line); 3.112—Detailed survey and design (trunk line); 3.21—Design (crossing); 3.22—Construction (crossing); 3.31—Design (subsea pipeline); 3.32—Construction (subsea pipeline) ; 3.12—Purchase; 3.14—Joint commissioning; 3.15—N_2 Inerting Service; 3.13—Construction; 3.16—Pipeline anticorrosion and pipe transport; 3.11—Design

Table 3-7 An LNG Project Work Breakdown Structure Directory

1 Jetty				
	1.1 Land formation			
		1.11 Design		
		1.12 Construction	1.121 Land Reclamation 1.122 Wave wall engineering	
	1.2 Hydraulic Engineering	1.21 Design		
		1.22 Procurement	1.221 Auxiliary berthing system 1.222 Harbor vehicle and vessel 1.223 Equipment for safety supervision	
		1.23 Construction		
			1.231 Jetty district	1.2311 Reef explosion 1.2312 Dredging 1.2313 Navigation system 1.2314 Sweeping the sea
			1.232 Main dock	1.2321 Work platform 1.2322 Berthing pier 1.2323 Mooring pier 1.2324 Steel footbridge 1.2325 Trestle 1.2326 Boarding ladder 1.2327 Joint Inspection Facilities 1.2328 Civil engineering and drainage works 1.2329 large temporary works
			1.233 Working Boat Quay	
			1.234 Water intake	1.2341 Hydraulic Structures for water intake
			1.235 Torch	1.2351 Torch trestle 1.2352 Torch tower
2 Recerving Terminal				
	2.1 LNG tanks			
		2.11 Preliminary Design		
		2.12 Contract Package of EPC		

Continued

		2.13 Owners procurement	2.131 Tank pump 2.132 The online instruments and valves	
	2.2 Process and utility equipment	2.21 Preliminary Design		
		2.22 Contract Package of EPC		
		2.23 Owners procurement	2.231 LNG high-pressure pump 2.232 Process sea-water pump 2.233 Electric Fire Pump 2.234 Diesel Fire Pump 2.235 Fire and gas detection system 2.236 Emergency shutdown valve and control valve 2.237 Terminal DCS 2.238 110KV airtight combination switch 2.239 Metering system (OFE)	
	2.3 Auxiliary Engineering			
		2.31 Building Design	2.311 Planning design 2.312 Architectural design	
		2.32 Construction	2.321 Building 2.322 Drainage 2.323 Lighting 2.324 Ventilation and air conditioning 2.325 Scenery	
	2.4 Outside Engineering			
		2.41 Design	2.411 10KV design of power supply 2.412 Design of site service facilities 2.413 Design of dispatch and control center	
		2.42 Construction	2.421 110KV outside power supplying engineering 2.422 Site service facilities 2.423 Dispatch and control center	
3 Pipeline				

Continued

	3.1 Trunk (branch) line (including station and valve chamber)			
		3.11 Design	3.111 Preliminary investigation and design 3.112 Detailed investigation and design	
		3.12 Procurement (including crossing and subsea pipeline)	3.121 Communications systems 3.122 Pressure regulatory pry 3.123 Instruments 3.124 Pipes 3.125 Valves 3.126 SCADA system 3.127 Gas power plant 3.128 Long distance power supply unit for valve chamber 2.129 Flowmeter metering skid 2.129a Heating furnace/splitter/receiver and launcher 2.129b Torch system	
		3.13 Construction	3.131 Trunk (branch) line 1 3.132 Trunk (branch) line 2 3.133 Trunk (branch) line 3	
		3.14 Joint commissioning of full range		
		3.15 N_2 inerting services		
		3.16 Pipeline anticorrosion and pipe transport		
	3.2 Crossing engineering			
		3.21 Design	3.211 Preliminary investigation and design 3.212 Detailed investigation and design	
		3.22 Construction		
	3.3 Subsea pipeline			
		3.23 Design		
		3.24 Construction		

3.4.2.5 Work List Instructions

For the bottom of the work package, it needs to give a detailed description, that is, to prepare the work list instruction. The instruction shall include detailed description of the activities, as far as possible for the management and operating staff in site knowing how to operate after reading. These work list instruction could be compiled together into a Book. The instruction should include the work package resource requirements, and all the instruction of assumptions and constraints. The details of the content vary by application domain. Such manual also provides a reference for future similar project management.

3.4.2.6 Project Work Breakdown Modification and WBS Checklist

After work breakdown structure compiled, it is necessary to carefully check the project WBS by checklist (Table 3–8), to avoid the project details missing or thoughtless. Any such modification must be reflected in the WBS related files (for example, cost estimation), the above changes usually happen when the technology is new or has not been verified in the project.

Table 3–8 Main Content of the WBS Checklist

Link	Constituent element	Level
Whether Links of the project are complete	Whether each component of sub-project is complete, reasonable in the project	Whether the levels of the project goal description are clear
Whether the main documents of each link are signed	Whether there are measurable quantity, quality, time parameter index values in the sub-projects	Whether the levels of the work breakdown structure are logical
Whether every link of the goal to achieve or not	Whether each of the sub-project engineering signed with contractor	Whether the packages in work breakdown structure are served to form a result of the project
Whether project goals are described clearly	Whether each stage deliverable's description are clear or not	Whether the work breakdown structure levels matched with the unified description of project target levels
Whether the logical relationship between each link is correct and reasonable	Whether all deliverables are served for the project goals	Whether measured indexes matched with performance of the project work
Whether resources each link required are clear and reasonable	Whether the finished sub-projects accord with the milestones	Whether there are the reasonable work quantity, quality and time metrics
Whether overall coordination of decomposition structure is reasonable	Whether sub-project goals are in conformity with the general target	Whether the work package are equipped with the corresponding resources

3.5 LNG Project Scope Verification

Scope Verification refers to the work process of the LNG project stakeholders on the formal approval and acceptance of the project scope and deliverables that have been completed. The object of project scope confirmation is the main document generated by the project scope.

3.5.1 LNG Project Scope Verification Basis

3.5.1.1 LNG Initiator and Local Government Cooperation Framework Agreement

On the basis of the market research of LNG initiator and the consensus with local government authorities, it is clear that the introduction of LNG project need to build the project, including jetty, receiving terminal and pipeline, at the same time including supporting construction of gas power plant projects, chemical industry, city gas projects, industrial fuel projects. Initiators should have technical, business and management experience and strength to develop LNG projects. Local government can agree the initiators in charge of the LNG project, and will recommend one or several powerful local enterprises to participate in the research and construction of the project. On the basis of initiators' market research, combined with local economic development planning, it needs to determine the scale of LNG project and the late LNG project development plan, and seriously study the feasibility of the project. The above is the important basis of LNG project scope management.

3.5.1.2 LNG Project Scope Documents

All files such as the project scope, WBS, and WBS directory and recognition and definition of the project scope compiled by the project company control departments and contractors, will be the basis of LNG project scope verification.

3.5.1.3 LNG Project Users

After the preliminary determination of LNG project scope, it is also necessary to communicate with each LNG user deeply to verify where and how gas delivery. For the moment, after LNG is vaporized, Project Company transfers the natural gas by the trunk lines to the gate stations, via the gate station distribution to each user. The LNG project company shall make full consultation with each user. All these work will also be included in the scope of the project management.

3.5.1.4 Deliverable Achievements

If LNG project completed, it must have hardware parts, including jetty, receiving terminal, storage tank, gas transmission trunk line and its supporting utilities. There are software parts, which are important documents that formed early in the project and in the project life cycle. Such as LNG project pre-feasibility study, feasibility study report, approval document of the project by government, HSE evaluation report and government approval letter, LNG resources supply and marketing contracts (agreements), and shippers to sign the contract of carriage, and users should sign "Take or Pay" contract. Above the hard and soft parts are indispensable, but also are the basis of the project scope verification.

3.5.2 LNG project Scope Verification Tools and Content

3.5.2.1 Project Scope Verification Tools

(1) Similar project scope specification.

The similar project scope specification can be used as reference for the new project. Even if the different types of projects have a relative reference to the contents of the process. We do like to draw Gantt chart in the project, how much work, which involves in the project, and all are defined in the scope statement.

(2) The work breakdown structure template.

As for LNG projects that are under the same category, template of their work breakdown structures can be used as reference for reviewing this project. Although every project is unique, because of their similarities, WBS of previous projects can always be used as template for the new projects. It is especially helpful for control department to learn and borrow from the deliverable achievements at different stages of previous projects. The project work breakdown structure is a kind of hierarchical structure system which is composed of the elements of the project. It is a detailed description of the project team to complete the project tasks, which constitute the scope of the project work.

(3) Expert review for hardware design.

Currently, industrial design in LNG project including FEED design, basic design and detailed design is completed through internal and external enterprises. These designs are one of the important parts in project scope. Project owner can audit the hardware design through expert review conference in order to check the completeness and reasonableness of the design. Also, the chosen design standard must have been proven successful and safe in international projects.

(4) The standard version of the software part.

The similar project research report, evaluation report, LNG resource contract, transportation contract and "Take or Pay" contract and other documents from other project company or the initiators, the standard specimens can be used as reference documents software review of the project.

3.5.2.2 Verification Contents of Project Scope

In consideration of all aspects of the project stakeholders, it's requested to verify project scope with the following contents.

(1) Whether project objectives are clear and in place.

Each LNG project must be defined with clear objectives including the sub goals and the total target, whether there is a logical relation between sub goals. We should analyze the degree of importance of each sub goal, and the influence degree of each sub goal to the total target, and seriously work out important sub goal execution scheme and a backup plan, etc. Especially when it could be formally put into production and this target could be accepted by LNG Company, national authorities, local government and each user enterprise.

(2) Whether indicators are reliable and effective.

It's also required to establish all kinds of reasonable project indicators, including quality, budget, schedule, HSE and etc., so that the project could be controlled based on a clear rule, try to define the quantitative indexes. For the indexes which could not be defined as quantitative ones, we can adopt method of expert grading and qualitative description.

(3) Whether the restrictive factors are changed.

It is also necessary to analyze the restrictive factors, including budget, milestones, quality, HSE and the terms of the contracts. According to the delay and change of time, the trend analysis is carried out for the objective constraints, in order to take into account the changes in the project implementation stage, to take countermeasures. For Project Company's own constraints, the degree of effect also need to be analyzed, when its constraints seriously affect the overall project objectives, Project Company should actively create conditions, and even seek the initiators and local governments to give the solutions.

(4) Whether important assumptions are reasonable.

For every project in its start-up stage, some important preconditions will be assumed. This is because the project initial grasp the extent of the project is not enough, and if these conditions are not considered, this project can't go on. For instance, we can assume that the project scale is controlled in (2.5~3) $\times10^6$t/a before resources are determined. As this quantity achieves economies scale, therefore, we can have negotiation regarding resources in the upstream of LNG and develop market based on this amount.

(5) Whether the risks are acceptable.

Project risk is objective existence. We should make qualitative and quantitative analysis to the possible risks in the project cycle in advance. The risks mentioned here include technical risk, economic risk, political risk, social risk and so on. In the stage of the project scope verification, it is required to make further risk analysis again, to find out which ones are acceptable and which are not acceptable. For those which are not acceptable and could bring subversive influence to the establishment of project, we should dare to stop the project.

3.6 LNG Project Scope Control

3.6.1 LNG Project Scope Control Level and Procedure

3.6.1.1 General Concept

LNG project scope management actually runs through all stages of the project cycle. Changes in the condition and environment of the project will change the project scope, and cause the change of the project duration, cost and quality, so changes in the project scope must be strictly controlled. This work has occurred many times in the project cycle, we must seriously deal with. The principles of the project scope control are: try to keep the integrity of the project's original performance measurement baseline; ensure the consistency of the project outputs changes and projects tasks plan update, coordinate all aspects changes requests.

3.6.1.2 Scope Control Level

It is necessary to make a scope control for project as the change of situation and the factors. But the strict management levels also should be followed. According to severity of the factor's changes, the situation can be divided into the two following cases.

(1) The shareholders' committee or the Board of Directors control.

This control generally involve the major issues of the project, for example, significant change of the LNG resources, a fundamental change of the downstream users, changes in government guiding principles, changes of the receiving terminal site and so on. These matters Project Company cannot make the decision, which must be made by the shareholders' committee or the board of directors.

(2) Project company control.

When some general matters of control involve the project itself, it can be determined by the project company themselves, such as project investment has not changed significantly, the pipeline routing or land acquisition costs change, users buy gas change, equipment and materials suppliers and the price change, but belongs to the controllable range, etc. The project company may make arrangements for the change to be submitted to the shareholders' meeting or the board of directors for the record.

3.6.1.3 Scope Control Procedure

As shown in Figure 3-9, in accordance with the degree of the scope control, the project company submits to the shareholders' committee or the board of directors or performed by the project company leadership. If the shareholders' committee or the board of directors does not pass the changes, then perform the original plan or terminate the project. If the significant change is passed, then it is referred for the project company to study the change of solutions. The project company can also according to its own jurisdiction to decide whether to change, and also according to its own technology and management ability to decide whether to borrow other brain, to hire consulting company or experts to propose the solutions. Referring to experts' opinion, Project Company asks its control department according to concrete situation to propose change scheme, including management plan, corrective measures and defect remedy, then reports to project leader or the boards of directors for approval or for the record. The above change file archive is also very important. Eventually, according to the changes, the LNG project company continues to operate the project. In the process, the company should try to make these changes toward the beneficial direction, and make efforts to eliminate the negative impact of the project scope changes.

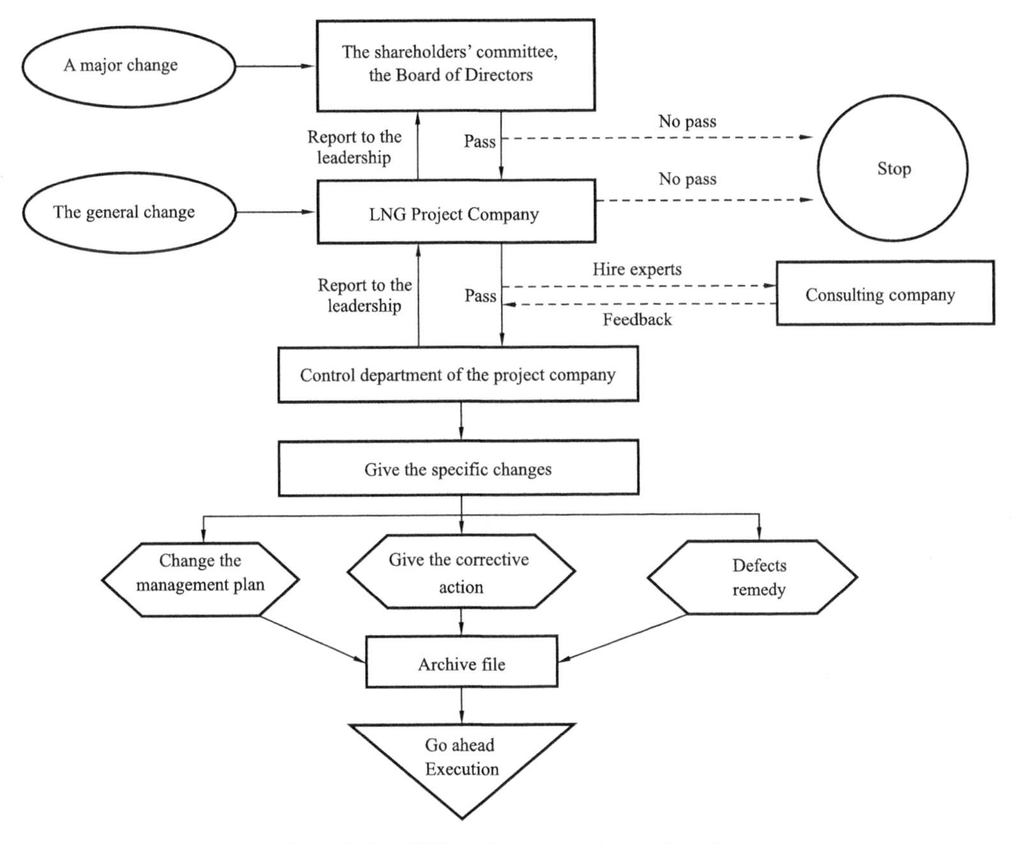

Figure 3-9 LNG project scope change flowchart

3.6.2 LNG Project Scope Control Results

3.6.2.1 Project Scope Statement

After the change control request was approved by the shareholders' committee or the board of directors or the project company, the planning department of the project company should modify the project scope statement at any time. The updated project scope statement will become the new benchmark for project scope.

(1) Further clarify project goals.

Project goals change with the change of the investment environment and conditions. It includes the change of goals, also includes the change of the ultimate goal, for example, Guangdong Dapeng LNG project, the original supply in stable year is 3.7×10^6t LNG, in order to meet the increasing market demand, and through hard negotiations, the stable supply is increased to 3.85t/a, the number of the storage tank is also increased to other two. In a word, the target must be adjusted in order to adapt to market changes.

(2) Changed LNG project deliverables.

Through the changes, LNG project deliverables will change too, for example, an LNG receiving terminal, with the expanding of market demand, the LNG supply obviously cannot meet the need for the market that will be supplied, then the tank numbers will be increased on the basis of the original tanks, and the design of the plane will have to be changed gradually to set aside the new tank location, this may need to change the diameter of gas transmission pipelines which were already designed, and the construction period will also be changed accordingly, and the new project milestone plan will be drawn up again.

(3) Changed LNG project milestones.

Milestones also change with the hardware aspects and software ones changed. Such as LNG projects using foreign tanks EPC construction contractor to build tank, because they don't understand Chinese national conditions, the construction plan they made is not entirely appropriate. Then Chinese partners need to negotiate with them and modify their plan to meet the project details. Another example is about LNG project commissioning, especially early LNG project in Chinese, generally to hire foreign LNG receiving terminal operation such as South Korea or Japan as an operation team, Chinese will send corresponding staff to learn the operation, this might delay the original milestone to ensure a smooth transition to the production period of the project.

(4) Changed project organization structure management plan.

The change of organization structure plan is the most frequent, this is because China's LNG project started shortly, it has not formed a complete, with the progress of the project and taken the change of organization structure mode, LNG Project Company gradually gropes organization structure change rule. LNG project company should carry on the organizational structure change in every stage and the change of human resource plan according to the early project management strategy and gradually master the proportion in the number among the size of the workload, organization structure and personnel changes, in order to explore the relationship between project change and organizational structure for the new LNG project.

(5) Changed LNG project technical standards.

LNG project technical standards are generally relative stable, but with the increase in global LNG terminal projects, technology and management experience are more mature, the new standard and modified technology, management standards have also emerged. In the process of the LNG projects, if there is a new standard for the proposed construction project does not change the original designed idea, and can meet the technical requirements, also can save cost, then the new standard can be considered to be used, such as earthquake design standard and so on.

(6) Changed project constraints and assumptions conditions.

Any project has its constraints and assumptions, LNG project is no exception. Such as Zhejiang LNG resources are not implemented early, as a result, the project delayed for two years before they get approved by the government. Although the land formation had been completed, the major sup projects can't be promoted according to the procedure. We must act according to the LNG project operation rule, to avoid working blindly and causing unnecessary waste.

3.6.2.2 New Work Breakdown Structure

According to the updated project scope statement, the project WBS must be modified at the same time, so that the project company and the contractor can make it as performance evaluation and guidance for the future work.

(1) Changed the project WBS.

These changes are bound to bring change to the WBS. According to changes in work tasks, new plan which is used to rearrange the work should be made, and be used to arrange human resources, financial and material resources to complete tasks change. This work is a process that must be carefully planned and reasonably arranged, and led by the project vice general manager, conducted with specific implementation by the Planning Department, especially in the critical path, and needs to carry out the WBS work under the general manager's attention to ensure the completion of the project.

(2) The new basis of performance evaluation of Project Company and contractors.

The new work breakdown is the new basis of performance evaluation of Project Company and contractors. Each contractor should strictly follow the WBS to arrange work and input resources. Project Company should fulfill their duties in accordance with the provisions of the contract, and supervise the contractor to work in accordance with the new work breakdown, and finally evaluate its performance under the new work breakdown.

3.6.2.3 The New Work Breakdown Structure Directory and Work Lists Instructions

Correspondence with the above mentioned, the new work breakdown structure directory and work lists instructions are prepared, so that the project company, contractors and supervision company can refer to them to make the specific work arrangement and plan, including new resource inputs.

(1) The new work breakdown structure directory.

The new work breakdown structure directory is a complete set of documents of work breakdown, it should be a more detailed list of contents, the project company must organize the experienced specialized person of contractor's to draw up the directory, pay attention to the logical work details and do not leak.

(2) The new work list instructions.

The new work list instruction is a list of supporting documents. To specific work as a starting point, it needs to arrange manpower, financial resources, material resources, time, and information resources, to ensure that the work could be carried out on the basis of investigation.

3.6.2.4 The Recommended Corrective Measures

Recommended corrective measure is working procedure in order to make the results of the project and the project management plan consistent with the project scope statement.

(1) Resources input.

Resource input is the foundation of project promotion, When a change occurs, the first from the resource input changes, such as human, financial and material resources, time and information resources, in terms of resources, it is necessary to ensure the quality, but also to ensure the progress, not only to meet the standards, should also be economically viable. Corrective measure in this regard is the support of other measures.

(2) The change of program.

When the changed work task has new goals, the original work program may be not suite. Project Company must supervise and urge contractors to find a more reasonable working procedure, and at the same time also pays attention to the changes in their work procedures.

(3) The new technology.

The new technology adoption is not only to pay attention to science, but also pay attention to the economy. For the changed work tasks, on the basis of improving quality, the new technology also should be used, but the project company should be more cautious, and if necessary, through expert demonstration to solve.

(4) Management innovation.

The novelty of LNG project brings opportunities for management innovation. Management innovation includes the coordination of the group working interface, the use of the bidding procedures, the reasonable coordination of the relationship between the contractors and supervision company, reasonable organization design and a complete set of human resources, the transformation of project acceptance and commissioning stage, and so on. Each project company has its own management methods. As the investors of a number of LNG projects, it is necessary to summarize the experience, in order to promote the new LNG project. CNOOC has been tried in this area.

3.6.2.5 Organizational Process Assets

The control and corrective measures, as well as other types of lessons learned in the project scope control, can be used as a project company and its parent company's process assets and input update process asset database.

(1) To increase organizational process assets.

The promoting and operating of the LNG projects provide a platform for LNG Project Company and its parent company to accumulate organization process assets. In particular, the parent company must supervise the project company to summary process assets, learn good management experience, working procedure, operation model and relative database of each project and form a complete set of documents, and finally submitted to the parent company of investment.

(2) The experience and lessons.

In some sense, in the process of LNG projects' preliminary studies, feasibility study, design, construction, acceptance and commissioning process, the experience and lessons are even more rare process assets. Different from the process assets documents mentioned above, experience and lessons are more attention to failure, the process of detours, records of the limitations and assumptions under which the results and conditions are not successful, and the results and conditions should be recorded at the same time, in order to learn valuable lessons from them.

3.6.2.6 Project Management Plan

If the approved change requests have an impact on the project scope, then the corresponding part of the file and the project management plan, including the budget plan, schedule plan, quality plan and HSE plan will also change accordingly.

(1) Budget plan.

The change of task will affect the change of budget and funding period. According to the changes, financial departments need to make new budget plan to adapt to the completion of tasks.

(2) Schedule plan.

The change of task will affect the change of schedule plan and work window. Planning department needs to make schedule plan timely, especially the reasonable arrangements for milestone plan and final production plan.

(3) Quality plan.

The change of task will also affect the quality, including the quality of the hard task and the quality of the soft task. Quality objectives are generally not reduced, but the need to re arrange the work of quality inspection agencies, inspection time and investment resources.

(4) HSE plan.

Changes of HSE plan may increase the difficulty of management. The original capital investment is likely to increase, and measures to improve HSE and personnel may be changed, test methods and equipment may have to be re-purchased. Tests procedures may have to be changed and preventive measures may need to be strengthened, and so on.

Chapter 4 LNG Project Human Resources Management

4.1 General Concept of Human Resource Management

4.1.1 Relevant Definitions

4.1.1.1 Definition of Organization Structure

Organization structure refers to the division of labor, grouping, coordination and cooperation.

4.1.1.2 Definition of LNG Project Organization Structure

According to the different stages of the project, different organization structures will be set up to meet the needs of the division of labor, grouping, coordination and cooperation in order to achieve the objective of the organization. To promote the process of LNG project in Chinese, three progressive forms of organization structure can be adopted generally, that is early project team-linear type, project office-the functional type and the project company-linear and functional type.

4.1.1.3 Definition of Human Resource Management

Under the conditions of certain time and space, human resources are the sum-up of the reality and potential labor quantity and quality.

According to the requirements of the organization development strategy, human resource management refers to allocate human resources in a planned way. Through a series of process including the recruitment, training, using, evaluation, motivation, deployment of the employees, to mobilize the enthusiasm of the staff, play to their potential capability, create value for the enterprise, and ensure the realization of strategic goals of the organization.

4.1.1.4 Definition of LNG Project Human Resource Management

LNG project human resource strategy should be based on project management strategy (As shown in the Chapter 2) to develop project human resource strategy, to allocate professional staff and set up project company organization, including recruitment and selection, training and development, performance management, salary management, employee mobility management, employee relationship management, employee safety and health management activities.

In view of the practical principle of this book, we can't introduce all the aspects of human resources management, but only elaborate the quality conditions of project general manager and staffs, project company organization structure, the job duties of the various departments, the project company's enterprise culture, etc.

4.1.2 The Role of Human Resource Management

4.1.2.1 Organization Structure to Serve the Project Stages

To implement the LNG project, an appropriate organization structure is the key to the success of a project. At different stages of the project, the tasks of the focus is different. As the early stage of the project, the main work is to analyze the local market development, the local macroeconomy, to study the conditions of natural environment, sometimes also participate in the LNG resource research activities. Next are the stages of pre-feasibility study and project proposal. Then are the stages of feasibility study, project application report writing, project design, construction, acceptance and commissioning, these are a substantial project operation stages. They correspond to the three kinds of progressive type organization structure respectively, those are Chinese LNG project widely used and considered the most appropriate form of organization.

4.1.2.2 Organization Structure Provides the Basis of Human Resources Allocation

Projects at different stages need to adopt different forms of organization structure, clear work objectives and key points are the foundation to formulate organization structure. According to the number of tasks, investors can draw up the human resources planning, staff deployment, staff recruitment and training work. According to its unique organization form to select project general manager and other technical management personnel to do the work of their responsible for to ensure the smooth completion of the project.

4.1.2.3 Appropriate Human Resources Can Accelerate the Progress of the Project

Any project is done by people, for LNG project, the first thing is to choose right project team or company leader, that is general manager of the project. If the person not only has a strong LNG project management and technical skills, but also has a high social sense of responsibility, sense of mission, and consciously resists the ugly social phenomena and corruption, and has the affinity to others and can command a powerful army commander force, which must be beneficial to push into the project. Of course, LNG projects not only by the general manager, but also depend on the each department manager and the general staffs, that is the team spirit and resultant force. putting the right people in right position can make the LNG project team or the project company work together and speed up the pace of the project, make the project as soon as possible to achieve the investors' strategy.

4.1.3 LNG Project Human Resource Management Workflow

4.1.3.1 Three Components of the Human Resource Management Process

The general flow chart of LNG project human resource management is shown in Figure 4–1. We can see from the chart, firstly, the LNG project human resource management bases on the contents of the project management strategy, correspond to the period of project organization structure and human resource planning. the middle part is the core part of human resource management, recruitment and selection from the start, the establishment to the system and work standards, salary management and performance the assessment, and is closely related to every employee team; and cultural construction is the main means of project organization to provide a harmonious and collaborative working environment. Human resource management is always a process of dynamic adjustment and control, which runs through the whole life cycle of the project.

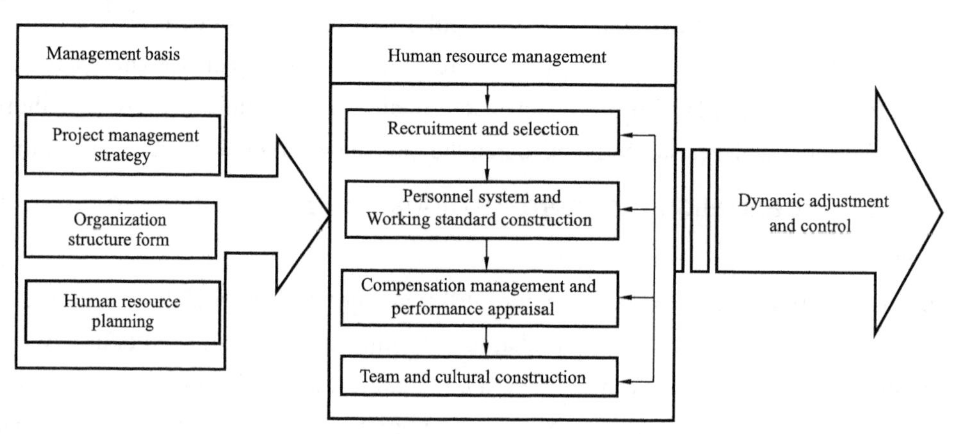

Figure 4-1 LNG project human resource management flow chart

4.1.3.2 The Relationship among the Three Components of the Process

(1) The relationship between the management basis and human resource management.

Starting from the LNG project, the project management strategy should be taken into consideration, and the human resource management strategy is the main aspect. If CNOOC has a long-term management and technology accumulation, the ability to complete the LNG project from the design, construction and operation of the task, you can use a totipotent strategy or self-dominated strategy, rarely hire external organizations and personnel, and other companies just entering the industry, although there is no direct operation of the LNG project, but has the operation of similar energy projects, you can use self-dominated strategy, or by means of appropriate outsourcing. Private enterprises can use the auxiliary or rely on dependent strategy.

Once the project management strategy has been established, the organization structure and human resource management tasks are clear. For totipotent strategy management, external recruitment and selection of the workload is very few, but must start internal personnel allocation preparation in advance. For self-dominated strategy, external recruitment and selection tasks is large, but must consider the internal redeployment of staff professional background and working experience, so that Internal redeployment staff and external recruiters are able to complement and coordinate. for auxiliary or dependent strategy, it needs to consider the scope of the project and the integrity of the project, hire a powerful external Project Manage Company or use EPC model.

For totipotent strategy or self-dominated strategy, it is necessary to fully consider the recruitment and selection, personnel system and working standard construction, salary management and performance assessment, and team culture construction, human resource management task is very heavy; relative to the auxiliary or dependent strategy, only to consider the more indirect management, a small amount of direct management.

(2) The relationship between human resource management and dynamic adjustment and control.

From the point of view of human resource management intensity, four management contents are closely related to the project management strategy, for the totipotent strategy or self-dominated strategy, four aspects must be exhaustive, and from top to bottom are progressive, each management process can feedback to any of the above level adjustment, in order to achieve the best matching of human resources. For ancillary or dependent strategy, the management of these four aspects can be weakened, but the supervision must be strengthened, such as outsourcing system and standardization, salary management and performance assessment and inspection team and cultural construction, to prevent the outsourcing of human resource management of the absence and omissions. For totipotent strategy or self-dominated strategy, the Project Company is responsible for dynamic adjustment and control, and it includes adjustment and control under the condition of organization structure determined and, also includes the adjustment and control under the different stages of the organization structure changed. For ancillary or dependent strategy, adjustment and control can be put forward by the Project Company, and by contractors to coordinate and solve.

(3) The relationship between dynamic adjustment and control and former two.

As shown in Figure 4-1, the dynamic adjustment and control of human resources run through the whole process of the project cycle, which is determined by the uncertainty of the factors in the project life cycle and the changeability of the social environment. Management basis is usually put forward by the project initiator, if decision-making layer understanding is in place, formulated project management strategy, the initial organization structure and human resource plan are reasonable, it is conducive to the project human resource management. Four aspects of human resource management process, is just for the organization and human resource planning and action, is nothing more than the recruitment qualified personnel, to formulate personnel system and working standard with LNG project characteristics, so that employees enjoy a reasonable salary and welfare and work hard in the positive team, together for the project to contribute their value. Thus, the dynamic adjustment and control of human resources workload is low. On the contrary, if the human resource management based on development stage is not in place, the task is not clear, it may affect the human resource management effect, also it will bring more workload to the dynamic adjustment and control, resulting in the increase of labor cost and waste. According to the dynamic development of the project phase needs to be adjusted and controlled in time, so that the LNG project is always in the human resources allocation and work in a coordinated environment.

4.2 Early Project Team

4.2.1 Organization Structure and Characteristics of the Early Project Team

4.2.1.1 Organization Structure of the Early Project Team

The early project team is the initial organization structure established by the LNG project initiators (Investment Company or Enterprise) for the development of its LNG project, that is, according to their own natural gas development strategy and planning to carry out LNG demand identification stage of the organization structure. According to the organization classification, it belongs to straight-line organization structure (Figure 4–2). Such as the organization chart shows that a project manager must be selected firstly, subordinates with relatively small staff, mainly from the cultivation and research of local natural gas utilization with the market, and is equipped with business, market development, government public relations, site selection, engineering early reconnaissance and other staff, usually the LNG project initiators arrange manpower, the organization structure is often tinged with the initiators' management traces.

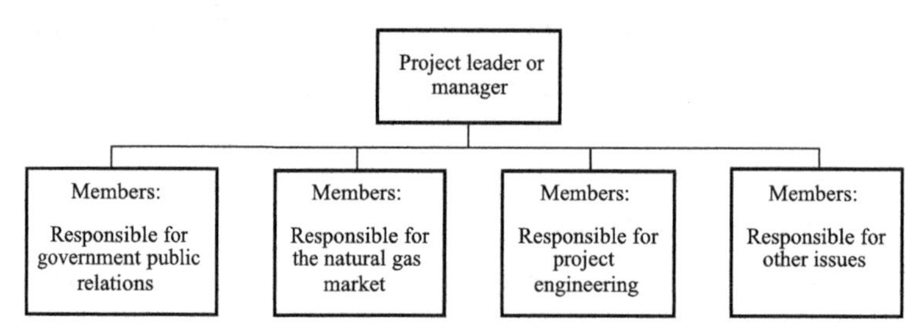

Figure 4–2 Project team - linear organization structure

4.2.1.2 Characteristics of Early Project Team Organization Structure

Early project team as a form of linear structure, it is mainly set up according to the project characteristics and the nature of the early work. The characteristics of the organization structure are followings.

(1) Clear hierarchical structure.

Project leader or manager is responsible for the work of the project, and the subordinate members only have one direct leader.

(2) Low cost.

Because the use of fewer personnel, generally no more than ten people, in order to avoid wasting too much in early investment of manpower, material resources and financial resources.

(3) Task clear.

Under the guidance of project manager to carry out the demand research for natural gas market and users, according to the characteristics of the project has a lot of uncertainty in the early stage, it needs to send a few people to check out the uncertainty factors, guiding the next step of work.

(4) High quality project director or manager.

The quality requirements of the project director or manager is to have a comprehensive understanding of LNG project, better to have a professional background, strong management and leadership skills, especially needs to coordinate the relationship between local parties and the staff of the organization of all kinds of ability.

(5) Members with broad knowledge.

Members are required to have strong ability in market, business, public relations, government coordination, and specific background, etc., especially to have team spirit, can get along with everyone.

4.2.2 The Main Work of the Early Project Team

4.2.2.1 Basic Work

(1) General requirements.

The main task of the early project team is the opportunity study of the LNG project, the investigation and study of the contents is very extensive, includes the macro factors of local economic development, the energy structure and the price level, the requirements of environmental protection, industrial policy and living standards, also includes the LNG specific to market conditions. The user also includes receiving projects; engineering aspects of the jetty, receiving terminal, pipeline natural environment etc.

(2) Natural gas publicity.

Before the use of natural gas in place, people is lack of understanding of natural gas, or just take it as a simple fuel energy to look at. LNG is not the original gas, but through the gas separation and treating, which has the characteristics such as high thermal efficiency, no pollution, widely application. People should have a process for understanding the advantages of natural gas used, propaganda natural gas advantages is the work of early project team. At this stage, the project investment initiators usually prepare for "LNG Brochures" and other related information, in order to carry out publicity and promote LNG awareness to the local government and the people. It will play a positive role in the advance of the project. In contact with local government officials, we should understand their basic attitude to promote the LNG project, but also judge the local government's investment policy and the environments. If the LNG market size and gas prices bear ability has not meet the requirements, it will suspend the pre-feasibility study of project, waiting for the market to grow or stopping the project.

(3) Project size estimation.

First of all, it is necessary to study the economic scale and development trend of the region which is likely to carry out the LNG project in-depth. In the early stages of the project, the target gas market in the region and the energy structure to be replaced will not be very clear, therefore, the project team must do a lot of research work, includeing preliminary study on gas supply market and technology economy, judgment of the gas price affordability industrial users and the implementation of specific user units, in combination with other gas supply and pipe network layout, preliminary judgment of the market share of LNG, a preliminary estimate the annual supply quantity of project planning needs. Put forward the proposed terminal scale of immediate and long-term, pipe route scheme, etc.

4.2.2.2 Major Items of Specific Project

(1) Primary user type.

Natural gas will be used as urban gas? Or used as gas power generation? Or used as natural gas chemical industry and so on. Right now, 70% of LNG is used in gas power generation, so, early development of gas power plant in gas market is the key to the establishment of the project. It is necessary to make a preliminary evaluation of the local electric power balance and technical economy, and to determine the layout, scale and annual gas consumption of the new power plant with the gas price bearing capacity. At present, most experts agree that LNG is applied in city gas, natural gas stations, LNG filling stations, industrial boiler will plays a better role in the economic capacity and have social environment benefit, therefore, to strengthen the development of non-gas power users is the focus in the future.

(2) Optional site ranking.

Through early reconnaissance, it will draw up more than two LNG project sites for comparison and selection. It includes the waterway, jetty access, terminal site, pipeline routing, etc.. Whether the coastal geological conditions can meet the requirements of building, is sedimentary type? Is flush type? Is stable type? The water depth condition is to meet the requirements of LNG ship to berth? Channel is beneficial to LNG ship to enter? How climate and waves, and currents affect? In order to answer the above questions, we must work to make in-depth investigation and observation, and even hire port design institute to undertake the work, to prepare for the future to determine the project site.

(3) Preliminary economic evaluation.

Combined with mature CIF (cost, insurance and freight) and the investment of LNG Project, the project investment is estimated, to predict the user's gas price, as a basis for judging the user's gas price afford ability, combined with the specific engineering conditions are preliminary economic evaluation of specific projects.

(4) LNG resource preliminary supply program.

With the market, the LNG resource supply has become an important problem that must be faced. From the current LNG resource supply method, it is from LNG project entity to seek itself LNG resources, point to point the supply mode, to change the project initiator to provide resources to the one or more of the LNG projects. In the project team stage, it needs to propose the preliminary LNG resource supply program.

4.2.2.3 Preliminary Results of the Project Team

(1) Signing a joint research agreement.

In order to promote the project progress, important goal of project team is to achieve investment company signed LNG project joint research agreement with the local government, namely "The Principles of Cooperation Agreement", in order to make clear of their respective responsibilities and tasks.

(2) Looking for partners.

After signing a joint agreement with local government, the government will not directly involve in investment projects. It requires the local government to recommend one or more local company or enterprises to cooperate with LNG Project Company, for laying the foundation to set up project office and push for substantive progress in the project.

(3) Completing opportunity research report.

Through the above work, the project team will understand the LNG market and the gas user types, determine the size of the gas market, complete the preliminary site schemes, from strategy, basically lock LNG resources. Then the early project team completes the project research results - opportunity study report.

4.3 Project Office

4.3.1 Owners Committee and Project Office

4.3.1.1 Organization

(1) Owners Committee.

Based on joint research protocol, usually to set up a LNG project owners committee (Figure 4–3), respectively, by the local government agencies such as the development and Reform Commission and the LNG sponsored investment company leaders served as director, deputy director, the members also are appointed by above-mentioned units.

(2) Project Office.

The project office is subordinate working group of the owners' committee, the agency is based on expanding and strengthening of the project early project team, mainly composed by the director or deputy director of the office and subordinates' work team, the members include marketing, resources, engineering, public relation, financial and administrative staffs.

4.3.1.2 Owners Committee Duties

(1) Determination of investment scale.

On the basis of the work of early project team, according to the market demand, to determine the project to import LNG volume, to ensure that users with gas, while to determine the size of the investment.

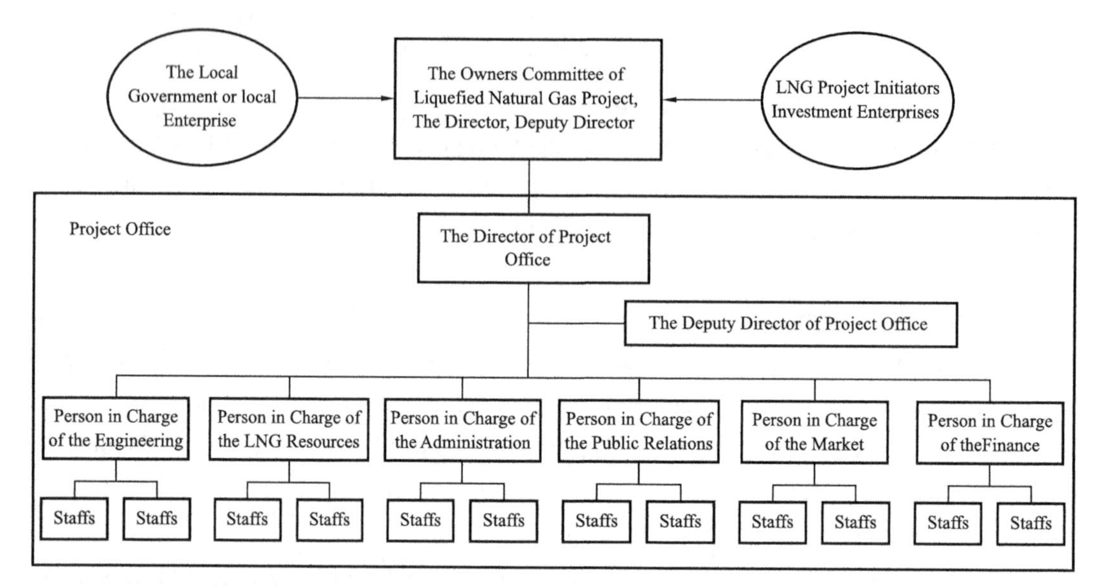

Figure 4–3 The Owners Committee, the project office relationship-functional organization structure

(2) Determination of user types.

Based on the in-depth investigation combined with the local economic scale of development, to determine the appropriate gas users, such as gas fired power plant, city gas, industrial users and so on. Generally, large gas users are ranked in a priority position, and sorting gas peak valley users and gas unit types, selecting the most likely users.

(3) Determination of local participants.

After the local government recommended, the local cooperation unit with positive response degree, will be confirmed by the initiator, both the local government and the initiator finally determine the local investment unit, entering the project substantive work stage.

(4) Project office establishment.

In the strong support of the local government, initiator company needs to conduct in-depth discussions with the local investment companies, in order to determine the quality requirements of office staff and allocation of all selected personnel, on the basis of the above work to set up the project office, which can be classified as functional organization structure (as shown in the Figure 4–3).

4.3.2 Project Office Organization Structure Characteristics

4.3.2.1 Advantages

(1) The function-centered management.

The advantage of the functional system is the administrative organization according to the function or nature of business management division, to select professional talents and play the role of professional expertise.

(2) Fine management.

Due to the function and the professional division of labor, it is conducive to employee business specialization, similar business is put in the same department with responsibility determination, it is helpful for setting up effective work order, preventing "Care for this and lose that" and "buck each other", to improve the management level because of the fine management work.

(3) To reduce the burden on direct leader.

It can give full play to the role of professional management of functional organizations, to reduce the work burden of line leaders.

4.3.2.2 Disadvantages

(1) Multilayer leadership.

It impedes with the necessary centralized leadership and unified command, formed a multilayer leadership. Middle management level tends to compete for contribution and buck-passing.

(2) Affecting work efficiency.

When the superior administrative leadership and the functional organization's instruction and the command conflict occurs, the lower levels don't know whose suggestion to follow, which can affect the normal operation of the work, easy to cause slack discipline, production management order chaos. It is not convenient for the overall coordination of administrative organization between departments, easy to form the phenomenon of lack of coordination between departments, so the administrative leader is difficult to coordinate.

4.3.3 Project Office Personnel and Responsibilities

4.3.3.1 General Requirements and Key Personnel Composition

(1) General requirements.

The office is in the early project executing agency. The main task is to carry out the important decisions of the owners committee and complete all tasks. Important responsibilities are: to decompose above tasks one by one which proposed by owners Committee, sorting and division of work. The work is based on the opportunity research results, to conduct further work, complete the pre-feasibility study report within the prescribed period of time.

(2) Key personnel composition.

Office staffs are expanded from early project team personnel, and mainly are sent by the LNG investment company, a few people sent by local enterprise or the company. At this time the personnel are relatively few, but they can carry on the division of labor, try to play each employee's professional skills, and do other work, but also pay attention to the team spirit and work overload, which is one of the characteristics of the work office stage. In most cases the staffs sent by investment initiator and the local company work in the office is a part-time job, they also need to take into account of the work by original company and the work of the project team. This issue should be coordinated between the project participants at early stage, to ensure that the office staff no matter where they come from, the work of the project team must be as high as possible commitment, and usually the work should be in the "Cooperative Principle Agreement" expressly stated.

4.3.3.2 The Main Responsibilities

(1) The implementation of market demand.

Combined with local economic development planning, and the local energy, electricity, gas planning, project office carries on a preliminary study on the competitiveness of the user of fuel and gas price mechanism replacement. In combination with other gas source supply, the office makes a clear positioning of LNG sources, market share and scale.

(2) Determination of the scale.

According to the classification of project users, the following works should be done, such as to determine the maximum users, especially to clear new gas-fired power plants, city gas and independent industrial users of the layout and scale etc., to clear LNG main users, initialed letter of intent to supply, to promote the pre-feasibility study of major gas user's project, to judge the main factors of gas project construction conditions, product market and the influence of gas project implementation. If necessary, to urge local governments to solve problems of gas project owner's policy, synchronously to start their projects, and lay the basic user groups for LNG project. On this basis, to find out the gas load center, develop a preliminary phase Ⅰ and phase Ⅱ project annual supply quantity.

(3) Carrying out the preliminary project plan.

According to the gas load center and coastal port conditions, the office selects one or more receiving terminal jetty sites, and makes preliminary study of jetty site. Through demonstration and selection by experts, local government prequalification, the jetty preferred site and alternative site should be identified, with matching between the sites and other conditions. The long period of meteorological, hydrological observation etc. need to start timely. The main contents are as follows:

① Jetty project pre-feasibility study work, including the jetty scheme comparison, engineering, hydro geological study at first, investment and operating cost estimates.

② Receiving terminal project pre-feasibility study, including the general layout, process scheme comparison, the engineering geological study, investment and cost estimates for operating.

③ According to the terminal address and gas load center, to carry out pipeline project feasibility research, preliminary planning of land pipelines, route engineering geological investigation, social infrastructure and environment evaluation, including routing, process scheme comparison, preliminary terminal site reconnaissance, investment and operating cost estimates.

(4) LNG resource suppliers selection.

The current LNG resources suppliers selection has the following two ways:

① By the project office independently or with the help of the parent company to carry out the foreign LNG resource suppliers selection.

② LNG resource is provided by the project sponsor parent company, project office has not a relationship with foreign LNG suppliers.

No matter what kind of solutions, all need to survey foreign LNG resources, including study on the international energy and LNG industry development trends, the existing liquefaction plants production capacity, new liquefaction plants construction conditions, potential LNG resources, market competition situation, the gas price and gas supply conditions. According to the selection strategy and preliminary scheme, to select several LNG resource countries, from political, economic, legal, technical research, geological resource conditions, to prepare for the final LNG resources selection.

(5) Preparing for LNG resources negotiations.

In order to ensure the buy and sell gas upstream and downstream of the contract, in accordance with the project owner committee's decision and authorization, the office participates in bidding and purchase agreement or organization LNG resources, sales agreement, transportation and other business contract negotiations, if initiated by the parent company to provide resources, the work is transferred to the parent company.

(6) Transport plan selection.

The office needs to research on domestic and foreign LNG transportation, shipbuilding development situation, the existing international excess capacity and excess LNG carriers, the potential transportation side and main conditions of shipyard, the formulation of transport options and preliminary scheme of strategy, preliminary determination of offshore LNG transport operators and transport modes, that is to use FOB? or DES? To study and judge the resources of CIF, the major issues affecting the supply of resources and transportation and solutions, etc.

(7) Research on financing and insurance.

The office needs to research on financing and insurance development situation at home and abroad, the existing conditions of financing and insurance, the insurance selection strategy and preliminary scheme, and also researches on pre-feasibility needed the financing, insurance, and cost structure.

(8) Pre-feasibility study.

According to the above mentioned opportunity research report, led by the project initiators to write the pre-feasibility study report or select qualified research units by the project office to do so. On the pre-feasibility study report to carry out the economic analysis, finally the office could make the local gas users, project sponsors, transportation side, foreign gas supplier all satisfy with the project's economic and social goals.

(9) Preparation before working start.

According to the budget approved by the project owners committee, the office could control the project investment in early stage, study project office structure and preparation for the project implementation stage and the joint venture company, offer some study results for the project investment initiators to prepare the principles of project cooperation agreement, organize engineering pre-phase research, and prepare for construction bidding, etc.

(10) Preparation for the project company establishment.

The office carries on the engineering contractor pre-investigation, prepares the project company establishment, starts the feasibility study and formulate of contractors' bidding strategy.

In short, if the pre-feasibility study report is recognized by sponsors of the parent company, local government and the cooperation units support the project forward actively, at the same time, then it will be transferred to the stage of preparation for project company establishment and the formulation for feasibility study report. If the pre-feasibility study reports proposed there are large uncertainties, supplement study should be done and make decision again. If the per-feasibility study report conclusion is not feasible, it will have to suspend projects to enter the stage of feasibility study, in order to make the decision for termination of the projects.

4.4 Project Company Organization

4.4.1 Project Organization Structure

4.4.1.1 The Basic Conditions for the Establishment of the Project Company

(1) The first case.

On the basis of the Project Office, the Project Company is an organization established with the increase of the workload of the project and deepened organization structure. The first case is: the project pre-feasibility study report has been approved by the initiators' company, and "Application for the Project to Carry out Preliminary Work Report" has obtained the consent reply by the State Competent Department, that is after the "Pass" obtained, then to set up the Project Company.

(2) The second case.

Similarly, the Project Company is on the basis of the Project Office, with the increase of the workload of the project and the deepened organization structure to be established. The second case is in the first case is satisfied, and the feasibility study report of the project has been approved by the initiator company, since then the Project Company is established. At present, the common LNG Project Company usually uses a linear functional organization structure, which is based on a single LNG project to establish the organizational form (as in the lower part of Figure 4–4). The General Manager of Project Company will control and take in charge of project.

4.4.1.2 Advantages

(1) Highlight the leadership responsibility system.

The General Manager is responsible for the project, and team members are responsible for the general manager. The General Manager of the project is the real project leader because he can mobilize all kinds of favorable factors inside and outside.

(2) Flat management.

Project management level is relatively simple. It makes decision-making speed and response speed of project management become fast. Establishing a project team, general manager of the project has control over projects and teams, at the same time, making down or break down the decision-making authority to the project team of each employee according to the needs of project.

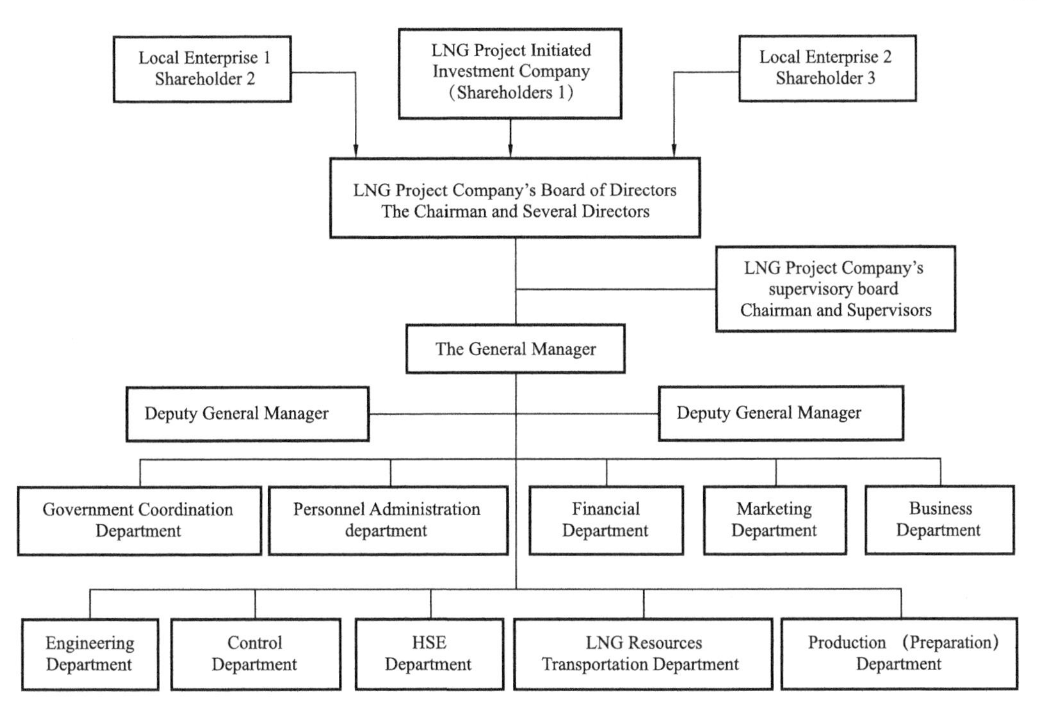

Figure 4–4 The Relationship of board of shareholders, board of directors, board of supervisors and project company—linear functional organization structure

(3) Clear objectives.

The team member's target is clear and single, the whole project team members work hard for the common goal of the specific LNG project construction, and create conditions for the team spirit, and can be relatively easy to play a synergistic effect of work together. Team members are independent of their functional organization, with no interference from the original work, they can put their heart into project work, also contribute to the formation of team spirit.

(4) Wider work interface.

For team members play potential as possible, pave the way for the generalist and versatile professionals.

(5) Convenient and concise communication path.

General Manager of the project can directly report to the top managers of the parent company, in order to avoid the distortion of information transfer.

4.4.1.3 Disadvantages

(1) Increasing the management cost.

Because of project to set up a company solely, the parent company has to invest and set up the posts, increasing manpower and equipment etc.. The stages of project work often cause employees uneven busy, project personnel are locked in a project, after the end of the project, the staff may be stuck in the project, which is not conducive to the overall allocation of human resources, and will increases the cost of management.

(2) Short-term Behavior.

As a result of the project life cycle, project staffs are often lack of long-term continuous work thinking, especially during the completion of the project, staff in an anxious state of mind, lack of career development opportunities and impact of the work. The decision-making layer is less considering operating demand, which may result in a large increase in operating costs due to the later cost of renovation.

(3) The formation of the concept of small groups.

Team members may be limited in this project work objectives, lack of overall situation consciousness and form a small groups, and restricting cross project technical communication and development.

4.4.2 The Main Responsibilities of the Company's Board of Directors and the Chairman of the Board

4.4.2.1 The Composition of the Board of Directors

From the affiliation of project company, usually the LNG project initiator investment company and the local investment company are shareholders, the general meeting of shareholders is established in accordance with the Company Law, which is the highest authority of LNG Project Company. In order to carry out the work easily, the shareholders appoint staff to set up the board of directors of Project Company, responsible for major matters of the company decision. Generally, the chairman of the board of directors is appointed by a large amount of investment party, the other party or several parties serve as vice chairman, the parties also appoint several directors (Figure 4-4).

4.4.2.2 The Main Responsibilities of the Chairman of the Board of Directors

(1) Status of chairman of the board.

The chairman of the board is one of the company directors as company's legal representative. Chairman of the board of directors is recommend and confirmed by the relevant shareholders or by the shareholders' meeting. Generally, the chairman is elected for a term of three years and can be re-elected.

(2) Responsibilities of chairman of the board.

① To convene and preside over the meeting of shareholders and preside over the meetings of the board of directors.

② To check implementation of the resolution of general meeting of shareholders and the board of directors, and report to the board of directors.

③ To sign the relevant documents according to law and the company "Constitution" on behalf of the company; to perform other functions and powers prescribed in the general meeting of shareholders, the board of directors and the Company Law.

④ When the board of directors is not in session, for the urgent matters would be decided by the board, the general manager will report to the chairman. The chairman of the board makes the decision after consultation with the vice chairmen, and report to the board of directors at the next meeting.

⑤ When serious natural disasters encounter, such as force majeure, not timely treatment will cause great damage to the company, the chairman shall have the right to exercise the special right of disposal in conformity with the law and the interests of the company to the company's assets, and will report to the board of directors in the post.

4.3.2.3 The Main Responsibilities of the Board of Directors

(1) Status of the board of directors.

The board of directors is responsible for the general meeting of shareholders, the board of directors complies with the company law, the articles of association and the provisions of the state laws, regulations and regulatory documents, faithfully performs their duties, to maintain the company's interests.

(2) Responsibilities of the Board of Directors.

① To convene and report on its work to the general meeting of shareholders, to perform the resolution of the general meeting of shareholders.

② To develop business plans and investment plans; to formulate the annual financial budget, final accounts plan of the company; to formulate plans for profit distribution of the company and plans for making up losses; to formulate company's annual and major financing plan; to develop company increase or reduction of the registered capital scheme.

③ To formulate the company merger, division, change of corporate form and dissolution of the scheme, to determine the company's internal management mechanism, the positions of senior management personnel setting, to set up the subcommittee of board directors; to appoint or dismiss the company's general manager, deputy general manager, financial controller, general manager of financial department, to decide on matters concerning their remuneration, to approve the company's basic management system such as compensation, recruitment and secondment.

④ To decide signing and revising LNG supply contract and the contract of carriage; to determine the LNG sales contract signed and any amendments.

⑤ During the construction of the project, it also exercises the following functions and powers: be responsible for establish construction funds financing, report to all shareholders for approval before implementation; to audit and report preliminary design and budget documents; to study and resolve major issues appeared in the process of construction.

⑥ Be responsible for the putting forward completion and acceptance of project application report; to validate the debt repayment plan and production management policy; to review and decide other major matters shall be decided by the board.

4.4.3 The Main Responsibilities of the Project Company Board of Supervisors and the Chairman of the Board of Supervisors

4.4.3.1 The Composition of the Board of Supervisors

The company also set up the board of supervisors (except for Sino and foreign joint venture company) and chairman of the board of supervisors, the supervisors are sent by the shareholders, but also by the workers democratic elections of representatives of the staff and workers to participation. The term of office of the board of supervisors shall be 3 years, and supervisors may be re-elected at the expiration of the term. The board of directors, general manager, financial controller and financial department manager shall not serve concurrently as supervisors.

4.4.3.2 The Main Responsibilities of the Chairman of the Board of Supervisors

(1) Status of chairman of the board of supervisors.

The chairman of the board of supervisors shall be the person in charge of the daily work of the board of supervisors, and shall be responsible to the shareholders of the meeting of the board of supervisors and the host.

(2) Responsibilities of the chairman of the board of supervisors.

① To convene and preside over meetings of the board of supervisors;

② Be responsible for the daily work of the board of supervisors;

③ To approve and sign the report of the board of supervisors and other important documents;

④ Other duties shall performed by the chairman of the board of supervisors.

4.4.3.3 The Main Responsibilities of the Board of Supervisors

(1) Status of the board of supervisors.

The board of supervisors is a permanent supervisory institution of the company under the leadership of the general meeting of shareholders. The board of directors and the board of supervisors are parallel.

(2) Responsibilities of the chairman of the board of supervisors.

① To check the company's finance;

② To supervise the board of directors, the general manager, deputy general managers and other senior management personnel in performing their duties in violation of laws, regulations or the articles of association of the Company Act.;

③ When the board of directors, general manager, deputy general managers and other senior management personnel behavior damage the interests of the company, ask for corrected;

④ To propose a temporary shareholders meeting;

⑤ To perform other functions and powers prescribed by the articles of association of the company;

⑥ Members of the board of supervisors attend the meetings of the board of directors;

⑦ The decision of the board of supervisors shall be agreed upon by all the supervisors or a certain proportion of the above.

4.4.4 The Main Work of Project Company

4.4.4.1 Human Resource Staffing

Generally, the general manager, deputy general manager and department managers are appointed by shareholder's companies. Under the terms of the shareholders agreement, subordinate staffs can be sent by the shareholders, also can be recruited by the project company in the society directly.

4.4.4.2 Determining the Downstream Users

Based on the earlier work, the Project Company should do in-depth study of alternative fuels competitive users, in consultation with relevant government departments, determine the gas price mechanism, implement gas power plants, city gas, gas for chemical industry and industrial gas users, and determine phase Ⅰ gas consumption volume, and prepare for signing the gas purchase contract.

4.4.4.3 Business Contract Negotiations

Combined with resources and financing structure, the Project Company carries on the downstream projects due diligence, implements the users' owner credit, financial investment and performance ability, to ensure that upstream and downstream signing gas purchased and sales contract, in accordance with the decision and authorization of the board of directors, participate in or organize LNG resources bidding (except sponsor shareholders to provide resources) and gas purchase agreement, gas sale agreement, transport agreement and other business contract negotiation.

4.4.4.4 The Preparation of Project Feasibility Study Report and Project Approval

(1) The preparation of project feasibility study report.

In order to carry out the work and promote the feasibility study report preparation, the following conditions must be met.

① The relevant government departments approve the jetty site, receiving terminal site and pipeline main routing scheme.

② The relevant government departments approve the main users, layout and scale.

③ The main users recognize the LNG terminal pre-feasibility study report, and agree to promote users' own project feasibility study synchronously, signe a letter of intent with LNG project side by gas supply.

④ The relevant government departments accepte the main conditions of land requisition, and signe a letter of intent of utilities supply.

⑤ After the argument with LNG project owners, the pre-feasibility study report is feasible, signe a project cooperation agreement in principle.

On the basis of the above work, the Project Company organizes and completes the preparation of project feasibility study report, to ensure that the pre-project design and feasibility report according to quality requirements and deadline (see Chapter 6).

(2) Project check and approval.

The LNG project is the basis and energy industry, in accordance with the national reform project approval system to the NDRC Energy Bureau check and approval, the procedure is based on the project application report on the feasibility study report of the project (see Chapter 6), after the project is checked and approved, the substantive work is carried on.

4.4.4.5 The Engineering Construction Bidding and Engineering Construction

(1) The engineering construction bidding.

When the LNG project is approved by NDRC Energy Bureau, according to the project budget approved by the board of directors, the company carries out the tender work of the design unit of the LNG project. Combined with the market, resources, transportation and pipeline network requirements, carefully examines the basic design data, organize the concept design of jetty, terminal and pipeline and project feasibility study report preparation, carries out engineering, geological and hydrogeological survey, program optimization, investment and operation cost estimation, and needs to get the approval of the owner, to implement water, power, road, communication and other public facilities supply, negotiates and signs supply contract agreement in principle, carries on detailed investigation of major equipment and materials supply and engineering contractors, formulates bidding strategies, and prepares project tender documents, organizes the FEED design, preliminary design, detailed design and tender evaluation.

(2) Engineering construction.

According to the project owner's committee approval, the project company carries on constructing project receiving terminal, jetty, pipeline and supporting project with required quantity and quality, and completes the project on time.

4.4.4.6 Project Acceptance and Commissioning

(1) Project acceptance.

According to the project feasibility study report and detailed design scheme, the company organizes experts to the project acceptance in different stages, write acceptance evaluation report.

(2) Commissioning.

According to the design requirements, in the commissioning stage, the company carries on the LNG receiving and unloading, tank storage, LNG gasification, natural gas transmission and distribution and other aspects of the trial running, finds the problems and rectification, and lays the foundation for the formally operation.

4.5 Human Resource Management

4.5.1 Project General Manager Post Qualification

4.5.1.1 Basic Requirements for Project General Manager (PGM)

(1) Summary.

With the appropriate organization structure, the required start-up capital, the maturity of the LNG technology, another is to have a competent PGM who plays a vital role in the whole project.

He or she must meet as general manager of the necessary general management and decision-making skills, at the same time, he should have relevant project knowledge and certain technique background of LNG project. Mentioned in MBA Course of Harvard Business School, the project general manager must possess three kinds of skills: one is technical skill, refers to the proficient and familiar with one of the major activities, particularly relates to a method, process, procedure or technical activities; The next is the personnel skill, refers to the ability of the general manager as a molecular of organizations, he should not only complete his own tasks excellently, but also lead employees to do their jobs in a cooperative atmosphere; and the last one is concept skill, that is to drive the whole project's operation, and he should realize the interdependence of various functions, one of them changed will affect the others. The lack of any one aspect of the basic quality of the above mentioned, will adversely affect the success of LNG project. The following parts are to explain the personal qualities and management capacities of the LNG PGM.

(2) The personal qualities of the general manager for LNG project.

① Leadership ability. According to the definition of scholar Stephen.P.Robbins, leadership is the capacity to make a team achieve its goal which is reflected in personal attractiveness and influence mostly. The characters of a leader are visionary, dare to think and to act and full of confidence, but he is not foist his opinions upon others, his charm lies in the use of ideal to inspire subordinates, good at the prospects of the project to show employees and motivate subordinates to work hard for the project goals.

② Team organizer. A project organization should be a team, because of the need for everyone to make concerted efforts to complete tasks in the limited time, financial and material resource, and the PGM is just the leader of the team. Meanwhile, he or she should play multiple roles. For example, he or she is innovator and explorer, estimator and pusher, controller and summarist.

③ Solid knowledge background. LNG project for China is an industry unprecedented, the PGM must be an expert in a certain area on technique and management fields, thus, he can keep advantages in management dominanted by developed and set up the authority of individual in a familiar field. Whereas one person cannot be perfect, it requires the PGM has the ability to mobilize employees service for the project by their skills and knowledge.

④ Excellent personal quality. In addition to the above mentioned qualities, as a PGM, in order to get others' respect and admiration, he still must have open-minded, cheerful character, firm and indomitable will, steadfast steady, anti-corruption, thoughtful, inclusive of others' dissent, care about their living and working environment, timely to solve the staff's personal and family difficulties, to think for others in front of benefits, to make friends with the subordinates etc.

(3) Requirements of LNG PGM's management ability.

① Planning ability. Planning is a premise for a project. As the LNG PGM must firstly have the overall perspective and strategic vision, on this basis, the PGM should play to the initiative of all team members, working together on the project plan, including the project's goals and milestones, also the detailed planning for every phrase. Let every team member understand their own goals and the ultimate goal of whole project.

② Organizational ability. From the perspective of the internal division of the project team, basing on the acquaintance of every team player's skills and working background, he should be an executor and cooperator. He should allocate tasks reasonably, and strive to arrange the right person in the right position. Meanwhile, he also equips office with corresponding equipments and creates a good working environment, and strives to achieve the character matching.

③ Controlling ability. The PGM should control the process of the project, and keep the key task in mind. He is familiar with the key issues and disruptive factors, in order to control the project process. Every job arranged must be completed in reasonable time and range, no waste of manpower, or not occupation of funds, but also accord with the financial system and the project audit requirements. Controlling is also included on the processing crisis event to occur, when it is not conducive to the progress of the project and to complete, to solve the problem timely and effectively, so as not to cause significant economic losses.

④ Decision-making ability. Decision-making plays a vital role in the process of project. An excellent decision represents the PGM's high theoretical level and comprehensive understanding to the project. Decision-making must be based on the deeply understanding to all information and developing rules of processes. Except the decision-making ability, the PGM should also pay attention to playing the subordinates' role under the decision, so as to avoid major decision-making errors.

⑤ Communication skills. Communication is the key to play in a team where the synergistic effect is good or bad. As a LNG PGM, who must firstly establish project team communication channel, especially in today's information society, there are so many communication channels, such as regular or irregular conference, weekly, monthly, annually and special reports, and electronic e-mail, phone, and fax, especially do not forget the oral report, talk and so on. Based on various communication channels, he must pay attention to the following aspects: firstly, in order to grasp the whole process and relevant problems of the project, it is necessary to set up a series of systems which include timely exchanging meeting about project progress and weekly, monthly and annually reports. Secondly, putting electronic e-mail, phone and fax as supplementary methods so that we can get the facts in time and discover and solve problems promptly. The last one is that oral communication which can make the PGM know the facts more clearly, if the use of oral talk properly, it can reflect more emotional exchanges and communication, to achieve the best effect.

⑥ Coordination ability. Coordination is the lubricant of communication among people, it includes internal and external ones. As a PGM, internal coordination refers to under no written rules to establish the cooperative relationship, handle the work link among the team members, usually after a relatively short period of time to straighten out. The most difficulty is the external coordination which refers to central and local government, local enterprises and local citizens. Facing an area where the people is not familiar to LNG, promoting the LNG project must have more understanding and accept LNG, make all parties mentioned above recognize that the implementation of the LNG project is the best way to win-win for all the parties. In order to make all parties understand and support the project on terminal site selection, pipeline land, tax concessions, the gas price determination, and user's enterprise cooperation. If the external coordination is not ready, may affect the progress of the project, even can't reach the expected economic and social effect. If it is a Sino foreign joint venture project, the staffs also need to have a certain level of foreign language, understand the international practice and cultural differences, so as to unite Chinese and foreign employees to work hard together for the project objectives.

⑦ Learning ability. The leader plays a vital role in the process of setting up a learning enterprise. We cannot ask each project manager to have foresight, but a lot of knowledge need to learn from the working, although sometimes feel that we have rich experience or knowledge, but there will be such a phenomenon: "when you use knowledge, you will find that you are not enough". Because new things emerge in endlessly, especially we are in the era of knowledge and information explosion, in order to make ourselves keep pace with the times, we must always, everything, everywhere to learn, have to be good at learning, good at digesting, combining with the current work, to solve practical problems.

⑧ Innovation ability. Innovation-oriented country needs numerous innovation-oriented enterprises which is the same for a project. As a PGM, should have the innovation in the foundation to absorb and learn from others and other project management experiences. For each project, even the same kind of project, such as Guangdong Dapeng and Fujian LNG projects, because of differences in the specific conditions of each project, and we can't copy mechanically other project experience, and to have the innovation based on the original work experience, the only way to make the project management to achieve the best effect.

4.5.1.2 PGM's Post Responsibilities

LNG Project Company is the PGM responsibility system under the leadership of the board of directors.

(1) Responsible for setting up the organization.

To organize the formulation of organization structure, labor force quota scheme, wage and welfare scheme and report to board approval.

(2) Formulating strategic goals.

According to the long-term development goals issued by the board of director, to organize and formulate the company's development strategy and medium and long-term development plan, establish business strategy and implementation plan, and carry out various resolutions and decisions that are issued by the board of director.

(3) Formulating annual work and investment plan.

According to the annual operating goal issued by the board of director, to draw up the annual business plan of the company, to organize and implement the annual operating and investing plan approved by the board of director.

(4) Company management.

Within the authority of the board, on behalf of the company he signs the relevant agreement, contract and handles relevant issues, responsible for improving the organizing, planning, coordinating, controlling and managing mechanism. He set up rules and regulations and workflow management system, and revises and improves according to the development situation of the project company.

(5) Five controlling systems.

The PGM has the responsibility to set up and implement five controlling systems which are schedule control, cost control, quality control, HSE management system and risk control system.

(6) Other tasks assigned by the Board of Directors.

4.5.2 Project Company Department's Post Responsibilities

The departments of current LNG project company in china include Administration and HR, Public Relationship, Engineering, Control, Market Development, Business, Finance, LNG Resource Transportation, HSE, and Production Preparation. Some project company divides finance into the accounting and financial department independently, some ones make the Administration and HR separately, some ones merge Marketing and Business into one department, etc.

4.5.2.1 Administration and HR Department

(1) The preparation of manpower planning.

The department is responsible for the preparation of human resources planning, formulates regulations of human resources management, preparation of project staff handbook, recruitment, remuneration, welfare, subsidy management, and company and employee's performance assessment according to the project progress and human resources planning.

(2) Salary and welfare management.

The department draws up the company salary structure and guiding principles, salary adjustment plan and budget. According to a staff position change, updating archives salary timely, answering staff salary calculation and payment of relevant problems. Based on the business development of the company, put forward the welfare, subsidy scheme, as a supplement to salary outside, complete the staff welfare subsidies calculation and payment of the work.

(3) Performance appraisal.

The department is responsibility to carry out the performance appraisal about the company and its employees, collects performance appraising data of all departments and performance appraisal index analysis at the same time, responsible for solving or participating in solving problems associated with performance management encountered in daily routine, putting up with reasonable and effective performance appraisal plan positively and proposing the performance appraisal linked to the reward and punishment system, and timely proposing and revising the personal career design, coordinating and solving the problem of human resources matching.

(4) Archives management.

The department manages all kinds of personnel files, so that the department can provide in-

formation support for the company's personnel decisions.

(5) Training.

The department is in charge of various training of the project company, such as comprehensive training, certificate acquired, new comer training, skill improving, short and long term, home and abroad training, etc.. And also has the responsibilities to give support to various special techniques training.

(6) Administrative management.

The department is responsible for the administration, secretary, documentary, supervision, meeting affairs and routine management, and also responsible for the approval of office facilities, equipment and other fixed assets, the purchase of office supplies.

(7) Office expense management.

The department is responsible for office expenses and related costs of the budget preparation, use, supervision and control.

(8) Other management.

The department is responsible for IT work, enterprise culture and propaganda work; responsible for foreign affairs and company and external liaison and reception, and responsible for the work of the party and the masses.

4.5.2.2 Government Coordination Department

(1) Land acquisition.

The department is responsible for land routing, land and sea land acquisition (permanent, temporary) application, forensics and compensation, also in charge of liaison and coordination with local government departments, such as customs, taxation.

(2) Policy research.

The department is responsible for national and local LNG tax, gas, environmental protection and other policy researches and making recommendations to the state and local government, responsible for writing the relevant policies and recommendations of the report, and constantly improving the revision until the approval of the competent authorities.

(3) Government coordination.

During the construction and production, the department has the responsibility to communicate and coordinate with various local government departments such as marine and fisheries, land resources, construction, transportation, environmental protection, price, fire control, water conservancy, industry and commerce, finance, taxation, customs, maritime affairs, public security, labor.

4.5.2.3 LNG Resource and Transport Department

(1) Resource and trade research.

The department is responsible for the organization and analysis of the world natural gas resources and LNG trade, and the possibilities of selecting several foreign LNG resource suppliers to supply LNG to China.

(2) Shipping research.

The department grasps the global LNG vessel's technological developing tendency and the shipping building industry change, appraise potential transporters from home and abroad, focuses on vessel fleet, route, operation, coordination ability and economic analysis of transport operators to provide transport service.

(3) The LNG resource purchase.

The department puts forward LNG purchase plan and recommendations, organizes to carry out international investigation for LNG resources supply and LNG shipping, and reviews potential LNG resource suppliers, including natural gas resources, liquefaction plant, port, jetty ability.

(4) The LNG resource negotiations.

According to the government's approval opinion the department, organizes the LNG resources and transportation supplier qualification, bidding, bid evaluation and negotiation, responsible for implementation of LNG purchase and sales contract and LNG carriage contract.

(5) Trade arrangements.

The department is responsible for the gas sales, payment guarantee, insurance, financing, etc., organizes to draw up or involve in checking the relevant feasibility research report of LNG supply and transportation.

(6) Natural gas utilization plan.

According to the requirements of natural gas users the department, coordinates the relationship of LNG chain, and puts forward reasonable LNG annual demand plan and shipping plan, signs trade contract with the gas users under the conditions of practical demand for natural gas, the annual plan, the method of settling accounts.

(7) Providing upstream opportunity.

The department creates conditions for the company or associated company having opportunity to enter the international natural gas upstream market.

(8) Relevant training.

The department formulates and implements the employees' training plan for LNG resource, transportation, and vessel, etc.

4.5.2.4 Market Development Department

(1) Market research.

The department is responsible for the research and demonstration of LNG project gas utilization plan, draws up the gas market developing plan and stage's goals, market developing strategy, organizes risk research under the quantity of gas required, the user undertakes the "Take or Pay" risk ability analysis, assures the sTableand reliable of quantity of gas required and cash flow of the project.

(2) Policy recommendations.

The department tracks the international and domestic natural gas market dynamics and de-

velopment trends, understands the relevant national policy orientation, combined with the practical problems existing in the development of the project, put forward the corresponding policy recommendations for government department as reference.

(3) Market cultivation.

The department is responsible for the user training, market promotion and LNG advertisement and propaganda, increasing users' awareness of natural gas.

(4) Natural gas utilization project coordination.

The department coordinates the local government, promotes the implementation of gas user's project market, makes the gas user's project keep in pace with the LNG project synchronously, contacts with all the owners of gas user's project, coordinates the relationship between the gas suppliers and gas users on the aspect of business and technology. Responsible for tracking progress by gas project feasibility study and for modifying and checking the part of marketing of the whole project.

(5) The sales contract negotiation.

The department is responsible for control of the market dynamics, organizes the drafting of natural gas sales contract, draws up the contract negotiation scheme, negotiation strategy, and responsible for organizing the sales contract negotiation and coordination.

(6) Research on the market competitiveness.

The department is responsible for the study of natural gas price and price adjusting mechanism, market competitive force of urban gas projects, gas turbine power competitive force, and local urban gas market development, surrounding gas market consumption, LNG tank car market study and exploitation.

(7) Information communication.

The department in accordance with LNG purchase and sale agreement, is responsible for providing the market data to LNG resource party, supports financial or business department to study and implement by gas project payment mechanism, responsible for coordination with Financial Department with financing work, and provides necessary market information and materials for financial agencies.

4.5.2.5 Engineering Department

The department is responsible for laying down company overall project and each monomer project schedule, according to such plan, organizes and supervises the feasible study, FEED design, EPC bid, preliminary design, details design, construction design, site construction, joint commission, up to operation and relevant work for the project. The department is the core of LNG project company. The department can be divided into the following several kinds of work.

(1) Pipeline category.

The category is responsible for the terminal and trunkline project bidding, design review, contract execution and construction, commissioning until the operation process of work.

① To be responsible for the review of the qualification examination and report of the feasibility study unit of the gas transmission line.

② To be responsible for trunkline bidding and design review, putting forward the work adjustment schemes and requirements to the pipeline contractors, based on the up-downstream project progress changes, interface requirements, pipeline project progress.

③ To coordinate and solve the pipeline contractors' problems at work. To communicate and coordinate with the local government departments effectively and timely when pipeline contractors meet local interference during designing and implement, and track and supervise the issue in advance and afterwards.

④ To understand and summarize the pipeline contractors' work situation and the work method timely, predict and judge the future possible problems, adjust the working way of making it more suiTablefor the requirements of the project.

⑤ To collect and sort relevant materials of gas trunkline project constructing and constructed from home and abroad, master the advanced engineering organization and mature application skills.

⑥ To organize and arrange the theoretical and skill training of relevant pipeline operators.

(2) Instrument and telecommunication.

The category is responsible for instrument control, security and communications professional work in LNG project implementation.

① According to the project progress, to draw up instrument and telecommunication professional work plan, responsible for instrument control, security and communications professional work in LNG project implementation.

② To understand latest technologic development of domestic and foreign instrument control, safety, and communications, according to the requirements of project implementation, participate in the development of new technology research, promotion and scientific research, complete the demonstration and research on the major technical scheme.

③ In accordance with the general requirements of automatic control system of LNG project, to determine the receiving terminal control level, communication requirement and trunkline control system level.

④ According to the project requirements, to review the concept design, feasibility study, FEED design, preliminary design and detailed design, supervise and manage the jetty, terminal and trunkline contractors.

⑤ During the implementation of the project, to check and appraise the achievement of instrument control, safety and communications in research and design stages, including design drawing, specification, statement and design report;

⑥ Be responsible for biding and purchasing, contract executing and changing of instrument control, safety and communications.

⑦ Be responsible for handling the engineering unit instrument control, electrical and communications professional technology management, to coordinate the relevant professions, relevant interface work and communicate with the related local departments.

⑧ To participate in site construction, implementation and appraisal, to participate and organize in control system, security system and communication system, pre-commissioning, commissioning and operating, ensure the achievement on the four controlling goals of the project.

⑨ Be responsible for organizing and arranging the theoretical and skill training to operation workers on the instrument control, safety, and communication.

(3) Jetty engineering project.

The category covers managing and controlling the research, design, implement, debug and running during the whole process of jetty project, ensuring the project implementation by its plan and technical and economic scheme reasonably.

① In accordance with the overall objectives and requirements of the project, to manage the jetty design, construction, commissioning of the whole process of sub-project, to ensure that the sub-project is implemented according to the plan with the technological and economical scheme reasonably.

② Be responsible for participating in the project bidding process of jetty engineering, design review, contract execution and construction of the whole process of construction management, commissioning and commissioning work, to coordinate and handle related work interface and interface.

③ Be responsible for drawing and appraising the technology requirements, taking in charge of the technology bidding and negotiation, and sorting out the bidding results.

④ Be responsible for tracking the process of contract execution, identifying the research results, ensuring that contracts are completed on time and according to quality, and collecting, arranging, reviewing and submitting the basic design materials.

⑤ To understand domestic and foreign technology trends timely, according to the needs of the implementation of the project, organize the special topics on new technology study, approval of drawing, calculations, design scheme, design documents, data sheets, and specifications, review the construction personnel quality, construction organization plan.

⑥ In conjunction with the chief contractor, to formulate the water intake and outlet and pipeline (box culvert) and terrestrial synchronous construction plan, and coordinate the general schedule, participate in the preliminary design and detailed design of jetty engineering.

⑦ To participate in the receiving terminal's land formation engineering contractor of dredging, hydraulic filling, vibratory compaction, dynamic compaction, bank-protection engineering construction scheme, construction organization plan, construction equipment, construction schedule and on-site management; if there is a matching power plant projects, participate in the coordination of water intake, drainage pipe gallery design and construction of the power plant.

⑧ To appraise and identify the design and site change under the certain authority, summarize and modify the professional works in suitable time continuously in all running stages of the project.

⑨ Be responsible for the theoretical and skill training for the operators working on the jetty.

(4) Electrical.

It is responsible for the relevant work in electrical professional of the jetty, receiving terminal, and gas trunkline, including design review, bidding, contract execution, construction, implementation, commissioning and operation.

① Be responsible for drawing up electrical professional's working goal and plan, and arranging the job of the employees in the electrical major.

② Be responsible for the demonstration and research on electrical technology program, responsible for electrical professional's bidding, design review of conceptual design, FEED design, preliminary and detailed design on jetty, receiving terminal and trunkline sub projects.

③ Be responsible for the feasibility study of the electrical engineering of the jetty, receiving terminal and gas trunkline, and fully responsible for the bidding and purchasing of electrical equipment, the implementation of the contract and the processing of the change.

④ Be responsible for the coordination and handling of electrical professional and local power bureau and power plant among users, the various contractors, the various engineering monomers, electric professional and other professional work interfaces, responsible for electrical professional work quality, cost, schedule, and risk management.

⑤ Be responsible for the supervision and management of the contractor's electrical professional work, understanding the latest technology development of electrical professional at home and abroad, and according to the needs of project implementation, proposing the promotion of new technology development, scientific research, suggestions and participating in research project.

⑥ Be responsible for the theoretical and skill training to the electrical operation workers.

(5) Process category.

The category is responsible for the management and technology of LNG treatment process of the receiving terminal project and pipeline process of the gas trunkline.

① Be responsible for organizing experts to review the process design, construction plan, approval, and solving the technology problems appearing in the construction, ensuring the implementation of the project smoothly.

② To supervise the contractor to complete the work according to the contract and the requirements of the owner, identify the contractor's technological achievements in each stage and put up correct measures and recommendations timely, coordinate the business relationship among contractors, enterprises and local government.

③ To prepare the bidding document of engineering technology, participate in the evaluation work, responsible for equipment and pipeline installation and commissioning work in receiving terminal and gas trunkline engineering implementation, coordinate and solve the various problems appearing in the construction.

④ To coordinate the relevant government approval, ensure the implement of the receiving terminal and gas trunkline project, forecast and analyze the key activities and technological difficulties which affect the implement of the project, and discuss with other professionals searching for solutions.

⑤ To track the technology developing trend of the LNG process and gas trunkline at home

and abroad, grasp advanced process simulation software, familiar with the relevant technical specifications, so that accumulating technology experience and document for the operation of LNG terminal and trunkline project and others in China.

⑥ Be responsible for the theoretical and skill training to the process operation workers.

(6) Machinery category.

The category is responsible for main mechanical equipment's design and selection, bidding, procurement and installation and debugging and other related technical work in LNG jetty, receiving terminal and gas trunkline projects.

① Be responsible for organizing and arranging demonstration and research of the major equipment selection scheme.

② To take part in drawing up the project biding document and bidding evaluation, responsible for the bidding and purchase, contract execution and change of relevant mechanical equipment.

③ To participate in on-site storage tank supervision, equipment inspection, installation and final debugging and commissioning.

④ Be responsible for and to participate in the determination of work interfaces between contractors and the data reconciliation; handle and organize experts to review various results of the contractor completed.

⑤ To understand the latest technology developing trend of relevant mechanical equipment at home and abroad, according to the needs of project implementation, put up suggestions and critical research issues of new developing and improving technology.

⑥ Be responsible for the theoretical and skill training to equipment operation and maintaining workers.

(7) Engineering economy category.

The category is responsible for monomer engineering project investment estimation, gathering and economic evaluation, assisting in cost control, project completion and acceptance and post-evaluation in the project implementation in receiving terminal, jetty and gas trunkline projects.

① Be responsible for undertaking engineering economy professional work, participating in project plan discussion, selection, assessment and acceptance at each stages.

② To examine and verify the valuation basis and method for estimation of each unit of project investment, select the economic parameters reasonably, establish the proper economic measuring methods for the economic evaluation, responsible for the preparation of project budget in quarterly and annual plans, responsible for budgetary estimate decomposition.

③ In the project implementation stage, to analyze and forecast the execution of the expense timely; participate in contract price negotiation, reviewing and implementation.

④ Be responsible for supporting project's economic and technology related financing, insurance, contract, procurement; participating in contract price negotiation, review and implementation, and process of contract change.

⑤ To collect, settle and study recently built in LNG engineering economic information and data at home and abroad, understand and grasp the estimation and evaluation method for project investment, provide reference opinion on economic evaluation and decision-making.

4.5.2.6 Control Department

The category is totally responsible for the schedule and expense control and quality supervision.

(1) Schedule control.

① In the stages of feasibility study, FEED design, preliminary design and the detailed design, the department is responsible for the basic materials and data collection, major program and technical route determination, and change management, achievement acceptance and related technical interfaces and the key links of the progress control, etc.

② Be responsible for drawing up the overall schedule of the project, based on it to prepare project schedule sheet and establish the key route map; using the professional control software, decompose the work unit to WBS.

③ In the different stages of the project bidding pre-qualification, responsible for checking contractor's the schedule plan.

④ In the implementation stage of the project, carrying on dynamic analysis to the labor hour, workload and work result, tracking the overall schedule plan of the project and drawing the schedule S-curve, and also tracking the schedule plan of the contractor's project and controling it.

⑤ To offer the project schedule plan to the technology department, prepare the cost and quality management plan, and participate in schedule management and coordination of work related to the project.

⑥ To organize and participate in the analysis of schedule change, coordinate and help to bring about contractors or manufacturers cohesion in between.

(2) Cost control.

In the stages of design, procurement, prefabricating, construction, commissioning and acceptance, the department is responsible for timely accurate cost control, support, analysis and supervision in the receiving terminal and gas trunkline project, so that the project can achieve the all-round control to the project expense.

① Be responsible for project cost estimation and budget plan, preparing the cost control scheme, making promises of payment management and account clearing.

② To offer the all aspects of expenses of the owner and EPC contractor, provide overall cost management and the estimated service; put forward cost control policies and procedures recommended.

③ Based on the WBS, to establish project cost breakdown structure (CBS).

④ To set up and run the project cost control system, proposal the budget which reflects the project plan, including remedial action plan, forecast the cost trend and the influence of potential change, guide and supervise the invoice procedure of contractors and suppliers.

⑤ To audit the invoices of contractor and supplier, sign payment suggestion; track and supervise expense the deviation between actual expense and budget or forecast expense, report the changing situation, offer the relevant suggestions.

⑥ To evaluate and coordinate the change affairs, advocate and keep the project original budget goal, offer the database and materials used in all stages of the project.

⑦ Be responsible for the whole expense management of contractors and other interfaces.

(3) Quality management.

According to the engineering quality management standards and norms, keeping all process quality management.

① According to the whole requirements and characters of the project, to study and draw up the documentary of project quality guarantee, quality control procedure and project management, supplementing, revising, perfecting and updating timely.

② To assist each sub project manager to track, check, prevent, correct the management of quality in the process of project.

③ Be totally responsible for quality management of the processes including unit each engineering bidding, design checking, contract execution, and construction, implement, debugging and operating of the jetty, receiving terminal, and gas trunkline; coordinating and dealing with related work interfaces.

④ In the pre-qualification of project bidding, responsible for qualification evaluation on contractor's quality system and program files; in the project implementation stage, responsible for reviewing, tracking each contractor's quality plan, program files.

⑤ Be responsible for its application in the process of interpretation, training, supervision, inspection work, ensuring the effectiveness of quality control procedures and management regulations and traceability of the records and files.

⑥ Be responsible for cultivating and improving quality awareness of all members and quality management training.

⑦ To understand the new development of quality management at home and abroad, in accordance with developing requirements of the project company, set up quality system and carry on certification work timely.

4.5.2.7 Finance and Accounting Department

(1) Drawing up planning.

The department organizes to draw up medium-long financial planning, annual plan budget and plan adjustment, organizes to develop the company financial management and accounting policies and system.

(2) Financial analysis and report.

The department organizes the company accounting work; dynamically analyzes investment and running activities from the aspect of finance, draws up the financial statements and report, so that to provide financial information support for the leadership of the company decision-making.

(3) Fund raising and financing.

The department organizes and draws up the company capital plan, responsible for the fund raising, fund revenue and expenditure, debt and risk management, organizes and draws up the financing strategy and the implementation of the program.

(4) Insurance and tax management.

According to the business characters of the company, the department formulates and implements the insurance strategy and plan, including insurance and claims etc.; responsible for the company's tax management and tax planning and tax management, according to related laws and regulations, pays tax.

(5) Auditing.

The department is responsible for assisting the shareholders and social firms to audit the project company. According to rectification from the audit report, straighten out the relationship among the government, shareholders, creditors and customers.

(6) Involving in purchase activities.

According to the limitation of financial management of the company, the department checks and approves accounting vouchers, payments and expense reimbursement, takes part in the decision-making process of the buy and purchase activities by the actual requirement, participates in major business contract financial part of the negotiations, guidance and coordination.

(7) Finance control.

The department participates in the project company's schedule, quality inspection and performance appraisal.

4.5.2.8 Business Department

(1) Planning.

The department is responsible for drawing up the work system and management procedures of the department, formulates acquisition strategy, purchasing plan and business documents.

(2) Market research.

The department carries out the market survey of goods procurement scope and procurement implementation at home and abroad, implements supply and determines the degree of competition in the market.

(3) Bidding purchase.

The department formulates the biding method, biding time, contract model, determining the package numbers, acquisition method (OFE, CFRE, CFE), the type of contracts, and in accordance with the provisions of the preparation of procurement plans for approval, responsible for organizing the implementation of the construction project procurement division of work, procurement package; ensuring the reasonable price on acquisition of equipment and materials, providing materials needed to meet the quality requirements of the standard of construction project on time.

(4) Contract management.

The department is responsible for negotiation, assignment, execution and management of purchase contract related in the project; responsible or participating in the purchase activities included in the project technology service contract and implement contract.

(5) Coordinating management.

The department assists engineering and technical personnel to implement and determines the owner's procurement scope and general contractor's procurement interface; responsible for expediting technological information, goods delivery and technical service, organizing goods receiving, handling customs clearance, commodity inspection, cargo insurance, inventory storage inspection, claims processing and completing the contract cost settlement with the finance department.

4.5.2.9 HSE Department

(1) Establishing the system.

The department organizes and draws up HSE policy, regulations and standards of the project company, responsible for promoting the setting up and implementing of HSE system.

(2) HSE management.

The department carries on the evaluation of environment affect, occupational disease and safety, and "Three Meantime" of safety and environmental protection management; writes or involves in reviewing the HSE contents in the LNG project feasibility study report.

(3) Routine management.

The department presides over and studies the technological problems of safety and environmental protection with the technologic departments of project, checks and draws up safety and technology measures. In the process of implementation and operation, finds hidden dangers of accidents, to formulate and implement rectification measures, organizes the training company employees on health, safety, environmental protection, improving employee health, safety, environmental awareness. According to the regulations, the department is responsible for the health, safety, environmental accident statistics, investigation, analysis, reporting, rectification and processing.

(4) Contractor management.

The department checks and examines contractor's HSE system and rules and regulations, pays more attention to site HSE management, improves contractor's HSE operating level.

(5) Certification management.

The department applies and obtains for the business license, dangerous chemical production running license, and keeps the validity of the licenses, responsible for the registration of dangerous chemicals and coordinates with government sectors related HSE work, implements the scientific and technical research related to health, safety of the project.

(6) Risk control.

The department organizes and takes part in hazard identification, risk assessment and risk control, and responsible for daily work of safe and health management.

(7) Communication and coordination.

The department communicates with government competent department positively, coordinates the related HSE affairs.

4.5.2.10 Production Preparation Department

The department is set up in the middle and later periods, it will have more advantages if it can be set up in the starting period through the project actual running. The department is responsible for the implementation of organization, employees, and management system of transitional period from trial production to normal one.

(1) Plan arrangement.

The department is responsible for HR and recruitment plan in the production preparation and trial production period, cooperating with the personnel administrative department for personnel recruitment.

(2) Production preparation.

The department is responsible for drawing up the production rules and regulations and running operating manuals of the project company, according to trial production running, revises and improves them.

(3) Training.

The department draws up the training plan of new employees; assists in organizing employee of LNG project post training, on-site training and certification training at home and abroad.

(4) The work handover.

The department involves in the acceptance of project machines, pre-commissioning and the supervision of quality standard and safe production, environmental protect, occupation safe sanitation supervision, lighting protection, pressure tank and fire protection; responsible for work taking over and connection with the department of project in the later period.

(5) Commissioning.

The department takes part in pre-commissioning, commissioning, trail production and operation adjustment in the period of commissioning adjustment, according to the actual situation, put up with related suggestions and measures to coordinate related department's improvement.

4.5.3 Organization Structure Adjustment in the Commissioning Stage

The commissioning stage is the last one in the life cycle of LNG project. Viewing from the several LNG projects in china, it exists the alternation period of trial production operation and engineering construction, such as the construction of the tank no.3 and 4 in the project of Guangdong Dapeng and Fujian LNG, at this point, the project company in terms of organization structure and human resources will be adjusted and changed.

4.5.3.1 Changes of Organization Structure

(1) Increasing power and function of production department.

In order to keep the two functions of commissioning and project construction, except the departments of original project in construction period, it must increase power and function of the production department which is formed on the basis of production service, the department should improve the operation function of receiving terminal and gas trunkline.

(2) Expansion and early intervention of the original function of the department.

Due to the trial production operation and construction in the same time, the original functions of some departments need to expand. Firstly, HSE department, in addition to the HSE management function in construction, it is needs to add the HSE management function with commissioning. Secondly, the market department should change its static management to dynamic one. Thirdly, resource transportation department needs to be changed from plan arrange to implement. Fourthly, HR department should change from salary, social insurance management to recruitment, training, stimulation stage. Lastly, the republic relationship department should pay more attention to communication with all users instead of related government departments.

(3) Canceling some original departments.

Adjusting and canceling the original departments' duties and sub project group functions are closely related to the completed sub projects. For the LNG project, after the jetty sub project completion acceptance, the original department will be adjusted partly and canceled, the same as the jetty, the storage tank and trunkline sub projects, some responsibilities will be adjusted and cancelled, and the land acquisition office will be canceled too. Taking the Guangdong Dapeng project as example, many functions under the construction have been transferred to the late new technical service department.

4.5.3.2 Re Allocation of Human Resources

(1) Some staffs Work adjustment.

Due to the completion of the projects, some departments' responsibilities will be adjusted or canceled, the existing staffs of the work will be adjusted too. For example, in the engineering department, some personnel will enter the operation department through the training, and some will be allocated into other project departments or comebacks to parent company.

(2) Strengthen training.

Before the period of pre-commissioning, with the adjustment of the organization structure, recruiting new staff for the operation department is an urgent and arduous work. According to the needs of production and operation, it will be recruiting the right staff filled to the new posts. Whatever new staffs in new department or old staffs on new occupations, they are all needed to be trained. The HR department should focus on the training to the above staffs by drawing up training plan and contacting training firm.

(3) Salary and welfare plan.

With the new department and occupation, salary and welfare directly influence employee recruitment and steady extent. During the production trial operation period, some personnel's work place is in the remote area, such as some on duty's station staff, some on shift personnel, if the project company ensures the higher level of compensation and benefits in the same position within the region, which will benefit the quality of staff and work at ease.

4.5.3.3 Project Company Organization Adjustment

In accordance with the dual tasks of commissioning and project construction, the organization structure of project company will also make adjustments, and the scheme is suitable for the period of operation and phase Ⅱ construction.

(1) Department setting.

Taking Fujian LNG as an example, the company has set up 13 departments (Figure 4–4), there are several responsibilities expanded than previous organization structure, such as: General Manager's Office, Finance Department, Resource Marketing Department, Government Affairs Department, HSE, Business Department, Engineering Department and Operation Department(On the basis of production preparation department to add function and personnel), and new departments are: Logistics Center, Dispatching Center, the Monitoring Center Project Team, Tank Project Team, Phase Ⅱ trunkline networks project team. The company sets up 189 positions, post: 389.

(2) Department responsibility division and position Management.

① General manager's office. It is a comprehensive management department, serve for the leaders and other departments, its main responsibilities are sectary, information and human resource management, party and masses'work, and also part-time functions: the management of daily work of the board of directors and labor and legal affairs of the company. Position: 11, post: 12.

② Logistics center. It is responsible for the logistic management and service, and the main functions are managing property, administration and security, and managing and maintaining the non-production buildings and equipment. And also part-time functions: estate and catering contractors'management. Position: 11 ; post: 12.

③ Finance department. It is responsible for budget planning and financial accounting management, main functions are planning, budget, accounting, capital, asset, tax and insurance management. Position: 16, post: 16.

④ Resource and market department. It is responsible for LNG resource, transportation, market management and development, and main functions are resource and transportation management, product sale and market development. Position: 15, post: 15.

⑤ Dispatching center. It is responsible for production scheduling. Main functions are: production planning and statistics, production monitoring, user's coordination, and also part-time functions: the company emergency treatment center. Position: 4, post: 7.

⑥ Operation Department. It is responsible for receiving, storage, gasification and transmission. Main functions are: the production technology management and equipment technology management, jetty coordination, vessel affairs management, pipeline protection, production and running organization, electric instrument maintenance, equipment check and repair, and production area building management. And also part-time functions: production and repairing contractors'management. Position: 48, post: 209.

⑦ HSE department. It is responsible for quality, health, safe (security guard), and environmental protection and standardization management. Main functions are: quality management (ISO 9000), occupational safety and health management (OHSMS 18001), environmental management (ISO 14000), standardization management (GB/T 15496), social responsibility management (SA 8000), security management, and also part-time functions: fire control management and public security management. Position: 8, post: 9.

⑧ Business department. It is responsible for legal affairs, contract and material management. Main functions are: legal affairs, contract management and material purchasing accepting, custody and checking management. Position: 11, post: 16.

⑨ Government affair department. It is responsible for coordinating the affairs related to the government. Main functions are: government affairs coordination, public relationship management and certificate management. Position: 5, post: 5.

⑩ Engineering department. It is responsible for the project planning and construction management. Main functions are: project planning and pre-engineering planning, engineering preparation, construction, and archives management. Position: 11, post: 22.

⑪ Tank project team. It is responsible for new tank project management. Position: 16, post: 30.

⑫ Monitoring center project team. It is responsible for management of monitoring center project. Position: 16, post: 20.

⑬ Phase-Ⅱ trunkline networks project team. It is responsible for management of phase-Ⅱ trunkline networks project. Position: 16, post: 18.

(3) Organization structure.

According to the above departments set up and HR arrangement, the organization structure chart can be seen in Figure 4-5, as reference for other companies.

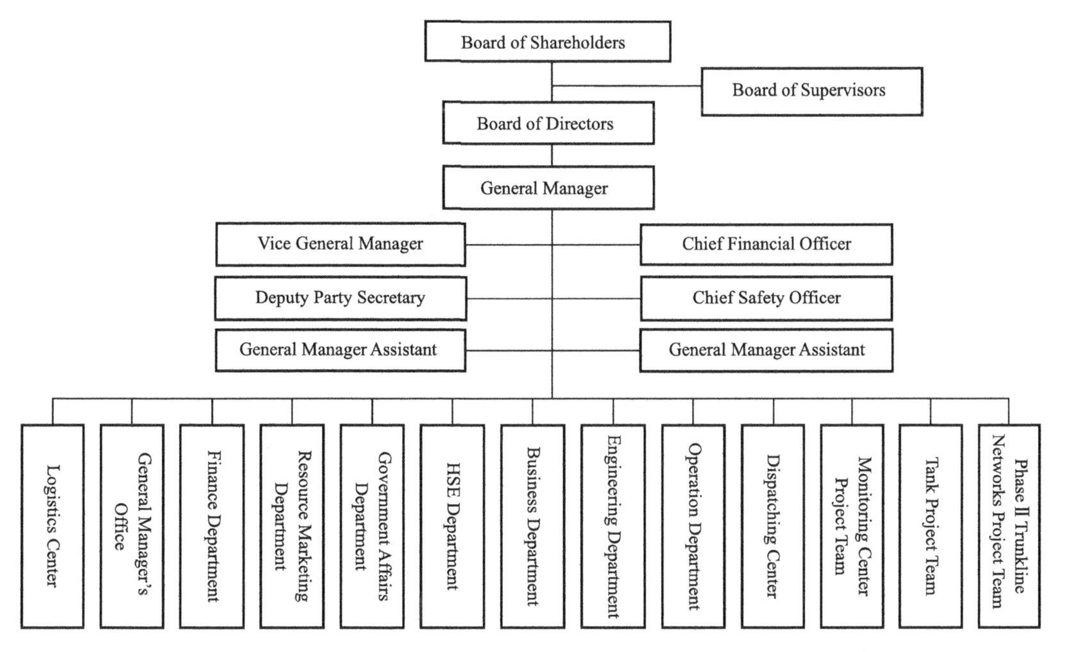

Figure 4-5 LNG project organization structure in trial production and operation adjustment stage

4.6 Corporate Culture Construction of the Project Company

4.6.1 General Concept of Corporate Culture

4.6.1.1 Definition of Corporate Culture

Corporate culture refers to the practice of enterprise in the market economy, and gradually formed by all staff for compliance, with the values of the enterprise characteristics, the operating guidelines, management style, enterprise spirit, business objectives, ethics, has followed the tradition and habits, etc.. It is also summarized into three levels: the concept level, material level, system level.

Organization framework, rules and regulations are important, but rooted in the heart of the enterprise employees values, ideas and behavior is the long-term survival and development of enterprises eternal power. LNG project companies in China are new enterprises, its culture is particularly important.

4.6.1.2 The Role and Significance of LNG Project Company's Corporate Culture

(1) Establishing behavior norms.

In reality, we find that some employees work resolutely, seriously and responsibly, while some employees are sluggish and inefficient. The reason is that corporate culture factors in the role, as a new company, we must firstly form the good behavior norms, and become the conscious action of employees. In the whole LNG project operation stages, with the attitude of ownership, work diligently and conscientiously, bear hardship without complaint, down-to-earth, step by step, do solid work steadily.

(2) Establishing moral standards.

LNG project management focuses on the professional staff, to open up a new path of clean energy, the benefit of mankind's creative activities. To achieve this goal, only by the mandatory rules and regulations is not enough, we must rely on the moral force to guide employees' thoughts, and consciously tighten the string of anti-corruption. The new company must firstly establish good moral standards. Moral power is not abstract, empty moralizing, and it will be conducive to the realization of the goal of project company.

(3) Promoting harmonious development.

The economic construction of our country is still at a stage of rapid development in the background of the global financial crisis, and energy efficient and environmental protection are the foundation of harmonious society. LNG project company is responsible for providing clean, efficient, environmental protection energy to the users, undertaking the significant social responsibility, also shouldering the glorious mission of providing value to shareholders. Harmonious development is the needs of the country, the needs of the community, also is the needs of the

staff. In a word, the corporate culture construction is a long way to realize the harmonious development of enterprises.

(4) Forming a common vision.

A project company is a team, only its members work efficiency is greater than the sum of their individual work efficiency, can be said that team spirit is embodied. If you want to achieve this, building the common vision of team is the only choice. LNG project company vision, that is under the human, financial and material conditions, to ensure the project put into operation as scheduled, make up for the lack of resources, change the regional energy structure, for the benefit of local people. This is a great career, a glorious career, the courage to create a daring and prospective new business, complied with the time development demand. LNG project company team only heart to a thought, make an effort to form a cohesive force, can make each individual contribution to the realization of a common vision.

4.6.1.3 The Characteristics of the LNG Project Company Corporate Culture

(1) The new corporate culture.

LNG project is new business in Chinese, the project company has set up a short time. The longest Guangdong Dapeng and Fujian project company is only 20 years, while other LNG projects such as Shanghai, Zhuhai, Shenzhen, only 7 ~ 1 year, especially the project into the trial production stage, at the same time, the most employees recruitment, and hire employees are just out of college and graduate students. Corporate culture construction has just started, most have a prototype, and the corporate culture has a long way to go, so the new corporate culture is the most significant characteristic.

(2) The variety of excellent corporate culture integration.

From existing LNG project companies, at the very least, composed of three parts: one is the initiator shareholders dispatched personnel, and the second is personnel sent by local enterprises shareholders, the third is project company recruitment of personnel in the society, including social personnel and new undergraduate and graduate students. Due to personnel sources, will bring the corporate culture characteristics of original enterprise unit, if it is a Sino-foreign joint venture projects, also brings the foreign culture. Different sources of culture sometimes will resonate, but sometimes there will be a collision and conflict. We cannot simply comment on what kind of corporate culture is superior than other cultures, it should be said that each culture has its own characteristics and advantages, through all kinds of background the collision and fusion of culture, the source of various cultural advantages and characteristics of centralized or reprocessing, continue to improve and perfect, can eventually form their own unique culture.

(3) The collision of Chinese and foreign cultures.

Under the jurisdiction of the CNOOC LNG project companies, Guangdong Dapeng LNG project company is working with BP as the operation of the company, Sino-foreign parent company's culture is remitted to the new company, with their respective development of social history. On the one hand, it may mean that Dapeng's corporate culture carries on the excellent culture inheritance of its own national history, but on the other hand, it will also have conflicts with western corporate culture, which will bring invisible pressure to work with foreign employees in the Chinese regional market conditions. The solution can only be more Chinese and foreign staff communication, understand each other norms and principles, overcome their pattern of behavior that cannot be accepted by the other party in the Chinese market operating environment, create new culture of Sino-foreign joint venture company.

4.6.2 The Establishment of the LNG Project Company Corporate Culture

LNG project company corporate culture establishment cannot be achieved overnight, but under the guidance of project manager and through the joint efforts of all staff, accumulate over a long period, and gradually improve. Its formation process is shown in Figure 4–6.

4.6.2.1 The Corporate Culture Construction Contributors

(1) The decisive role of PGM.

In the process of corporate culture construction, the general manager of a project company is a person who has a set of rights and responsibilities, the general manager's influence is huge, often the general manager values, management style, moral standards, will affect the staff around him consciously or unconsciously, general manager advocates what, likes what, hates what, will be passed to the subordinates, and gradually form the corporate culture, will play a decisive role to form the corporate culture. Therefore, investment initiators select the candidate of qualified general manager with higher ideological, moral, technological, management qualities is the key to form the excellent corporate culture of the project company.

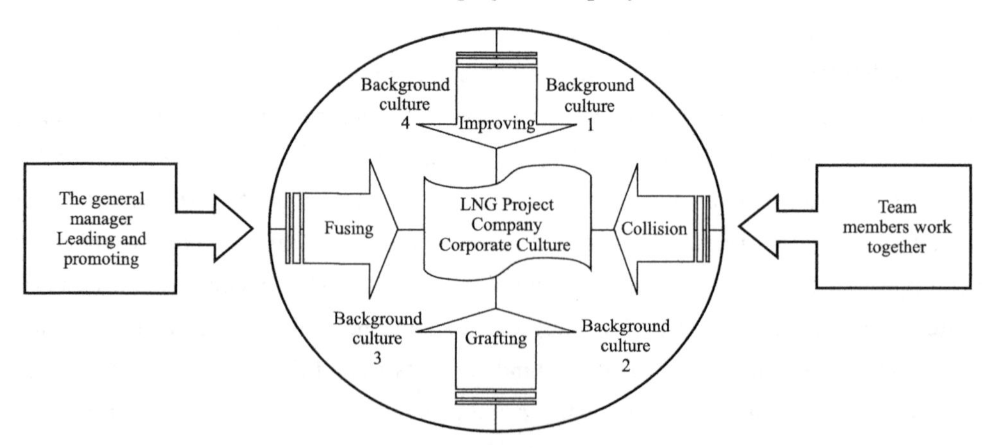

Figure 4–6 LNG project company corporate culture forming process

(2) The role of other staffs.

In the process of corporate culture construction, in addition to the leading role of the general manager, also cannot ignore other leaders of each level and the general staff. The cultural idea of the general manager depends on the levels of leadership to implement, to enrich and develop, but also depends on general staff to perform and support. Sometimes, the lower the cultural idea of corporate culture idea may be formed to "The fire burns high when everybody adds wood to it" effect.

(3) The role of the parent company.

In the process of corporate culture construction, the parent company, especially holding company's corporate culture, will affect to form a new company culture by sentting general manager, at the same time to play the role of deviation correction and strengthening the construction of corporate culture, the other parent companies will play a corresponding role to the new company culture.

4.6.2.2 LNG Project Corporate Culture Construction Steps

(1) Culture audit.

The above mentioned LNG project company personnel from a wide range, in order to set up the company's corporate culture prototype in a relatively short time, we must clear each bringing different cultural background, which is consistent with the new project company's goals and values, which deviate from, it is a cultural audit stage which the problem must be solved. The second is how the LNG project company general manager orientation on corporate culture? How the high-level contribute to the enterprise culture? How the middle and the general staff identity of culture? How the new company's social responsibility, organizational goals, the staff demand, business philosophy, team spirit status, etc.

Culture audit can be responsible for the competent administrative department of the project company, through the questionnaire survey method, interviews, seminars etc.; also can employ social consulting institution to do so. The achievement is corporate culture present situation investigation report.

(2) Cultural design.

Based on an investigation report on the present situation of corporate culture, and can put forward several design ideas, design scheme of the new corporate culture, through the staff to discuss and feedback, by the project general manager confirms final approval. Its contents include: concept level, material level and system level.

(3) Culture implementation.

According to the cultural design scheme selection, it will be implemented in project organization, project construction and project operation, generally divided into the following stages.

The first stage: to establish their own culture in the foundation of the project company culture basic blank, there is a process of filling in the blanks and practice. The time for the project start to the project put into operation before the acceptance.

The second stage: summing up. Through the joint efforts of all staff and practice will find their deficiencies, to modify the original program, to achieve all the degrees of approval. Generally fits the project final acceptance.

The third stage: the corporate culture accepted by most of the staff, to accelerate the development of the new company, also in the community have a certain influence, then there is a relative curing period. Generally it is consistent with the project company to enter the normal production stage.

(4) The corporate culture cycle.

The above three stages is a cycle of culture construction. After practice or new general manager takes over, it will be revised and innovated to the original culture, so a new cycle begins again.

Some famous enterprises both at home and abroad, their long-term existence and development is an important reason why there are a good corporate cultures. The formation of corporate culture is not once a day, but needs several generations of leaders to advocate and the common effort of the staff. LNG project as the newly established company, in addition to a cultural framework of the early prototype, more need to have a long-term process of accumulation and perfect. Only in the practice of the daily work, absorbing essence, eliminate bad piece, in each operation and action, extraction and selection step by step, can eventually form a distinctive LNG project company corporate culture.

4.6.2.3 Introduction of the Prototype of LNG Project Corporate Culture Construction

(1) Concept level.

Concept level includes project company enterprise mission, purpose, values, management philosophy, organization spirit, ethics, business ideas, etc.

① Guangdong Dapeng LNG project company's mission: to provide China with clean, efficient, reliable and secure support for rapid economic growth.

② Fujian LNG project company's mission: to offer the mostly clean energy, to create a better life.

③ Fujian LNG project company's core values: respect for employees, happy work, interactive growth, dedication to the community.

④ Guangdong Dapeng LNG Project Company's vision: to become a well-known international energy company, beyond the stakeholders, including shareholders, the government, employees and the community expects of us. As the leader of China LNG industry, we will adopt the world advanced technology and management, with our rich professional knowledge, keen business opportunities and rapid market response is famous in the world.

⑤ Fujian LNG project company's goal: To uphold the business principles of legal, justice, equality and mutual benefit, on the basis of competition and profit operating LNG and other clean energy, to promote the local energy consumption structure, adjustment and sustainable development of social economy, and get commensurate with its risks assumed ideally each shareholder to obtain economic benefits and investment return.

⑥ HSE policy is the central theme of the project company HSE management system, must have the correct policy guidance. The followings are a recommendation of the HSE management policies:

a. Safety first, prevention first, prevent the occurrence of major accidents

b. People-oriented, care for life, ensure employee's health and safety

c. From good science, follow the rules and regulations, provide quality services

d. Saving energy and reducing consumption, prevent pollution, protect living environment

e. Full participation, continuous improvement, striving to be first-class enterprise performance

⑦ Guangdong Dapeng LNG company is Sino-foreign joint venture company, composed of 11 shareholders, in order to promote integration of corporate culture, Dapeng President Thoms M.King puts forward "One Dapeng" management idea. After several years of Chinese and foreign personnel running in, form their own corporate culture core. Figure 4-7 shows Guangdong Dapeng culture framework.

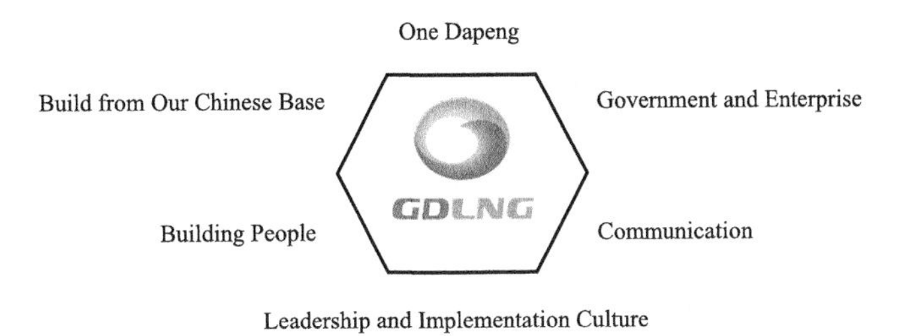

Figure 4-7 Guangdong Dapeng LNG Project Company Culture Framework

The framework is listed "One Dapeng" as the core of the Dapeng 6 principles. For a number of shareholders of cross-cultural enterprises, the formation of "One Dapeng" enterprise culture is particularly important, embodies the Chinese and foreign equality, cooperation, mutual benefit and win-win. Government support is the guarantee of enterprise success, provides users around the clean energy, contributes to the government energy conservation and emissions reduction. Chinese and foreign employees should strengthen communication, it is the premise of understanding and learning from each other. Only communication can eliminate misunderstanding, to reach a consensus. In order to make Guangdong Dapeng LNG Company bigger and stronger, we must have a strong leadership and hundred-percent execution culture, to the smooth implementation of government decrees and come to the success. Employees are valuable wealth to the Guangdong Dapeng LNG company, set up "One Dapeng" award in the company, make outstanding contribution to employee motivation. Guangdong Dapeng is located in Chinese, from the national conditions of Chinese, respect Chinese outstanding traditional culture, into the foreign advanced management concept, a joint venture to create a new culture. It should be said that the implementation of One Dapeng culture, will promote the projects, which plays a great role in promoting the efficiency.

(2) Material level.

Material level includes the company logo, standard color, standard architectural style, standard medals, commemorative book etc.

① The company logo: such as Fujian LNG project company, invested by CNOOC and Fujian development corporation, take the two parent company's logo as a company logo, reflected in the two shareholders, facing the common risk and development (Figure 4–8a). Guangdong Dapeng LNG Company for many shareholders, the Sino-foreign joint venture company was set up, which formed a harmonious corporate culture, and the logo is shown in Figure 4–8b.

(a) Fujian LNG logo (b) Dapeng LNG logo

Figure 4–8 LNG Project Company Corporate Logo

② Company standard color generally is dominated by striking colors, such as blue or red as some companies' standard color, the company uses standard color uniformly prints in covers of documents and reports, embodies the unique company tone.

(3) System level.

System level includes the company system, method, and model (Table 4–1). The rules and regulations is the project company of various functional departments according to the scope of their duties, carry out the normative documents described self-rights and obligations, the project company according to their own characteristics, can develop their own, there is no uniform model, the following is a classification of some LNG project company regulations.

Table 4–1 LNG Project Company Rules and Regulations Classification Table

Department	Content	Remark
The Corporate Level	The procedure rules of board of shareholders, board of directors and board of supervisors. The articles of association of the company, general manager's working rules, general manager's meeting system	
Administration Department/Office	Office, secretarial, confidential, secret, letters and visits, seal management regulations, employees wear and behavior specification during work time etc.; the company foreign affairs management regulations; Staff dining, welfare, housing management regulations; Public relations management regulations; Document management regulations	
HR Department	Staff health management approach, the provisions of major epidemic management; Safety and environmental protection management regulations; emergency plan; accident analysis report template; Provisions on environmental protection	
Business Department	Pre-qualification management, Tendering regulations, model, performance guarantee model; Opening agenda, open label word list, quotation, bid evaluation method, evaluation, the bid evaluation report template record; The bid winning notice template, contract approval regulations, contract models; Procurement regulations	
Government affairs Department	Regulations for the government department examination and approval	

Continued

Department	Content	Remark
Marketing Resources Department	The measures for the administration of market development; LNG resource contract model; transport contract model	
Finance Department	Measures for the management of financial management way, funds, accounting rules, tax insurance regulations; Accounting vouchers, account books, reports and statements and other accounting archives management regulation	
HSE Department	Employee health management method, major epidemic management regulation; Safety and environmental protection management regulations; The emergency plan; Accident analysis model; Environmental protection regulations	
Engineering Department	Provisions on the administration of bidding documents, contract template, land acquisition and project management; Provisions on the administration of design, construction, supervision, completion and acceptance of project construction; Provisions on the management of the purchase of equipment and instruments for the construction of the project; and the provisions on the administration of the installation and commissioning	
Control Department	The regulations for the schedule control; Quality management regulations; Cost control management regulations	
Operation Department	The regulations of human resource management during operation; Operation rules and production operation manual; Ensuring the operation of the contract template; Late work handover and cohesive arrangement regulations with engineering department	
Archives Management Room	The provisions on the administration of the original materials, such as text, tape, disk, audio and video, which are involved in the project; Archives library management process; Provisions on the administration of archives destruction	

4.6.3 The Main Content of LNG Project Company Corporate Culture

Different LNG project companies have different corporate culture, there is no fixed corporate culture model, and LNG project company generally respected several corporate culture elements as follows.

4.6.3.1 Honest and Trustworthy

The honesty is the original meaning of "true, real", that is honest, sincere, true justice, neither the old nor the young will be cheated. Faith is the original meaning of "truth-seeking, keep sincerity", namely the pursuit of truth, stick to the truth, keep its promise and abidioy his agreement.

The first is putting loyalty to the country, to establish a reliable, responsible corporate image, in order to ensure the authenticity of information disclosure, abide by the laws and regulations, adhere to the legal business, pay taxes according to law, maintain the social stability and contribute to local economic development.

The second is putting loyalty to the contractors, all in accordance with the contractual rights and obligations to do business, to create safe construction environment for contractor's frontline staff, and make concerted efforts to the completion of the project, if the contractor is in violation of the law, whom shall be punished by law and contract, maximize the interests of both sides as the premise.

The third is putting loyalty to the suppliers, in the sales contract as the basis, to fulfill the obligations of the owners, on-time payment, provide valuable information to the supplier, and to cultivate long-term cooperation relationship.

The fourth is putting loyalty to the customers, in accordance with the sales contract, to supply gas or LNG with quality and quantity and on time. If force majeureacurs, inform customers honestly with mutual understanding and mutual love, work together to overcome difficulties, don't take the resource supply as bargaining chip, to damage user's interest, and continuously improve service quality for customer's satisfaction.

The fifth is putting loyalty to the employees. To maintain staff and staff's legitimate rights and interests; employees also put loyalty to the company, provide personal information truthfully, responsible to perform, be loyal to their duties, keep commercial secret, don't do anything harmful to company, always consider for the company, grow together with company.

4.6.3.2 People-Oriented

The former general electric CEO Jack Welch is a master in the field of business management, his motto is: "All we can do is to bet on our choice of people. Therefore, all our work is to choose the appropriate people." Former president of Tsinghua University, Mr. Mei Yiqi words "University, non building that also, but the master of that also."

The project company will take employee's development in an all-round way as one of the goals of enterprise development. And through the management of human nature, cohesion of staff management and the need for staff management, to achieve the common development of employees and the company, the goal of the both is the integration and interdependence, that is, the company will be the development of employees into its own goals, the staff will also be the company's development into their goals.

"People-oriented" as a major goal of the corporate culture is the employee satisfaction. Through the satisfaction of employees to achieve external customer's satisfaction, obtain the company's competitive advantage. The influence factors of job satisfaction include: the work itself, fair wages, supportive work environment, harmonious relationship among colleagues, and personal character and ability matching with work. The enterprise only forms the "people-oriented" corporate culture which can really put people in the center position management, pay attention to employee needs, to meet the needs of employees.

It has always devoted to the study of the establishment of an effective incentive system, external competition and internal fairness of the overall pay levels, improve employee satisfaction and loyalty to enhance staff morale and efficiency, so as to increase the company's benefit and value.

It should be to establish an effective training system, there are plans to carry out the training of employees, improve their job related knowledge, skill, ability and attitude of quality, in order to adapt to the need for and now the survival and future work needs and organizational development. Improve the quality of the staff at the same time improve the value of employees.

It should be to establish a full range of welfare, security system and the relief mechanism concerning the workers and groups with difficulties, to establish physical examination, vacation system, to carry out recreational and sports activities in a variety of forms, to emphasize the HSE, pay attention to employee health and safety.

4.6.3.3 Innovation

Famous American economist J.A.Schumpeter thinks innovation is essential to development. Innovation is the individual according to certain goals and tasks, using all the known conditions, to produce new, valuable results (mental, social, physical), cognitive and behavioral activities.

Innovation is essentially a positive and effective way of thinking and working methods, and also can be said to be a kind of work ability and attitude towards life. Innovation, needs us not to bound the traditional thinking; innovation, requires us to ignore all the sacred rules of the religious order; innovation, needs to release a spark of thought; innovation, needs to keep on carving of the spirit; innovation, needs to tolerance and courage; innovation, needs to be prepared to fail.

It is required to renew the management concept, break through the traditional management mode, to develop employees with technology, improve efficiency organically; implement target management, clear the owners contractors supervision of responsibility, establish common interests directly linked with effective incentive mechanism, the formation of network management affected each other and connected; apply the modern management science, the "additive effect" into a "multiplier effect", greatly improve work efficiency.

LNG project is new to our country, first of all, we must have the innovation based on introduction, digestion, absorption, rely on progress of science and technology, by supporting technology, seek the best combination. The introduction of new technologies do not seek "leading", but for "the most suitable", and exert its maximum efficiency. According to the actual project practice, to carry on the scientific research, technological reform and technological innovation; from project management perspective, play in one step ahead of the advantages in Chinese, to develop national, industry and enterprise standards, and make due contributions to the development of LNG industry.

4.6.3.4 Win-Win Cooperation

Win-win concept itself is the meaning of: considering the problem from the perspective of interdependence to solve problems, seek solutions of mutual respect and reciprocity. So as to achieve the various stakeholders to pursue consistent success, achieve the mission, share the achievements.

The LNG project company based on the cooperation of all parties on the basis of complementary advantages, operate strictly in accordance with international practice, regulate their own behavior, through sincere cooperation and effective communication, and to maximize the overall interests of the parties, all parties will be the winner.

Win-win and mutual benefit is not the traditional sense of the only making profit or loss with vicious competition, it is at the core of all parties work together to make big "cake", to obtain the larger income unchanged in their respective shares.

The first, the relationship between LNG project company and the country, needs to speak win -win. As the management of state-owned assets, the government is necessary to consider the sustainable development of enterprises, but also pay attention to preserve and increase the value of state-owned assets.

The second, the LNG project company is composed of two or more shareholders, needs to speak win-win. Any investor wants to get a good return, realize capital appreciation. As a joint-stock enterprise, to ensure that the shareholders should have good return is as unalterable principles. Internally, the maintenance of the shareholder's interests will be put in the supreme position. Externally, needs to establish a win-win image, gain the trust of the majority of shareholders.

The third, the relationships among users need to adhere to a win-win situation. To ensure user's energy supply, for the sake of users, to flexibly treat with gas volume adjustment for user reason, to maintain smooth communication channels. At the same time, according to the government's gas prices to obtain compensation, make the project company benefit.

The fourth, the project company and the staff, advocate a win-win situation. Through scientific human resource configuration, enable employees to do their best at work, maximize its advantages and realize their own values, thus to enhance the benefit of the company by increasing work efficiency and value.

The fifth, among the employees, also need to strengthen the spirit of advocacy win-win cooperation. From the point of view, the staffs consider the long-term cooperation in the work for each other. That is the win-win cooperation is a long-term accumulation of mutual aid and win-win, rather than short-term benefit. To bring this spirit into one, form a habit, reflecting in the daily behavior and work.

4.6.3.5 Environment Friendly

LNG project is not as a chip development, manufacturing a new sports car as simple, LNG project involves many aspects of upstream and downstream enterprises and local societies. It must firstly consider the jetty coastal environmental protection and ecological protection, also consider the animals, plants and humanities' environment coordination in receiving terminal area. But also concern for the pipeline along the excavation backfilling and vegetation protection and concern the safety of gas users, city residents and occupation disease prevention etc. So the LNG project company HSE philosophy should become an important content of corporate culture.

Project company should not only consider the operation of the project from the point of view of economy, but also establish a kind of social responsibility, to do with the environment friendly and harmonious development of the society.

4.6.3.6 Team Spirit

Today, people often refer to a famous saying of American steel magnate Andrew Carnegie: "If my entire plant, equipment, materials were burned, but as long as you keep my entire team, a few years later, I will still be a steel king." Henry Ford said: "coming together is a beginning; keeping together is progress; working together is success."

LNG project is a group engineers, alone a person or several people can't complete the task, it must rely on the joint efforts of every person, carry on the reasonable personnel gradient collocation, combinatorial optimization, explore the minimal resource input, to obtain the maximum benefit of enterprise management mode, form a joint force to make the project more successful.

Corporate culture construction LNG project company is another important content of project management. It requires the command role of general manager of project company, needs the joint efforts of all staff, needs the support of shareholders, and more needs a long-term corporate culture to build, exercise, supplement and perfect. At present, Subordinate project companies of CNOOC have a prototype of corporate culture, but the whole corporate culture are also far from perfect. In order to realize corporate social responsibility and economic value, it requires project company hardware engineering quality, more requires the software protection, and the corporate culture is a magic weapon for prosperous LNG project company. From beginning of the project, the general manager of project company should pay attention to corporate culture construction and project construction at the same level. In the joint venture company, taking the employees' sources as the focus, it should be paid attention to the fusion of multi-culture, from the perspective of cultural audit, focusing on cultural design, gradually to establish corporate culture with enterprise characteristics, to create the foundation of corporate culture for a hundred years old corporation.

Chapter 5 LNG Project Site Selection and Land Requisition Management

5.1 Concepts of the LNG Project Site Selection and Land Requisition Management

5.1.1 Relevant Definitions

5.1.1.1 The Definition of Site Selection and Management

Each LNG project is composed by the jetty, receiving terminal and supporting pipeline engineering. The first two sub-project's locations are called the LNG receiving terminal site. In order to select the appropriate LNG receiving terminals, based on satisfied with LNG jetty for ship berthing, unberthing, loading and unloading and jetty channel conditions (jetty part) to select reasonable layout of receiving terminal for docklands, tank farm, process areas, public works, sea water intake and drainage area and administrative area (receiving terminal part), then combined with the natural gas pipeline transportation system and the distribution and network market conditions, through a large amount of reconnaissance and survey work, to construct the matching long distance pipeline or to use of the existing pipeline system, and then by choosing a variety of one-one corresponding to the jetty, the receiving terminal and pipeline layout schemes in different geographic locations of the same area, the experts appraise and elect them and sort them from geography, safety, technical, economic and other factors, in order to determine the best site scheme. This process control is called site comparison and selection management.

In the LNG project, when the market demand condition has been confirmed, and LNG resources, transport conditions have also been available, then LNG receiving terminal site selection is becoming a top priority project construction. However, LNG receiving terminal site is subject to constraint of natural conditions and construction conditions, we should consider the three parts — LNG jetty, LNG receiving terminal, gas pipelines, not only are separate projects, but also are integrated system projects with assistance mutually, interlaced mutually, restraint mutually, and associate mutually. Starting from the whole, in the early stage of small adjustments to the single construction project, will greatly reduce the cost for the project. So in the project construction scheme selection, should plan as a whole to LNG overall project, focus on the optimization scheme of the overall project construction, in order to maximize the overall efficiency of LNG project as the premise, and ultimately elect the best scheme of LNG project site.

Usually this work is different from the feasibility study, it only emphasizes the comparison and selection of nature and geography as the main contents of the construction conditions, which is completed before the LNG project feasibility study, and offers the target site for the feasibility study report.

5.1.1.2 Definition of Land Requisition Management

The land within the PRC territory belongs to the state or collective ownership, no private land ownership.

In order to obtain the LNG project construction land, through legal procedures such as application → review → approval → announcement (1) → land acquisition → announcement (2)→ compensation → land amendment registration and so on, the project company obtains the two kinds of permanent and temporary land use rights. This process of coordination is called land requisition.

Receiving terminal, gas distribution station and block valve chambers need permanent land, and the pipeline construction needs temporary land. For permanent land may be occupied by the state or collective land, some coastal sites may be non-occupied, such as the intertidal zone, beach, reef dam, even some need to through the land reclamation. In general, the lands occupied by LNG receiving terminal and pipeline are composed by above three parts, but the jetty construction needs to occupy the coast, channel and associated waterways, these must be approved by the local government land and maritime administration, to obtain the right to use the land.

5.1.2 Significance and Principles of Site Comparison and Selection and Land Requisition

5.1.2.1 Significance of Site Comparison and Selection

(1) Rational use of land and shoreline channel resources.

LNG project not only involves land resources of permanent occupation such as the jetty, receiving terminal, gas distribution stations and valve chambers and temporary occupation such as pipelines laying, but also involves the jetty, sea coastline and waterways resources, and these resources are very precious to our country. Through the experts site's comparison and selection, from the point of conservation of land and shoreline channel resources, we will use the limited resources wisely.

(2) To promote the rapid development of the local economy.

LNG project often drives downstream users, including the rise of city gas, gas power generation, natural gas chemical industry, fuel industry, LNG filling, cryogenic energy utilization and so on. Such as Guangdong and Fujian LNG project have driven dozens of enterprises production and extension in industrial chain. LNG project site selection will take LNG project as a leader, focus on clean energy advantages, form industrial parks and industrial clusters in a region, make the local industry development rapidly and promote economic growth as soon as possible.

(3) Accord with national environmental protection policy.

As natural gas is low-carbon, clean and efficient green energy, industrial users will not have a negative impact on the local environment, the industry except make contribution to environmental protection, but also will give local residents a blue sky, and promote economic and social benefits in same steps, so as to form a virtuous circle.

5.1.2.2 The Principles of Site Comparison and Selection

(1) Principle of balancing the interests of all parties.

Site comparison and selection involves state, local governments, enterprises and individual interests, consideration should be given on co-ordination, weigh the advantages and disadvantages, especially in the area to consider building a LNG project for the first time. It's need to under the policy guidance of state competent department, deeply investigatation of LNG project initiators and survey of the regional economic development, industrial layout, environmental protection, energy structure adjustment and price endurance factors and so on, offer the study report of the use of LNG project planning, in order to let the experts make a choice based on the report and submit to the state for approval.

(2) Principle of technical economy and the layout optimization.

Site comparison and selection should consider not only the optimum from the point of view of technology and economy, but also consider the economic development from the angle of planning optimal layout. Technology and economy are based on specific LNG projects, planning and layout are based on macroeconomic development planning of the local government, combined with technical economy and the layout, overall consideration, make choice of site to achieve overall optimization.

(3) Principle of system optimization.

LNG gas market and gas consumption decide the initial scale of LNG terminal, and the trunkline direction and diameter, pressure and other design parameters. Jetty, receiving terminal, trunkline and LNG gas market constitute a complete engineering system, receiving terminal site not only is the end of unload LNG carriers, but also is the start of the gas systems to transport gas. Site selection is appropriate or not, depends on whether the system is optimal, the optimal system is mainly reflected in the system provides user's the lowest average gas price.

(4) Principle of gas load center.

Starting from the local overall layout, economic construction and development, planning the receiving terminal site should be selected in the vicinity of the backbone of user's location, in order to facilitate the shortest pipeline route and installation cost is the lowest, and that is the LNG receiving terminal should be constructed in gas market load center.

(5) Principle of prioritize.

In a multi sites selection for ranking, it should adhere to the principle of open, fair, justice, according to expert argumentation, the owners choosing, government approval procedures, the first, screened on a number of basic LNG project construction sites, to select more than two from them, then through the comprehensive comparison, and recommend the preferred site and alternative site.

5.1.2.3 Significance of Land Requisition

(1) Legal significance.

Land requisition should be carried out according to legal procedures. Land occupied includes crop land, non-crop land and marine reclamation land. Only obtain above lands by the law, can we avoid legal disputes later or has the initiative in the dispute, which is the prerequisite for the construction of LNG projects.

(2) Social significance.

By legally obtaining the land use rights, we not only take into account the interests of local government and the farmers whose lands occupied, but also take into account the interest of the project company, especially land requisition, removing and resettlement at communities where the receiving terminal located, can better improve the living conditions of local people, and generally improve the quality of life of the residents, all these can contribute to the realization of a harmonious society.

5.1.2.4 Principle of Land Requisition

(1) Principle of land requisition by law.

The construction of LNG project belongs to the scope of public interest, and therefore, land requisition by the law is the first principle to solve this problem. Currently in Chinese there are "Property Law" and "Land Management Law" involving in land requisition. Land requisition should be in accordance with legal procedures, which is often the most intense conflict, should be according to the national and local policies, without prejudice to the interests of land users, to give the original land user's reasonable compensation, especially the interests of farmers, and make full payment of reasonable compensation.

(2) Principle of serving the overall project.

LNG project involves the land occupied by receiving terminal, jetty, long distance pipeline, and shoreline and waterway resources, from the land requisition work, it must serve the whole project, the key is to grasp the reasonable land expropriation principle, and don't because of a local land requisition delay or difficulty to affect the progress of the whole project.

(3) Principle of saving land resources effectively.

Since china is a country with great population and less arable land, the construction of LNG project should minimize the occupation of agricultural land. Project company should according to the "Property Law" special provisions, pay attention to the special protection of arable land, strictly limit agricultural land turn to construction land, and control the total of construction land, not violate the law and procedures to impose state and collective-owned land.

(4) Principle of environmental protection and beautification.

Construction project can't destroy the local environment, and try to keep the construction of LNG project in a friendly interface with the surrounding natural environment. In the late completion of the project, we should create the project covering an area of within the scope of the green, beautify the environment, and strive to achieve the harmony and unity of the work scene and the natural environment, especially on the open ditch pipeline backfilling and landscaping, to prevent soil erosion.

5.1.3 Site Selection and Land Requisition Management Process

5.1.3.1 Three Parts of Management Process

(1) Basis and organization preparation.

The general flow diagram of LNG project site comparison and selection is shown in Figure 5–1. From the figure, the first is basis and organization preparation, its contents include regional economic planning which are the basic documents of LNG project site selection. Only the LNG project is in line with the local economic development, and promotes the role of local industrial layout, it will be got the support from the local government. In order to carry out site selection, generally the jetty, receiving terminal, pipeline design and research institute with the national Class-A are entrusted to complete the work, this requires the project company to sign a contract with these institutes to do the three single project selection based on their condition. Finally, the overall design and research institutes to carry out the portfolio selection, select the best site, no doubt, the commission contract is the base. Selecting the location of land requisition is the premise of the construction of a project, in order to legal compliance of land requisition, project company must carry out investigation including national and local laws and regulations and procedures of land expropriation. Project company must establish a leading group or strong land office at the beginning, usually the general manager of project company is appointed as the head of the leading group, deputy general manager of project company who is dispatched by local investment company serves as director of the land office, in order to play the role of local enterprises in the land expropriation.

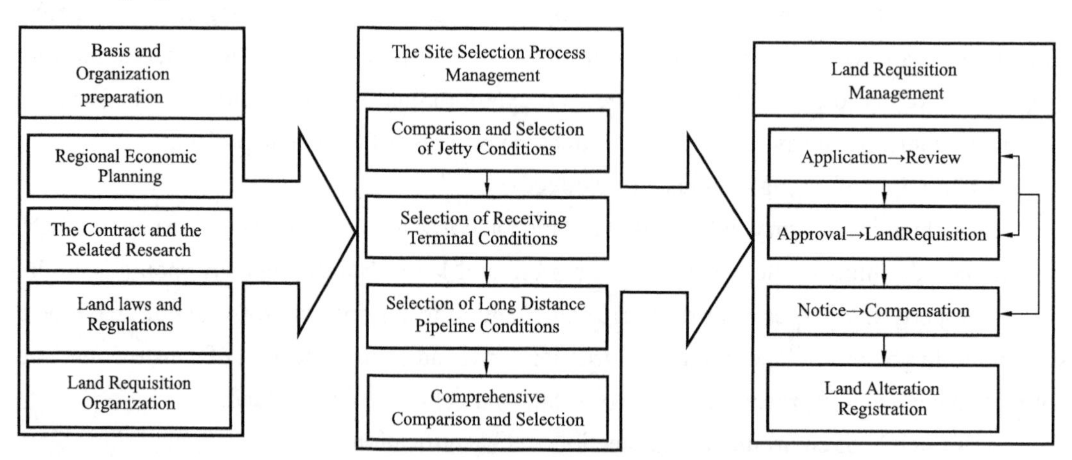

Figure 5–1 LNG project site selection and land expropriation management flow diagram

(2) The project site selection process.

The project site selection process includes two stages. The first is the jetty, receiving terminal and pipeline individually selection stage, selected from their independent conditions, such as Fujian LNG project's jetty location has been selected three alternative ports, receiving terminal also has been chosen three alternative sites, pipelines has been chosen different route lines according to local circumstances, the purpose is to provide alternative schemes for final choice. The second stage is comprehensive selection, which is generally completed before the feasibility study, through expert recommendation, project company proposal, the steps of government decision-making.

(3) Land expropriation management.

Just in this way to select jetty and terminal sites and pipeline route, but it's on the paper work, only through the process of land expropriation, which can realize the real land use right under the LNG project company and the company can start substantive work. While the land requisition process is very complex and lengthy, and through the operation of CNOOC several LNG projects, these all reflect the difficulty of land requisition, and sometimes it even become the biggest bottleneck which affects the completion of project.

5.1.3.2 The Relationship of the Three Components Process

(1) Relationship between basis and organization preparation and project site selection process.

Seen from two parts, the basis and organization preparation and site selection process management, the local regional economic planning frame is the choice of LNG project. Only adapting to the local economic construction pace and scale, can we talk about the LNG project settled in certain place. Laws and regulations collection of land requisition is very important pre-project work which is the data preparation for land requisition. While the commission contract and the land requisition institution belong to the project company organizational preparation, usually it's better to select the employees who understand the local language and culture to participate in the work. Only the basis and organization preparation work have done well, can we talk about site selection work is steadily. According to the commission contract, Design and Research Institute carries out multi-schemes selection for jetty, receiving terminal and pipeline sub projects. If the commission contract does not reflect the project company working intention, there would be no best selection from multi-schemes, also won't get both economical and reasonable, but also environmental protection, not only to facilitate the construction and protection of cultivated land of optimal location.

(2) Relationship between selection process and implementation management of land requisition.

Site comparison and selection is the basis of land requisition, single jetty, receiving terminal and pipeline route selection, are just as the sites selected to provide alternatives, the government eventually determines the site, through land requisition to achieve the use rights of land, coast, and sea areas. If the site selection work is not solid, it is necessary to choose a site again, without doubt which will increase the workload for land requisition.

(3) Relationship between the land requisition management and above two.

Land requisition implementation management is the last link of site selection, and also the actual operation end of implementing the land expropriation. Through implementing of land requisition, we can look back the commission contract signed which is complete and accurate, and through practice, we can revise commission contract which provides better for future LNG project. Land requisition implementation can also reflect the organization if it meets the land requisition workload or not, and if workload is huge, it needs to replenish the manpower, and even the local government should play its greater responsibility for the work of land requisition, and through land requisition, can also see that the site selection process is whether careful and meticulous, in particular, whether the pipeline route is examined in detail along the way. If it does not meet the requirements, it will be unable to implement land requisition, or increase the cost, or not conducive to the future pipeline construction, or serious damage to the environment, and contrary to the principle of land use. Of course, land requisition implementation is a work has a strong policy consideration, heavy workload, mass interpersonal conflicts, and huge social impact, so the project company has to guarantee various resources in order to complete the land requisition.

5.2 LNG Jetty Condition Selection

5.2.1 General Concepts of LNG Jetty Condition Selection

5.2.1.1 Work Scope

LNG jetty is a building that provides LNG vessel for channel, berthing and unloading or transshipment LNG, also is an important component of the receiving terminal, and generalized also includes ancillary facilities. According to the "JTS 165—5—2016: Code for design of liquefied natural gas port and jetty" in the division of the scale of the LNG jetty, China's coastal LNG jetty built belongs to the large jetty (LNG vessel capacity of more than $8m^3$).

Through the selected jetty analysis, such as the natural environment, climatic conditions, construction conditions, ship berth capacity, jetty site construction plan and safety influence and so on, after preliminary economic evaluation and investment estimation, the advantages and disadvantages of each proposed jetty are obtained, for multi jetty comprehensive sites selection to provide the preferred and alternative sites, it is also the premise of receiving terminal and pipeline selection.

5.2.1.2 General Requirements

Generally, the project owner entrust the jetty engineering design institute and the relevant special research institute with the corresponding qualifications to carry out the LNG jetty site

selection work. LNG jetty site should be connected with the local city planning and overall shoreline planning, as a part of LNG project, LNG jetty should be integrated with the layout of LNG receiving terminal. The first, according to the general requirements for jetty site selection, as an independent jetty unit to compare the shoreline, dock, channel, anchorage, meteorology, hydrology, berthing, environmental protection and other conditions, complete the jetty plane and vertical layout and construction economic evaluation, the schemes are recommended as the preferred and alternative solutions.

5.2.1.3 Basis of Documents

LNG jetty site selection should be based on the following documents.

(1) Laws, regulations and standards.

① Relevant national port and jetty construction laws and regulations.

② Domestic and international standards and norms in port and jetty construction.

(2) Relevant planning and research results.

① Long-term national development planning and industrial policy.

② Local social development planning, urban planning, the overall development plan of the harbor area and the sea (or rivers) shoreline using plan.

③ Commission contract.

④ Relevant requirements and documents provided (or approved) by owner.

⑤ Gas market provided by owner and the preliminary determination of the construction scale.

⑥ Terminal scale and vessel type design approved by owner.

⑦ Relevant research results cited by this study.

5.2.2 LNG Jetty Construction Conditions

5.2.2.1 Jetty Location, Surrounding Status and Planning

(1) Port and jetty site selection principles.

LNG jetty includes the main loading and unloading pier and the auxiliary wharf, channel, anchorage, etc., which is an important node in the industrial chain. In order to select suitable sites, some principles should be mentioned as followings.

① LNG jetty location should be coordinated with the overall layout of receiving terminal, which should be combined with the comprehensive determination of receiving terminal location, user layout and external transmission mode. Natural conditions including geomorphology, topography, geology, meteorology, hydrology and earthquake safety in this area should meet jetty construction conditions.

② Natural water depth should be appropriate. In topography, geological changes and water depth too deep and the hydrological conditions of the complex lot, where the natural water is too shallow with coastal sediment movement actively, where maintenance dredging is too large, should be fully justified. The mutual influence between the proposed jetty and the sediment movement should be considered, which can avoid the serious deposition of the jetty and the evolution of the coast.

③ It is suitable to be placed in the place easy for evacuation, and convenient for transportation. At the same time, the water should satisfy the condition of good natural cover, and natural water depth also meet the requirements of LNG ship is never going by tide navigation, if difficult to satisfy, a special demonstration is required. LNG jetty should be selected in the area of better access to sea water use.

④ It needs to consider whether the selected jetty site has the basic conditions with LNG ships entering or leaving, berthing and departing, loading and unloading, such as suitable water area, water depth channel conditions for LNG ship normal operation (or through appropriately channel excavating, harbor basin dredging and breakwater building to meet the requirements).

⑤ Consideration should be given to the mutual influence of LNG jetty construction and operation on the activities of other peripheral facilities. At the same time to consider the number of LNG transport ships and the number of days of continuous operation, to achieve the annual LNG transport, unloading requirements.

(2) Jetty site selection matters needing attention.

① Special jetty site should be connected with the city planning and the overall planning of the port, terminal land and water area status shall meet the government requirements for land and water area planning.

② Safety evaluation should be made for the alternative site locations to the surrounding environment of existing factories, schools, public facilities, underground pipelines, and high-voltage wire on the ground, etc.

③ Special jetty should be far away from the coastal recreational areas and densely populated residential areas, the distance should be determined by safety evaluation. The jetty should not be arranged in upwind side of the constant wind direction.

④ Special jetty should be selected in the convenient transportation, easy evacuation sites. If the LNG jetty or receiving terminal constructed in isolated island, external traffic should be solved, to ensure staff safely evacuate in all-weather condition.

⑤ Without the specialized demonstration, LNG jetty is strictly prohibited to be built in the complex geological structure and unfavorable quake-proof area with late and recent existence of active faults.

5.2.2.2 Natural Conditions

(1) Meteorological conditions.

Meteorological data, such as temperature, precipitation, relative humidity, wind (including typhoon, cold wave), thunderstorms, fog conditions, ice, weather disasters and so on.

① Temperature: including the annual maximum, minimum temperature, and average temperature etc..

② Precipitation: obvious rainy and dry season or not, precipitation concentrated in that month, the annual maximum, minimum, average, daily maximum precipitation and so on.

③ Humidity: including the influence of an oceanic climate, air humidity, the annual average relative humidity, the humidity in summer and winter.

④ Fog and haze: local fog and haze occurred season, the average annual fog and haze days, continuous fog days and haze days, visibility etc.

⑤ Thunderstorms: including the average annual thunderstorm days, thunderstorms occurred season.

⑥ Wind conditions: based on observation data over the years, obtaining the statistics of wind direction and speed, and drawing up the wind rose diagram, and obtaining the data of frequent wind season, wind direction, wind strength and so on.

⑦ Typhoon: typhoon multiple seasons, the largest in the history of the typhoon frequency, the effect of waves, tsunami, tide level.

(2) Hydrology.

Related hydrological data, such as tides, waves (including wind storm and swell), seawater temperature, salinity, silt content, water depth, ice etc.

① The tide: tidal datum relations, tide characteristic values, including the highest, the lowest, annual maximum, annual minimum, average high and average low tide level, average sea level, maximum and average flood tidal range, maximum, average ebb tidal range, average flood tide lasting time, average ebb tide lasting time. Tide design, including design of high and low tides, extreme tidal which includes extreme high and low tide.

② The wave: wave condition, according to the general wave observation data for statistical analysis, such as common and strong wave direction, common wave level, surge, etc. Design wave elements, including deep water and jetty area elements.

③ The tidal current: including periodic reciprocating current fluctuation, the trend of tide, flow speed and so on.

5.2.2.3 Engineering Geology

(1) Beach and underwater terrain.

Through on-the-spot investigation and data collection, it needs to describe the selected LNG jetty site where the beach and underwater terrain, such as muddy or sandy beach or reef beach, or above all kinds of combination of ocean currents, tides, waves; sediment siltation, deposition and variation of shallow marine species, and growth condition seawater cleanliness, coastal stability etc.

(2) Stratum.

If there is engineering geological drilling, it should give the description of depth of rock-soil layers and distribution.

(3) Earthquake.

Regional earthquake zone attribution should be in accordance with the "GB 18306—2015: Seismic ground motion parameters zonation map of Chinese", the location of earthquake activity, destruction records, earthquake intensity and peak ground acceleration to be concerned, and in accordance with the national standard "GB 50011—2010: Code for seismic design of buildings", for building earthquake fortification level analysis. It is also possible to collect large-scale regional geological maps as large as possible, to understand the major faults in the region, especially the active faults and their activity data.

5.2.3 Ship Form Analysis and Berth Capacity

5.2.3.1 Matching Analysis of Ship Form and Jetty

(1) Main dimensions of ship type.

According to the project scale in first phase, LNG transport ship size should be initially selected, at the jetty site selection stage, mainly focus on ship's main dimensions (Table 5-1). Such as, under the prevailing circumstances of Guangdong Dapeng and Fujian LNG project, selected membrane LNG carrier in cargo type, choose the mainstream type with length of 292 m, breadth of 43.35 m, depth of 26.25m, speed of 19.5 knots, and loading capacity of $14.7 \times 10^4 m^3$. For the new project according to the specific circumstances, select the LNG transport ship scale, and should be considered for LNG transport ship development. Please refer to the relevant sections of Chapter 11.

Table 5-1 LNG Vessel Main Dimensions

Loading Capacity	Length L (m)	Molded Breadth B (m)	Molded Depth H (m)	Load Draught T (m)	Note
$8 \times 10^4 m^3$	239	40	26.8	11	
$13.5 \times 10^4 m^3$	280	45.8	25.5	11.9	
$14.7 \times 10^4 m^3$	292	43.4	26.3	11.7	
$16.5 \times 10^4 m^3$	298	46	26	11.5	
$21.6 \times 10^4 m^3$	315	50	27	12	
$26.6 \times 10^4 m^3$	345	55	27.2	12	
3200HP Fire and Tug Boat	45	10	4.5	3.8	

(2) Jetty waters.

Through the initial selection of LNG ship size, it is necessary to make the LNG transport ship into and out of the jetty smoothly, and the jetty should have the following basic requirements.

① LNG carrier braking section advised by linear arrangement in the direction of the jetty. When with difficult layout, can be arranged in a curve, but shall not be less than 5 times the radius of curvature of the design carrier length. LNG carrier braking distance can take 4 to 5 times of design carrier length.

② LNG carrier turning basin should be located near the jetty and convenient for ship docking and departing. Swing diameter of swirling waters shall not be less than 2.5 times of the carrier length. By the water flow impact of jetty waters, the turning basin can be arranged in oval layout, length along the flow direction can be extended to length that is not less than 2.5 to 3 times of carrier length.

③ Depth of turning basin designed should not be less than the design depth of jetty apron.

(3) Finally to determine the size and type of ship.

The following factors will be used to determine the ship size: the annual gas supply volume, storage tank volume, shipbuilding capacity of China or foreign, the diameter of the pipeline, the backup gas supply, jetty conditions including channel, route, anchorage and other factors. It is also necessary to analyze the matching between the ship and jetty, to see whether the ship meets with the berthing conditions? The ship type selected needs to be change? All of these questions should be answered at the end of the site selection.

5.2.3.2 Channel and Anchorage

(1) General requirements.

When a LNG carrier is sailing in and out of the channel, the front should be piloted by patrol boat and rear should be escorted by fire and towboat escort. When there is difficulty setting, it needs for specialized demonstration. In addition to the pilotage and escort boats, the distance from LNG ship to the channel sideline is not less than 100 m, or in accordance with the relevant standards.

(2) Channel sharing.

The channel of LNG carrier in and out can be shared with other ships when under traffic control. LNG jetty artificial inbound and outbound waterway can be designed as single-track, the effective channel width can be in accordance with the relevant provisions of "JTS 165—2013: Overall Design Code for Sea Ports", and should be given by corresponding calculation method. LNG jetty effective width of artificial bi-directional jetty channel is axis distance between two boats, should not be less than 6 times the maximum design beam.

(3) Channel design depth.

The computing datum of LNG jetty in and out channel depth should be designed from the local theoretically lowest tide level. The design calculation of the each rich depth of water should be according to the relevant provisions of the current industry standard "JTS 165—2013: Overall Design Code for Sea Ports".

(4) Special anchorage.

Anchorage and LNG jetty, the channel and other anchorage safety net distance should be greater than 1000m. According to the current industry standard for anchorage of scale should be confirmed as the relevant provisions of "JTS 165—2013: Overall Design Code for Sea Ports". In addition to the normal pilot anchorage, the quarantine anchorage and lying anchorage set up, if the ship in and out channel is longer, the emergency anchorage zone should be set up at the proper place of the jetty channel side.

5.2.3.3 Hydraulic Structure, Revetment and Breakwater

(1) Hydraulic structures.

It is necessary to select from multi schemes, such as the LNG jetty, trestle, work boat dock, receiving terminal revetment, water intake and drainage building structure, torch base, and trestle structure, to recommend the best solution.

(2) Receiving terminal revetment.

In order to meet the safety requirements of the receiving terminal, it should be according to the natural conditions, construction conditions and local material supply etc., to determine the revetment structure.

(3) Breakwater construction.

It needs to integrate the influence of the construction of the breakwater to the number of operating days and its investment and construction period, to determine whether to construct the breakwater.

5.2.3.4 Berth Throughput Capacity

LNG annual handling capacity can be pressed as the following formula to estimate:

$$P_t = \frac{TGt_d}{t_z + t_f + t_p + t_h}\rho \tag{5-1}$$

Where:

P_t——designed annual throughput capacity, t/a;

T——the number of calendar days, take 365d/a;

G——designed actual ship cargo capacity , t;

t_d——day and night hours, taking 24h/d;

t_z——time required of unloading a designed ship (h), can be considered according to the similar berth operating data and ship loading, the number of unloading arms, efficiency and other factors. If there is no data, net unloading time can take 14 ~ 24h;

t_f——auxiliary operation time of loading and unloading (h), refers to the operation time that can not be performed simultaneously with unloading on berth, consider according to the statistical data of similar berth into operation.

t_p——not in the night, additional time (h) of LNG ship in and out of the channel, berthing and unberthing, according to the ship from inbound to outbound of each link of the whole process, can draw a flow chart to determine;

t_h——waiting the tide, waiting flow time (h);

ρ——berth utilization rate (%), should be based on annual volume, ship type, ship unloading efficiency, the number of berths, ship in the jetty charges, the jetty investment, operating costs and other factors, and the overall economic benefits for the incoming target to determine. If there is no data, can take 55%~70%.

5.2.3.5 Jetty Working Days

According to "JTS 165—5—2016 : Code for Design of Liquefied Natural Gas Port and Jetty" provisions of article 4.2.1 : LNG ship in various stages of the process of operation, the wind speed, wave height, allowing visibility and flow rate should be consistent with the specified in Table 5-2.

Table 5-2 Standard for LNG Vessel Operating Conditions

No.	Operation stage	Allowed wind speed	Allowed wave height (m)		Visibility	Flow speed (m/s)	
		(m/s)	beam sea $H_{4\%}$	Following sea $H_{4\%}$	(m)	Transverse flow	Following flow
1	Sailing in and out of jetty	≤ 20	≤ 2.0	≤ 4.0	≥ 1000	<< 1.5	≤ 2.5
2	Berthing operation	≤ 15	≤ 1.2	≤ 1.5	≥ 1000	< 1.0	<2.0
3	Loading and unloading operations	≤ 15	≤ 1.2	≤ 1.5	—	< 1.0	<2.0
4	Mooring	≤ 20	≤ 1.5	<2.0	—	≤ 1.0	<2.5

Note: ① The beam sea means it and the ship's angle is greater than or equal to 15° waves, less than 15° is called following sea; The transverse flow means it and the ship's angle is greater than or equal to 15° flow, less than 15° is called the following flow; ② The allowed average period of wave is 7s; For more than 7s large periodic waves need to make a special demonstration; ③ H_4% is wave height when frequency calculate to 4%

Effects of natural factors in operation, due to the wind, thunderstorm, wave etc. are likely to occur simultaneously, so the actual impact of the cumulative time of operation should be less than the cumulative value of the individual statistics. It should be considered the effect of various natural factors overlap time, such as Guangdong Dapeng Chengtoujiao jetty site, the overlap influence time of natural factors is 4 days, and according to measured data of 8 years from 1993 to 2000, the days ≥ 7s wind last longest are two days, but the observation period is much shorter. Because the longest impact working days of Southern China coastal are generally caused by typhoon, a typhoon phenomenon is considered as the its path location, each typhoon impact working days counted as 5 days. So the actual non-operating days are 34.88 days, the annual operating days are 330 days.

5.2.3.6 Summary Tableof Jetty Conditions

(1) Qualitative evaluation.

The single jetty conditions are summarized in the following table (Table 5-3), multi sites selection provides the basis. When collecting the above information, it is advisable to collect the information of the location of the selected jetty; if not, in the jetty selection stage, the data can also be used or referenced to the nearby existing information in the construction project.

Table 5-3 Jetty Site Conditions Qualitative Score Sheet

Classification		Item	Score
The geographical position	Land topography	Topographic map	
	The underwater terrain	Sea map	
Coastal Geology	The soil type	Sand, clay	
	Buried depth of bedrock	Bedrock elevation, bedrock properties	
	Soil properties	Penetration Number, Physical and Mechanical Index	
	Earthquake	Intensity, faults	
Meteorological	Wind	Wind speed and direction rose diagram	
	Typhoon	Through the frequency, path	
	Other	Temperature, water temperature, number and days of precipitation, foggy days and visibility	
Marine conditions	Tide/current	Tide types, characteristics tidal level, tidal current, residual current, flow direction	
	Wave	Wave rose diagram	
	Sediment	Sediment concentration, particle size, sediment movement characteristics	
Jetty operation days analysis		According to the "Code for Design of Liquefied Natural Gas Port and Jetty (JTS165-5-2016)" to analyze jetty working days	
Construction conditions		Marine environment construction conditions	
		Construction and materials	
Total			

(2) Investment Estimation.

The contents of the investment estimation table can be selected according to the actual situation of the jetties, Table 5-4 can be used as a reference. It is general requirements for each candidate jetty at the same factors, no items missing, in order to make a comprehensive comparison of scores.

Table 5-4 Jetty Sub-project Investment Estimate

No.	Items	Investment	
		RMB	U.S. dollar
A	Engineering cost		
1.0	LNG pier		
2.0	Work boat wharf		

Continued

No.	Items	Investment	
		RMB	U.S. dollar
3.0	Trestle		
4.0	Revetment/retaining wall		
5.0	Sweeping sea		
6.0	Reef explosion		
7.0	Channel dredging		
8.0	Temporary Works		
B	Auxiliary engineering cost		
1.0	Dock drainage		
2.0	Auxiliary Building		
3.0	Demolition		
4.0	Research and test fees		
5.0	Local materials fee		
C	Reserve funds		
1.0	The basic reserve funds		
D	The static investment		

5.3 Condition Selection of Receiving Terminal

5.3.1 General Concept of Condition Selection of Receiving Terminal

5.3.1.1 Work Scope

The work scope includes the receiving terminal overall planning and layout, traffic conditions (mainly land), power supply, communication, gasification water and fresh water supply, geological and earthquake data collection and analysis, the receiving terminal site and foundation treatment, and through preliminary economic evaluation, investment estimation, get a single terminal's strengths and weaknesses of the conditions to provide multi alternative sites for comprehensive comparison and selection.

5.3.1.2 General Requirements

Receiving terminal site selection work is undertaken by qualified engineering design institute. The main consideration is link with the jetty, geographical location, land formation, the receiving terminal construction conditions, social support conditions, preparation of project investment estimation of the receiving terminal and so on.

5.3.1.3 Basis Documents

LNG terminal site selection should follow and comply with following documents.

(1) Laws, regulations and standards.

① Relevant national laws and regulations.

② Relevant national, industry and enterprise's standards.

(2) Other documents.

① Local social development planning, urban planning, overall development planning of the port and shoreline use planning.

② Commission contract.

③ The relevant requirements and documents offered (or approved) by owners.

④ Gas market and the preliminary determination of the construction scale approved by owners.

⑤ Cooperation agreement signed between owners and local government.

⑥ The government departments can provide the social conditions for receiving terminal site.

⑦ Based on the investigation data of jetty engineering.

⑧ Current LNG terminal project specifications and standards.

⑨ Geological and engineering survey and research results quoted by this report.

5.3.2 Receiving Terminal Geographical Position and Surrounding Environment

5.3.2.1 Principles of Receiving Terminal Site Selection

(1) Saving land.

Not occupied or less occupied fertile farmland, to avoid a lot of remove and demolition, when the condition which can make full use of dredging earthwork or near earth reclamation.

(2) Relying on the superior social conditions.

Power and water supply, communication, railway, highway, water transport and other social support conditions are good.

(3) The receiving terminal near the gas load center.

Receive terminal selection should be selected and combined with jetty address, but in addition to the construction of the jetty conditions, also have its own construction conditions and requirements, such as receiving terminal should be as close as possible to the gas load center, it can be beneficial to the laying of the pipeline is short and extended, beneficial to expand the user market, reducing investment and energy saving.

(4) Meeting the safety requirements.

According to the requirements of terminal facilities, it should be fully considered the safety

distance requirements in the plane and vertical design.

5.3.2.2 Engineering Geology and Earthquake

We should collect the regional engineering geological data, including the surface sediments, shallow sediments, whether there is a suitable bearing stratum, and the bedrock geologic age and lithology, but also through field survey, to describe type of land site landform, collect site basic earthquake intensity, dynamic peak ground acceleration, have or not have recent or late earthquake, if near the old and new geological fault zone, particular attention is paid to the relationship between adjacent active faults and so on.

5.3.2.3 The Receiving Terminal Land Formation

(1) Foundation treatment.

Foundation treatment scheme should be determined based on site requirements, natural conditions, terminal safety requirements, sources of material and construction conditions and other factors. For example, some terminals' land formation is required to cut into a mountain, and some needs reclamation, some foundation piling are shallow from bed rock or supporting layer, some are deep, and some sites' bedrock is smooth, some have big change of depth. These all should be determined by technical and economic appraisal. It should be foundation treatment when receiving terminal may generate soil liquefaction or soft soil seismic subsidence. Liquefaction judgment should comply with the relevant construction, structures and roads seismic requirements of terminal.

(2) Elevation and subsidence.

Receiving terminal elevation should be formed based on site requirements, earthwork balance, site area surrounding terrain elevation and flood proof requirements and so on, the receiving terminal land area residual settlement, uneven settlement, after the treatment of standard value of foundation bearing capacity should meet the structures requirements in different regions of terminal.

(3) The receiving terminal area extendibility.

From the perspective of the existing LNG domestic receiving terminal constructions, during a period of completed or been completed has increased investment in LNG storage tanks and other facilities construction, therefore, the receiving terminal site has room for expansion, is one of the selected conditions, if there is space to expand, it should be described as the expansion area, foundation processing difficulty and surrounding situation.

5.3.2.4 The Other Terminal Construction Conditions

(1) Less occupation of farmland and land conservation.

Base on the policy of conservation of land and shoreline resources, we should minimize to occupy the farmland, and as much as possible make full use of coastal beaches and abandoned land.

(2) General layout.

In accordance with the basic requirement of LNG unloading — storage — gasification —

outside transport process and public facilities, the entire work places can be divided into 6 regions: jetty area, LNG tank area, process area, regasification area, utility area and factory front area. In accordance with the principles of convenient construction arrangement, meet the safety standards and facilitate future staff access, we should propose the best terminal layout with reasonable optimization, and calculate the amount of civil engineering program and the corresponding investment.

5.3.3 Receiving Terminal External Cooperation Conditions

Terminal external collaboration conditions include transportation, communication, water supply, fire-fighting, demolition and resettlement, and other related external conditions to terminal.

5.3.3.1 Transportation

(1) Highway and planning.

Highway conditions include the existing roads distance from the receiving terminal, road grade, road extension distance of the city, and the recent expansion of the transformation. If there is no road now, describe the future highway planning.

(2) Railway and planning.

Railway conditions include the existing railways distance from the receiving terminal, the railway level (whether it is electrified railway, with or without intercity rail), railway extension distance of the city, and the recent expansion of the transformation. If there is no railway now, describe the future planning.

(3) Waterway and planning.

Waterway conditions include the terminal itself as jetty, or whether there are nearby waterways, waterway extension distance of the city, and the recent expansion of the transformation. If there is no waterway now, describe the future waterway planning situation.

(4) Aviation and planning.

Whether there is an airport near the receiving terminal, the distance from the receiving terminal, the distance from the airport to the city, and the future of the airport construction plan.

5.3.3.2 Communication, Water and Power Supply

(1) Communication.

The receiving terminal nearby whether adopts locality PBX (private branch (telephone) exchange) total capacity, digital communications network, local telephone module capacity, and switches margin and so on.

(2) Water supply.

Water supply system nearby whether is perfect or not, the difficulty degree and length of water lines access to terminal, the water supply capacity, and if it is affected by seasonal climate.

(3) Power supply.

What about the existing power plant and substation, the voltage level, near the inlet and the degree of difficulty, the construction of electricity access difficulty degree; second power options and so on.

5.3.3.3 Fire Fighting, Demolition and Resettlement and Other

(1) Fire fighting.

Since receiving terminal stores LNG, gasified LNG is flammable and explosive products, therefore, nearby there is fire station to use or not, is an important aspect in the future in case of emergency incidents and saving investment in the future. The investigation includes the existing fire station and its distance, level, equipment and facilities, staff number. If there is no fire station can be borrowed, then consider investment of self-built firehouse.

(2) Demolition and resettlement.

Whether have demolition project near the receiving terminal site, is also one respect to measure the difficulty degree of LNG project construction. The demolition includes residential property, collectively-owned enterprises, state enterprises, historic and civilization sites and military installations demolition, at the same time, also includes the demolition of the resettlement project, the final demolition and resettlement costs and other calculations.

(3) Other.

It includes whether there is near hospitals, schools, public services places and facilities.

5.3.3.4 Scoring the Conditions of Receiving Terminal Construction

(1) Qualitative evaluation.

According to the single receiving terminal conditions, by expert qualitative scoring, as shown in the following table (Table 5-5), provides the basis for multi sites selection.

Table 5-5 Receiving Terminal Site Conditions Qualitative Score Sheet

Classification		Item	Score
Land formation	Land topography	Topographic maps	
Geology	The soil type	Sand type, clay type	
	Bedrock depth	Bedrock elevation, bedrock property	
	Soil property	Penetration number, physics mechanic index	
	Earthquake	Intensity, fault	
External collaboration conditions	transportation	Highway, railway, river way and aviation and planning	
	communication	Communication status and planning in the region	
	Power and water supply	Power and water supply situation and planning in the region	
	Fire fighting	Fire-fighting situation and planning in the region	
	Others	Utilities and matched auxiliary facilities	

Continued

Classification	Item	Score
Cultural and environmental conditions	Housing demolition, vegetation, cultural relics, dangerous chemical plants to the LNG terminal, etc.	
Construction condition	Onshore construction of environmental conditions	
	The acquisition of local construction materials	
Total		

(2) Investment estimation.

The investment estimation of receiving terminal can be chosen according to the actual situation of each terminal. Readers can take table 5-6 as reference. Similarly, each candidate terminal address has the same investment items which cannot have missing, in order to score when comprehensive comparison and selection.

Table 5-6 Receiving Terminal Sub-project Investment Estimate

No.	Engineering name	Investment	
		RMB	US dollar
A	Engineering cost		
1.0	Storage tank region		
2.0	Process region		
3.0	Utility region		
4.0	Site and civil engineering		
5.0	Other direct costs		
6.0	Demolition/relocation costs		
B	Other costs		
1.0	Upfront costs		
2.0	Design costs		
3.0	Land expropriation		
4.0	Off-site construction costs		
C	Reserve fund		
1.0	Basic reserve fund		
D	Static investment		

5.4 Selection of Gas Trunkline Conditions

5.4.1 General Concept of Condition Selection for Gas Transmission Trunkline

5.4.1.1 Work Scope

According to the users area and the gas load center, to plan the LNG project of natural gas pipeline system, including the distribution of trunk lines and branch lines, the gas transmission station, block valve chambers. After a preliminary economic evaluation and investment estimation, it needs to choose the most appropriate trunk line route.

5.4.1.2 General Requirements

The pipeline route selection work shall be undertaken by the pipeline engineering design institute with the appropriate qualifications. The main consideration is cohesion of trunk line and receiving terminal and the main users, or according to the optimal pipeline route selection as the main users of the user site to provide advice; on the advantages and disadvantages of conditions to a single receiving terminal, to select multi trunk gas preparation routes, to determine the final line trend, the pipeline technique, the line investment estimate, and to provide alternatives.

5.4.1.3 Basic Documents

The following documents should be complied with and conformed to when choosing the pipe route.

(1) Laws, regulations and standards.

① Relevant state laws and regulations.

② Current pipeline engineering specifications and standards.

(2) Related documents.

① Local project program of natural gas network.

② Commission contract.

③ Gas market and preliminary determining scale of construction provided by the owner.

④ Engineering geological survey data of pipeline route.

⑤ Pipeline route relying on social conditions.

5.4.2 General Concept of the Gas Trunkline Route Selection

5.4.2.1 Selection Principle of Pipeline Routing Scheme

(1) It is close to the large users.

In the market research of LNG project users, there are always some large users, such as in Guangdong Dapeng LNG project and Fujian LNG project, the power plant users account for

60%～70% gas total consumption, therefore, close to the large gas users will be conducive to the layout of the entire pipeline route, thus reducing the total investment.

(2) The total line is the shortest.

It is required with straight lines and striving to shorten the line length; as far as possible by passing mountains, waters and bad engineering geological zones; making efforts to reduce the cross to natural and artificial barriers.

(3) It is conform to the line main direction.

The location choice of large or medium-sized underground crossing or spanning river project, gas station and block valve chambers should be consisted with the main direction of pipeline. Local trend should be adjusted based on the large or medium-sized underground crossing or spanning river project.

(4) Relying on the best conditions.

Considering power, transportation, water supply and building materials supplying condition, the pipeline should be laid along the existing road as far as possible for easy pipeline construction and pipeline maintenance in the future.

(5) Conservation of resources.

Considering the farmland, water conservancy project and the planning development of towns, industrial and mining enterprises, the pipeline route should be designed to try to avoid through towns, industrial and mining areas, airports, train stations, sea wharves , river ports and so on. And the pipeline route should be designed to save land and social resources as far as possible. The design should avoid the difficult construction areas and adverse geological location with selecting favorable terrain and the design should protect the national natural resounes.

(6) Taking into account the follow-up projects.

It is necessary to consider the cohesion of related projects and follow-up projects; in one phase of the project, it needs to take into account the two phase of the project.

5.4.2.2 Topography, Engineering Geology and Earthquake

(1) Topography.

Based on the pipeline route determined by the selection principles of pipeline route scheme, it needs to describe the topography of the zones crossed by the pipeline, such as plain of accumulation, lagoons lacustrine, coastal marsh, water network of benchland, hills, mountains areas. The difficulty level of the pipeline engineering can be seen from the whole topography.

(2) Engineering geology.

① Stratum. It is mainly described pipeline through the sections of the surface covering layer, thickness, lithology and base rock, and crossing section of rock lithology.

② Hydrogeology. Through the region, it is mainly described the buried depth of phreatic aquifer, whether the atmospheric precipitation recharge is mainly, or lateral discharge is mainly or vertical (evaporation) excretion mainly; stratum area tectonic fracture or weathering fracture occurrence of water; and the relationship between the depth of buried pipeline range.

③ Adverse geological condition. It will be mainly described the adverse geological section, such as the position of soft silt seam, the section of low bearing capacity, shallow fault, lithology mutation zone and so on.

(3) Earthquake.

Considering the pipeline through the region of strong earthquake activity level, the basic earthquake intensity and the dynamic peak ground acceleration, it should be evaluated according to "SY/T 0452—2004: Seismic design code for oil and gas buried steel pipeline".

(4) Climate.

It is necessary to carry on the climate investigation along the pipeline, such as climate zone, rainfall, monsoon, typhoon; water network density, rainfall caused mountain torrents, flooding frequency and influence on buried depth of pipeline.

5.4.2.3 Engineering Quantity and Investment of Pipeline Laying

(1) Overall requirements.

According to the recommendation of LNG receiving terminal as the starting point of pipeline to many users, especially to main users place extension, it should be met above pipeline of route selection principles. The investment estimation will consider the engineering quantity statistics, including the pipeline trend, trenching, landfill, vegetation recovery, pipeline length, crossing the road (highway), crossing the railway, and crossing large and middle rivers and so on.

(2) Pipeline block valve chamber.

According to the "GB 50251—2015: Code for design of gas transmission pipeline engineering", it should be calculated the block valve chamber engineering quantity and investment, considering the certain distance along the pipeline, in special sections on both sides, and the population density of area level, and set two block valve chambers on both sides of large river crossing.

(3) River crossing.

The crossing type includes excavation, direction drilling and shield, according to the actual situation to decide in what manner, to assess the impact of the limitation of project duration, investment, technical condition and season, through the methods of selection, to calculate the engineering quantity and the investment.

(4) Pipeline anticorrosion.

By field electric measuring the soil resistivity along the line, according to "SY/T 0087.1—2012: Standard of steel pipeline and tank corrosion assessment-Internal corrosion direct assessment for buried steel pipeline", the anticorrosion coating and cathodic protection scheme is put forward, calculating the anticorrosion workload and the corresponding investment.

5.4.2.4 The Utilities of Gas Transmission Station

The utilities include power supply, water supply and drainage, firefighting, general layout, architecture and so on.

(1) Power supply.

The power supply for each station should include the substation and distribution station, air conditioning and lighting, power electricity, lightning protection, anti-static grounding and so on. Station field power supply investment may be affected by the degree of difficulty for power supply near the station accessing to and the selection of online UPS auxiliary power supply for the primary demand, such as station communication, automatic control, emergency lighting and so on.

(2) Water supply and drainage and firefighting.

The water supply for each station is mainly used for living water of operator in charge, equipment cleaning water, site flushing water and so on, the investment is affected by the local water supply conditions. The water drainage includes industrial waste water discharge, domestic sewage and so on. Station firefighting is established to the fire control system in accordance with the fire hazard category in unit area, including fireproof distance, road set, fire- fighting equipment configuration, fire water and so on.

(3) Communication.

Station near the lines of communication and access conditions need to be considered, such as the use of the board with the total capacity of the telephone exchange, digital communications network, telephone module board capacity, switch margin.

(4) General layout and construction.

According to the requirements of urban planning and environmental protection and fire departments, national standards and specifications, it needs to conduct the station process layout, such as process equipment area, instrumentation control room, shift dormitory, sewage reservoir, and also including the area of land expropriation and so on.

5.4.2.5 Other

In order to ultimately compare the relative investment size, it needs to consider the above engineering quantity and investment, if there are other effects in the future pipeline scheme selection, should also be considered, such as: gas pipeline and pipe technology, material, energy saving, environmental protection, which involves in environmental impact and recovery measures etc.

5.4.2.6 Engineering Quantity and Investment Summary Sheet

(1) Qualitative evaluation.

According to the construction of the pipeline involved in the engineering geological conditions, external conditions, human, geographical environment and construction conditions, the Table 5–7 is given, by experts scoring, for the preparation of comprehensive selection.

Table 5-7 Pipeline Route Selection Qualitative Score Sheet

Classification		Item	Score
Topography	Land topography	Topographic maps	
Engineering geology	Stratum	Sand type, clay type, bedrock elevation, bedrock property	
	hydrogeology	Density of rivers network, mountain torrents, flooding occurrence frequency, depth of phreatic aquifer, the impact on pipeline buried depth and so on	
	Adverse geological condition	the position of soft silt seam, the section of low bearing capacity, shallow fault, lithology mutation zone and so on	
Earthquake	earthquake activity	Intensity, fault	
Climate	Along the line	Climate zone, rainfall, monsoon, typhoon	
External collaboration condition	transportation	Highway, railway, water way and aviation and planning	
	Communication	Communication status and planning in the regions	
	Power and water supply	Power and water supply situation and planning in the regions	
	Firefighting	Firefighting situation and planning in the regions	
	Others	Utilities and matched auxiliary facilities	
Cultural and environmental conditions		Housing demolition, vegetation, cultural relics, dangerous chemical plants to the LNG terminal, etc.	
Construction conditions		Onshore construction of environmental conditions	
		On the local acquisition of construction materials	
Total			

(2) Investment estimate.

The changes in the pipeline route direction will inevitably bring about changes in the investment of each item in the Table 5-8, the summary of the total investment through the following table is the preparation for the comprehensive selection.

Table 5-8 Pipeline Sub-project Investment Estimate

No.	Engineering name	Investment	
		RMB	US dollar
A	Pipeline cost		
1.0	Trunk line cost		
2.0	Branch line cost		
B	Crossing project cost		
1.0	Large rivers		

Continued

No.	Engineering name	Investment	
		RMB	US dollar
2.0	Medium and small rivers		
3.0	Highway/railway		
C	The process station/block valve chamber Construction cost		
1.0	Process station		
2.0	Pipeline anticorrosion		
3.0	Building cost		
4.0	Land requisition cost		
5.0	Off-site construction costs		
D	Reserve fund		
1.0	Basic reserve fund		
E	Static investment		

5.5 Receiving Terminal Comprehensive Comparison and Selection

5.5.1 General Concept of Comprehensive Comparison and Selection

5.5.1.1 General Requirements of Site Comprehensive Comparison and Selection

(1) In accordance with local economic development planning.

The local economic development planning is the foundation of the LNG project, in terms of the three main single sub-projects as a composition of LNG project, the beginning is more considered from the technical factors, or considered single sub-project as the least amount of investment from the economic, however, often ignoring the industrial layout and development direction of local industry, hence the selected scheme may not fully meet the requirements of economic development planning. Only with a minimum overall project investment, and in accordance with local economic development planning, the site selection is the best place.

(2) Overall layout optimization.

The comprehensive comparison and selection is not standing in the single project such as jetty, terminal or pipeline standpoint, but from the overall LNG project as a starting point for selection. The combination of three optimal single sub-projects cannot represent the best overall project comparison and selection. Our ultimate goal is that the combination of three single sub-projects is the best though the three single sub-projects are not their best each.

(3) Relatively less investment.

The work of comprehensive comparison and selection is generally carried out before the project feasibility study. Given the depth of research and the grasping of information, the comparison and selection work is mainly commenced with qualitative scoring of construction conditions and estimation of project investment. The combination of jetty, terminal and pipeline with good construction condition and less investment will be selected. The work does not have to have a complete economic evaluation. Of course, a complete economic evaluation is better if the condition is permitted.

5.5.1.2 The Principles of Terminal Site Comprehensive Comparison and Selection

As mentioned above, the selection of LNG terminal site is a high technical and powerful policy work, which must adhere to the principles of expert argumentation, owner choose and government approval. Experts focus on the technology, while owners concern about the benefits and government grasps the policy. Only the combination of the three, the interest of all parties can be balanced.

(1) Expert argumentation.

The LNG project site selection is a technical work. It must be based on the technical conditions for the construction of LNG jetty, terminal and pipeline, which cannot be contrary to scientific principles. Project company or project team must firstly hire the experts in these three majors respectively to compose the expert's group, let them intervene the sites selection work early, tell experts real intentions of the project company site selection, provide single scheme data to three engineering design institutes, so that the experts can investigate and study on-site with problems. Sometimes the associated experiment and computer simulation are necessary, in order to verify the original plan. The project company should pay great attention to the advice of the experts.

(2) Owners' choice.

As the main body of investment, the project company or project group should proceed from the overall project, hire qualified design institutes to carry out single project site selection work, carefully check the quality of the work. When the design institute assumes the terminal site comprehensive comparison of the work with the above expert group opinion reached consensus, the project company may submit its proposal to the competent authorities for approval; when the two opinions appear at odds, to urge the design institute to do more work, or do the full argument, at the same time the project company will ask the expert's group and the design institute fully to exchange views, using scientific data to validate the comparison conclusions, so that making a solid comprehensive comparison work in place, tending to form a unified opinion. Finally the project company - owner take out the tendentious solutions.

(3) Government approval.

Government approval is the last stage of the sites comprehensive comparison, generally taking the form of site comprehensive comparison deciding meeting, presided over by the competent department of the government. In the meeting leaders and experts from other government departments, sites comparison group of experts, single comparison design institutes and comprehensive comparison design institute, project owners and so on will be invited. Through introducing the scheme of single sub-project and comprehensive comparison, combined with government departments' macroeconomic policy and planning of the project, after extensive discussion and full consultation, eventually forming the government approved the preferred site and alternative site.

5.5.2 Comprehensive Comparison and Selection Process

5.5.2.1 Flow Chart

General site comprehensive comparison and selection process is shown in Figure 5-2.

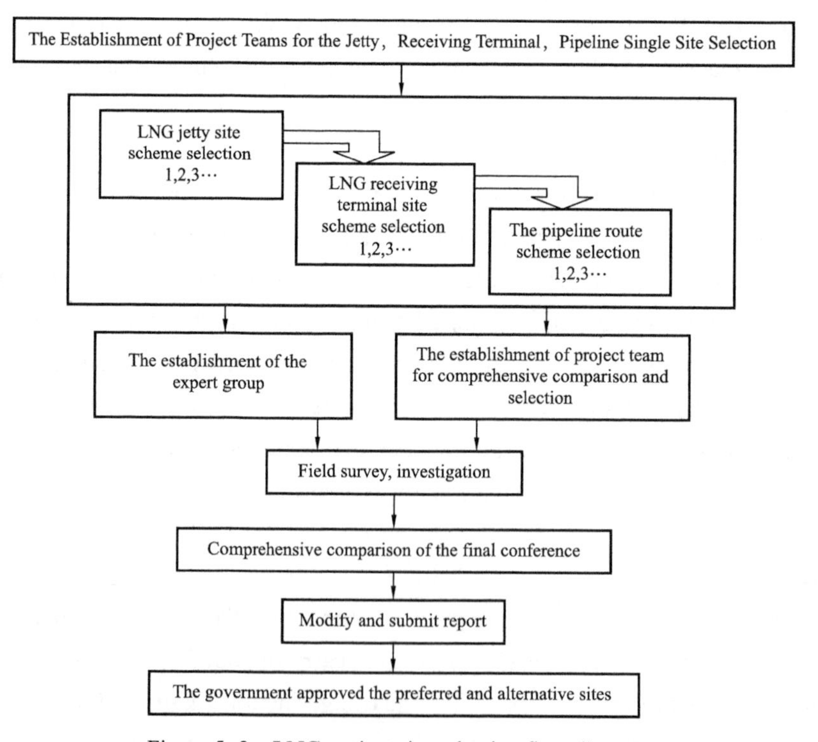

Figure 5-2 LNG project site selection flow diagram

5.5.2.2 Process Operation

(1) Multiple single sub-project schemes are the foundation.

From the above introduction, the project company firstly hires design institutes respectively for the jetty, terminal and pipeline to carry out field surveys and data collection, to select the sites and routings for the single sub-project. Under the circumstance which the local data cannot be obtained, the peripheral data can be learned from, for qualitative evaluation of each single project and project investment summary.

(2) Establishment of expert group.

On the basis of single sub-project site selection, LNG project company independently hires jetty, terminal and pipeline experts, establishing the panel. The panel will study the design institute report of single sub-project site selection, to inspect and survey on the site with questions, forming a preliminary opinion.

(3) Establishment of the project site comprehensive comparison project team.

Then, the project company chooses an engineering design institute to conduct the comprehensive sites comparison, the concrete task is to study the integrated project plan combined with jetty, receiving terminal and pipeline three single sub-projects portfolio. Generally it is necessary to start from the jetty site, select the receiving terminal, together with the supporting pipeline routing, to form a combination, similar to other portfolio analysis. The type of combination is a combinatorial mathematics problem, that is to say, the final combination of each LNG project is the product of the number of combinations of each scheme.

(4) Comprehensive assessment.

From the existing LNG project scheme selection practice, the design institute bears the comprehensive comparison, based on the single sub-project construction condition qualitative evaluation and investment estimation, to implement the jetty, terminal and pipeline combination comparison and selection, get the site sorting, and submit a comprehensive comparison report.

(5) Expert scoring.

Comprehensive scoring by the panel, the scoring methods can be used as collective discussion method, and or Delphi method as well, and the former is easily affected by authoritative expert, while the latter can play the subjective initiative of each expert without the interference of others.

① Comprehensive scoring based on the above qualitative evaluation of single sub-projects and investment estimation, sorting by the comprehensive scheme of each expert.

② On the basis of single sub-project qualitative evaluation and investment estimation, if adjustment for the engineering program, original design institute can be arranged to re-compiled and calculated, and then comprehensive scoring by experts, and sorted the comprehensive scheme by the experts.

(6) Submitting the final report.

On the basis of expert reviews, the site comprehensive comparison institute modifies reports, and submits the final project site selection report, accompanied by expert collective opinion.

(7) Government approval.

According to the principles of expert argumentation, owner choose and government approval, finally the project company submits the result to the competent department of government for approval, to determine the preferred site and alternative site.

5.5.3 The Introduction of the Key Work in Sites Comprehensive Comparison and Selection

5.5.3.1 Comprehensive Comparison Tableof Site Factors

CNOOC's operation of the Guangdong Dapeng, Fujian, Shanghai and Zhejiang LNG project sites comprehensive comparison and practice, can be summed up the site factors comparison table (Table 5–9), which can be used for reference to other new projects.

Table 5–9 Project Site Comprehensive Factors Comparison Table

No.	Comparative factor	Combination site 1	Combination site 2	Combination site 3
Jetty				
1	Water domain, water depth	Text description and scoring	Text description and scoring	Text description and scoring
2	Jetty plan layout			
3	The length of trestle			
4	Hydraulic structure of jetty			
5	Breakwater			
6	Channel			
7	The dredging volume of the water domain			
8	Annual back siltation (intensity/quantity)			
9	Annual average operation days			
10	Construction conditions			
11	Shoreline planning and the impaction of surrounding environment			
12	Others			
Terminal				
13	Land formation (Fill/extraction)	text description and scoring	text description and scoring	text description and scoring
14	Geology condition			
15	Transportation			

Continued

No.	Comparative factor	Combination site 1	Combination site 2	Combination site 3
16	Communication			
17	Power and water supply			
18	Firefighting			
19	Cultural and environmental conditions (demolition/environmental treatment)			
20	Onshore construction and environmental conditions			
21	Obtaining the local materials			
22	Others			
		Pipeline		
23	Topography	Text description and scoring	Text description and scoring	Text description and scoring
24	Engineering geology			
25	Earthquake			
26	Climate			
27	Transportation			
28	Communication			
29	Power and water supply			
30	Firefighting			
31	Cultural and environmental conditions (demolition/environmental treatment)			
32	Onshore construction and environmental conditions			
33	The local availability of construction materials			
34	Others			
35	The final score			

5.5.3.2 Project Sites Comprehensive Investment Comparison Table

Comparison of investment estimation (Table 5-10) is a very convincing work. It combined with the qualitative evaluation of the above together, laid the foundation for sites comprehensive comparison.

Table 5-10 Project Site Investment Estimation Comparison Table

unit: ten thousand Yuan

No.	Comparison of elements	Unit	Combination site 1			Combination site 2			Combination site 3			Notes
	Total investment		quantity	RMB	US dollar	quantity	RMB	US dollar	quantity	RMB	US dollar	
	Jetty											
1	LNG pier	Seats										
2	Trestle	m										
3	Work boat	m										
4	Harbor basin and channel dredging	$10^4 m^2$										
5	Sweeping sea/reef explosion	m^3										
6	Revetment/retaining wall	m										
7	Temporary engineering	Items										
8	Jetty drainage	Items										
9	Auxiliary building	Items										
10	Demolition/resettlement	Items										
11	Research and test	Items										
12	Local materials	Items										
	Receiving terminal											
13	Storage tank region	Items										
14	Process region	Items										
15	Utility region	Items										
16	Site and civil engineering	Items										
17	Other direct costs	Items										
18	Demolition/resettlement	Items										

Continued

No.	Comparison of elements	Unit	Combination site 1			Combination site 2			Combination site 3			Notes
	Total investment		quantity	RMB	US dollar	quantity	RMB	US dollar	quantity	RMB	US dollar	
19	Other costs	Items										
20	Upfront costs	Items										
21	Design costs	Items										
22	Land requisition	m^2										
23	Off-site construction costs	Items										
	Pipeline											
24	The length of trunkline	m										
25	The length of branch	m										
26	Crossing large river	Items										
27	Crossing medium and small river	Items										
28	Crossing highway/railway	Items										
29	Construction of stations/block valve chambers	Seats										
30	Process station	Seats										
31	Pipeline anticorrosion	m										
32	architecture	m^3										
33	Land requisition	m^2										
34	Off-site construction costs	Items										
35	Reserve fund											
36	Basic reserve fund											
37	Total											

5.5.3.3 Comprehensive Scoring Table

Comparison and selection is the system engineering, on the basis of single sub-project site strengths and weaknesses analysis, making a multi sites comprehensive combination comparison, recommend the preferred site and the alternative site. The following is a recommended comprehensive score table (Table 5-11) in this book, offering readers as reference. The weights in the table can be used to weight scoring according to the experts comparing the options. The recommended weight below is a total of 200 points, namely: jetty 60, receiving terminal 40, pipeline 50, investments 30, other factors 10 (note: The superior condition scores high, the inferior condition scores low) and uncertain and potential risks 10 (note: low risk scores high, high risk scores low); the weight of the item is the highest score, scoring according to each site factor. Combination of the highest total score is the preferred project site scheme, and next score is the alternative solution.

Table 5-11 Project Site Comprehensive Comparison and Selection Table

No.	Item	Sub-items	Weight (the highest score)	Combination scheme 1	Combination scheme 2	Combination scheme 3	Combination scheme···
	Jetty		**60**				
1	Geographic location	Land topography	3				
		The underwater terrain	6				
2	Coastal Geology	The soil type	4				
		Bedrock depth	6				
		Soil property	2				
		Earthquake	7				
3	Meteorology	Wind	3				
		Typhoon	5				
		Others	1				
4	Marine conditions	Tide/current	4				
		Wave	2				
		Silt	3				
5	Analysis of jetty operation days		7				
6	Construction conditions		7				
	Terminal		**40**				
7	Land formation	Land topography	6				

Continued

No.	Item	Sub-items	Weight (the highest score)	Combination scheme 1	Combination scheme 2	Combination scheme 3	Combination scheme···
8	geology	The soil type	2				
		Bedrock depth	4				
		Soil property	1				
		Earthquake	6				
9	External collaboration conditions	Transportation	2				
		Communication	1				
		Power and water supply	3				
		Firefighting	5				
		Others	1				
10	Cultural and environmental conditions		2				
11	Construction condition		7				
	Pipeline		**50**				
13	Topography and geomorphology	Land topography	6				
		Underwater topography	2				
14	Engineering geology	Stratum	4				
		hydrogeology	3				
		Adverse geological condition	6				
15	Earthquake	Earthquake activity	5				
16	Climate	Zone along the pipeline	2				
17	External collaboration condition	Transportation	3				
		Communication	2				
		Power and water supply	1				
		Firefighting	5				
		Others	2				
18	Cultural and environmental conditions		2				
19	Construction condition		7				
20	Investment		30				
21	Other factors		10				
22	Undetermined potential dangerous		10				
23	Comprehensive score						

5.5.3.4 Project Site Comprehensive Comparison and Selection Report

The following is the recommended LNG project site selection report basic contents (Table 5–12), the report contents can be set according to the actual site selection of different factors, the conditions of the natural environment and the external construction of different conditions.

Table 5–12 The Chapter Contents of LNG Project Site Comparison Report

1 summary 1.1 comparison and selection basis and methods 1.2 comparison and selection principles 1.3 comparison and selection conclusions	4 Comparison of site construction economic investment 4.1 Summary of the investment cost 4.2 Summary of operation cost 4.3 Gas price calculation
2 The basic conditions of site comparison 2.1 Market gas consumption volume and size 2.2 the design basis for project site	5 The comprehensive comparison and conclusion of project site construction 5.1 The total investment cost of preferred and alternative project sites 5.2 The pipeline transportation cost, gasification cost and gas price of preferred and alternative terminal sites 5.3 Conclusion of sites comparison and selection
3 Comparison of site construction 3.1 Comprehensive comparison of the jetty engineering 3.2 Comprehensive comparison of the terminal engineering 3.1 Comprehensive comparison of the pipeline engineering	6 The drawings, accessories 6.1 The drawings 6.2 The accessories

5.6 Land Requisition Management

5.6.1 Land Requisition Management Overview

5.6.1.1 Related Units Involved in Land Requisition

The units involved in the LNG project land requisition include government departments, investors and operators, designers, contractors and supervisors (Figure 5–3).

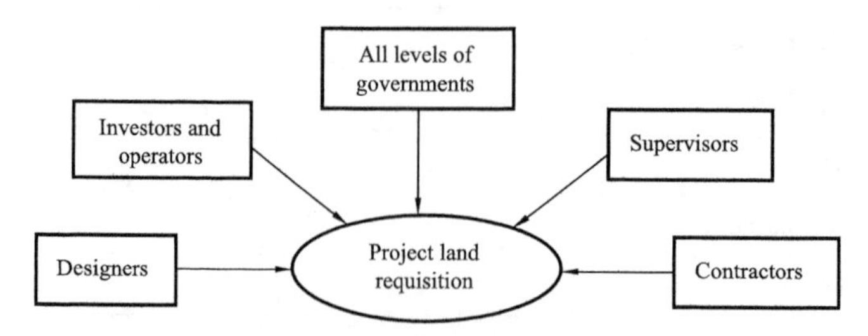

Figure 5–3 Land requisition relationship diagram of LNG project

(1) Government departments.

The LNG project settled in a certain place, it must be approved by the competent authorities. From the point of land requisition administration, we must get the strong support and cooperation

of all levels of governments, such as Land Resources, Programming, Planning, Commerce, Forestry and other sectors, especially in terms of land requisition and demolition, we must get the full support of local government. Without the support of local government, we can't do anything.

(2) Investors and operators.

The project investors and future operators are the main parts of the project land requisition. We must firstly set up a strong leadership group and a land requisition office. General Manager of the company serves as the director of the office, deputy general manager sent by the local enterprises serves as the deputy director of the office, in order to give full play to local advantages. Land requisition staffs are generally familiar with the local folk customs, so that easily communicate with the people in land requisition area, and also need some public relation skills of coordinating with governments, contractors, supervisors. Under the conditions of not violating the policy, some disposition right must be authorized to the land expropriation staff, in order to accelerate the process of land expropriation.

(3) Designers.

Pipeline route designers must fully investigate on the spot. If this is not done, it will seriously delay land expropriation. For example, the pipeline route designers of Guangdong Dapeng LNG, their original route design in Guangzhou, Shenzhen and Dongguan did not get the local planning department approval, because the amount of land requisition and housing demolition is so large that the department cannot be carried out. The land requisition department has to organize design personnel to visit the planning department of local government in each city for re-route. The LNG project pipeline route should be optimized as far as possible in the compliance with design safety regulations, to reduce the amount of housing demolition.

(4) Contractors.

In order to speed up the progress of the project, the project company in terms of temporary land often pays attention to make the contractors function, adopting incentive policy which the contractors replace the project company in temporary land requisition, stimulating the initiative of each contractor to support matching land requisition and housing demolition work. So at the same time the contractor would pay attention to conservation while using land in construction, avoiding unnecessary waste, to achieve the "Win-Win" objective.

(5) Supervisors.

In general, the supervision work should run through the whole process, it is helpful for the owner reasonably to divide bidding blocks, and the supervisor reviews the bidding documents and put forward advisory opinions for owner as reference in the bidding process. Before the construction company entering the site, the supervision company can assist the owner in accordance with the requirements of the contract to provide entering conditions for the construction company. For example, the land requisition and resettlement, local relations, four connections and one leveling, and construction drawings supplying and so on. The supervision company should issue the "Notice to Proceed" under the construction condition required. During construction, the supervision company should audit, sign and issue the design drawing, and organize the design exchange, especially the modification for pipeline route caused by land requisition. An important pipeline route must have verified in the presence of supervisors. Supervisors must often pay close attention to the project cost, including the rationality of land requisition and so on. To carry out independent research, identify problems and timely urge the constructor to take corrective action, supervisors should handle the design changes and claims caused by land requisition change, so that the project can be proceed smoothly.

5.6.1.2 Laws and Regulations Related to Land Requisition

(1) The state laws and regulations.

It includes the State Land Administration Law and Implementation Regulations; Forestry Law and Implementation Regulations; Maritime Space Use Administration Law; Urban Planning Law; the Provisional Regulations on the Sale and Transfer of Urban State-owned Land Use Right; Farmland Occupation Tax Provisional Rules; Interim Regulations on Resource Tax; the Provisional Regulations on Land Value-added Tax; the Specific Provisions of the Arable Land Occupation Tax Policy; New Construction Land Use Fee Collection Management Approach; the Fee Collection Interim Measures for the Administration of the Forest Vegetation Recovery; Notice of the Ministry of Construction on the Issuance of "City Housing Units Valuation Guidance" ; Notice of National Price Bureau, Ministry of Finance Issued on Land Management System Part of the Charging Items and Standards; the Ministry of Land and Resources Issued Notice of Guidance about Improving the System of Land Expropriation Compensation and Resettlement; Notice on Implementing the "Decision of the State Council on Deepening the Reform of Strict Land Management" ; the State Council on "Deepening the Reform of Strict Land Management Decisions", etc.

(2) Local laws and regulations.

It includes Local Farmland Occupation Tax Measures for Implementation; Land Requisition Administration Fee Collection Regulations; Notice on the Relevant Issues Concerning the City Housing Demolition and Resettlement Grants and Temporary Placement Subsidy Standard; the Measures for the Implementation of the Forest Law; Temporary Land Use Management Approach; Notice of the Approval of the State Council of the Construction Land Review Issues Related to the Work Submitted for Approval; Pre-examination of Construction Project Land Use Management Approach; the Measures for the Administration of Examination and Approval of Forest Land Requisition and Occupation; the Agreement Transferring State-owned Land Use Rights Provisions; the allocation of Land Use Management Approach; Notice of the Implementation of Land Purchase and Reserve System and so on.

5.6.2 LNG Project Land Requisition Characteristics and Difficulties

5.6.2.1 Characteristics

(1) Many spots and wide range.

LNG project involves the jetty, channel, anchorage, the receiving terminal in many par-

tition, pipeline transportation stations, block valve chambers and the hundreds of kilometers of pipeline, so the permanent and temporary land requirements on many spots and in a wide range, each LNG project involves a number of cities and more than a dozen of counties (cities, districts), and even the trans-provincial area. Such as Guangdong Dapeng LNG project ran through 175 villages, 21 large-medium rivers, 90 times crossing highways, 10 times crossing railways, $52km^2$ of woodland, 60 houses and 12 workshops to demolish, this is the best result by repeatedly demonstrated.

(2) Large amount of land used.

In general, a LNG project will be the expropriation of all land thousands of acres, including several hundred acres of permanent and temporary land, such as Guangdong Dapeng and Fujian LNG project with a total land area of 130 acres of permanent (Table 5–13).

Table 5–13 Guangdong and Fujian LNG Projects' Land Used Statistical Table

unit: acre

LNG project	The total land area (no temporary land)	Terminal land (permanent land)	Pipeline land (temporary land)	Block valve chamber land (permanent land)	Gas station's yard land (permanent land)
Guangdong	136.6	98.8	1221.9	8.9 (24 Seats)	28.8 (11 Seats)
Fujian	133.6	98.8	1248.9	4.0 (20 Seats)	30.8 (12 Seats)

(3) In the face of local governments at all levels and individual people.

Due to land requisition on many spots and in a wide range and having large amount of land used, involving local governments at all levels, we need to get their support and help, especially when dealing with individual people, under the cooperation of local governments to do the ideological work of the people, while taking full account of the interests of local people, but also reducing the cost of land for the project.

(4) Too many documents submitted for land use approval.

According to the existing projects, LNG project for examination and approval has strict policy requirement, requirements for document preparation is in a huge amount, including dozens of text documents and nearly 10 kinds of drawings.

5.6.2.2 Difficulties

(1) Land requisition compensation for the specific operation is difficult.

In order to complete LNG project permanent and temporary land requisition work, generally it needs to sign permanent and temporary land requisition compensation, permanent and temporary land requisition fruit crop compensation, permanent and temporary land requisition demolition compensation and other special demolition works. For example, Guangdong Dapeng LNG project signed the above agreements reached more than 1000 copies.

(2) No uniform standards for temporary land use.

According to the current operation of the project, land requisition compensation and resettlement fees, ground attachments and young crops compensation fee, farmland reclamation deposit, because of the different economic levels around the city and there is no uniform standard, the specific operation is difficult.

(3) Bigger pipeline route change.

The selection of pipeline route needs to take into account the technical requirements and local planning, sometimes, in order to comply with local development plan, needs to modify for many times. The task of coordination with the local government and relevant company is heavy with large back and forth. China's southern densely populated, scarce land resources, the complexity of the underground pipe network, which gives the LNG pipeline selection and demarcation of the invisible and increased the number of times the difficulty.

(4) Long pipeline restricts the construction period.

The entire pipeline is up to tens of thousands of square meters of demolition, demolition work will likely become one of the restrictive factors of pipeline construction, while in the process of construction, especially in the coastal southern water network, numerous rivers, the need for crossing, plus the southern rainy season also affects the construction, so that the entire construction period is prolonged.

(5) Approval process is complex and for a long time.

By the LNG land use features of a wide range, too many documents required by approval, it is determined that land examination and approval procedure are complex and require a long time. Before line delimitation, a lot of communication and coordination with relevant units (such as water, electricity, roads, cable, railways, environmental protection, fire protection, forestry and other related departments) are required. Such as the route selection and line delimitation of Guangdong LNG had to be in twists and turns so that to run through the line, as well as that bring difficulty for the planning department for site planning and handling of land planning permits, affecting planning approval.

5.6.3 Methods and Types of Temporary Land Requisition

5.6.3.1 Temporary Land Requisition Methods

The temporary land use needs to submit a written application to County Land and Resources Bureau, the use of non-cultivated land should be approved by the County Land Bureau and issued a "Temporary Land Use Permit". The flexibility and skills of temporary land requisition are relatively large, there is no fixed pattern. Temporary land use practices can be summarized into two.

(1) Legal procedures.

In accordance with the legal procedures, land users and land owners sign a temporary contract, and then report to the relevant departments for approval.

(2) Negotiation.

When selecting the pipeline contractor, if the temporary land expropriation task is given to the contractor, and is written in the construction contract, it will be completed by the contractor and the owner to negotiate, the project company will coordinate with the contractor, and the key steps need the tripartite of owner, contractor and supervisor at presence and witness simultaneously.

5.6.3.2 Temporary Land Requisition Types

(1) Below 0.49 acre.

The use of basic farmland protection area or outside the basic farmland protection area of 0.49 acre or less, it should be approved by the City Land Bureau, and the County Land Bureau issues a license.

(2) More than 0.49 acre.

The use of basic farmland protection area or outside the basic farmland protection area of 0.49 acre or more, it needs to be approved by the Provincial National Land Agency, and the County Land Bureau issues a license. Guangdong Dapeng LNG pipeline occupied temporary and permanent land and demolition work could be seen in Table 5-14.

Table 5-14 The Statistics of Guangdong Dapeng Pipeline Project Land Expropriation

Pipeline section	The length of pipeline (km)	Temporary land expropriation (acre)	Permanent land expropriation (acre)	The demolition area (m^2)
Shenzhen section	128	379.6	103.8	300
Huizhou section	38	112.7	29.1	230
Dongguan section	88	65.2	10.9	1000
Guangzhou section	91	444.5	52.7	6000
Foshan section	44.5	220	1.7	

(3) To sign temporary land use contract.

Temporary use of state-owned or collective owned land, it should be signed temporary land use contracts with the original state-owned land management units, collective land ownership units or the original land contracting operators respectively. The temporary use of the state-owned unused land, it should be signed a temporary land use contract with the County (city, district) Land Bureau.

5.6.3.3 Cost and Process Management for Temporary Land Use

(1) Cost composition.

The temporary land-use fees involve land compensation fee, young crops and ground attachments compensation fee, reclamation fee.

(2) Matters needing attention.

Firstly, in accordance with the laws and regulations and standards of compensation of the local government, in the specific operation to face local residents and the people, it may occur disputes involved in the negotiations over the cost compensation, and even a lot of aggressive behaviors, the project company and contractors should avoid direct conflict with land owners, rely on the local government, do meticulous work, and strive to satisfactorily resolve the problems.

5.6.4 Permanent Land Requisition Preparation and Process

5.6.4.1 Land Use Pre-examination and Data Preparation

(1) Before the feasibility study report is submitted.

① Project company qualification certificate.

② Land application (description of land use, quantity, location and so on), attached construction land application table.

③ After determining the location of the valve room and station of the pipeline, to handle site selection proposal.

④ Entrusting the relevant companies to write geological hazard assessment report.

⑤ Involving the use of cultural relics, forest land, land use permits should be proof material has approval by authority department.

⑥ Pre-examination opinion of each district and city land competent departments, and to submit them to Land and Resources Office progressively.

⑦ The construction land boundary survey report and land boundary survey map.

⑧ Planning to take up 1 ∶ 10000 current land use map.

⑨ The general layout drawing and linear engineering plane diagram of the construction project.

⑩ Land use pre-examination results as supporting documents for the feasibility study report.

(2) In the stage of preliminary design.

① The document of approval of feasibility study report or other project approval documents.

② In preliminary design process, if there is a change of site location, the site selection submission and blue line graph need to reapply. The blue line graph is the matching accessory in the "Process of Planning and Site Selection", which is painted by the Urban and Rural Planning Bureau.

③ After all the site selection ends, statistics about the land use type.

④ Completing statistics on the occupation of basic farmland, consult with the Provincial Land and Resources Office to determine land use planning adjustment and handle the adjustment procedure of farmland protection area planning.

(3) After the preliminary design review approval.

① Preliminary design approval documents or other design approval documents.

② For planning permits and red line map. The red line graph is the matching accessory in "Construction land planning permit", which is the required document for land approval, painted by the Urban and Rural Planning Bureau. Handling these two important documents is the premise that there must have site selection submission, red line graph, preliminary design review and approval and other project approval documents.

③ Whether or not to cover important mineral deposits to prove material or to cover important mineral deposits evaluation report.

(4) Application materials.

① The main documents include: agricultural land conversion plan, the cultivated land supplement plan, the land requisition plan and land supply plan, the construction land used presentation.

② Accessories include: construction project site survey records, basic farmland identification field survey records, new construction land use fees paid for the preparation of the case description, land units for land requisition program views, land ownership certificate, the construction unit qualification certificate, land acquisition funds proof, the report of geological hazard risk assessment and cognizance opinion, survey report of the boundary points and other materials.

5.6.4.2 Approval Procedures for the Permanent Land Requisition

(1) All levels of government approval.

① Construction projects shall be approved by the State Council or the Provincial Government in accordance with the law.

② Construction company makes an application for construction land use to the land administration departments of city, county government.

③ Land administration department of city or county government issues land requisition program after examination.

④ Land requisition program is reported to the provincial or state government step by step after the approval by the city, county government.

⑤ Land requisition and other programs are approved by the State Council or provincial government in accordance with the law.

The government at all levels approval process is shown in Figure 5–4.

(2) LNG project permanent land expropriation process.

The LNG project company should pay attention to the work of land requisition from the start, arrange strong manpower to set up the Land Acquisition Leading Group and Office, to complete declaration, negotiation, implementation of compensation, land change and registration. If the use of social forces, to sign a contract with the commission unit, land requisition and its procedures are as follows (Figure 5–5).

5.6.4.3 Permanent Land Acquisition Procedures

(1) Declaration.

The project office makes permanent land application for approval to the local government, fills in application form for approval of land, and prepares relevant materials and drawings (general layout, route map). According to the Ministry of Land and Resources issued "Notice on the Relevant Issues Concerning the Examination and Approval of Construction Land for Approval by the State Council", for the provincial office of the overall land use planning partial adjustment

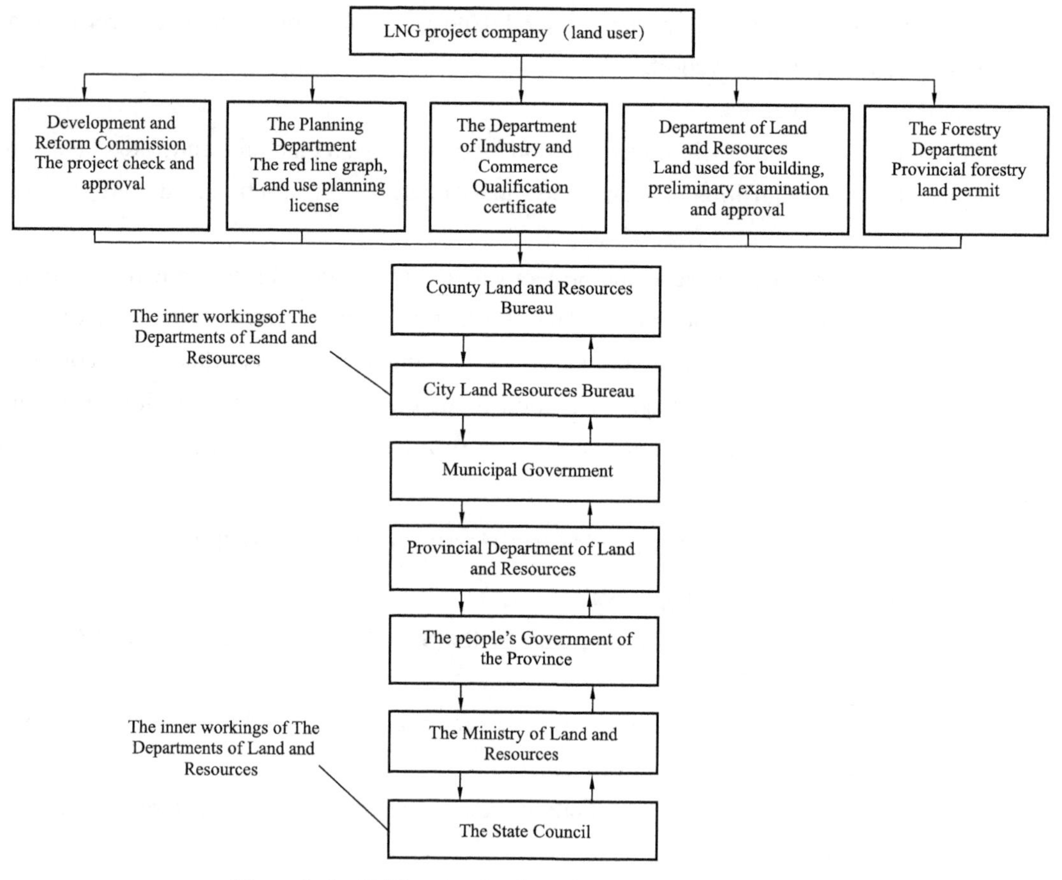

Figure 5–4 LNG project land requisition approval flow chart

scheme and application and approval procedures reported at the same time, in order to simplify procedures, and reduce costs.

(2) Issuing the announcement of land requisition.

① Issuing authority: city and county government.

② Issuing scope: appropriated land of township (town) and villages.

③ Announcement content: land requisition authority for approval, license number, the use for land requisition, scope, area and land requisition compensation standards, measures for the resettlement of agricultural personnel, land requisition compensation deadlines and so on.

④ Announcement consequences: after the release announcement planting crops or constructing buildings are not included in the scope of compensation.

⑤ Boundary survey: application to the Department of County Land and Resources for boundary survey.

(3) Hearing of witnesses.

Cooperating with each Land and Resources Departments surveying in land ownership, to formulate land requisition scheme (including compensation standard), to inform the villagers, if there is a request for a hearing of witnesses, the Land and Resources Department will organize the hearing.

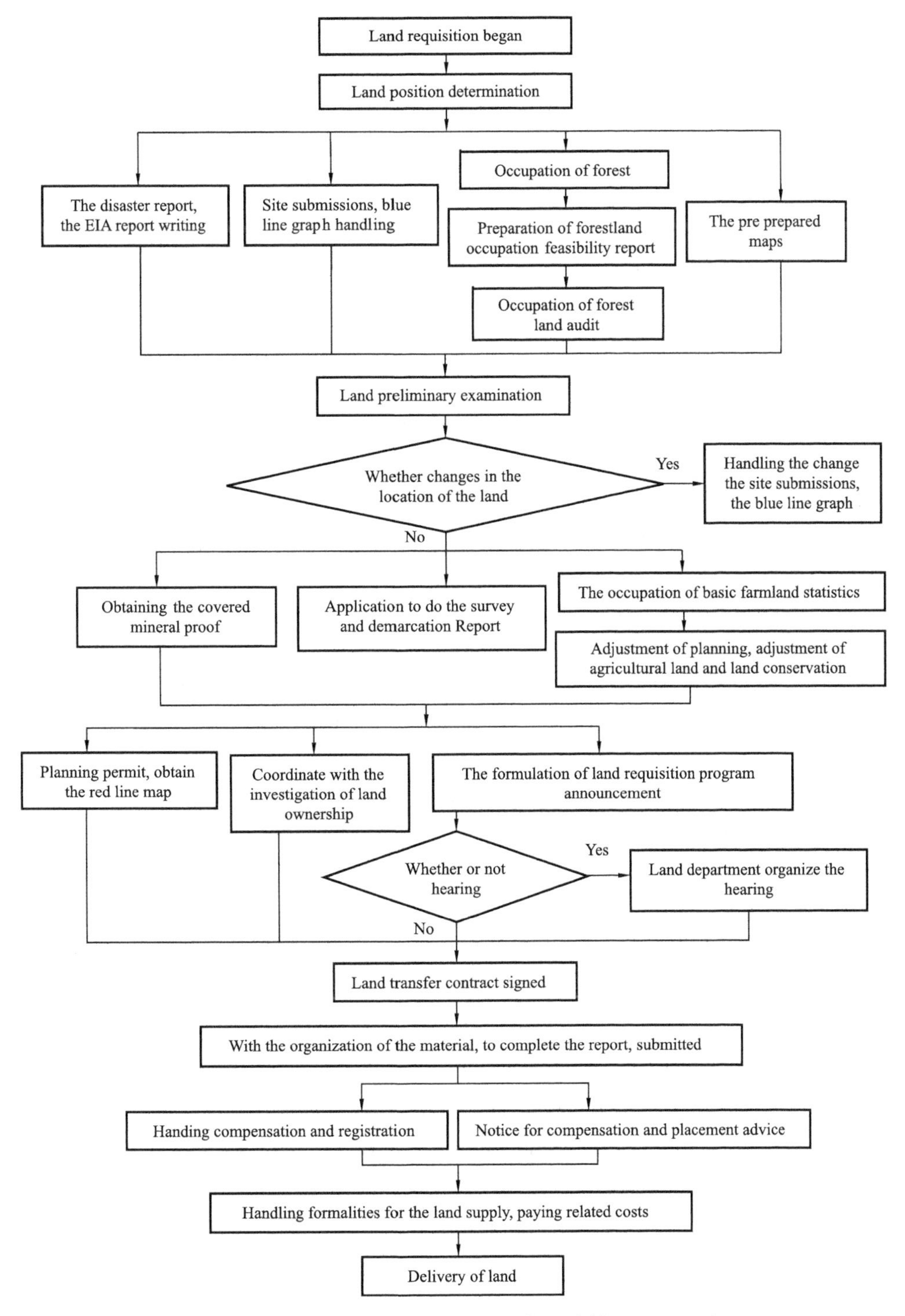

Figure 5-5 LNG project permanent land requisition process chart

① According to the decision of the State Council on deepening the reform of strict land management, article of the Nation [2004] No. 28 fourteenth "before the examination and approval of land requisition in accordance with the law, the use of land requisition, location, compensation standard, and resettlement measures should be told to the farmers whose land are expropriated; the survey results about land status must be approved by the rural collective economic organizations and farmers; if necessary, Land and Resources Department shall organize a hearing in accordance with the relevant provisions. The material about informed farmers with the land requisition, the identification of the relevant materials should be applied for land acquisition approval".

② According to the communication with local land management departments, the hearing of local planning and adjustment of agricultural land and land conservation can be combined together at the municipal level, but the compensation standard hearing of witnesses can only be carried out at the County level, in order to ensure the smooth development of land requisition work, it is recommended to rely on local government as far as possible.

(4) Registration for land requisition compensation.

① Registration authority: land requisition announcement designated government land administration department.

② Registration applicant: the owner, user of the expropriated land.

③ Registration deadline: stipulated period in land requisition announcement.

④ Registration required materials: land ownership certificate, certificate of property rights and other documents attached to the ground.

⑤ The consequences of not registered: not included in the scope of compensation.

⑥ With the local Land and Resources Bureau signing the initialed land paid transfer contract.

(5) Elaboration of land acquisition compensation and resettlement program.

① Draft authority: city and county Lands Department in conjunction with the relevant units.

② Draft basis: land registration data, the results of the field exploration, the checked registration of land requisition compensation, land requisition compensation standards by law.

③ Program content: land compensation, resettlement subsidies, young crop compensation, compensation fees of attachments matters.

④ Program announcement: program announcement of city, county land administration department at the location of the land requisition township (town) village, listen to the opinion of the rural collective economic organizations and peasants expropriated land.

⑤ Progressively approval: after report submitted, step by step approval, after examination and approval of the people's government at the same level, report to the people's government with agricultural land conversion, land requisition rights for approval step by step, until the approval of the State Council for agricultural protection zone adjustment and approval of the land use.

(6) The implementation of the program for land compensation and resettlement.

The implementation and approval authority: city, county government (and provincial government departments for the record).

(7) Implementation of compensation and resettlement program.

① Organization and implementation of authority: government lands departments at or above the county level. Project company should coordinate with the County Land Department for land requisition (issuing announcement of land requisition, compensation for registration, notice for compensation and placement suggestions, the tax payment and the payment of compensation and resettlement costs).

② Payment: within 3 months from the date of the announcement of the program, the project company pays the units and individuals of land requisition, failing to pay the cost according to the provisions, the units and individuals have the right to refuse to give the land.

(8) Land delivery.

Land requisition units and individuals shall, in accordance with the provisions of the procedures for the period, organize the ground material cleaning and deliver the land.

5.6.4.4 The Amount of Land Compensation Fees and Standards

(1) Land requisition compensation tax items

① Land compensation fees. The land used company should pay an economic compensation to the rural collective economic organizations with land requisition for the economic losses in accordance with the law.

② Young crop compensation. Land used company should pay young crops on the requisitioned land due to land requisition being destroyed with a compensation fee for the young crops planting companies and individuals.

③ Attachments compensation. The land used company pays compensation to the owner for attachments on the expropriated land such as housing, other facilities destroyed causing by land expropriation.

④ Resettlement subsidy. Land used company pays compensation to the land requisition companies for resettlement of surplus labor causing by land requisition.

⑤ Other costs include land reclamation fees, new construction land paid use fee, land management fee, and restore forest vegetation cover costs, infrastructure supporting fees and so on.

⑥ Taxes include farmland occupation tax, deed tax and so on.

(2) The amount of compensation standard for land requisition

① The land compensation fees specific criteria, the amount acts as the land requisition compensation and resettlement program regulations approved by cities and countries government in accordance with the laws.

② Determination of the average annual output value of the land having been expropriated in 3 years before (about the compensation standard for land compensation fees, resettlement grants) : according to annual statistical report of the most basic companies audited by Local Statistics Departments and price approved by the Price Department.

③ The land compensation and resettlement subsidies paid in accordance with the provisions cannot allow resettled farmers to keep original living standard, the resettlement fees can be increased. But the sum of the land compensation fees and resettlement fees shall not exceed 30 times of the average annual output value of the expropriated land in the previous 3 years.

5.6.4.5 Compensation Management and Ownership

(1) Treatment of the each compensation cost

① Land compensation resettlement fees paid to the collective in accordance with the law, young crop compensation at collective land and attachments compensation, are managed and used by land acquisition unit.

② Young crop compensation and attachments compensation are owned by the owners of young crop and attachments.

(2) Ownership and use of resettlement fees.

① Where a rural collective economic organization is located, it shall be paid to the rural collective economic organization, which shall be managed and used by the rural collective economic organization.

② Where the arranged by other units, the resettlement fees are paid to the other units for resettlement.

③ No need for a unified resettlement, the resettlement fees are paid to the resettlement personnel or after the consent of the resettlement staff for the payment of insurance premiums for the insured.

(3) Allocation of compensation expenses for collective ownership.

① Set up account deposit in local financial institutions.

② Open the public use, accept supervision by the villagers.

③ Allocation by a majority vote of "all villagers meetings" or "villagers representatives meeting", submitted to the township government for the record.

(4) Land expropriation compensation disputes and settlement modes

① Compensation standards controversy. Firstly coordinated by the government at or above the county level, the coordination failed, ruling made by the People's Government who makes land expropriation decision.

② Disputes over compensation cost allocation. The nature is a civil dispute, the parties are the village committee or rural collective economy and the villagers, the parties can be resolved through civil litigation.

③ Land information publicity disputes. It belongs to administrative dispute, and the parties can resolve the dispute through administrative review and administrative litigation mode.

5.6.4.6 The Land Permit Handling

(1) Issuing the "Construction Land Ratification".

"Construction Land Ratification" is issued to the land use company after approved by Land and Resources Bureau, a legal certificate can be used in land requisition, and the certificate will be recovered by land departments in the processing of handling land permits.

(2) Handling the land certificate.

The full name of land certificate is the land certificate of the People's Republic of China, and it is issued by the competent authority in charge of the land to the owner of the land after the

acceptance of the project.

5.6.4.7 Construction Process Permit

After the project land requisition contract is obtained, all of government permit work, we are collectively referred to as the construction process permit, including planning permit, land planning permit, construction permit and other engineering permits. In order to obtain a license in a timely manner, it should be fully mobilized all resources and advantages of the contractors and owners, when project in the implementation stage, according to the permission, consent, authorization, government approval and other applications and obtain parties (the company or contractor) are different, the license is divided into three categories.

(1) Class I.

It includes the approval of the project land, sea use certificates, temporary operating permit and so on, mainly responsible for the application by the owners. The contractors shall cooperate with the owners to ensure that the company timely get the license and the contractors shall provide all necessary assistance to the company, mainly including documents and technical support and so on.

(2) Class II.

It includes the necessary licenses and applications required by the government in the construction process, including: project planning, operating license, judicial verification of imported equipment, the use of a certificate of special equipment, company project quality inspection certificate and so on. The contractors are responsible for this kind of licenses, in the name of the owners, the owners need to provide the necessary support, including authorization support, the company file support and so on.

(3) Class III.

It includes qualifications obtained by contractor to engage in this project, tax registration, temporary construction power, temporary water use certificate, and the contractor supplier manufacturer qualification certificate and so on. It is responsible by the contractor independently.

Chapter 6 LNG Project Feasibility Study and Project Check and Approval Management

6.1 General Concepts of LNG Project Feasibility Study

6.1.1 Definition and Significance of LNG Project Feasibility Study

6.1.1.1 Definition of LNG Project Feasibility Study

Before the LNG project investment decision (PID), it needs to demonstrate on the relevant construction proposal, technical proposal and production management proposal in order to obtain the economic and technical solution. Starting from the overall system, the feasibility study aims to analyze and research various aspects, such as the LNG resource, market, engineering (including jetty, receiving terminal, pipeline and matching engineering), technique, economy, finance, business, health, safety, environmental protection and energy saving, etc., to determine whether the project is feasible and provide the scientific basis for PID.

6.1.1.2 Definition of LNG Project Feasibility Study Report

LNG project feasibility study report is the final text results of the project feasibility study. It is mainly based on the previous opportunity research and pre-feasibility study, then by the commissioned units and project team further to implement the basic data and analysis of market, resources, engineering, economy and other factors, with multi-proposal selection, and overall consideration to complete the deep processing LNG project overall written report.

6.1.1.3 The Significance of LNG Project Feasibility Study Report

(1) It is the basic conditions for the competent department of the state approval

LNG project is whether able to obtain the record and registration of the national competent authorities and its basis is still the project feasibility study report, because the report gives the reasons of the necessity and feasibility of project construction. Although the formal approval system by the national authorities is substituted by the record and registration, the basic condition for the application report is still the project feasibility study report. The report can reflect the full view of the project from the engineering technology, social development, economic benefit and people's life, etc. From the perspective of LNG project investment company, which on the economic benefits of the shareholders, from the perspective of national authorities, which focus on the social benefits much more, therefore, the report writers must give considerations on both aspects.

(2) It is an important link to the internal approval procedures of Project Investment Company.

As the initiated investment company to invest in a LNG project, must demonstrate whether the project meets the investment company's development strategy, whether rely on or promote the existing industries, whether it is the business development direction and new economic growth point for the initiated investment company, all those above must be demonstrated in the feasibility study report, and also is an important link to the internal approval procedures of project investment company.

(3) It is the basis for the bank to provide loans.

Like other projects, after the project being approved will enter a substantive resources investment stage, it needs for financing. It is necessary for the investor to provide the project feasibility study report to the bank for the evaluation of the project. The feasibility study report is the basis for the bank to provide loans, so as to arrange the fund delivery from the banks.

(4) It is a guiding document for the next step of the project.

For a LNG project, the project feasibility study report is the guidance document for FEED design, preliminary design and detailed design. So, the report should strive to complete comprehensively, clearly describe all kinds of risks, and put forward corresponding measures. For project sponsor investment companies, it should be made comprehensive economics analysis in feasibility study report, to find out the main factors that affect the project economy, including the economic sensitivity analysis on the project, the important factors how to influence the operation of the project, and put forward the solution. The conclusions on the economic analysis are also the economic evaluation benchmarks after the project is put into operation in the future. The report also demonstrates the overall scheme of arrangement, environmental assessment, safe operation and energy saving, etc.

6.1.2 Guiding Ideology and Basis for the Feasibility Study of LNG Project

6.1.2.1 The Guiding Ideology of Feasibility Study

(1) The introduction of LNG needs to obey the national energy development of the overall strategy.

With the increasing pressure of environmental pollution, the national demand for low carbon, clean energy is also getting more and more intense, especially for China's industrial developed coastal provinces, natural gas will be included in the environmental protection, high efficiency, and high-quality energy, always ranked in the priority development status. LNG investors should seize the historic opportunity to echo the overall strategy for national energy development, and timely start the LNG project, the project must comply with the national industrial and investment policy.

(2) The LNG project should be consistent with the overall planning and layout of regional

energy.

In the case of the investment objective, LNG projects should be consistent with the regional energy planning. A lot of research work on market needs to be done, including a full argumentation on feasibility study before deciding to introduce LNG to the area. The construction of the engineering must comply with the overall energy planning in the region.

(3) It should be paid more attention on affordable users of gas market.

LNG is used in coastal city in priority, replacing the liquefied petroleum gas, the existing turbine oil and other industry fuels. The main reason is that the economy is relatively developed in coastal areas, which has higher bearing capacity in the import price of natural gas. Moreover, natural gas power generation capacity will reach 70%, in favor of a period to achieve economies of scale, at the same time, the construction of a new LNG power plant, considering the amount of LNG and peak power requirements, the recent planning of new LNG power plant and layout should be properly controlled.

(4) Foreign investors are allowed to play a role in the project introduction.

If the project has taken the introduction of foreign capital, it should strive to achieve China holdings, and increase the voice of China in the project. On the other hand, it is also important to inspire foreign company's role, to make full use of foreign direct investor's experience in the design, construction and operation of the LNG project. In order to accelerate LNG projects in China and avoid detours, to achieve the desired purpose. In the future, LNG projects can be flexible and consider the introduction of foreign businessman.

(5) Overall planning and synchronous construction of industrial chain.

The whole project includes LNG jetty, receiving terminal, pipeline and the main user's projects. It should be tried to make a unified planning, approval, simultaneous construction of upstream and downstream projects, to select important users, to optimize construction scheme, to determine the scale of construction, peaking methods, economic analysis and so on. All of those need to be compared and demonstrated, in order to facilitate the decision in selection.

(6) To improve the overall economic benefits of the project.

It should take economic benefits as the center, strengthen the project market research, conscientiously implement the national principles on less input, more output, timely investment, rapid development, make everything possible to compress the project investment. In the stablepremise, to seek truth from facts optimization of project cost factors, to minimize the project targets of cost and improve the economic benefits of the project, to enhance the competitiveness of the project.

(7) To pay attention to the project social benefits in the local people's livelihood.

The LNG project feasibility study should pay attention to this project which brings to the local government social, economic and environmental protection benefit, for the improvement of regional environment contribution etc. Adhere to the construction principles which there are the market, competitive, sources of capital and benefit.

6.1.2.2 The Basis of the LNG Project Feasibility Study

(1) The principles of using relevant laws, regulations and appropriate standards

At present, the LNG industry is an emerging industry in our country. LNG project management is also a new thing. In order to successfully complete the design and construction of LNG project, must be based on relevant regulations and standards at home and abroad, especially in China, ISO, Europe and the United States and Japan and other countries. The relevant laws, regulations, standards adopted in the design and construction of LNG projects in our country are as follows.

① Strictly abide by the relevant laws and regulations of China.

② Strict implementation of the mandatory standards of the state, local and industry and the mandatory provisions.

③ Strict implementation of state and industry LNG standards.

④ Domestic LNG standard has not been prepared, then take the equivalent of the existing international LNG standard.

⑤ No technical standards and regulations have been set up at home and abroad, and the standards and regulations of the similar industries are adopted.

⑥ Foreign acquisition equipment, materials, manufacturing, inspection, the main use of international standards (such as ISO, the United States, the United Kingdom, etc.).

⑦ Domestic procurement of equipment, materials, manufacturing, inspection, will adopt international standards (such as ISO, the United States, the United Kingdom, etc.) or the relevant standards of our country.

(2) The corresponding industry engineering standards.

The LNG project engineering includes: receiving terminal, jetty, storage tank, pipeline, and power supply, communication, heating and ventilation, measurement, construction, water supply and drainage and the HSE environmental protection engineering. All kinds of section have their own standards and regulations. The industry standards will update gradually with the discovery of new materials and new technologies. It depends on the specific circumstances to adopt certain standards in the process of implementation so as to reach safety and less cost requirements. Specific engineering standards include the following contents.

① Receiving terminal-LNG industry standards.

② Port-Jetty industry standards.

③ Pipeline-Natural gas pipeline industry standards.

④ The power supply-Power industry standards.

⑤ City gas project-Municipal engineering industry standards.

⑥ Industry gas project-Natural gas industry standards.

⑦ Chemical gas project-Natural gas and chemical industry standards.

6.1.3 Feasibility Study Unit Qualification Requirements and Research Procedures

6.1.3.1 Feasibility Study Report Preparation Unit Qualification Requirements

(1) General requirements.

The compiling unit qualification of feasibility study report must be the professional design institute who has a Class-A aptitude. The LNG project includes 3 most important professions: jetty, receiving terminal and gas pipeline, which can be respectively completed by 3 professional design institutes.

(2) Itemized requirements.

The feasibility study report shall also include contents on LNG project construction market, LNG resources, economic evaluation and the HSE. As a whole project, it should also include the feasibility study report such as gas power generation project, city gas project, industrial gas project, and chemical project and so on. Finally, select a head institute to compile the final report. At present, the units to participate in the preparation of the LNG project feasibility study report as the following categories.

① The overall feasibility study report — Chemical Engineering Institute and Owner's Institute.

② The feasibility study report of jetty project — Port Engineering Institute.

③ The feasibility study report of gas pipeline — Pipeline Design Institute.

④ The feasibility study report of power generation project — Electric Power Design Institute.

⑤ The feasibility study report of city gas — by the Municipal Engineering Design Institute.

⑥ The feasibility study report of industrial gas — Relevant Industrial Engineering Institute.

⑦ The feasibility study report of chemical engineering — Relevant Chemical Engineering Institute.

6.1.3.2 Feasibility Study Procedures

(1) Signing contract with commissioned institute on feasibility study.

If the owners or subordinate institutes have the ability to design qualification, can be directly assigned tasks, let it take the feasibility study work; if without this ability, the owners can entrust the design institutes of the society, but whether internal or external institutes to do the work, it needs to take signing the contract and executing contract mode.

(2) Forming the project team for feasibility study.

The entrusted unit, according to work requirements and tasks, sets up the project team, and appoints the person in charge of the project, the project team is demand to make work plan and schedule. All above arrangements must be approved by the owner.

(3) Information investigation and collection.

Feasibility study is a large task with technical and complicated management, project data collection including prophase opportunity research and prefeasibility study report; basic data of jetty, receiving terminal, pipeline and utilities project involved, there will be some field observation and engineering geological drilling. It requires cooperation from the owner of the project.

(4) Construction design and scheme optimization.

According to the requirements of the contract, combined with market and LNG resources, on the basis of the basic design data and technology based on certain economic data, the project team proposes a number of alternative construction schemes, by comparison and evaluation, and selects and recommends the best construction scheme.

(5) HSE evaluation.

After the basic engineering program introduced, it is necessary to carry out the HSE pre-evaluation, and entrust the corresponding qualification units. When the HSE evaluation report is completed, the owner will report to competent department of the government, in order to get the HSE evaluation report approved by the government. The HSE evaluation is always the important content for feasibility study.

(6) The financial and economic evaluation.

In accordance with the requirements of the construction project economic evaluation method, the recommended construction plan needs to carry out the detailed financial and national economic analysis and calculate the corresponding evaluation index, to evaluate the financial benefits of the project and the rationality of national economy and efficiency. Only above two have reached the minimum requirements of the investment company's internal profit rate and the national social benefits, can the project be established.

(7) Preparation of feasibility study report.

① General requirements.

The feasibility study report of LNG project is a comprehensive introduction to the overall situation of LNG resources, marketing, engineering, economic and social benefits, which reflects the research results, corresponding to each part of the scientific calculation and analysis and forecasting, is the programmatic document of project investment decision. From the research content and the professional point of view, the feasibility study is usually divided into the following parts: LNG resources research, LNG and natural gas market, jetty, the receiving terminal, pipeline sub-projects and LNG overall project feasibility study.

② Special researches and report preparation unit.

It can be commissioned by a number of professional research units, the main research contents include: the project of LNG resources, the market and the implementation of ocean shipping; jetty, receiving terminal, gas transmission trunkline three of the sub-projects technology scheme selection; research and evaluation of HSE and seismic safety; project organization and personnel, land use, project the bidding and implementation plan; investment estimate and financing and financial and economic evaluation; the introduction of supporting gas use projects, including city gas, gas power plant and other gas projects.

③ The overall report preparation unit.

Based on the above research work and special reports, firstly, it needs to describe the suitability of LNG resources, gas market and marine transportation, focus on the 3 major sub projects: jetty, receiving terminal, the best combination of pipelines and engineering process optimization on a comprehensive demonstration; pay special attention to the safety fire control, earthquake safety and environmental water and soil conservation research, put forward the implementation of equipment and personnel arrangements; carry out investment estimation and financial analysis and sensitivity analysis of the project, find out the most influential factor of the economic benefits of the project, emphasize the relationship among users and special users and small users in the gas market; evaluate the social impact of the project on national economy, put forward the main risk factors of the project and the circumvention measures, finally draw the project recommendations and conclusions.

(8) Submitting feasibility study report.

After the overall feasibility study report is completed, the report compiling unit reports to commissioned side and listens to the opinions of the experts and leaders from commissioned units, and then revises and improves the report, finally, submits to commissioned units with satisfied report.

6.1.4 The Relationship between LNG Project Research and Check and Approval Management

6.1.4.1 Three Components of LNG Project Research and Check and Approval Management

(1) The basis and foundation.

Investors in the investment of LNG project before, it is necessary to seriously study the national natural gas utilization policy, at the same time to find the answer in the strategic direction of the investment on their own, when both have a great fit, it should be considered in the LNG project investment, such as the relation of LNG project research and check and approval management (Figure 6-1) shows, it is necessary to conduct a more extensive regional project opportunity study and a preliminary feasibility study of the project within the region, if the first two studies continue to support the investor to invest the LNG project, the investor will make decision to set up the project organization and LNG design and construction standard study, so as to lay the foundation for the next step of implementation.

(2) The feasibility study of the project.

The feasibility study of LNG project is based on the research of project opportunity and the preliminary feasibility study. It is mainly for specific LNG projects. The local natural gas market demand, international LNG market can provide the corresponding volume of LNG. It is necessary to carry out LNG receiving terminal, jetty location and pipeline route selection work timely, on the above basis, to conduct the 3 LNG sub projects scheme selection and design and construction demonstration, at the same time, it needs special research work including LNG and HSE, as well as the comprehensive economic evaluation and social benefit evaluation.

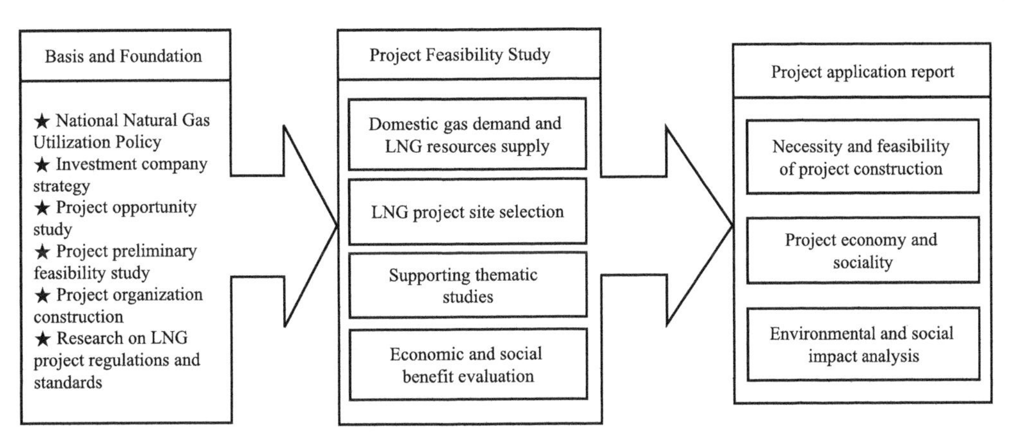

Figure 6–1 Relation graph of LNG project feasibility study and project check and approval report

(3) Preparation of project application report.

Feasibility study and report provides the basis for the project application report. For LNG, this major energy project, it has been directly approved by the NDRC, and the project application report is prepared according to the project feasibility report, and its focus is more concerned about the status and role of the project in the industry, the necessity and feasibility of the project construction. It pays special attention to the maximization of social benefits and to prevent the emergence of industry monopoly, to minimize adverse environmental impacts and get friendly along with the ecological environment.

6.1.4.2 Relationship among Three Components

(1) The basis and foundation Provide basis for feasibility study.

With the support of the national clean energy policy and the investment company's strategic guidance, LNG project investment will become a possibility. Project research opportunities and a preliminary feasibility study are the basis of the feasibility study, and project organization construction and the natural gas regulations and standards research both are to promote the organization guarantee and technology base of LNG project. Only above work has been done well, in order to lay the foundation for the project feasibility study. If the preliminary feasibility study is not up to the depth of research, and organization construction and technical standards are not in place, then the feasibility study work will have to rework, so both the funding will be wasted and the progress will be severely affected.

(2) Feasibility study is the basis for the project application report.

Project feasibility study is to further deepen on the basis of the preliminary feasibility study, and it focuses on the implementation of the specific LNG receiving terminal site, jetty location, pipeline routing, natural geographical environment, construction conditions, and environmental protection, under the above conditions, together with the economic and social impact of the project, to give the overall concept of the project. In addition to the necessity and feasibility of LNG project outside, project application report pays special attention to the regional economic and social benefits, environmental impact, community contribution. Therefore, only the project feasibility research achieves the depth requirements, and also pays attention to its social macro, such as environmental impact and social stability evaluation, in order to ensure the project application report is fully convincing, and avoid supplementary materials and some work continued to deepen and repeat national approval process, so as to improve the approval rate of the competent authorities of the state.

(3) Project application report is comprehensive reflection of the former two.

Successful approval of each LNG project application report means the basis and foundation are solid work and feasibility research work is in place, in turn, if the first two works can't reach the depth, such as improper LNG standards selected will affect future production safety and environmental protection, then the State Department will not approve the project. It is necessary to sum up the application process and provide experience for future project approval.

6.2 LNG Resources and Transportation Research

6.2.1 LNG Resources Research

6.2.1.1 General concepts

LNG resource is the most important preliminary work in the feasibility study of LNG project. It is to solve the problem about providing products to downstream users. Although there is a single LNG project to find a single resource, to change the trend of initiated investment enterprise unified coordination to supply LNG resources, but from the integrity of the project, where resources come from is still to be answered? The supply of LNG resource is single point to single point? It is single point to multipoint? Or it is multipoint to multipoint? The resources can meet the scale of the project or not? The resources can cover the LNG production cycle or not? This is the first question to be answered by an investment company when it is engaged in a LNG new business or expansion project.

6.2.1.2 Analysis on the Situation and Development Trend of LNG Resources Supply and Demand

(1) Analysis on the situation of natural gas supply and demand.

The first analysis work of LNG project is to confirm supply and demand of natural gas in the world, which is the basis of resource supply. The analysis work is based on the distribution of natural gas reserves, focusing on the related petroliferous basin, tectonic and reservoir types and the main gas producing countries and areas will be put into the development of gas fields and natural gas trade, focusing on the analysis of natural gas reserves to provide LNG resources.

(2) The analysis on the existing LNG production capacity.

The analysis focuses on the existing, construction, and expansion production capacity of international LNG liquefaction plants, but also the planning, potential international production capacity of LNG liquefaction plants and target market, targeted LNG projects related to LNG

manufacturers needs to be analyzed.

(3) LNG demand analysis.

The relationship between LNG consumption and supply and demand of LNG resources is the main point of research content, which in details includes analysis of the international market demand, the demand for LNG range, especially the requirement analysis of LNG market in the Asia Pacific region, including Japan, South Korea, India, Taiwan, China LNG market demand in the influence and proportion of regional demand, but also the existing LNG contracts and potential demand of equilibrium of supply and demand needed to be analyzed.

(4) LNG price and trend forecast.

LNG price mechanism, price system and the trend are the focuses of each LNG project concerned, it decides whether the project will be launched in a certain extent, but also is the most sensitive and core issue of the proposed project. It will focus on the analysis of Japan, Indonesia price mechanism and the price formula, especially the emerging exporter Qatar LNG price mechanism and the price formula impacts on the world LNG price.

(5) LNG trade development trend.

The analysis focuses on the dynamic development of LNG trade, including long-term, short-term and spot trade proportion, development trend, and then the LNG trade in natural gas prospects for long-term development of the overall trend in the allocation of resources and supply and demand are needed to be analyzed.

6.2.1.3 Analysis on the Relevant Situation of LNG Producers

(1) Natural gas reserves.

The analysis and demonstration focus on the LNG producers, which include target gas field reserves, investor's composition, and gas production period, output historic data, residual reserves and development plan, gas production could cover the LNG project production cycle should be particularly concerned.

(2) LNG production capacity and auxiliary conditions.

The analysis focuses on the LNG producers' processing capacity, production capacity, the other by-products manufactured, include: process flow, storage capacity, auxiliary production facilities, production management, staff and support conditions, the HSE system establishment and implementation, and also jetty and terminal throughput capacity and management etc. of host country should be analyzed.

(3) Market development of producers.

The analysis focuses on the LNG producers' country social development and stability, the legal protection, manufacturers marketing situation, including sales model—long-term, short-term, spot trade proportion, the supply of objects, supply contract period, the remaining supply capability and residual reserves and so on.

6.2.1.4 Analysis of LNG Resource Supplier

The analysis of LNG resource supplier is about its resources supply capacity, including as follows.

(1) Supplier's LNG resource supply requirements.

According to the characteristics of LNG resource providers (whether production and sales are combination), analysis of the basic situation of suppliers includes the ownership's structure, investment, production capacity, market development, the number of existing and potential users, transportation capacity, jetty conditions.

(2) Supplier's LNG resources business requirements.

The analysis focuses on the relationship between suppliers and manufacturers, suppliers and purchasers, including contract mode and contract period, delivery and payment of insurance and guarantee, etc.

(3) Other influencing factors.

The analysis focuses on whether the LNG supply has a significant impact on natural factors and the degree of influence, such as natural disasters, political unrest, terrorism, piracy and corresponding measures.

(4) The relationship with the local government

The analysis focuses on the relationship between LNG suppliers and the local government, such as the situation about implementation of the LNG supply contract, credit conditions. Also it needs to analyze the relationship between the Chinese government and the government of the country where resources are located, so as to minimize the political risk and ensure the supply.

6.2.1.5 LNG Resources and Overall Strategy

(1) Resources and transportation.

For the emerging LNG industry in Chinese, the analysis focuses on the transportation conditions of LNG resources, including bilateral jetty conditions, the ability to receive the ship, ship construction, delivery time, transitional arrangements, normal shipping plans, arrangement coincide.

(2) Resources and receiving.

The analysis focuses on matching analysis between LNG resource and receiving conditions, including storage capacity, regasification capacity, transmission capacity, standby supply condition, etc.

(3) Resources and users.

The analysis focuses on natural gas market conditions, including users, user amount and proportion, stable users'proportion, the demand for natural gas, peak shaving and reserve requirements, alternative energy arrangement, etc.

(4) Resources and policies.

Generally, LNG resource acquisition contract requires the approval of the competent authorities of the Chinese government, and the project company should fully listen to the guidance of government departments, but also obtain reasonable tax preferential policies.

(5) Resources and business.

The analysis focuses on the overall business arrangements, including bidding, invitation to negotiate, business strategy and proposals. After full argument, it should strive for the favorable LNG resource procurement contract terms, price terms, the supplier's transport capacity and shipping capacity, on the basis of full investigation, to propose the project LNG resources trade and transportation methods.

6.2.1.6 LNG Resources Selection

After LNG purchase bidding, negotiation, due diligence and signing, final results are obtained: to finalize the LNG supply of resources, implement the project's LNG resources volume, LNG trade mode, business contract, gas price model, effective conditions, implementation key points, potential risks and countermeasures, etc.

6.2.2 Research on Marine Transportation of LNG

6.2.2.1 The Overview of Worldwide LNG Transportation

The analysis focuses on the worldwide LNG transportation industry characteristics and the status quo, including the relationship between supply and demand of LNG transportation market, the utilization of ships, the current worldwide situation of LNG transport fleets (basic LNG transport fleets, size and safety condition) and the relationship between the LNG mode of transport and LNG business. Finally, the development trend of international LNG transportation, LNG trade and shipbuilding are obtained.

6.2.2.2 To Select the Project LNG Transportation Mode

(1) Comprehensive analysis on the basic condition of transportation.

The analysis focuses on the both domestic and international LNG ship building, transportation status and future development, and also bases on the project jetty, hydrological and meteorological conditions, combines with the LNG supply of resources, export jetty conditions, LNG resource procurement contract characteristics, trade amount and market implementation situation, to carry out the comprehensive analysis of LNG transport capacity demand.

(2) To make transportation plan.

① The analysis focuses on the study and to formulates the basic transportation scheme of the project and the transitional period of transport plan, and to put forward the overall strategy of LNG transport under the trade of FOB (Free on Board, by the buyers to provide transportation), CIF (Cost; Insurance and Freight, supplied by the seller transport) or DES (Delivered Ex Ship, supplied by the seller without the import customs clearance of goods), and to carry on the related due diligence.

② The transport scheme includes but not limited to: the basic situation of ship-owners and ship management companies, transports party credit and operational performance, ship source, ship personnel and the basic situation, the ship in the transitional period and the ship crews' arrangement.

(3) Transport ship selection.

① If the adopted ship needs a new order, it should be given the basic situation of shipyard, including the performance of technology, equipment, shipbuilding facilities, safe operation management, financial credit, contract fulfillment and delivery of the ship as expected and so on.

② If the ship is on loan, it needs further investigation of the owner, including financial credit status, the status of fulfilling a contract, lease contract, ship performance, capacity and the project matching condition.

6.2.2.3 LNG Transportation Risk Measures

(1) To deal with business risk.

① Deal with business risk through the transport business contract, freight mode, effective conditions, implementation, terms of the contract.

② Deal with the production risk by scheduling and shipment arrangement, pipeline network, storage, peak-shaving and emergency safety analysis.

(2) To deal with the risk of transportation.

① Deal with the risk of transportation through the opening of jetty, routes and channel conditions, not continuous operation days (meteorological conditions, jetty operating conditions etc.), shipping quality and performance guarantee, shipping management analysis.

② Deal with the ship shore matching risk through the LNG ship and jetty matching, the unloading rate and receiving system matching, ship capacity and tank capacity matching.

6.3 LNG Market Study

LNG resources and market research, can be called LNG project feasibility study of the most important two preliminary work, the former is to solve the problem of product, while the latter is to solve the problems of purchaser, only these two have been implemented, the LNG project can be successful.

6.3.1 Demand for Natural Gas Market and its Characteristics

6.3.1.1 Chinese Natural Gas Market Status and Development Prospects

The analysis focuses on the current situation of Chinese natural gas market, including reserves, production and sales volume, pipeline construction and operation, natural gas production and consumption difference. The development of China natural gas market, including natural gas demand of regional distribution, natural gas consumption structure, the oil companies to participate in the international trade of natural gas.

6.3.1.2 The Project Demand for Natural Gas Market and its Characteristics

(1) Analysis of the industrial gas demand.

① Natural gas demand on existing industrial projects.

② Natural gas demand for natural gas replacement industrial projects.

③ Natural gas demand on new projects.

④ Natural gas demand on the planning projects.

⑤ Natural gas demand on the potential industry.

(2) City gas demand.

City gas users include the urban residents, industry and business customers, LNG filling, and gas supply for car and gas air conditioner. According to the population and the economic development of the cities, it can determine the rate of gasification, the ratio of consumption of different kinds of customers and the amount of gas use in each city.

(3) Natural gas demand of power plant.

① The existing gas power plant of natural gas demand.

② "Oil to gas" power plant of natural gas demand.

③ The planning of new LNG power plant of natural gas demand.

(4) Other potential natural gas demand.

In addition to above users, the analysis focuses on other users and potential users.

(5) The utilization characteristics of natural gas.

Based on the data of gas consumption provided by the general customers, then calculate the monthly uniformity coefficient, daily uniformity coefficient and hourly uniformity coefficient and analyze its characteristics.

(6) Comprehensive analysis of characteristics.

According to the special users and big users, such as gas power plant, it is necessary to conduct a comprehensive analysis, to obtain the special user market uneven coefficient, the special user market monthly uneven coefficient, daily uneven coefficient and hourly uneven coefficient, to provide the basic data for the entire supply chain capacity arrangement.

(7) The total market demand.

The natural gas demand of all kinds of users is summarized, and the scale of the proportion of each user type and the characteristics of market structure are analyzed.

6.3.2 LNG Price Competitiveness Analysis

6.3.2.1 Analysis of LNG competitiveness

(1) Analysis of the LNG and the existing natural gas competitiveness.

Introduction of LNG with the existing competitiveness of natural gas needs total cost analysis, including gasified LNG to the gate station price, due to the difference of gas price and calorific value conversion and environmental protection effect and social benefit.

(2) Analysis of LNG and other alternative energy competition.

According to the LNG quality provided by the upstream resources side, based on calorific value, equipment transformation, labor cost, social benefits, and other alternative energy sources, such as coal, petroleum, LPG, oil, etc., it needs to carry out the comparative analysis of competitiveness.

6.3.2.2 Analysis of User's Price Bearing Capacity

(1) Power plant price affordability analysis.

According to the feasibility study of the power plant, the feed-in tariff and gas price bearing capability analysis should be carried on, including the economic comparison of LNG for power generation, the peak price, base load price, price bear limit, and the analysis of the relationship between feed-in tariff and LNG price.

(2) The analysis of price affordability for industrial users.

According to the industrial users of alternative fuels as well as production management, the analysis focuses on the natural gas price affordability, including the transformation cost, natural gas and calorific value and benefit of alternative fuel, environmental benefits, energy saving and land saving and comprehensive benefit.

(3) Urban user's price affordability analysis.

According to the situation of city gas users of alternative fuels, the analysis focuses on the price difference between the two, including the city residents, commercial users with gas, natural gas vehicles, gas air conditioning, combined with the local actual situation, to analyze the user's price affordability.

(4) Potential user's price affordability analysis.

According to the situation of potential users, the analysis focuses on price affordability, based on the calorific value and price data on the market of LNG and potential users of energy.

6.3.3 LNG Receiving and Natural Gas Purchase and Sale

6.3.3.1 LNG Purchase Agreement

If the LNG resources project has been implemented, it needs to analyze the main clause LNG purchase agreement signed, including the annual supply, LNG component, LNG transportation, LNG price formula, effective conditions, "Take or Pay" clauses, etc., as well as the impact of these clauses on LNG project company gas distribution and downstream users.

6.3.3.2 Natural Gas Sales Agreement

If the project has signed a gas sales agreement, the project's main items of the gas sales agreement should be analyzed, including main users of gas consumption amount, gas prices, supply and demand terms, "Take or Pay", peak shaving, gas quality, and other core business and technical provisions, as well as other constraint conditions to upstream suppliers and the LNG project company.

6.3.3.3 The Both Match Analysis

As a middleman LNG project company, needs to analyze the following key elements: how to create storage solutions, including storage capacity, gasification capacity; how to choose the best LNG and gas reception, distribution schemes; how to deal with the season peak shaving, month peak shaving, day peak shaving; and how to build emergency and backup schemes.

6.4 The Jetty Sub-project Feasibility Study

6.4.1 General Concept of Jetty Study

Jetty sub-project is an indispensable part of imported LNG project, including anchorage, inbound and outbound channel, mooring and berthing jetty facilities for loading and unloading operations. Whether it is new construction and renovation projects, the proposed optimal jetty solution needs full argument on the basis of comparison and selection.

6.4.1.1 Basic Work

(1) Basis.

① Preparation of contracts and the related minutes signed by the project owner and the entrusted institute.

② Project pre-feasibility study report.

③ Assessment submissions of expert review.

④ Jetty planning in project area.

⑤ Compilation method of feasibility study report of construction project (Ministry of Communications).

(2) Basic data.

It includes: engineering survey, engineering geological investigation, measurement scan detection in jetty area, hydrological survey, earthquake safety evaluation, the navigation environment safety assessment report etc., and the information about selected site near the adjacent unit, the main industry standards and design specifications.

(3) The scope of the study.

It includes the LNG terminal piers, trestle, revetment, land formation, foundation treatment, water intake and outlet structure, torch and pipe rack foundation, dock basin and waterways, as well as the required power for the ship docked, water supply and drainage, fire protection, communications, control, environmental protection and other facilities and temporary works.

6.4.1.2 Proposed Jetty Status

(1) The basic situation.

Based on full demonstration and comparison, it needs to propose the jetty and terminal location, status quo and development planning of the jetty berths, jetty anchorage, channel, turn round water region and anchorage status.

(2) Regional advantage analysis of the proposed LNG jetty and port.

① According to the port and the main users of geographical distribution, from the construction difficulty, land, shoreline resource utilization efficiency, economic benefits and other aspects, the analysis focuses on the jetty location advantages and port construction necessity.

② It needs to demonstrate the advantages of the LNG jetty region from location, shoreline conditions, investment environment, facilities, throughput prediction, ships'type analysis and construction scale.

(3) The development status of local national economy

The research focuses on the local basic situation of economic and social conditions, including land area, population and economic characteristics, mineral, forest, land, water resources and plant species resources.

(4) LNG and natural gas consumption analysis of local area.

The analysis focuses on the market status and development of natural gas in local area, especially for user's load center position, the main users, including city gas, industrial users, distribution patterns and potential users of gas power plants forecast.

(5) Jetty construction scale.

The scale of construction of jetty sub project is proposed through comprehensive analysis.

6.4.1.3 Analysis of LNG Ship Type

(1) Analysis of the main world LNG ship type and LNG routes.

To analyze the world LNG ship type and development trend, the main LNG routes of the word , and focuse on the sea condition and navigation condition of this project.

(2) The characteristics of LNG transport ship.

① The analysis focuses on the design and construction of LNG ship, new type of ships and the relationship of price with tank capacity, as well as the type of LNG transport ship and manufacturers, according to the characteristics of LNG transport ship.

② According to China's LNG shipyard ship building experience, the analysis focuses on the possibility of building LNG ships, proposes the construction of LNG ship in domestic shipyard proposal.

(3) The ship type selection of design and ship source.

The tank capacity range and the main type of LNG ship are listed, and to explain the source of the ship, and propose the main ship type used in this project and the scope of the ship type berth.

(4) Ship freight analysis.

The analysis focuses on the main factors, including transport distance, age of the ship, ship type, ship numbers, charter period and recommend reasonable transportation plan, and analysis of the ship freight constitute, etc.

6.4.2 Study on Project Design of Jetty Technology

6.4.2.1 Layout

(1) The general layout scheme comparison.

It should be based on the selected jetty site, carry out the multi schemes comparison for the LNG jetty, receiving terminal, work boat berths, by the multi layouts of economic and technical comprehensive comparison, to recommend the best layout plan.

(2) Jetty layout.

① It should be combined with the water area condition, the receiving terminal, land formation, work boat dock layout, and various influencing factors to select the best jetty location.

② It should be combined with the surrounding water depth, the wind, wave and current conditions, conveniently berthing and unberthing operation, and the analysis focuses on the constant wind and secondary constant wind angle, harmful wave direction and harmful secondary wave direction angle, current direction angle, and contour angle and channel angle to determine the jetty axis.

(3) Jetty length.

According to "JTS 165—2013: Design Code of General Layout for Sea Ports", and other relevant standards, the size of the ship berthing, loading and unloading, mooring conditions and other operation requirements should be taken into account, to determine the LNG jetty length and work boat wharf length.

(4) The work platform of LNG jetty.

LNG jetty platform scale is mainly composed of unloading arm, power distribution room, control room, fire cannons and other facilities, it is strived to meet the requirements of reasonable layout and safe distance.

(5) Dolphin arrangement.

According to the "JTS 165-5—2016: Code for Design of Liquefied Natural Gas Port and Jetty, 5.4.5 section" and related specifications, to determine the dolphin arrangement layout type and setting distance.

(6) Mooring Dolphin layout.

It should be able to meet a variety of design ship mooring angle and the length of the line.

(7) Jetty mooring.

The length of berth shall meet the safety of ship berthing, unberthing and mooring operation requirements, according to 1~1.2 times of ship length to estimate the berth. Pier type LNG jetty should set two dolphins. It should meet the LNG ship in the local maximum tidal range and wave range requirements for the safe operation.

(8) Hydraulic structures.

The multi scheme comparison and selection should be done in the LNG jetty and trestle, workboat dock, receiving terminal revetment, take and drainage buildings, the torch base, trestle structure, recommend the best solution.

(9) The waters in front of LNG jetty berth.

According to the "JTS 165—2013: Design Code of General Layout for Sea Ports", it needs to determine the width of the waters in front of LNG jetty berth and workboat berth, and the water depth and bottom elevation should be designed and calculated based on "JTS 165-5- 2016: Code for Design of Liquefied Natural Gas Port and Jetty".

(10) Harbor basin and connecting waters.

In the premise of ensuring the ship safety and handling, it needs to determine the swirling water diameters of LNG harbor basin and workboat harbor basin. Usually the LNG berth harbor basin, connecting waters designed bottom elevation and jetty designed bottom elevation are the same.

(11) Channel.

It should be contrasted the design length of the ship to determine channel width, bottom elevation of design channel. Harbor basin and jetty bottom elevation are the same.

(12) Anchorage.

According to local port planning and after soliciting the opinions of administrative departments, the anchorage range should be basically determined based on "JTS 165-5-2016: Code for Design of Liquefied Natural Gas Port and Jetty", the LNG ship should set up special anchorage, and the distance with other anchorages shall be greater than 1000m.

(13) Load design.

Fixed and mobile equipment, pipelines, LNG jetty work platform, work boats dock, mooring force, impact force, ship breasting force load, etc. should be analyzed.

(14) Receiving terminal revetment.

The revetment structure should be determined based on natural conditions, local conditions and construction material prices, etc.

(15) Breakwater construction.

The comprehensive construction with the breakwater operation days, breakwater construction investment and construction period should be analyzed, to determine the construction of breakwater.

(16) Boats for the jetty.

According to the provisions in the design code of JTS 165-5-2016 LNG jetty, LNG vessel berthing should be equipped with three or more tugboats assistance, while unberthing should be equipped with two or more tugboats assistance. According to the needs of ship operation, also it is required to equip with guard boats, mooring boat, traffic boat etc.

(17) Navigation facilities.

In order to improve the safety of navigation of LNG ship in and out of jetty, as the pilot assist pilotage, it shall be equipped with high precision overlay chart differential positioning system, monitoring system and setting auxiliary berthing at the jetty, in jetty and workboat pier, shall be set up with bankheadlight, on both sides of the channel and the harbor basin border shall be set up with navigation buoy etc.

6.4.2.2 Longitudinal Design

(1) Water level design.

Designed high water level, designed low water level, extreme high water level (return period of 50 years, 100 years of annual extreme high water mark) should be listed, extremely low water level (return period of 50 years, 100 years of annual extreme low water).

(2) Jetty top elevation.

① Top elevation of LNG jetty: top elevation shall be determined according to the "JTS

165—2013: Design Code of General Layout for Sea Ports" or other relevant provisions of the standard.

② Top Elevation of workboat pier: top elevation shall be determined with the comprehensively consideration of the effects of surrounding waters as well as the waves.

(3) The receiving terminal land elevation.

The receiving terminal land elevation shall be determined by extreme high water level in history, and terminal drainage and cut and fill balance in the project region.

(4) The wave wall revetment elevation.

The wave wall revetment elevation shall be determined according to JTS 165-5—2016 LNG terminal design specifications and project region.

(5) Trestle width and its top elevation.

The trestle width shall be determined according to the requirements of process and the road. The trestle's top elevation shall be determined according to the connection of jetty work platform elevation and terminal bank revetment elevation.

6.4.2.3 Supporting Sub Projects

(1) Land formation and foundation treatment.

According to the natural geographical conditions of practical jetty, it needs to put forward the feasible scheme of land formation, and to determine the design elevation. According to the site engineering geology and drilling, determine the standard and the foundation treatment scheme.

(2) Power supply and lighting.

① It is necessary to study and determine the power, lighting design and design demarcation point relevant to LNG ship operations, and to calculate electricity load and select the power supply as well as the major electrical equipment relevant to LNG vessel operations.

② It needs to compare and select low-voltage distribution lines and cable types, determine laying way and set warning lights, lighting systems as well as grounding system.

(3) Water supply and drainage and fire control.

According to the standard, it needs to determine the water supply system of LNG ship and jetty surface water drainage system, equipped with fire extinguishing system and facilities.

(4) Communications and traffic control.

It needs to equip with ship-shore communication, including very high frequency (VHF) coast stations, shortwave single sideband (SSB) radio, etc. according to relevant specifications. It needs to select guide and navigation systems, including differential global positioning system (DGPS), vessel traffic services (VTS), etc. according to specifications.

(5) Control and computer management.

It should be installed including ship fire monitoring, inspection, operation, management and computer control system, equipped with ship programmable logic controller (PLC) control system, ship fire interactive television (ITV) system, fire control cabinet, on-site fire detection device and control system of auxiliary equipment etc.

6.4.2.4 HSE Design

(1) Environmental protection.

According to the results of environmental monitoring, it needs to sum up the status of air environment quality, the water quality investigation and analysis on the present water quality status of the overall area. According to the monitoring results of environmental noise, to summarize the status of the sound quality of the environment through the area; the main pollution sources and pollutants on the construction period and operation period, analysis of the environmental impact; the various pollution sources and pollutants, to take control and protection measures; finally, to put forward the scope of environmental management and environmental monitoring responsibilities and requirements, the existing problems and suggestions.

(2) Occupational safety and health.

In view of the natural conditions of the operation of the wind, fog, waves, rain, lightning, sunshine and other safety and health hazards, It needs to list the main effects and preventive measures. The various occupational hazard factors in the production process are required to be analyzed, in order to put forward the measures for the protection of occupational safety and health, and put forward the problems and suggestions.

6.4.2.5 External Collaboration Conditions

External conditions include construction and port planning, from the location of the project of highway, railway traffic engineering and planning, surrounding facilities and institutions, and it needs to put forward available social life and engineering support facilities.

6.4.2.6 Construction Conditions

(1) Summary.

The analysis focuses on the project facilities and ancillary facilities, land and water conditions, weather conditions, the amount of demolition, road traffic, sand and gravel and other building materials supply and transport capacity, transport distance, construction of water, electricity, telecommunications and etc.. It needs to list the major engineering and the amount of hydraulic engineering.

(2) The construction organization plan.

After comparison and analysis, the following optimization scheme is proposed:

① The scheme of the reclamation and external dredging earthwork amount and transportation distance.

② The main structure and construction scheme of the LNG jetty and trestle engineering.

③ The structure and construction scheme of workboat pier.

④ The revetment structure, construction scheme and the foundation treatment construction scheme.

(3) Other supporting engineering.

It includes jetty fire control, power supply and lighting, control, communication, ladder, quick release hooks, navigation aids and other facilities with engineering quantity, construction scheme and the reasonable arrangement of construction schedule.

(4) Large temporary works.

It includes the caisson, concrete post tensioned prestressed beam, steel bridge fabrication and large temporary construction yards and facilities and construction scheme.

(5) Construction period.

It is necessary to list the major project construction sequence and construction period, and make appropriate adjustments based on project progress.

(6) Investment estimation and economic evaluation.

It is required to work out the project investment estimatation, including the analysis of unit costs, unit fee rates and operating costs, and carry out the economic evaluation on the basis.

6.4.2.7 Comprehensive Demonstration and Recommendations

(1) Recommending scheme.

Based on the different land, jetty, berths, water intake and outlet arrangement, it needs to propose several general layout schemes, and carrys on the analysis, put forward recommendations and existing problems and suggestions.

(2) Preparation of attached drawings.

In order to explain directly, the jetty engineering of the LNG project needs the following figures: geographical location and jetty plan, general layout, working platform structure, breasting dolphin structure, mooring pier structure, trestle structure, workboat pier section, revetment structure section, etc.

6.5 Receiving Terminal Sub-project Feasibility Study

The LNG receiving terminal is the work station for LNG receiving (including ship unloading), storing, vaporization and transportation (including tanker loading station). This section is suitable for investment in new and expansion of imported LNG terminal sub-project.

6.5.1 The Basic Research of the Receiving Terminal Engineering

6.5.1.1 The Main Basis and Research Scope and Content

(1) Main basis.

① Contract documents signed by the project owner and entrusted institute and relevant conference minutes.

② The concept design or FEED design of the LNG receiving terminal.

③ The pre-feasibility study report.

④ The expert assessment review comments.

⑤ The overall planning of the project area.

(2) Research scope and content.

The research scope includes the study of plan layout, storage tank, equipment selection, equipment sparing, technical process, LNG ship to shore uploading transfer, utility and off-set utilities, the optimal design of the safety and sanitation, firefighting, environmental protection, as well as the investment estimation of the receiving terminal sub-project.

6.5.1.2 Receiving Terminal Size and Composition

(1) Function and size of the receiving terminal.

According to market research, it needs to predict gas demand and potential demand of LNG project, based on the receiving terminal function, to determine receiving terminal project size and project construction schedule. With annual turnover as the foundation, to comprehensively consider the parameters of LNG storage tank (design pressure, design temperature, daily evaporation rate, the inner tank material, inner tank diameter and height, the outer tank diameter and height, etc.). With the factors affecting gas market change, determine the receiving terminal in dynamic storage capacity.

(2) The receiving terminal composition and optimization.

The size and composition of process, utility and off-set utilities should be optimized and selected, the compatibility of the constituent parts (including jetty, receiving terminal, gas transmission line) needs to be further demonstrated, then the cohesive relationship among main facility, process, planning, market, resource and transportation need to be analyzed.

(3) Equipment configuration and standby principles.

The analysis focuses on the configuration of main process equipment of receiving terminal, including unloading arms, LNG tanks, LNG low pressure pumps, LNG high pressure pumps, BOG compressors, BOG condensers, sea water pumps, vaporizers etc.. According to users' characteristics, gasification capacity, pipeline gas transmission capacity, pipeline gas storage capacity, it needs to optimize the composition of the device and configure the proportion of standby equipment.

(4) Project overall process and control.

Based on optimization, the research focuses on the process and control of the receiving terminal, including LNG ship unloading system, LNG storage system, BOG process system, LNG transport system, LNG output system, metering system, tanker and ship loading system, flare system, utility system and so on.

(5) The gas product specifications.

It is necessary to list the selected LNG resource supply contract or letter of intent, to give the scope of the corresponding LNG component and proposed LNG main components, including the calorific value adjustment etc.

(6) The maximum hourly flow rate of natural gas output

Analysis provides hourly peak gas consumption of large industrial users and power users and the average gas consumption in peak months or peak days. According to market research, it needs to consider vaporization facilities ability, gas storage ability of pipeline, LNG transport capacity and operating efficiency, to determine maximum LNG output ability and peak-load ability. According to the annual production plan and maintenance design, it needs to calculate LNG project operation in hours. Finally the main facilities design capabilities and whole project design ability is to be summarized.

(7) LNG ship type selection.

The project size and jetty characteristics should be fully considered, to determine the minimum and maximum of ship volume range, under the requirements for the safe operation of the ship's berthing conditions, to recommend the size of the major marine vessel. The study includes LNG transport ship capacity, onshore LNG storage tank capacity, onshore natural gas trunkline transporting and its gas storage capacity and matching with LNG transport ship during the voyage, LNG transport ship in time delay, to meet the peak-valley changes of gas market.

(8) The maximum output volume of the LNG tanker.

If there is LNG tanker system designed, the research focuses on LNG tanker loading platform and plane layout of LNG terminal, the safe operation and the maximum output volume.

6.5.1.3 Layout

(1) Device partition.

Layout bases on the equipped facilities, research results about environmental influence and safety evaluation and the latest measured data etc. According to the division of functions, it can be divided into 7 zones: jetty, LNG storage tank, processing, utility, sea water intake area, administrative office and tanker loading.

(2) Layout specification.

The layout design should be follow the existing national and international industry laws, standards, regulations, natural conditions and mature production experience. The zoning should be based on different function of each unit. It should be based on the simulation results on the requirement of fire and heat radiation, to determine the spacing between each functional zone. At the same time, the flow path must be considered reasonable direction, reasonable pipe arrangement, safe operation, equipment maintenance and installation convenience channel etc.

6.5.2 Technology Scheme and Automatic Control

6.5.2.1 LNG Ship Unloading System

A matching LNG unloading system must be set in order to safely and quickly unload LNG from ship to storage tank in the receiving terminal after LNG ship berthing.

(1) Unloading arm number and BOG return arm.

According to the matching of LNG ship and jetty, under the conditions to satisfy the LNG storage scale, the numbers of matching unloading arms complied with the unloading rate and BOG return arms need to be equipped through research.

(2) Emergency release joint and interlock system.

In case of emergency the LNG ship can be safely discharged from the unloading arm as soon as possible, the emergency release joint and interlock system need to be equipped on the unloading arms.

(3) Thermometer set.

For the control, detection of pipeline pre cooling temperature, thermometers should be set at different positions on unloading pipeline.

(4) LNG collection tank.

In order to collect remnants of the LNG at the branch pipe end after ship unloading, LNG collection tank and heater should be set in jetty, to gasify the LNG and return to evaporator manifold.

(5) Transport process.

At the beginning of the unloading, a small amount of LNG is used for cooling unloading arm and auxiliary facilities. Under the normal unloading, LNG will be conveyed by unloading arms via the transfer pumps on the ship, and collect LNG to main pipeline through the branch lines, and transport LNG to the storage tank through the main pipeline. After LNG entering the tank, it replaces the BOG, and BOG is transported to the LNG ship cabin by return pipe. In the absence of unloading operation, a small amount of LNG flows from main pipe to re condenser through a circulation pipe from the low pressure output main pipeline, to keep LNG unloading main pipe in cold standby.

6.5.2.2 LNG Storage System

(1) Types of storage tank.

① Single containment tank.

A single containment tank shall consist of only one container to store the liquid product(primary liquid container). This primary liquid container shall be a self-supporting, steel, cylindrical tank. The product vapours shall be contained by: either the steel dome roof of the container; or, when the primary liquid container is an open top cup, by a gas-tight metallic outer tank encompassing the primary liquid container, but being only designed to contain the product vapours and to hold and protect the thermal insulation. Depending on the options taken for vapours containment and thermal insulation; several types of single containment tanks exist. A single containment tank shall be surrounded by a bound wall to contain possible product leakage. For examples of single containment tanks, see figure 6–2.

② Double containment tank.

A double containment tank shall consist of a liquid and vapour tight primary container, which itself is a single containment tank, built inside a liquid-tight secondary container. The secondary container shall be designed to hold all the liquid contents of the primary container in case it leaks. The annular space, between the primary and secondary containers, shall not be more than 6.0 m. The secondary container is open at the top and therefore cannot prevent the escape of product vapours. The space between primary and secondary container can be covered by a "rain shield" to prevent the entry of rain, snow, dirt etc. For examples of double containment tanks, see figure 6–3.

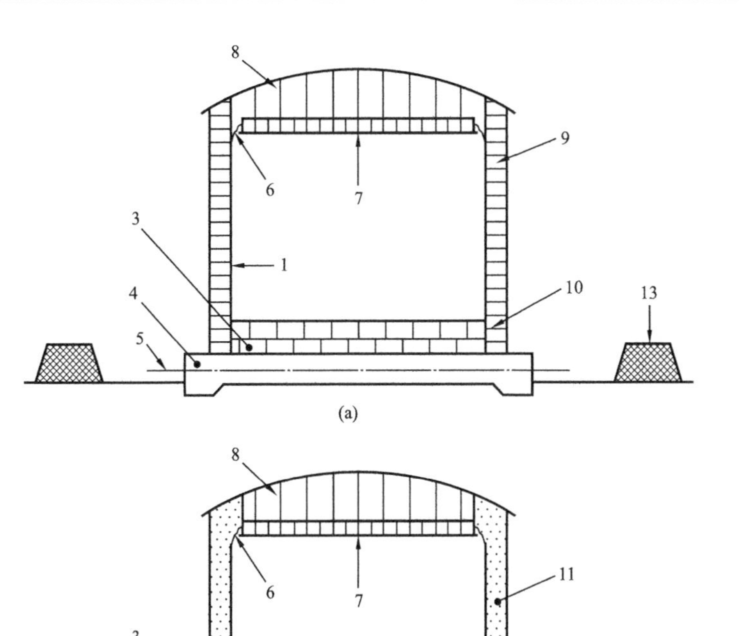

Figure 6–2 Examples of single containment tanks

Key

1—primary container (steel) ; 3—bottom insulation; 4—foundation; 5—foundation heating system ; 6—flexible insulation seal; 7—suspended roof (insulated) ; 8—roof (steel) ; 9—external shell insulation; 10—external water vapour barrier; 11—loose fill insulation; 12—outer steel shell (not capable of containg liquid) ; 13—bound wall

③ Full containment tank.

A full containment tank shall consist of a primary container and a secondary container, which together form an integrated storage tank. The primary container shall be a self-standing steel, single shell tank, holding the liquid product. The primary container shall: either is open at the top, in which case it does not contain the product vapours or equipped with a dome roof so that the product vapours are contained. The secondary container shall be a self-supporting steel or concrete tank equipped with a dome roof and designed to combine the following functions: in normal tank service: to provide the primary vapour containment of the tank (this in case of open top primary container) and to hold the thermal insulation of the primary container; in case of leakage of the primary container: to contain all liquid product and to remain structurally vapour tight. Venting release is acceptable but shall be controlled (pressure relief system). The annular space between the primary and secondary containers shall not be more than 2.0 m. Full containment tanks with thermal insulation placed external to the secondary container are also covered by these requirements. For examples of full containment tanks, see figure 6–4.

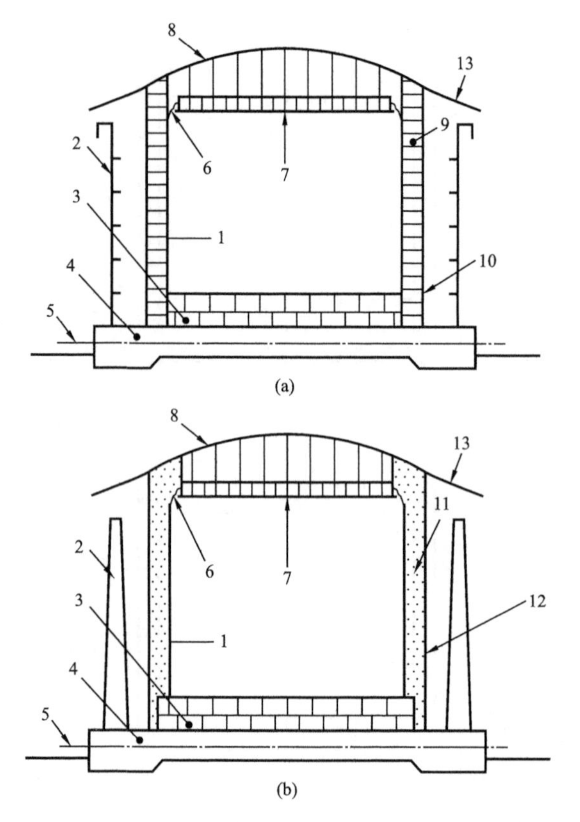

Figure 6–3 Examples of double containment tanks

Key

1—primary container (steel) ; 2—secondary container (steel or concrete) ; 3—bottom insulation; 4—foundation; 5—foundation heating system; 6—flexible insulating seal; 7—suspended roof (insulated) ; 8—roof (steel) ; 9—external insulation; 10—external water vapour barrier; 11—loose fill insulation; 12—outer steel shell (not capable of containg liquid) ; 13—cover (rain shield)

At present, the full containment tanks constructed in China, the tank bottom and tank wall structures are as follows (Table 6–1) :

Table 6–1 The tank bottom and tank wall structures

The tank bottom, from the bottom to the top	The tank wall, from outside to inside
Carbon steel lining plate	Prestressed concrete wall
Concrete leveling layer	Carbon steel lining
Foam glass brick	TCP (Thermol corner protection) back glass brick
Asphalt felt	Elastic mat
Concrete leveling layer	The inner tank (Ni 9% steel)
TCP The two layer (Ni 9% steel)	

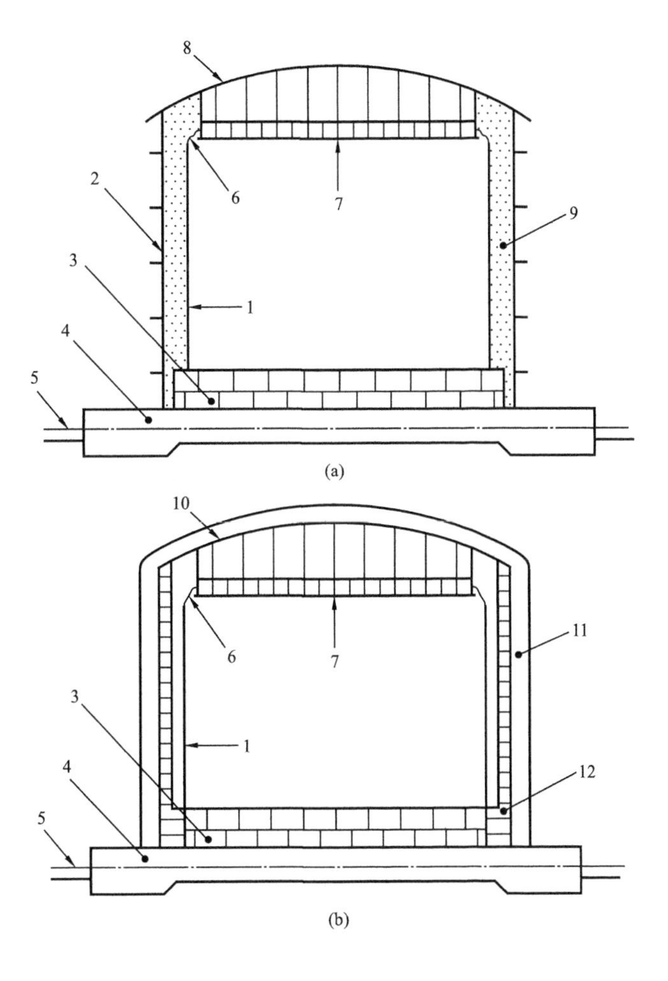

Figure 6–4 Examples of full containment tanks

Key

1— primary container (steel) ; 2—secondary container (steel) ; 3—bottom insulation; 4—foundation; 5—foundation heating system; 6—flexible insulating seal; 7—suspended roof (insulated) ; 8—roof (steel) ; 9—loose fill insulation; 10—concrete roof; 11—pre-stressed concrete outer tank (secondary container) ; 12—insulation on inside of pre-stressed concrete outer tank

④ Membrane containment tank.

A membrane tank shall consist of a thin steel primary container (membrane) together with thermal insulation and a concrete tank jointly forming an integrated, composite structure. This composite structure shall provide the liquid containment. All hydrostatic loads and other loadings on the membrane shall be transferred via the load-bearing insulation onto the concrete tank. The vapours shall be contained by the tank roof, which can be either a similar composite structure or with a gas-tight dome roof and insulation on a suspended roof. For an example of a membrane tank, see figure 6–5.

In case of leakage of the membrane, the concrete tank, in combination with the insulation system, shall be designed such that it can contain the liquid.

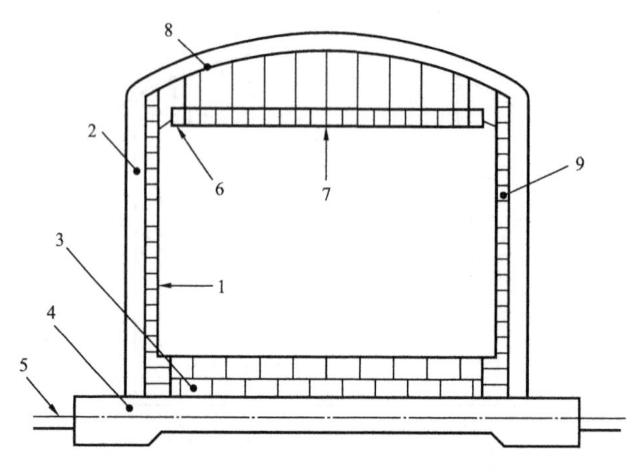

Figure 6-5 Example of membrane tank

Key

1—Primary container (membrane); 2—Secondary container (concrete); 3—Bottom insulation; 4—Foundation; 5—Foundation heating system concrete outer tank; 6—Flexible insulating seal; 7—Suspended roof(insulated); 8—Concrete roof; 9—Insulation on inside of pre-stressed

(2) Tank process system selection.

① Type of storage tank. According to the specific circumstances of the project, full containment tank were chosen in China's LNG projects at present. In order to determine the receiving terminal storage capacity, through the receiving terminal scale and gas transmission capacity, to calculate the number of tanks needed to be built. In accordance with the ISO 14620 international storage tank standards to construct, it is necessary to consider the late reserve increased storage tank location.

② Feed lines. For LNG storage, it needs to set the top and the bottom of the feed lines, according to the differences between the specific gravity of LNG for feed and the LNG in original tank, if the former is bigger than the latter, it can be fed from top line, otherwise it can be fed from the bottom, thereby preventing tank LNG stratification.

③ Liquid level meter, density meter, thermometer setting. In order to keep track of LNG liquid level, density and temperature at any time, set up the level meters, density meters, thermometers, in the inner tank, annular space and the bottom of the tank in different parts for continuous measurement of LNG related parameters and to select suitable instruments according to the requirements of the site and the overall detection. When the differences of temperature and density occurred, the low pressure pump should be start to cycle, thereby preventing LNG stratifying.

④ Low pressure pump. In order to output LNG from storage tank and prevent LNG layered, low pressure pumps need to be set. According to the output capacity, it needs to calculate and determine the number of low pressure pump in the tank. For the protection of low pressure pump outlet pipeline, the flow control valves should be set for each pump outlet pipe.

⑤ Tank Foundation heater. To prevent tank foundation frozen and concrete foundation damaged, it needs to set several heating systems on the tank foundation, by installing the temperature

detection facilities in the bottom of the tank to control the heating system.

⑥ To prevent over pressure and negative pressure. It is necessary to make primary and secondary pressure protection by pressure controlling.

6.5.2.3 BOG Processing System

Due to the difference in environmental temperature, BOG must appear in LNG tank. BOG processing system is born at the right moment.

(1) BOG compressor.

According to the amount of BOG produced during unloading and not unloading, it needs to determine the number of compressors. Step by step to adjust the flow control and the compression ability through the adjustment of tank pressure, the control of compressor can be automatic and manual. If the BOG flow is over the compressor capacity, the excess BOG can be discharge to the torch through the pressure control valve.

(2) Re condenser.

In order to save energy, BOG should be avoided to discharge to the torch or atmosphere, and the condensation as far as possible. Based on matching with BOG compressor, to select the number of condensers, re condenser has two functions, one is in the condenser, after the pressurization of the BOG which is mixed with the sent LNG and condensed into liquid, and the second is the re condenser can be used as the buffer container entrance LNG high pressure pump output.

6.5.2.4 LNG Transport and Gasification System

The main function of the receiving terminal is transporting gas or LNG to downstream. So the LNG transport and gasification system is the most important process system.

(1) LNG transfer pump.

According to the amount of transmission, it needs to determine the number of LNG transfer pumps and ratio of high pressure pumps and medium pressure pumps. Some receiving terminals only select high-pressure pumps. Flow control valve is set at inlet pipeline of vaporizer to control the amount of transmission of high pressure pumps and medium pressure pumps. The control valve has both automatic and manual selection.

(2) Vaporizer.

According to the amount of transmission, it needs to determine vaporization capacity, then to determine the number of LNG vaporizers and types. Usually in our country, three types of the vaporizer are selected, such as open rack vaporizers (ORV), submerged combustion vaporizers (SCV) and intermediate fluid vaporizer (IFV). The gas after gasification is measured and sent to downstream users.

6.5.2.5 Natural Gas Transportation and Metering System

(1) The natural gas transportation system.

According to the specific circumstances of the project, it is necessary to allocate the natural gas pipeline with the numbers of trunkline and branchline, the pressure control valve is arranged on the upstream gas pipeline of the metering equipment, to prevent gas transmission pipeline overpressure.

(2) Metering system.

It needs to install matching metering facilities for the production on the high pressure gas pipeline in receiving terminal area, and the trade measurement should be carried out by the metering facilities in the supplier point of city gas station, power plant and other users.

6.5.2.6 Torch System

When the evaporation pipe produces gas at low pressure due to overpressure, it needs to set liquid separating tank at low point in the upstream of the torch system, and set electric heater in the outer of liquid tank, the purpose is to cause the liquid to be evaporated in the liquid separating tank, and finally the part of the gas is burnt.

6.5.2.7 Tanker and Tank Ship Loading System

When the LNG receiving terminal based on the LNG transmission and distribution scheme, the tanker and tank ship loading system and transportation is adopted, according to the LNG loading amount, equipped with LNG loading platform or berth number, each loading platform or berth is provided with a liquid loading arm and gas return arm, also is provided with a control panel, weighing and control room etc.

6.5.2.8 Fuel Gas System

When the receiving terminal is designed with SCV and torch ignition device using gas as fuel, the fuel gas system must be set up, the gas can be high-pressure gas outside the gas after decompression, therefore to set up air heater, to heat the gas to the normal temperature.

6.5.2.9 Fuel System

When the receiving terminal is designed to fuel the fire pump and the accident emergency generator, the installation of fuel tanks and fuel pumps must be considered, the fuel for the use of equipment.

6.5.2.10 Nitrogen System

It is necessary to use nitrogen for equipment sealing and purging in receiving terminal, based on continuous and discontinuous with nitrogen and the amount of scale, therefore it is necessary to equipped with two sets of nitrogen system.

(1) Membrane nitrogen device.

For continuous using, membrane nitrogen device is equipped with air compressor to control the pressure of nitrogen gas pipe. The equipment needs continuously using nitrogen are: low pressure and high pressure pump, BOG compressor, rotating joint with nitrogen sealed in unloading arms for BOG return arm.

(2) Liquid nitrogen complete sets of equipment.

For discontinuous using, liquid nitrogen complete sets of equipment are equipped with air

heating evaporator and electric heater. It is used for peak intermittent nitrogen gas source and alternate source of membrane nitrogen device. The equipment needs discontinuously using nitrogen are: sealing and purging of unloading arms, BOG return arms, pipe, containers and other equipment.

6.5.2.11 Automatic Control

The central control system consists of seven parts, and each part has connection between hardware and software. The central control room is set in a non-hazardous area.

(1) Distributed control system (DES).

The DES system constitutes the core of the monitoring and control, communicating with all the other system. Its basic functions include: remote operation and control; process and operating window; provide alarm, process flow and parameter tracking record and display; automatic calibration and monitoring; automatic continuous, timing and logic control; calculation, advanced control and optimization control; report; signal display.

(2) Operator training system (OTS).

OTS system is equipped with simulation software of DCS system which is used for training new operators and for the optimization and simulation of control and regulation system.

(3) Emergency shutdown system (ESD).

ESD system is triple-redundant fault-tolerant. Redundancy concept: programming control processor redundancy, power communication redundancy, and I/O module redundancy. All signals of ESD sensors, control valves, local ESD switch and local control panel put into the ESD cabinets through the ESD terminal. ESD system commands three levels of shutdown: ESD1: whole terminal shutdown; ESD2: process shutdown; ESD3: unit shutdown.

(4) Fire and gas detecting system (FGS).

FGS can detect and report leak of dangerous gas or fire. Detecting and alarming equipment: combustible gas detectors, flame detectors, smoke detectors, infrared detectors, temperature detectors, fire alarm button, sound and light alarm devices. Signals get into the FGS cabinets through the FGS terminal. FGS trigger event can be invoked by DCS system, in order to take timely and appropriate measures such as starting the fire pump valve.

(5) Level, temperature, density system (LTD) in storage tank.

The monitoring facilities of liquid level, temperature and density are set in LNG storage tank. These signals can be monitored by LTD system and be invoked via redundant communications can be used by the DCS system.

(6) Skid equipment control system (PACKAGES).

The skid equipment suppliers will conFigure control devices of independently operation and the corresponding maintenance equipment, such as BOG compressor, the vaporizer control panel and the local control system of jetty unloading (PLC). Skid equipment control and other systems is connected with the status signal of the communication interface, turn off signal is hard wired. It is required that the skid equipment control design is compatible with other design of the receiving terminal.

(7) CCTV security system.

This system monitors the whole terminal of safety in production, mainly consists of a camera, safe cabinet, control panel, a recorder and a monitor.

6.5.3 Pipeline Material Selection and Design

6.5.3.1 The Principles of Material Selection

The universal standard criterion for purchasing pipelines and components at home and abroad need to be listed through the standard and specification research. The various pipes minimum corrosion allowance, various pipe grades and material selection table need to be given through the study. On the basis of comparison, the insulation designs for the equipment and pipes in the receiving terminal (not including LNG storage tank) need to be given. According to the environmental characteristics of the project site, the external anticorrosive materials design needs to be conducted.

6.5.3.2 Pipeline Design

It needs to list the principles of pipeline design, and explain the pipeline layout through research; considering the convenience of construction and maintenance, economic and security factors, list minimum headroom pipeline under various conditions and minimum clearance requirement. According to the specification, the parallel pipeline layout spacing of all kinds of buried pipeline need to be designed. The high vent points, low drain points, and sampling tube valve positions and grades need to be designed on the pipeline.

According to requirements of the specification, it is necessary to conduct the system planning for the common material pipeline, and carry out arrangement on the device within the safety valve, regulating valve, check valve and other valves for layout. According to the specifications, it needs to arrange pipe fittings, pipe accessories, instrument interface, location, size etc. According to the regulatory requirements, it is necessary to design the LNG pipeline size, flexibility and location, water pipeline in the device. The standards adopted in the pipeline design need to be listed and the requirements and methods of pipeline connections shall be specified.

6.5.4 General Layout, Utilities and Auxiliary Facilities

6.5.4.1 General Layout, Transportation

(1) General layout.

According to the principles of general layout and the layout of LNG jetty and workboat pier, combined with hydrology meteorological data and geological conditions, it is necessary to put forward a number of general plane layout schemes, choose the final scheme after the comparison and selection, and describe the receiving terminal scale and land components. The green area should be planned under the standards of national fire control rules.

(2) The receiving terminal vertical layout scheme.

Baseel on required water depth of jetty channel and (or) the jetty design elevation and the local theoretical lowest tidal level and design of the highest tidal level and considering cut and fill balance and outside road connecting factor, it is necessary to carry on the design to the site elevation.

(3) Transportation.

The transport condition of the receiving terminal and the inside roads should be designed based on the standards.

6.5.4.2 Utilities

(1) Water supply and drainage.

① Water supply and drainage system for producing and living include fire water supply system, technological seawater system, industrial waste water and sewage drainage, seawater drainage, rain water drainage etc.

② Combined with the project of local surrounding water, the water supply source is determined. Water supply and drainage system should be classified according to the properties of water and decontamination triage principles.

③ It is necessary to select the parameters for the main water supply and drainage system equipment name, specifications, quantity etc.

(2) Power supply and communication.

① Power supply. Combined with project where power status, the comprehensive analysis should be done with project electric load level, power consumption, the distribution principle and scheme, design standards, main equipment of power supply system.

② Communication. According to the project location communication status, the matching analysis should be done for the structure of the communication system according to relevant design specifications, including administrative telephone system, program-controlled scheduling/broadcasting system, UHF/VHF wireless intercom systems, access control system, local area network system, fire automatic alarm system, emergency backup system and TV monitoring system etc., and the communication of major equipment and materials need to be listed.

(3) Civil engineering, heating and ventilation engineering

① Construction. The building waterproof, fireproofing, freeze-proofing and fireproofing of the steel structure, corrosion prevention, etc. should be studied according to the relative design principles.

② Structure. The building structure type, seismic fortification, foundation treatment and foundation form should be studied based on the relative structure design principles and standards.

③ Heating and ventilation engineering. The heating and ventilation scheme of auxiliary facilities and utilities' building are designed based on the relative refrigeration design standards, and list the main equipment.

(4) Maintenance.

Based on the general equipment status of receiving terminal, it needs to set the service shop and equip with appropriate machining and welding facilities, electrical testing equipment and repair tools.

(5) Warehouse.

It needs to equip with warehouse in the receiving terminal.

(6) Laboratory.

It is necessary to set a laboratory in the terminal.

(7) Compressed air and instrument air supply.

It needs to study the use of compressed air for public works and the amount of instrument air used in instruments and control valves and the related supporting equipment in the receiving terminal.

(8) Nitrogen supply station.

It needs to study the amount of nitrogen and instruments in the receiving terminal.

(9) Office facilities and other auxiliary facilities.

According to the receiving terminal environment and function, it is necessary to study the need for office and auxiliary facilities.

6.5.5 Energy Saving and Water Saving

6.5.5.1 Energy Saving

(1) Energy consumption analysis.

All kinds of energy consumption in receiving terminal are described , and the energy consumption should be listed, including electrical power and fuel consumption, also the local energy supply status, the comprehensive energy consumption results are need to be analyzed, calculated and listed.

(2) Energy saving measures.

According to the technical features of the receiving terminal, it needs to put forward the adopting energy saving measures, and list the main energy consuming equipment and heat exchange efficiency, describe the thermal insulation of the cryogenic system, including insulation material design in project pipeline and tank cooling design etc.. The energy saving measures and requirements of the electrical, heating and ventilation, energy measurement system are need to be described. The design advice of the construction layout, facade modeling, lighting and ventilation are need to be provided.

(3) Cryogenic energy utilization.

Cryogenic energy utilization is one of the contents of the project examination and approval. It is necessary to survey the cryogenic energy utilization and develop the utilization of cryogenic energy market, including air separation, refrigeration, cryogenic shatter, power generation and other cryogenic energy utilization.

6.5.5.2 Industrial Water Consumption Index and Water Saving Measures

It is necessary to research the industrial water consumption situation, and put forward the main water saving measures and solutions of the receiving terminal.

6.5.6 HSE

6.5.6.1 Environment Protection

(1) Local environment status.

① It needs to describe the topography, geomorphology, geology, and climate, marine and land hydrology, soil and land or wetland and coastal ecological environment status of the project location.

② It is necessary to describe the local administrative divisions, population, industry, economic development planning, tax policy, transportation, etc.

③ It needs to describe the present situation of local atmospheric environment quality, ground water environment, sea/fresh water quality, terrestrial/planktonic/benthic organisms, and sound environment.

(2) Major pollution sources and major pollutants.

Combined with the scale of the project and drawing lessons from the major pollution sources and major pollutants produced in other LNG project construction, production, operation and life, it is necessary to analyze the fuel, emissions generated by the main pollutants including CO_2, SO_2 and NO_2 and production wastewater, domestic sewage discharge problems in the receiving terminal.

(3) Design of environmental protection measures.

In view of the major pollution sources and pollutants above, it is necessary to respectively formulate the corresponding environment protection measures, calculate the environment benefits.

(4) Greening.

Under the condition of not affecting safe distance, it needs to make full use of the space and both sides of the road for greening design.

(5) Environment management and monitoring.

It needs to arrange specific person to do the environment supervision work, make his responsibilities, and set up an environment data monitoring system to test the environment pollution, and analyze the environment impact of the atmosphere, water, noise and so on in terminal area.

(6) Environment investment estimation.

Capital investment is the fundamental guarantee for the implementation of environment protection. According to the results of environment monitoring, it needs to make the environment investment estimation of manpower, equipment and instruments.

6.5.6.2 Labor Safety

(1) Analysis of the risk factors in the process of project construction and production.

It needs to analyze the hazard factors such as pressure injuries, bumps, high altitude fall, asphyxia, electric shock, arc damage hazards in the construction of the receiving terminal. Also it needs to analyze the LNG or gas leak, freezing, combustion, explosion and the surrounding environment influence caused by the production of LNG storage and transportation and gasification.

(2) Labor safety and protective measures.

According to the above analysis, it is necessary to propose the labor safety and protective measures, including the general layout, structure, process equipment, electrical, mechanical, fire, explosion safety, physical, chemical damage protection, natural disasters, falling, personal protection, and the influence to the person and surrounding environment in the production of LNG storage, transportation and gasification. it also needs to analyze the safety and reliability of the LNG project chain system.

(3) Labor safety investment estimation.

It needs to implement the capital investment, including the labor protection, labor safety facilities, staffing and so on, and estimate the labor safety investment.

6.5.6.3 Occupational Health

(1) Analysis of the occupational hazards in the production process.

It needs to analyze the occupational hazard factors in the production process, such as cryogenic, asphyxia, toxicity, noise, production dust and other hazards.

(2) The occupational health protection measures and facilities.

According to the occupational hazards mentioned above, it is necessary to put forward the protective measures and facilities, including personal protection, anti-cold, anti-heat stroke, noise control, assistant hygiene and daily necessities and other measures and facilities. At the same time, it needs to propose the safety and health monitoring, accident prevention during operation and maintenance, medical first aid facilities, prevention and control of occupational health and management program. It needs to establish the health management mechanism, health education training and management system and develop a major epidemic, food poisoning and other emergency plans and measures.

(3) Occupational disease prevention measures and facilities investment estimate.

It needs to estimate the occupational disease prevention measures and facilities investment.

6.5.6.4 Fire Control

(1) Status quo of fire control environment of project.

It needs to analyze the current fire control environment, and design the fire control facilities according to the characteristics and actual situation of LNG fire control.

(2) Analysis of the risk factors of fire explosion in the production process.

It is necessary to analyze the hazard factors such as LNG and gas leak, fire explosion risk factors in the process of production.

(3) Fire prevention measures.

According to the characteristics of the project, it needs to put forward the fire prevention measures, including general layout safety, building structural safety, fire prevention of process equipment, process control safety, electrical safety, fire and explosion prevention.

(4) Fire control facilities.

It is necessary to analyze the facilities and process layout of the receiving terminal, put forward the fire control facilities in each area, including fire alarm and combustible gas detection, fire-fighting water, high-expansion foam, dry powder system, set fire control equipment and (or) set up a fire station.

(5) Fire control facilities investment estimation.

It is necessary to estimate the fire control facilities investment of the project.

6.5.6.5 Earthquake Safety Evaluation

(1) Earthquake geology.

According to the influence of historical earthquake, earthquake activity, regional earthquake tectonic features, near-field seismotectonic characteristics in project site, it needs to analyze the site engineering geological conditions of the earthquake evaluation, probabilistic seismic hazard analysis.

(2) Ground motion parameter design.

It is necessary to describe the calculation results of the designed ground motion parameters.

6.5.6.6 Enterprise Organization and Personnel

(1) Enterprise operation system and management system.

According to the needs of the project construction and operation, based on the experience of the existing LNG project company, it needs to design the personnel organization of the company in the construction and operation period and clear their respective responsibilities.

(2) Personnel training.

According to different stages, it needs to put forward the training methods and goals, and train the company staff.

(3) Implementation plan.

According to the project engineering proposal, it is necessary to make the plan for the implementation of the project.

6.5.7 Investment Estimation

6.5.7.1 Basis of Compilation

It needs to make the investment estimation according to the contract and related regulations, as well as the relevant basic information.

6.5.7.2 Compiling Methods

It needs to make the investment estimation according to the current price, and make adjustment based on the actual project situation and price differentiation.

6.5.7.3 Project Investment and Analysis

It needs to list the total investments of the project, including fixed assets, deferred assets and preparatory fees, etc.

(1) The total cost can be estimated as follows.

① Production cost plus period expense estimation method:

$$\text{Total cost} = \text{Production cost} + \text{Period expense} \quad (6\text{–}1)$$

Where:

$$\text{Production cost} = \text{Direct material cost} + \text{Direct fuel and power cost} + \text{Direct wages} + \text{Other direct cost} + \text{manufacturing expenses} \quad (6\text{–}2)$$

$$\text{Period expense} = \text{Management fees} + \text{Operating expenses} + \text{Finance charges} \quad (6\text{–}3)$$

② Production factor estimation method:

$$\text{Total cost} = \text{Purchased materials, fuel and power cost} + \text{Salary and welfare} + \text{Depreciation charge} + \text{Amortization} + \text{Repair charge} + \text{Financial expenses (interest expense)} + \text{Others} \quad (6\text{–}4)$$

(2) Financial internal rate of return (FIRR).

It refers to the discount rate that can make the cumulative present value of net cash flow equal to zero in the calculation period, namely FIRR as the discount rate, which is the basis of the:

$$\sum_{t=1}^{n}(\mathrm{CI}-\mathrm{CO})_t(1+\mathrm{FIRR})^{-t}=0 \quad (6\text{–}5)$$

Where CI——Cash inflow;

CO——Cash outflow;

$(\mathrm{CI}-\mathrm{CO})_t$——Net cash flow in t period;

n——Project calculation period.

Project investment financial internal return rate, project capital financial internal return rate and investment parties financial internal return rate are calculated on the basis of the above formulas, but the cash inflow and cash outflow are different.

When the financial internal rate of return is greater than or equal to the set of criteria i_c(often referred to as a basic return rate), the project financially can be considered to be accepted. Project investment financial internal return rate, the project capital financial internal return rate and investment party's financial internal return rate can have different criteria.

(3) Financial net present value (FNPV).

It refers to the setting discount rate (general use of benchmark yield i_c) to calculate the net cash flow in project calculation period and the sum of present value of the project, it can be calculated:

$$\mathrm{FNPV}=\sum_{t=1}^{n}(\mathrm{CI}-\mathrm{CO})_t(1+i_c)^{-t} \quad (6\text{–}6)$$

Where i_c——Setting discount rate (equals to benchmark yield).

Generally, financial profitability analysis can only calculate the financial net present value of the project investment, which can choose to calculate the income tax before or after income tax net present value of net present value as needed.

If the financial net present value calculated is greater than or equal to zero according to the setting discount rate, the project plan financially can consider to be accepted.

(4) Investment payback period (P_t).

It refers to the time required to recover the investment of the project by the net income of the project, generally in the year as the unit. The payback period of the project investment should be from the beginning of the project construction, while, it should be given clear indication if it is calculated from the year of operation. The investment payback period can be calculated as follows:

$$\sum_{t=1}^{P_t}(\mathrm{CI}-\mathrm{CO})_t=0 \tag{6-7}$$

(5) Sensitivity analysis.

It refers to the analysis of the influence on financial or economic evaluation index the uncertainty while factors are increasing or decreasing. And calculate the sensitivity coefficient and the critical point, and find out the sensitive factors.

Sensitivity coefficient (S_{AF}) refers to the ratio of rate of evaluation index change and the rate of uncertain factors change, which can be calculated as follows:

$$S_{\mathrm{AF}}=\frac{\Delta A/A}{\Delta F/F} \tag{6-8}$$

Where $\Delta F/F$——the change rate of the uncertainty factor F;

$\Delta A/A$——when the factor F changes ΔF, to evaluate the corresponding rate of change of A.

6.5.7.4 Other work

(1) Some descriptions.

It needs to clarify other descriptions, such as the scope of investment estimates, foreign exchange rates, tariff rates, value-added tax (VAT) rates, basic reserves, fixed asset investment direction adjustment tax, etc.

(2) Project investment estimation table.

It needs to list project investment estimates.

(3) Attached drawings.

The receiving terminal project research should be at least attached by: regional location map, a process flow diagram, vertical layout and general layout, etc.

6.6 Gas Transmission Pipeline Sub-project Feasibility Study

This chapter especially refers to the matching gas transmission pipeline sub-project of LNG receiving terminal, including gas pipeline, gas transmission station, pipeline crossing and public facilities etc. This part also can be applied to general natural gas transmission pipeline engineering.

6.6.1 Research Purpose and Scope

6.6.1.1 Research Purpose

(1) Providing gas transmission function for the receiving terminal.

Gas transmission pipeline is the auxiliary engineering of the LNG receiving terminal, which services for the LNG receiving terminal. One of the research purposes is to meet the terminal's the largest gas transmission capacity and transport range, including the seasonal, month and day peak shaving capability.

(2) Pipeline layout for saving land resources.

In addition to the pipeline construction takes up land, forming a complete set of gas transmission station (seat)/truncation, distribution valve chambers seat) also need to occupy land. To save land resources as the principle, through the research and comparison, it should be chosen as less as possible to occupy the permanent land resources.

(3) Economic and reasonable technical schemes.

On the basis of ensuring the gas transmission function of the receiving station and saving the land, it also needs to satisfy the safety of gas transmission, and the gas processing equipment should be chosen reasonably, in order to achieve the economic and reasonable investment benefit.

6.6.1.2 Research Scope

Research scope includes engineering scope, main body and supporting engineering.

(1) Technical schemes.

It is necessary to select the appropriate process flow of gas transmission pipeline engineering, control and communication systems, and adopt technical solutions according to the terminal layout, production operation and gas, energy saving, environmental protection and safety.

(2) Adopting the new process, technology, equipment and materials.

It is necessary to compare and make reliability analysis of the new process, technology, equipment and materials, especially demonstrate the effect on the energy saving, environmental protection, safety, transport efficiency and investment.

(3) The adoption of patent technology.

Patents and proprietary technology are needed in special case.

(4) Engineering technology level.

① The overall technical level of gas transmission process.

② Large pass through or crossing engineering technology level.

③ Automatic control and communication technology level.

④ The new management system and labor productivity.

(5) The research conclusion.

It is needs to describe from the resource/market situation, engineering survey, the main engineering quantity, the main technical and economic indicators, etc.

6.6.2 Gas Transmission Pipeline

It needs to analyze the pipeline routing, explain the permanent and temporary requisition of land demand, and recommend technologically feasible, economic and reasonable gas transmission lines.

6.6.2.1 Gas Pipeline Trend

(1) Pipeline selection principles.

According to the natural environment and engineering geological conditions, such as topography, geomorphology, meteorology, hydrology, etc., combined with the starting point and the location of the user and the city, humanities, transportation, water conservancy and other aspects of the overall planning of the development of a comprehensive comparison, it needs to choose safe, economical, convenient construction and maintenance of the pipeline trend.

(2) Gas pipeline trend.

The multi schemes design of the gas transmission pipeline is carried out, and the advantages and disadvantages of these schemes are analyzed.

① It needs to describe the dense areas of the population, natural and heritage protected areas, industrial and mining areas, construction planning areas, reservoirs, rivers, railways, highways, important facilities, and underground concealed objects along the line.

② It needs to describe the starting point of the line (the first gas station), the middle of the inlet, outlet point, end point (the end gas station) and regional location. It is necessary to draw the routing trend (macro and local) plan chart on the administrative divisions map or topographic map, and describe the line trend and natural survey along the line.

③ Outstanding engineering difficulties, the difficulty factors need to be introduced, such as land requisition, cultural relic, military facilities, complicated geological conditions, etc.

④ It needs to briefly describe the economic development and social relying conditions, the traffic, water supply, power supply, communication, building materials supply situation along the line, and list the agreements related to the line.

6.6.2.2 Routing Trend Scheme Comparison

(1) Comparison.

It is necessary to carry out the comprehensive comparison and selection of macro and local trend schemes of each line, including the crossing, construction management, engineering quantity and investment of the route.

(2) Pipeline trend recommendation scheme.

The scheme should be recommended based on the humanities, economic, traffic conditions, terrain, landform, vegetation, meteorological, hydrological, seismic, geological and other natural conditions, including statistics of length of administrative divisions, statistics of length of surface vegetation status, statistics of length of the landscape, statistics of length of the difficult location, statistics of the length of stone, meteorological climate along the line, hydrological characteristics, the climate characteristics, basic earthquake intensity and engineering geological zoning etc. along the route.

6.6.2.3 Pipeline Crossing

(1) Large rivers passing through (crossing).

It needs to study and describe the number of times the pipeline is required to cross the large rivers, according to the preliminary survey results, choose the crossing area of the river, location and the scheme, give a brief description of the recommendation, including design pressure and pipe size, tunnel structure, pipeline laying methods, etc.

(2) Medium and small rivers and water network passing through (crossing).

It needs to study and describe the numbers of pipeline passing through or crossing the medium and small rivers and ditches, and choose the crossing way based on the engineering geology, hydrogeological and geological conditions, such as directional drilling, tunnels and the excavation trench, etc.

(3) Railway and highway crossing.

It needs to study and describe the numbers of pipeline crossing the railway, highway and other county or township roads, the design of pipeline crossing rivers, railways and highways shall comply with the relevant provisions of the "SY/T 0015—1998: Design code for oil and gas transportation pipeline crossing engineering—Aerial crossing engineering".

(4) Other crossing.

It includes communication optical cable (military and civilian), heating and water supply piping, power cable, sewage and water pipeline, oil and gas pipelines and other crossing which should be met their special requirements.

6.6.2.4 Pipeline Block Valve Chamber

(1) Setting principles.

The pipeline block valve chamber should be set in the place where the transportation is convenient, open terrain and on the high ground. The design distant between the block valve chambers should meet the standard in 4.4 of "GB 50251-2015: Code for design of gas transmission pipeline engineering".

(2) Valve chamber setting.

It needs to study and set up the block valve chambers according to the above setting principles, combined with the engineering station distribution, the actual situation of the line passing through the region, and the distribution of the existing pipeline block valve chambers.

(3) Valve chamber function and equipment type selection.

It needs to study and list the function and selection of equipment in line with the valve chambers, including remote monitoring (RTU), cutting off functions and releasing etc.

6.6.2.5 Pipeline Laying

(1) Laying methods.

It needs to give full consideration to the matching degree of the design output and the actual working output, route trend and process optimization degree, pipe type and the main process equipment selection, anticorrosion technology level and adopted standards, clarify the requirements of pipe laying, including casing, buried depth and the methods, the elastic laying, backfill material and hydraulic protection and other protective measures.

(2) Main engineering quantityies.

It needs to list the main engineering quantities of the recommendation, including the specifications of the pipes, steel grades, length, the number of crossing, new highway length and the number of the line valve chambers.

6.6.3 Pipeline Anticorrosion

6.6.3.1 Basic Data

It includes soil erosion along pipeline, interference distribution of AC (alternating current) and DC (direct current) stray current along the pipeline, pipeline steel grades, pipe diameters, wall thickness and way of manufacture, process conditions for conveying gas, such as the transmission medium of H_2S and CO_2, temperature, pressure, water and hydrocarbon dew point, process station distribution and the power supply condition.

6.6.3.2 Pipeline Anticorrosion Scheme

Pipeline anticorrosion scheme should be determined according to the standard "GB 50251-2015: Code for design of gas transmission pipeline engineering".

6.6.3.3 Selection of Anticorrosion Coating for Pipeline

(1) Selection of outer coating for pipeline.

It needs to evaluate the various performances of outer coating which can be selected of pipeline anticorrosion, material sources, prices, construction equipment and the difficulty degrees of the mending and repairing technology. It should be based on the comparison of natural conditions, construction conditions along the pipeline and the economy to choose the outer anticorrosion layer.

(2) Selection of internal coating for pipeline.

It needs to analyze the using reasons, type and thickness of the adopted internal coating, and list the reference standards.

6.6.3.4 Cathodic Protection

(1) Scheme for determining cathodic protection.

The cathodic protection scheme should be made based on the comparison of the economy.

(2) Calculation of the cathodic protection.

When impressed current cathodic protection is used, the cathodic protection radius of stations protection should be calculated and determine the number and distribution of all cathodic protection stations. When using sacrificial anode protection, the anode material consumption should be calculated.

(3) Cathodic protection station current scheme.

It is necessary to study and develop the power supply scheme of the forced current cathode station, recommend specifications for forced current cathode station equipment and types of auxiliary anodes.

(4) Auxiliary facilities installation position and principles.

It needs to explain the auxiliary facilities (test pile) installation position and principles.

(5) Cathodic protection scheme in special area.

It is necessary to study and recommend the cathodic protection methods used in towns, mountains, crossing area and other special sections.

(6) Temporary cathodic protection.

It needs to make sure if the pipe needs the temporary cathodic protection, and explain the protection way.

6.6.3.5 Main Engineering Quantity Table

It needs to list and explain the main engineering quantities of the cathodic protection.

6.6.4 Gas Transport Process and Pipe

6.6.4.1 Gas Transport Process

(1) Main technical parameters

It needs to describe the gas transmission scale(annual or daily capacity), and basic data(gas composition and main physical properties, gas pipeline introduction, the first station exit gas pressure, temperature, environmental temperature, pipeline design pressure, the supply pressure of end users, the maximal and minimal transport capacity etc.)

(2) Process calculation formula and the software.

According to the standard in GB 50251–2015, it needs to simulate the transient state of gas pipeline and the transient of gas pipeline network, and describe the name and version of the software. The important gas pipeline engineering should adopt the internationally accepted software calculation and check.

(3) Process calculation.

① The sTableflow of the pipeline should be calculated, when the supply and the use of gas are unbalanced, it is necessary to calculate the typical dynamic state of the pipeline.

② It needs to seek the best combination according to the composition of each process, involving pipe diameter, gas pressure, gas transmission way.

③ When pressure conveying, based on capacity, transmission pressure, the pipe diameter, the number of gas station and pressure ratio optimized, it should be even refined to compressor model and compression ratio, etc.

④ It needs to calculate processing parameters of each process scheme, and explain the scheme composition and calculation results by the form of chart.

⑤ It needs to calculate the annual consumption of water, electricity, fuel gas of each process scheme.

⑥ It needs to estimate the engineering investment and operation costs of each process scheme.

(4) Gas transmission process.

It needs to analyze and compare the recent or long-term gas transmission capacity, steady or dynamic output gas adaptability, reliability and safety accident self-help ability, construction and management characteristics, project investment and operating costs etc. The recommended scheme needs to be put forward.

(5) Pipeline gas storage capacity analysis.

It needs to calculate the gas storage capacity under various working conditions, which use pipeline storage capacity for peak shaving, adopt joint peak shaving way by LNG receiving terminal and pipeline jointly to meet the user's peak month, peak days and average hourly gas.

6.6.4.2 Pipe Material

(1) Steel pipe type selection.

It needs to select type of steel pipe for gas transmission pipeline engineering according to the needs of engineering, the diameter of pipeline, the gas composition, gas transmission pressure, pipeline area categories and environment conditions. All kinds of steel pipe availability, quality, price, etc. need to be analyzed, finally to propose the recommendation of steel pipe types.

(2) Steel grade selection.

Various steel grades of steel pipe performance, quality, consumption and cost should be compared, to analyze the adaptability in engineering, put forward recommendations to choose types of steel grade. Steel material should comply with the relevant requirements in "GB/T 9711–2011: Petroleum and Natural Gas Industrial—Steel Pipe for Pipeline Transportation System".

(3) Pipeline strength calculation.

According to the recommendation of pipe type, it needs to choose a variety of steel grades to calculate the thickness, on the basis of calculation results, determine the wall thickness, calculate the steel consumption, select the type of pipe diameter, design pressure, regional coefficient, radial stability coefficient, strength parameters, etc. The calculation formula of wall thickness should be listed, and the name, unit and coefficient of each parameter in the formula.

(4) Consumed of steel pipes.

According to the recommended line trend scheme, the selection of pipe diameter, steel grade, the length of gas pipeline through the various regions, steel elbow and other steel pipe dosage, it is necessary to calculate the weight of all kinds of steel pipe used in pipelines and the consumed of steel pipes.

(5) Pipeline strength and stability check.

① Strength check. According to the needs of the gas pipeline and gas line pipe strength, the strength of pipes need to be checked.

② Pipe radial stability check. For the pipeline which outer diameter and wall thickness ratio are more than 140, buried deep and large external load pipe, stability checking, each parameter of name, unit and coefficient should be taken, and the checking results explanation are need to be given.

(6) Pipeline seismic strength design.

In accordance with the relevant requirements of the SY/T 0450-2004: Seismic Design Code for Oil and Gas Buried Steel Pipeline", it is necessary to study and recommend the pipeline through a variety of complex geological units, seismic hazard zone of a variety of processing and material selection, structure, filling, erection, laying, connecting and crossing engineering facilities and emergency shutdown measures.

6.6.5 Gas Transmission Station

6.6.5.1 Station Setting

According to the length of the pipeline, across the region, the downstream supply, it needs to determine the number of stations, categories and locations.

6.6.5.2 Station Process

(1) Process selection.

According to the location of each stations (the first station, intermediate station, end station, compressor station, etc.), it needs to design capability, design parameters, process flow, propose the protection measures to adopt new techniques, new technologies and security.

(2) Selection of main equipment.

According to the requirements of station process, it is necessary to choose the main equipment involving filtration device, pig send and receive unit, water jacket furnace, all kinds of valves, etc.

(3) The main engineering quantity.

According to the station design, process and equipment characteristics, it needs to calculate and list the main engineering quantities of each station.

6.6.5.3 Automatic Control

(1) Automatic control level.

It needs to set gas transmission pipeline engineering automatic control level in accordance with the requirements of gas, and put forward the necessity of the automatic control system construction.

(2) System scheme.

The design of automatic control system for each unit includes the new or expanding station, valve chamber, dispatch center, station control system, RTU, instrument selection, etc. The automatic control system is recommended on the basis of the scheme selection and the selection of the instruments and equipment are carried out.

(3) Main engineering quantities and equipment materials.

According to the automatic control level and the best system solutions, it is necessary to calculate the main engineering quantities and list the equipment materials.

6.6.5.4 Communication

(1) Main communication status along pipeline.

It needs to Figure out communication network status, planning and construction conditions along the pipeline and surrounding through research.

(2) Communication service demand.

According to the communication requirement of pipeline station, it needs to divide the automatic control and power supply system, access to the project other communication network and public communication network engineering interface, plan the main contents and scope of communication engineering construction, in order to meet the needs of gas transmission pipeline network system of communication services.

(3) Communication service prediction.

According to the forecast of the future demand of the pipeline and the station, it is necessary to determine the scale and reserve of the communication system.

(4) Technical plan.

According to the communication service demand forecast and the technical and economic comparison result, the best communication system plan should be recommended.

(5) Networking scheme.

According to the gas pipeline communication engineering of the system structure, functions and mutual relations, it needs to formulate networking schemes for transmission systems, telephone switching systems and other communication systems.

(6) Equipment selection and materials.

According to the communication needs of the business and networking solutions, it needs to select the matching equipment and list the major equipment and materials purchased at home and abroad.

(7) Main engineering quantity.

It needs to classify the main engineering quantities of the recommended schemes.

6.6.6 Utility Facilities

6.6.6.1 Power Supply

On the basis of the full demonstration, it needs to put forward the power load grade and its load, power supply source and its scheme, main equipment and main engineering quantities.

(1) Power load grade.

Based on the relevant provisions in the GB 50251—2015, the station is listed as level 2, and valve chamber is listed as level 3.

(2) Power load.

According to the station field with technology, automatic control instrument, cathode protection, water supply and drainage, communication and lighting, air conditioning and so on, it is necessary to calculate the statistical power load.

(3) Power supply and power supply and distribution scheme.

It needs to determine the source of station power supply and the distribution scheme through analysis and comparison.

(4) The main equipment selection.

It needs to select the main equipment, including power transformer, low voltage distribution panel, power box, power cable, control cable, wires, explosion-proof electrical equipment, etc.

6.6.6.2 Water Supply and Drainage

(1) Water supply.

According to the requirements, it needs to select the water source, water quality, water supply mode, calculate the main equipment, materials and engineering quantities.

(2) Drainage.

It needs to select the discharge direction, the type of sewage and the choice of disposal (discharge) ways, and calculate the main equipment and materials and the main engineering quantities.

6.6.6.3 Heating, Ventilation and Air Conditioning

Based to thermal load, thermal parameters and conditions of use of each station, it needs to recommend heating scheme. According to the requirements of the equipment and working environment, the heating, air conditioning and ventilation scheme are proposed. According to the requirements of the original water and heating equipment for water quality, the water treatment program, main equipment models, specification and quantity, and main consumption indicators are needed to be recommended.

6.6.6.4 General Layout for Road and Transportation

It needs to compare the station site selection, geographical location, general layout, road and transportation, calculate the main technical parameters and engineering quantities.

6.6.6.5 Building and Structure

According to suggestion for the seismic fortification intensity and seismic safety evaluation of building structure located, it needs to adopt seismic fortification measures, choose to conform to the requirements of the building and structure system, explain its use, fire prevention, health, fire protection, energy saving, anti-corrosion, insulation, and other forms of architectural design and structure, foundation treatment, material structure design, and list the quantities of the building and structures.

6.6.6.6 Maintenance

It needs to analyze society supporting condition, aiming at the maintenance content and the workload, set maintenance organizations, team personnel, and equipped with main machine, equipment, automobile, etc.

6.6.7 Energy Saving and HSE

6.6.7.1 Energy Saving

It needs to analyze and compare gas pressure conveying power consumption, engineering comprehensive energy consumption, unit energy consumption, consider energy saving project implementation plan according to the relevant state and other relevant policies and regulations requirements, analyze project plan in energy saving and consumption reducing, the specific measures taken by reducing gas leakage and loss of energy saving measures, emphatically analyze economic benefit, social benefit and environmental benefit.

6.6.7.2 Environment Protection

In the light of the major pollutants and sources of pollution name, type, quantity, discharge mode and the influence of pipeline construction on soil and ecology, etc., it needs to put forward the treatment technology of all kinds of pollutants, processing method and measures of controlling soil erosion and protecting the ecological environment. According to the emission standard configuration of main environmental protection facilities, it is necessary to calculate and list its cost. In accordance with the approval of environmental impact report from state and local environmental protection department , the environmental impact needs to be analyzed.

6.6.7.3 Fire Control

According to the guidelines and norms of fire-fighting design, combining with regional fire station cooperation strength and the equipment situation, it needs to put forward a recommendation for the firefighting and its explanations, list the main fire control equipment, material name, specification, quantity, etc.

6.6.7.4 Labor Safety and Health

(1) Occupational hazard analysis.

The analysis focuses on natural disasters, environmental hazards and occupational hazard factors in the process of production.

(2) Occupational hazard protection.

To protect occupational safety and health and occupational hazards in the process of production, it needs to list labor safety and health costs.

6.6.8 Organization and Implementation

6.6.8.1 Organization and Personnel

(1) Organization.

According to the pipeline and transmission stations, valve chamber distribution, it needs to work out the corresponding organization.

(2) Staffing.

According to the determination of the organization, it needs to list all the number of the staff of departments, and put forward the cultural degree, professional requirements for different types of work.

(3) Training.

According to the different stages of the pipeline operation periods, the training plan of pipeline need to be prepared.

6.6.8.2 Project Implementation Schedule

(1) Implementation stages.

It needs to list the the various implementation stages of the project, the main contents include construction project preparation mechanism, preliminary design and its approval, equipment and materials ordering, construction drawing design, construction preparation, construction, production preparation and commissioning, pressure test, drying, completion and acceptance and so on.

(2) Implementation schedules.

According to the work content in different stages of the project, the implementation schedule needs to be established.

6.6.9 Economic Evaluation

6.6.9.1 Investment Estimation and Financing

(1) Investment estimation.

It needs to explain the scope and the basis of the investment estimates, give the results of investment estimation, list the total investment summary and the total investment estimation Tableof the project.

(2) Financing.

On the basis of the scheme demonstration, it needs to propose the project legal person establishment and financing scheme.

6.6.9.2 Financial Evaluation

(1) Basis and basic data.

It needs to list financial evaluation basis and basic data, including the evaluation of the calculation period, gas pipeline operation and production data, the benchmark yield, the price, production capacity, salary, welfare funds, depreciation rate and amortization period, repair rate, all kinds of taxes and fees, etc.

(2) Cost and fee estimation.

It needs to analyze the cost and expense of the components, the estimation methods and cost indicators.

(3) Financial analysis.

It is necessary to analyze and explain the annual operating income, sales tax and additional profit and income tax, financial profitability, project solvency, transmission cost.

(4) Uncertainty and risk analysis.

It needs to analyze and explain the sensitivity, benchmark balance, profit and loss balance and other risks according to the different factors, and propose the countermeasures through risk analysis.

6.6.9.3 The Economic Evaluation Results and Suggestions

It needs to list the economic evaluation results and suggestions.

6.6.9.4 Appendix

For visualization of pipeline engineering research, it needs to prepare the appended drawings, including route diagram, process flow chart, station site general layout and natural gas pipe network layout sketch map, etc.

6.7 Preparation of the Feasibility Study Report of the Overall LNG Project

6.7.1 Status and Function of the Overall Feasibility Study Report

6.7.1.1 It is the Concentrated Expression of Research Results

As mentioned above, LNG project feasibility study involves many aspects of an industry, such as existing resources, the market, engineering, business. To get approval from the competent authorities of the state and the investment side, various problems must be in-depth studied. The LNG resource requisition, transportation mode selection, market analysis, engineering optimization, HSE research and the response measures are completed on the basis of in-depth analysis and research, and the overall feasibility study report is focused on the results of those study reports.

6.7.1.2 It is the Full Display of the Best Solution

Weather the LNG resources, market business contract mode, or receiving terminal, jetty and pipeline, in order to demonstrate the contract mode and the engineering economic and technological rationality, multiple schemes must be carried out, the best solution, and overall optimization and sorting in the total report need to be picked out. Therefore, it is not a simple solution to list and overlay, there may be some scheme optimal in the sub divisional works, but may not be optimal in general, this needs more in-depth research in the overall report preparation stage, to make every effort to achieve the overall optimal and reasonable.

6.7.1.3 It is the Optimal Combination of Economic Benefit

In each report, the economic evaluation generally starts from their respective areas, LNG resources report making economic evaluation with the most reasonable price and supply cycle, LNG and natural gas market used to evaluate the price to bear ability and consumption flexibility. Jetty, terminal and pipeline are made decisions based on investment and output balance, and LNG project economic evaluation of the overall feasibility report must be the least of the overall investment, to achieve the best economic and social benefits. When their respective economic evaluations are optimal, but the overall is not optimal, it is going to have to re-evaluate and adjust; when economic benefit is the best, but lacking of social benefits, we should strive to find the reasonable balance between economic and social benefits.

6.7.1.4 It is the Theoretical Basis of Design Work

After the overall feasibility study report approved, a preliminary design and detailed design will be conducted, the former is the foundation of the latter. Therefore, the overall feasibility study report, including resources, market, such as engineering, business and economy will be reflected in the design work. The quality of the overall feasibility study report is a key to a successful project.

6.7.2 Basic Content of Overall Feasibility Study Report

6.7.2.1 Compilation Basis and Necessity of the Construction of LNG Project

(1) Compilation basis of the construction of LNG project.

① Contracts and the relevant meeting minutes of project owner and the entrusted institutions.

② The composition reports, including the jetty engineering, receiving terminal engineering, pipeline engineering, markets and resources reports, transportation reports.

③ Basic data collected at different stages, including hydrometeorological investigation analysis research report, the engineering geological investigation report, measurement technology report, etc.

④ Special assessment or evaluation studies, such as labor safety, occupational health, earthquake safety assessment report, etc.

⑤ Background information of receiving terminal site location that needs to be considered or applied.

⑥ Expert preassessment meeting and (or) expert written evaluation comments.

(2) The necessity of the LNG project construction.

It is necessary to analyze the present gas supply and demand situation for the location of receiving terminal, elaborate the need for rapid economic development, the need for energy security, energy structure optimization, energy supply reliability and safety, and the needs of the energy consumption structure optimization and adjustment, the needs of environment protection, the needs of the natural gas market, the need of the development of relevant industries, etc. Finally, it is necessary to draw the conclusion that the construction of the LNG project is necessary.

6.7.2.2 The Scope and Content of the Overall Feasibility Study Report of LNG Project

(1) The scope of the overall feasibility study report of LNG project.

The research of the general report should be based on the summary analysis and comprehensive evaluation on the basis of the research results of jetty, receiving terminal, gas pipeline, market, resources, and transportation and so on. The main research scope should include:

① To analyze the energy and the supply market with natural gas as raw material in the project location, and propose the competitiveness analysis opinions;

② To analyze the LNG resources, trade mode and shipping, and provide potential resources (or the specific resources) and propose suggestions of trade and transportation mode;

③ To recommend and determine the composition and framework of LNG project through comparison and demonstration of market, the scale of construction, peak shaving, economic benefit analysis;

④ To determine the scale of the jetty, receiving terminal, gas pipeline and the construction scheme, as far as possible definite the LNG key user's projects, the purpose of gas using and scale;

⑤ According to the general requirements of the project, to make reasonable planning and setting of general layout, transportation, utilities and supporting facilities;

⑥ To discuss the safety technology of the project, and put forward the safety and reliability analysis;

⑦ To study the environment protection project, and purpose environmental impact analysis of the project of the province or region;

⑧ To clear the structure of project financing, research and implement the financing channels and financing methods;

⑨ To estimate the project investment, calculate, analyze and evaluate the financial benefit and the sensitivity of the project;

⑩ To research and put forward the need for state and local governments to help coordinate the relevant issues and give relevant preferential policies through policy analysis;

⑪ To discuss the feasibility of the project construction through engineering technology, finance, comprehensive evaluation.

(2) The content of the overall feasibility study of LNG project.

The feasibility study report is a most important task after the establishment of the project company, which agglomerates a lot of labor and effort, including the project company, jetty, receiving terminal, pipeline design institutes and general report preparation institute. The detailed content of the feasibility study are given by the above chapters, which is a milestone to promote the LNG project substantive progress. The LNG project will be successful or not, often depends on the feasibility study report (Table 6–2) quality and thoughtful risk factors' consideration.

Table 6–2 The Content of LNG Project Feasibility Study Report

1 Pandect	
1.1 Overview 1.2 Compilation basis, principles and guiding ideology	1.3 Project construction background and necessity 1.4 Study scope 1.5 Research conclusions
2 LNG Resources and transportation	
2.1 LNG resources	2.2 LNG transportation
3 The market demand and competitive analysis	
3.1 China gas market status and prospects for development 3.2 Natural gas status in provinces and cities (or regions)	3.3 Provinces and cities (or regions) gas market status quo 3.4 Market competitiveness analysis
4 Project scale and design capability	
4.1 Project scale and composition 4.2 Project overall process 4.3 The determination of project design capability 4.4 Operation hours	4.5 Maximum hourly gas output capacity and peak shaving of if receiving terminal 4.6 The gas product specifications 4.7 Summary of project design capability
5 Jetty engineering	
5.1 Summary 5.2 Jetty status 5.3 Ship hull form analysis and construction scale 5.4 Natural conditions 5.5 Loading and unloading process	5.6 General layout and scheme comparison 5.7 Hydraulic structure 5.8 Auxiliary engineering 5.9 Comprehensive demonstration and recommendation scheme
6 Receiving terminal engineering	
6.1 Summary 6.2 Receiving terminal function and size 6.3 Main technology process scheme 6.4 Technological process 6.5 The operation of receiving terminal 6.6 Automatic control	6.7 Equipment layout 6.8 Selection of pipe materials 6.9 Piping layout principles and instructions 6.10 General layout transportation, utilities and auxiliary engineering
7 Gas pipeline engineering	
7.1 Summary 7.2 Gas transmission line 7.3 Pipeline anticorrision 7.4 Gas transmission process 7.5 Line piping	7.6 Gas transmission station 7.7 Automatic control 7.8 Communication 7.9 Public facilities

Continued

8 Energy saving, water saving measures	
8.1 Energy saving	8.2 Water saving
9 Environmental protection	
9.1 Environmental status in the construction region 9.2 The main pollution sources and pollutants 9.3 The design of environmental protection measures 9.4 Greening	9.5 Environmental management and monitoring 9.6 Preliminary analysis of the environmental impact 9.7 Environmental protection investment estimation 9.8 The problems and suggestions
10 Labor safety	
10.1 Compilation principles 10.2 Compilation basis 10.3 The analysis of the hazard factors in the project construction and production process	10.4 Labor safety and health protection measures 10.5 Safety and health management 10.6 Safe reliability
11 Fire control	
11.1 Compilation Basis 11.2 Project fire control environment status quo	11.3 The analysis of risk factors for fire explosion in the process of production 11.4 Fire-fighting facilities
12 Occupational health	
12.1 Compilation Basis 12.2 The analysis of occupational hazards in the project production process	12.3 The occupational health protection measures and facilities 12.4 Occupational health management
13 Earthquakes and other safety evaluation	
13.1 Summary 13.2 Geologic tectonic 13.3 Terminal and jetty of seismic geology 13.4 The evaluation of the site earthquake engineering geological conditions in receiving terminal and jetty	13.5 Seismic hazard probability analysis results of receiver terminal and jetty 13.6 Design ground motion parameters of Terminal and jetty 13.7 Basic earthquake intensity of gas transmission pipeline 13.8 Other safety evaluation
14 Enterprise organization and personnel	
14.1 Enterprise operation system and management system	14.2 Personnel training
15 Investment estimation and financing	
15.1 Investment estimation	15.2 Financing
16 Financial evaluation	
16.1 Summary 16.2 National economic evaluation 16.2 Basic assumptions and data	16.3 Financial evaluation 16.4 Others 16.5 The main calculation report
17 Introduction of Supporting gas projects	
17.1 City gas 17.2 Gas power plant	17.3 LNG filling station 17.4 Other gas projects

Continued

18 Policy recommendations	
18.1 The necessity of policy support 18.2 The main goal of policy support	18.3 Policy suggestion
19 Risk analysis	
19.1 Classification of risks	19.2 risk analysis
Appendix	
A (informative) figures	B (informative) Enclosure

6.7.2.3 The Appended Drawings and Accessories

(1) The appended drawings.

LNG project feasibility study report at least should be attached to:

① Receiving terminal process flow diagram;

② Receiving terminal regional locations;

③ Receiving terminal general layout;

④ Natural gas pipe network layout diagram.

(2) Attachments.

The project feasibility study report as the attachment of the project approval reports, at the same time the following files should be completed.

① The document of import LNG for projects approved by the NDRC.

② The construction project opinion paper of local provincial and municipal government investment department.

③ The project site approved document issued by the provincial and municipal people's government departments.

④ The relevant urban planning views issued by the provincial and municipal urban planning department.

⑤ The project land grant review opinion issued by the land resources administrative department (issued by the competent department of the same level, if approved by the National Investment Department, provided by the Ministry of Land and Resources).

⑥ The environmental impact assessment document approval opinion issued by Environmental Protection Administrative Departments (provided by the competent authorities at the same level).

⑦ The approval document about the use of coastline, sea, waterways issued by the provincial and municipal authorities.

⑧ The soil and water conservation plan approved by the Ministry of Water Resources.

⑨ The local power sector commitment documents to the project for electricity use.

⑩ The local water supply department commitment to the project for water use.

⑪ The file of relocation of housing and other facilities within the region of project issued by the local people's Government.

⑫ Seismic safety evaluation and approval documents (approved by the State Seismological Bureau).

⑬ Geological disaster evaluation and approval documents.

⑭ The approval document for the cultural relics protection issued by the provincial and municipal department or above.

⑮ The approval document for covering mineral deposits issued by provincial and municipal mineral department level or above.

⑯ The document that the project does no harm to military facilities issued by the Corps level or above.

⑰ The project construction approval documents issued by provincial and municipal civil aviation authorities.

⑱ Bank loan commitment document for the project.

⑲ The cooperation agreement of the project investment parties, including the investment parties to issue an effective credit material and formal funding commitment documents.

⑳ Upstream natural gas purchase and transport agreement and downstream gas sale agreement.

㉑ Approval opinion on the route of the laying pipelines issued by the marine authorities.

㉒ The pre-assessment occupational hazards report and approval by the public health department.

㉓ The project safety assessment report and file for record in National Safety Supervision and Administration Bureau.

㉔ The forestry department approval of the receiving terminal.

㉕ The approval document for highway crossing issued by the Transportation Department.

㉖ The approval document for safety pre-assessment of navigable environment issued by Marine Department.

㉗ The approval document for flood control assessment report issued by the Water Conservancy Department.

㉘ Other related agreement and approval documents, etc.

6.8 LNG Project Check and Approval

6.8.1 Reform of the Examination and Approval System for National Major Projects

6.8.1.1 From the Examination and Approval System to the Check and Approval System

(1) National investment system reform.

In order to promote the separation between government and enterprise, and expand the pace of enterprise autonomy, since 2004, according to the "State Council on the decision on the reform of investment system", the NDRC has abolished the examination and approval system of the investment projects, and thorough reformed regardless of investors, regardless of investment sources, regardless of the nature of the project, according to investment management the scale of investment by governments at all levels and the relevant departments for approval of the enterprise, but in accordance with the "who invests, benefits, who bears the risk" principle, the implementation of business investment autonomy.

(2) "The Directory of Investment Projects Checked and Approved by the Government".

In order to clear the government investment for projects management, "The Directory of Investment Projects Checked and Approved by the Government" (hereinafter referred to as the Directory) has been formulated by the NDRC. Investment projects are divided into encouraging, allowing, restricted and prohibited category. "The Directory" clears the scope of the implementation of the check and approval system of investment projects, the check and approval authority of the project approval authority, and in accordance with the economic situation and the need for timely adjustment of macroeconomic regulation.

(3) The implementation of the check and approval system and record system.

For enterprises that do not use government investment projects, it will no longer implement the examination and approval system, as an alternative, adopt check and approval system and the record system according to different circumstances.

① Check and approval system: the government only for major projects and restricted projects, from the perspective of social and economic public management to audit the investment projects of enterprises. The project in "The Directory" promulgated by our nation needs to be checked and approved.

② Record system: the investment projects of enterprise outside of "The Directory" are changed to record system regardless of the sizes.

6.8.1.2 The Distinction between Check and Approval System and Examination and Approval System

(1) Difference in the applicable scope.

Examination and approval system is only applicable to government investment projects and enterprise investment projects using governmental funds; check and approval system is applicable to the construction of major projects and restricted project without using government funds.

(2) Different contents of the examination and approval system.

Past examination and approval system, the government audits enterprise investment projects not only from the perspective of social managers, but also from the view of investment of the owner. Check and approval system, the government audits enterprise investment projects only from the perspective of social and economic public management, and audit content is mainly "Maintain economic security, rational development and utilization of resources, protect the ecological environment, optimize the layout, protect the public interest, to prevent the emergence of monopoly", etc. Instead of replacing investors to audit the projects from market prospects, economic benefits, financial sources and product technical solutions, etc. The enterprises make decision independently and take their own risk.

(3) Different procedures for audit.

Examination and approval system generally needs to approve "Project Proposal", "Feasibility Study Report" and "Commencement Report" three links, but check and approval system only needs "Project Application Report" one link.

6.8.2 LNG Project Check and Approval Procedures

6.8.2.1 Implementing Agency for Check and Approval System

According to the "The Directory" by the government recently, the project for the reception and storage of imported LNG facilities shall be checked and approved by the NDRC in charge of investment under the State Council.

6.8.2.2 Project Check and Approval Authority Agency

(1) General requirements.

If application data is not complete or does not comply with the relevant requirements by project check and approval agency, the agency should inform the project declaration company at a time within 5 working days after receipting the project application report. Project declaration company needs to be asked to clarify and supplement the relevant information and documents, or to adjust the related content. If the required data is complete, the project check and approval agency shall be official acceptance, and issues a notice of acceptance to the project declaration company.

(2) Check and approval contents of the project check and approval agency.

The project check and approval agency shall review the project in accordance with the following conditions:

① Whether it is accord with national laws and regulations;

② Whether it is accord with the national economy and social development planning, industry planning, industrial policy, industry access standards and general land use planning;

③ Whether it is in line with the national macroeconomic regulation and control policy;

④ Whether it is conducive to the rational layout of the region;

⑤ Whether LNG products monopolize the domestic market;

⑥ Whether it affects China's economic security;

⑦ Whether it is the reasonable development and effective utilization of the resources;

⑧ Whether it protects the ecological environment and the natural and cultural heritage effectively;

⑨ Whether it has a significant adverse impact on public interest, especially in project area.

6.8.2.3 If Necessary to Entrust Assessment

After accepting the application for check and approval within 4 working days, the project check and approval agency shall, if necessary, entrust the qualified advisory body to evaluate the application report.

The advisory body entrusted by the project check and approval agency shall, within the time prescribed by the project check and approval authority, put forward the assessment report, and shall bear the responsibility for the conclusion of the assessment. Advisory body in the assessment, the project declaration unit can be required to explain the relevant issues.

6.8.2.4 Check and Approval Agency to Seek Advice from Relevant Departments

The project check and approval agency shall, in approval and review, if it involves the functions of the other departments in charge of industries, the relevant departments' opinions should be consulted. Relevant departments should put forward written opinions to the project check and approval agency within 7 working days after receiving comment letter (attached to the project application report). If audit written opinion is not feedback beyond the date, it shall be deemed agreed.

For projects that may have a significant impact on the public interest, the project check and approval authority shall take appropriate measures to solicit public opinion when conducting the approval review. For special projects, expert evaluation system can be implemented.

6.8.2.5 Check and Approval Agency Audit Opinion

Project check and approval agency shall put forward audit opinions on the project application report within the 20～30 working days. After acceptance of project application report, project check and approval agency will decide whether to approve the project application report and announce to the public, or offer audit opinions for the approval of the high level authority within 20 working days. If check and approval decision is really difficult to make within 20 working days because of special reasons, the date could be extended 10 working days by the person who in charge of the agency, and the project declaration unit should timely be inform with written notice and give the reasons for extension.

The project check and approval agency entrusted consultation assessment, the solicitation of public opinion and expert review, the time required for the period is not calculated in the preceding paragraph.

6.8.2.6 Check and Approval Documents Issued

To consent to the check and approval project, the project check and approval agency shall issue the project check and approval document to the project declaration unit, also copy to the relevant departments and subordinate project approval organ. To do not agree with the check and approval project, the project check and approval agency should issue a disapproval document to the project declaration unit, and explain the reasons for disapproval and copy to the relevant departments and subordinate project check and approval agency. The project is checked and approved by the State Council, while the project check and approval document should be issued by the NDRC.

6.8.2.7 Objection Handling

If the project declaration unit has objections to the decision made by the project check and approval agency, can put forward administrative reconsideration or administrative litigation in accordance with law.

6.8.2.8 Follow-up Work to the Project Declaration Unit

(1) Related procedures handling.

The project declaration unit shall, according to the project check and approval document, handle the procedures of land use, resource utilization, city planning, safety production, equipment import and tax relief and so on.

(2) Validity period.

Project check and approval document is valid for 2 years, from the date of publication calculated. If the project fails to be started construction within the validity period of the check and approval document, project company will apply for an extension to the original project check and approval agency within 30 days before the expiration of the validity of the check and approval document, and the original project check and approval agency will make a decision on whether to allow delay before the expiration of the validity of the check and approval document. If the project is not either started construction within the validity period of the check and approval document, or apply for an extension to the original project check and approval agency, the original project check and approval document will be void automatically.

(3) The adjustment of the check and approval project.

If the stipulated contents of the check and approval project need to be adjusted, the project company shall promptly report to the original project check and approval agency in written form. According to the specific circumstances of the project adjustment, the agency shall issue a written confirmation opinions or go through the formalities for check and approval again.

(4) Unauthorized project.

For the project which should be reported to the project check and approval authority for check and approval but not be reported, or the project application for check and approval but is not approved, land resources, environment protection, urban planning, quality supervision, securities supervision, foreign exchange management, safe production supervision, management of water resources, customs and other departments are not allowed to handle the relevant procedures, financial institutions may not issue loans.

6.8.3 Preparation of LNG Project Application Report

6.8.3.1 Preliminary Work for Preparation of Project Application Report

(1) Application report preparation institute qualification.

The institute which in charge of project application report preparation shall have the relevant engineering consulting qualification, among the projects which shall be subject to approval by the department in charge of investment projects of the State Council, the institute which in charge of project application report preparation shall be with a Grade-A engineering consultation qualification.

(2) The cooperation and coordination of project application unit.

① The project application entity shall not split the project, provide false materials and other illegitimate means to obtain the check and approval document of the project, and the check and approval agency shall revoke the check and approval of the project.

② The pre work cannot be weakened. The basic starting point of investment system reform is to give full play to the basic role of market allocation of resources, separate government functions from enterprise management, reduce administrative intervention; that is to establish the main body status of enterprises in investment activities, implement the enterprise independent investment, self-financing, bank credit audits independently, and take their own risk. The construction projects in business investment have been changed from examination and approval to check and approval, which increases the responsibility and the risk for investors. It is more important to do a good job in the construction of the project, to avoid the project decision-making mistakes, it is more important. The project pre-feasibility study and feasible research link are essential.

③ The main content of project pre-feasibility study is to make preliminary estimation and suggestions about the market of construction project, the scale of production, construction conditions, production conditions, technical level, environment protection, capital source, economic benefits, and so on, which mainly discusses the necessity and possibility of a project from the macro. After completing the work mentioned above, the investors can decide whether to set up projects, and whether to continue to do preliminary work. Therefore this stage is necessary.

④ The feasibility study is an important content of the project construction preparation, which is the core of the early stage of the work, and is an important part of basic construction program. The task of feasibility study is to conduct a comprehensive analysis, argumentation, and more comparison evaluation on whether construction project is technology, engineering and economic is reasonable and feasible, and is to provide reliable basis for project decision.

6.8.3.2 Matters Needing Attention in the Preparation of Project Application Report

The role of the project application report, the basic equivalent to the current feasibility study report and approval for a project belong to the government's permission. Therefore, the project application report must be well prepared, and gradually standardized.

(1) Entrusted preparation institution.

Firstly, in accordance with the project application report requirements of the NDRC, combined with actual project, it is necessary to write a project application report outline, and then entrust a qualified design and consulting organization. The project application company shall submit the application report of the project to the project application in 5 copies.

Generally, the project application report is completed in later period of the project feasibility study (or written at the same time).

(2) Preparing attached files earlier.

First of all, it needs to clarify documents that should be attached (see Table 6-2), and documents should be provided by institutions; apply to relevant units in advance, so as not to affect the approval.

(3) Working depth.

The check and approval system which has only one link "Project Application Report", is bound to increase the risk of pre working capital investment. Therefore, in the project that is not sure, we must do more work in the early stages, in order to achieve the "Project Application Report" depth requirements.

6.8.3.3 The Content of Project Application Report

According to the relevant requirements of the state to prepare the project application report, it needs to submit "Project Application Report" to the project check and approval agency which will examine and approve project application report, mainly from the maintenance of economic security, reasonable exploitation and utilization of resources, to protect the ecological environment, to optimize the major layout, to safeguard the public interest, to prevent monopoly, etc. The project application report is based on the project feasibility study report.

(1) Application report chapters.

Chapter 1 Application Company and Project Profile

Chapter 2 Development Planning, Industrial Policy and the Analysis of Industry Admittance

Chapter 3 Resource Utilization Analysis

Chapter 4 Energy Saving Scheme Analysis

Chapter 5 Analysis of Construction Land, Land Expropriation, Demolition and Resettlement

Chapter 6 Environmental and Ecological Impact Analysis

Chapter 7 Economic Impact Analysis

Chapter 8 Social Impact Analysis

Chapter 9 The main Risk and Countermeasures

Chapter 10 Conclusions and Suggestions

(2) Accessories.

Annex of the application report is given by Table 6-3.

Table 6-3 Annex to the Application Report

No.	Document
1	Province (or local) people's government, enterprises reported to the NDRC on the proposal of the LNG project letter
2	Reply of the NDRC on the approval of LNG project to carry out the work
3	Province (or local) planning administrative departments on the project site views
4	Preliminary comments of Ministry of Land and Resources on the LNG terminal and jetty engineering construction land
5	Official reply of the State Environmental Protection Administration on the project environmental impact assessment report
6	Official reply of the State Seismological Bureau on the earthquake safety evaluation report of the project site
7	Review record Tableof State Administration of Quality and Technical Supervision on the project safety pre-assessment report
8	Official replay of Ministry of Health on the project occupational disease hazard pre-assessment report
9	Reply letters of Ministry of Water Resources on the project of soil and water conservation scheme report
10	Review record Tableof Provincial Department of Land and Resources on the project geological disaster risk assessment report
11	Certificate of Provincial Department of Land and Resources on no important mineral deposit overburden in the project area within the scope of land
12	Preliminary comments of National Bureau of Oceanography on LNG project in the use of offshore area
13	Opinion of Provinces (or local) port and maritime authorities on the coastline and channel used by LNG project
14	Official replay of Provinces (or local) department in charge of maritime affairs on the LNG project navigation environment safety assessment report
15	Attached letter of project construction in line with the local flood control planning by the department in charge of water conservancy provinces (or local)
16	Reply letter of Province (or local) forestry authorities on the forest land used by the LNG project (If the project involves the occupation of forest land)
17	Reply letters of Provincial (or local) cultural authority on the archaeological investigation, exploration results within the scope of LNG project site
18	Reply letters of Provincial (or local) transportation authority on LNG project site selection
19	Reply letters of Military Corps level and above on without influence each other between projects and military installations
20	Investment commitment document of project's shareholders
21	Project financing bank issued loan commitment document
22	Upstream Natural gas purchase agreement
23	Liquefied natural gas transport agreement
24	Downstream Natural gas sales agreement

Chapter 7 LNG Project Schedule Management

7.1 General Concept of Schedule Management

7.1.1 The Definition, Significance and General Practice of Schedule Management

7.1.1.1 Definition of LNG Project Schedule Management

LNG project schedule management is also known as time management. It is based on the LNG project scope management, through further refinement, identification and definition of project work breakdown structure, specific tasks that must be carried out, in order to achieve the goals of the project, and give specific arrangements for the concept of time. According to the basic situation of the existing project company, finance and material resources, information resources, the time should be reasonably distributed, sorted and controlled. Schedule management is a combination of static management and dynamic management of the management process, at the same time, the project general manager and the schedule control department manager must have a solid schedule management skills and rich experience in project management.

7.1.1.2 The Significance of LNG Project Schedule Management

(1) The schedule management is the premise to obtain benefits.

The purpose of schedule management is to make the project completed on schedule. LNG project construction requires investors to invest a lot of money, and bear enormous financial pressure and risk of financial recovery, so the quality of the project, the amount of time to complete, and finally put into use is a prerequisite for producing benefit. Under the condition of market economy, if not successfully completed the project objectives, will cause the enterprise funds extended occupation, interest increased and additional wage expenses and other phenomena occurs, resulting enterprises in the expected profit in LNG construction project cannot be achieved, so no economic benefits will be obtained, it must through the scientific and reasonable method to carry out schedule management, to make the project put into operation as scheduled.

(2) The schedule management is a necessary condition for cost saving.

As an independent project, in the human, financial, material, information resources under certain circumstances, the shorter the time, the lower the cost, the more reasonable time to use, the better the effect, here include: the human resources department to provide with reasonable amount of labor force, technical department to provide reliable technical assurance measures, procurement department to supply enough qualified quantity of equipment and materials, the financial department to pay the full amount of equipment materials and personnel costs on time, the comprehensive departments in advance according to the requirements of the government management department to declare the project procedures, etc.. If each job is completed on schedule, it will bring about a lower cost.

(3) The schedule management is the direct embodiment of the capacity of project management.

Schedule, quality and budget are the three main factors of traditional project management. Generally speaking, under the premise of reasonable arrangements for the project schedule, the schedule delay will increase investment, while ahead of schedule may improve the benefit of investment; to speed up the schedule could affect the quality, and strict quality control may affect the schedule; but if the quality control in place to avoid rework, will accelerate schedule. Schedule, quality and budget three goals are the unity of opposites, and the schedule management is to solve the contradiction among the three, not only to schedule quickly, but also to invest saving, good quality. Therefore, the reasonable arrangement and control of the construction schedule are direct manifestations of the strong management ability.

7.1.1.3 The General Practice of Schedule Management

(1) Schedule management is closely linked to project goals.

Effective schedule management, first of all, must have a clear long-term, medium-term, short-term goals. The second, it is based on the goals of long-term, medium-term and short-term plans, and then divided into year, monthly, weekly and daily plans. The schedule management and the project goal setting, the goal implementation complement relations each other, and the schedule management and goal management are inseparable. LNG project to supply gas to the downstream on schedule, in fact, is a total target. However, the total target is composed by a series of small goals. Sharpening your knife won't waste your time for cutting your faggot. Before the project started, project general manager must clearly understand LNG project involving the scope of work, the logical relationship among small goals, the logical relationship between small goals and big goals, and formulate the year, monthly, daily goals and plans, and gradually achieve daily goals and plans until closing to the final target.

(2) Schedule management should grasp the principle of 80 : 20.

A project general manager should learn how to use principle of 80 : 20 in project management. That is, dividing the things that you are going to do into chronological order according to the priority level, to distinguish 20% of the most valuable things, and then arrange time in line with the value size; 80% is generated by the top 20% of thing, another 20% comes from 80% of the thing. From the perspective of LNG project, you can use 20% time for critical thinking and work preparation. To distinguish between urgent matters and important matters, urgent matters are often short-term, while important matters are usually long-term, set a deadline to all the list of things. To all of the things that have no meaning, apply the skills that are deliberately ignored will be removed from the list of things that are not significant.

(3) Schedule management should pay attention to its flexibility.

Schedule management also should consider its flexibility, here are the following several meanings, one is cleverly delay time, learn to say "no", and once you determine what is important, you ought to say "no" to those unimportant things. The second is to maintain a certain amount of time to maneuver in order to cope with emergencies and have to deal with things, including those seemingly insignificant things, but must do. The third, the project leader should cultivate the arts of authorization. For those things do not have to come forward by yourself, you can take full authority to subordinates, so as you take the precious time to do what should do.

(4) Schedule management should distinguish dominant work and recessive work.

Each task of the project should be carefully analyzed, to distinguish the some work is dominant, some work is recessive. Such as a sub project—LNG storage tanks built is dominant, and the thinking process of an engineering project plan is recessive. Therefore, it should be based on experience, list complete project necessary to complete the work, while the expert validation process, as a basis to develop a feasible project time plan, a reasonable time management.

(5) Schedule management needs to base on the team characteristics.

General Manager should fully understand the unique characteristics of the project, according to the different project organization and project scope, formulate the project management strategy, and determine the management scope and intensity. The project owner is responsible for the distribution of different roles in the project team, to take a direct management of the work, and the other work is to take indirect management. Team members in the work need to coordinate with each other to do a good job in the schedule of the project management. If the project team has sufficient human resources, for the same level tasks that have no logical relationship, and can ensure the working face requirements, we can take the concentration of forces to fight a war of annihilation, arrange several teams parallel to complete these tasks, in order to shorten the construction period; for the tasks that have logical relationship in between, we can only take in turn operation or in assembly line, according to the stage to complete one by one. If the project team is equipped with limited human resources or temporarily not in place, should consider making full use of contractors, supervision units, even design units of human resources.

(6) Schedule management should use computer aided tools.

Currently, the project management software is becoming more and more mature, which is widely used in the work including LNG project management. Its advantage is the clear expression way which is more convenient, flexible and efficient in the project time management. After inputting the following parameters into management software, such as task list, estimated task duration, logical relationship between tasks, human resource involved in the task, cost and so on, project management software can automatically perform mathematical calculations, balance the allocation of resources, deal with cost calculations, and quickly solve the schedule cross problems, and also can print schedules. In addition to the project schedule generation function, project management software also has a strong project execution record, tracking project plan, the ability to record the actual completion and timely give the actual and potential impact analysis.

7.1.2 LNG Project Schedule Management Process

7.1.2.1 The Three Parts of the Schedule Management Process

LNG project schedule management flow chart is shown in Figure 7–1.The figure shows that the first is the LNG project schedule management basis, which includes scope description, historical data, constraints and assumptions; the middle part is the core of schedule management, its working depth and corroboration each other directly affect the success or failure of schedule management; schedule control is throughout the entire project life cycle activities.

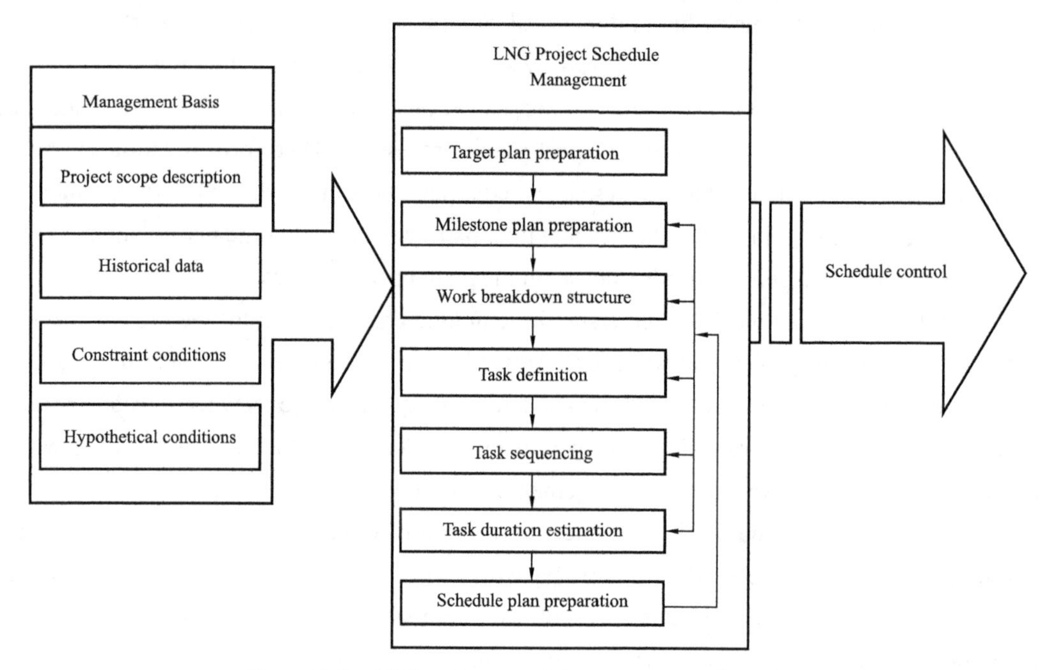

Figuer 7–1 LNG project schedule management flow chart

7.1.2.2 The Relationship between the Three Components of the Process

(1) The relationship between the management basis and schedule management.

Schedule management is based on the basis, the management basis is sufficient can avoid the project schedule management in the work of less detours, which is the biggest time saving. The first is the scope of the project which gives the LNG project composed of the soft and hard task, only all the tasks are clear, can we talk about on schedule, then collect the historical data, if there is a similar project LNG decomposition, task definition, task scheduling, task duration estimating and schedule data which have very high reference value for static time management. The schedule of the preparation is generally carried out in accordance with the seven steps of the middle box in Figure 7–1, according to the actual situation of the project to prepare. At present, Guangdong, Fujian Dapeng LNG project have been put into production, the schedule management basis and schedule data accumulation and summarization have great reference value for new projects.

(2) The relationship between the schedule management and schedule control.

From Figure 7-1, the first step of the schedule management is to formulate the general goal of the project, usually proposed by the LNG project sponsor or by the project team; the milestone plan is formulated jointly by LNG project company or the contractor; work breakdown is the premise of schedule management, task definition, task sorting, task duration estimation are the basis of the next step of work; in the middle box from the second to the sixth step work finished, enter the schedule plan preparation, but also can reture to adjust the above. The work in the middle box section is also called the component of time static management, which is the foundation of the schedule control.

(3) The relationship between the former two and schedule control.

Schedule control is the content of time dynamic management. It should be based on changes of internal and external environmental factors, timely adjustment and coordination, so that the LNG project has always been under effective control of the operation. It is necessary to strictly implement the system of project schedule report, but also master schedule adjustment causes, control the milestone and control the critical path of LNG project in accordance with the schedule management program, and strive in the overall schedule unchanged, re arrange and adjust the project schedule, or extend the time, but still within the allowable range etc.. If the schedule control method is appropriate, it will confirm the schedule management of the method is correct, whether to consider the project objectives, but also to take into account the milestones; whether to consider the work of decomposition and scheduling, but also to take into account the overall project schedule, it will also provide feedback on schedule management, making the basis for more adequate preparation.

Only the above three parts of the schedule management orderly, with mutual confirmation, combination of dynamic and static, timely adjustment, can be ensured that the project is completed in a reasonable time.

7.2 LNG Project Task Definition and Management

Task or work package (also known as activity) is the smallest unit of the project. It represents a relatively independent, single content, and is easy to manage and control, easy to cost accounting and inspection of the work unit. According to the nature of work, tasks can be divided into two categories: soft and hard tasks. The following two types of tasks and their characteristics are introduced, and focus on the definition of the building and process flow in hard tasks and the general practice of schedule planning.

7.2.1 Soft Task of LNG Project and Its Characteristics

7.2.1.1 Soft Task Definition

The soft tasks in the LNG project is expressed as the work of the brain, and the product is the non-visual form. Such as all sorts of business and contract negotiation, project report and file writing, software writing, purchase, installation and use, project internal and external relations of contact and coordination, various outcome review meetings, inspection and acceptance, all kinds of training classes, visit learning, consulting activities etc..

7.2.1.2 Characteristics of Soft Task Products

(1) Non-intuitive of the product.

Different from hard task products in the physical characteristics, soft task products generally have non-intuitive features, for example, the result of contract negotiations is in the form of text paper, it can also be WORD documents or other form of electronic documents in the PCs; another example, after consulting, exchanging, discussing, communicating, visiting, and learning, it can be impressions and concepts which exist in the personal brain. In short, the product form is generally-non-intuitive, but also some of the experts said the tacit knowledge.

(2) The input of the product is mainly mental work.

In the completion of the soft task, it is mainly based on mental work, such as negotiations with foreign suppliers of LNG resources in LNG sales and purchase contract; meeting of each sub-project completion or trial production acceptance review; HSE training for construction team; consultation activities from the consulting firm or expert, etc.. They are based on mental inputs, the subject and object in the discussion, exchange and thinking to produce a certain ideas and conclusions, etc..

(3) The quality of the product is closely related to the quality of the participants.

Soft task products have the characteristics of the mental labor input, which quality is determined by the participation of the person's quality, professional background, skills and experience. Such as a project for final acceptance, if the experts who involved in the acceptance have experience in similar projects design, construction of leadership and management, it is very easy for experts to grasp the key points of the project completion acceptance, find its shortcomings, put forward rectification opinions.

(4) The value of the product cannot be quantified.

The value of tangible products and the use function of the hard task are relatively easy to quantify, such as the pipeline which is completed and put into operation, the investment and the total cost can be quantified by the financial and audit methods, and its use function is to transport natural gas. But soft task products, such as results of business and contract negotiations, software installation and use, a variety of outcomes assessment meeting, inspection and acceptance, a variety of training courses, study tour, consulting and other activities, which form ideas and concepts, are difficult to quantify the value.

7.2.2 Hard Task of LNG Project and Its Characteristics

7.2.2.1 Hard Task Definition

The hard tasks of the LNG project are the output of the tangible products (including intermediate products), based on a variety of resource inputs. Such as the construction of the LNG

receiving terminal, jetty, storage tanks, gasifiers and pipe sinks, gas transmission trunk, sub gas stations, public engineering buildings and process facilities, etc..

7.2.2.2 Characteristics of Hard Task Product

(1) Fixity of product.

As mentioned above, the building products are fixed in one place, those foundations and the buildings on the ground need to have a FEED design, preliminary design and detailed design, and then construction and mechanical completion acceptance and handover acceptance, trial production acceptance.

(2) Products with huge dominant resources input.

LNG project building products, such as receiving terminal, storage tanks, jetty and gas transmission lines, have reached several billions Yuan. In addition, large amount of human, such as designers, construction contractor team, purchasing personnel, other management personnel, etc. up to hundreds or even thousands of people; moreover, great deal of construction machineries and tools. Like Guangdong Dapeng LNG project, in stable production period, the annual capacity of receiving and gasification LNG up to 3.85×10^6t, and total capital investment is over 7 billion Yuan.

(3) Product diversity.

Besides receiving terminal, storage tanks, jetty and gas transmission lines in a LNG project, the supporting projects need to be built, such as utilities, natural gas station, office buildings, etc., each kind of building products vary between the building structure and construction, and they have their own respective architectural style in the art and decoration, but they must achieve its rugged quality requirements according to their architectural characteristics.

(4) Long construction period of product.

In LNG projects, the construction cycle is generally longer, such as Guangdong Dapeng and Fujian LNG project, construction period of a 160,000 m^3 of storage tank is about 2 years, nearly 400 km pipeline is divided into 6 tenders and construction, construction period to in more than a year. A LNG project construction period of annual receiving and gasifying 3×10^6t/a is up to almost 3 to 4 years.

7.2.3 LNG Project Construction Task Management

From the point of view of the current implementation of CNOOC's LNG project construction tasks, in general, project contracting method should be adopted, the detailed construction site management content won't be discussed in this book. This book is only from the point of view of project company (owners) for the project schedule management, to introduce the general conception of architectural construction.

7.2.3.1 Construction Preparation Classification

(1) Classified by work scope of the construction preparation.

① The whole field construction preparation is based on a construction site for the object of the construction of the preparation. Aiming at the purpose and content of the whole field construction, to create favorable conditions for the whole construction activities, such as the LNG receiving terminal site and jetty site preparation belong to this category, and also taking into account the construction unit works preparation conditions.

② The unit works construction preparation is taken a building as the object of the construction preparation. According to its purpose and content, to create the favorable conditions for unit works construction activities, such as the construction of storage tanks are prepared belong to this category, and taking into account the construction conditions of divisional works and itemized works preparation.

③ Divisional works and itemized works construction preparation are based on divisional works and itemized works as the object for the construction preparation. According to their purposes and contents, to create favorable conditions for the divisional works and itemized works activities, such as one of a site preparation of gas trunk line belongs to this category.

(2) Classified by the construction stage of the construction preparation.

① Preparation before the project starts is to offer construction preparations for proposed project. Its purpose is to create necessary conditions for the construction of the intended commencement of the project. It can be whole field construction preparation; it also can be unit works construction preparation.

② The construction stage preparation is after the commencement of construction projects, and all construction preparation work is before the commencement of each construction stage. Its purpose is to create necessary conditions for the construction stage of the formal start. For example, the preparation works include: the foundation work of storage tank, platform work, bound wall work, the storage tank roof lift engineering. Each construction stage has different content, the required technical conditions, organizational requirements and site layout and other aspects are different, and therefore, before the start of each construction stage, it is necessary to do the appropriate construction preparation work.

7.2.3.2 Construction Preparation Content

LNG project is a group project that includes multiple sub projects, and its construction preparation work must be adequate. Construction preparation content is shown on Table 7-1.

(1) Construction investigation.

① Information on the characteristics and requirements of the LNG project. LNG project company must provide construction contractors with the feasibility study report and its background information, and the project address selection, etc.. Project company must provide the design companies with detailed design purposes, tasks and design intent and other information, including the design scale, the engineering characteristics, and production technical process and technical equipment. Project company also must provide construction companies with engineering stages and construction batches, supporting deliverables sequence requirements, drawings delivery time, as well as the standards and specifications of the construction, quality requirements and technical difficulties and so on.

② Information on the natural conditions of the construction site and the surrounding area. Project company must assist the construction company to obtain and understand the topography and environmental conditions, geological conditions, earthquake intensity, engineering, hydrogeological conditions, and weather conditions, etc. in construction site and nearby.

③ Technical and economic conditions of construction region. Project company should help the construction company to coordinate local water, electricity, supply conditions, transportation conditions, communication conditions, and local materials supply and collaboration conditions.

④ Social and living conditions near the construction site. It includes providing food, housing, transportation conditions for construction in the surrounding areas; in medical conditions, business services, post and telecommunications, fire control, security, etc.; also includes local residents, office, enterprises distribution, working time, living habits, etc.; safety problems and fire prevention problems caused by construction hoisting, transportation, piling, fire and explosion during construction as well as the impact of vibration, noise, dust, garbage and transportation scattered on the social life and the surrounding environment; also considering greening within construction site, and cultural relics protection.

Table 7-1 LNG Project Company Construction Preparation Contents

Construction Investigation	Construction organization preparation	Technical preparation	Material preparation	Site preparation
Information on the characteristics and requirements of the project	Establishing the organization and management corresponding to the project contractors	Familiar with and review the construction drawings and relevant design information	Construction materials preparation	"four connections and one leveling", Field control network measurement, supplementary exploration, temporary facilities erection
Information on the natural conditions of the construction site and the surrounding area	Providing support to the contractor, and giving a variety of education to construction team	Signing sub-contract for project	Processing preparation of accessory, components and pre-products	The on-site storage and stacking of building materials and components
Technical and economic conditions of construction region	Introducing organizational design, planning and technology to construction team and technical workers	Preparing construction technology design	Construction, equipment installation preparation	Organizing construction equipment approach, installation and test
Social and living conditions near the construction site	Supervising contractors to establish and improve construction management system	Preparing the construction drawings and construction budget	Production process preparation	Construction site preparation, setting fire security facilities under adverse climatic window

(2) Supporting construction labor organization preparation.

① To establish the organization and management corresponding to the project contractors. The project company according to its personnel and organizational status, after establishing the strategy for the management of a project, once through public bidding selected construction contractor, must establish corresponding and supporting construction of project management organization, in order to carry out the project management. The general contractor management agency chart is given below (Figure 7-2).

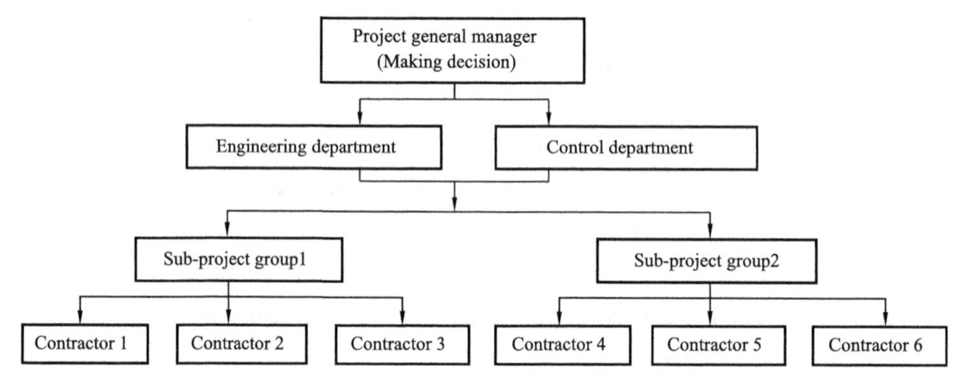

Figure 7-2 Relationship between project company and contractor construction management

② Requirements of project engineering and control organization. The responsibilities of Engineering Department and Control Department or the sub project team are to achieve the project objectives, carry out the project schedule control work. The corresponding working team and organizational forms are required.

a. Project general manager or deputy general manager makes decision for the entire project planning and control tasks.

b. Engineering Department and Control Department or sub project groups implements the sub project planning and control.

c. Specific responsible person-engineers manage the sub project planning and control.

d. Contractors set up the control room of bidding plan, compile and control the bidding plan, make the project management process, standardize the implementation of the project plan.

③ Providing support to contractor to start the engineering. The owner should provide necessary support to engineering contractors, and participate and understand project resources inputs plan from the contractors; also introduce the project company's rules and regulations to the construction team, especially emphasize on HSE regulations, site management, rewards and penalties, ensure that the construction team carry out the construction work within the framework of the project company's management, and complete the project on time, quality and quantity; and offer a variety of training to construction team, etc..

④ Providing the design, planning and technical disclosure to the construction contractors. Relevant departments of the company and the detailed design unit should introduce the design, planning and technology to construction contractors, including implementation and guarantee proposals about construction techniques, standards, safety precautions, new structures, new materials, new technologies and new processes.

⑤ Reviewing contractors to establish and improve the construction management system. The project company should carefully examine the management system of the construction company, technical category including: project quality inspection and acceptance system, engineering technical file management system, the inspection and acceptance system of building materials (components, accessories, prefabrication), technical responsibility system, construction drawing examination system, the technical disclosure system; management category including: attendance system, construction site and team economic accounting system, the materials in and out of storage systems, equipment use and maintenance system and the HSE system, etc..

(3) Technical preparation.

① Organizing and participating in reviewing construction drawings and related design information. It is a very important construction management preparatory work for organizing experts to review construction drawings and relevant design information, including reviewing the basis of construction drawings, purpose of design drawings, and content of design drawings, etc.. After reviewing constnution drawings and related design information, the experts put forward the questions, the project company should supervise the design company to supplement and change, and all of the above data will be archived, and submitted to the construction company.

② Assisting the construction company to prepare construction organization design. According to the scale of the project, the structural features and the specific requirements, project company should help the construction company to deal with relationship between people and objects, subject and auxiliary, technology and equipment, professional and collaboration, supply and consumption, production and storage, usage and maintenance as well as the relationship between the spatial arrangement and the time ordering, and help the construction company to prepare a scientific program, which can effectively guide all construction activities.

③ Assisting the construction company to prepare construction budget. Based on the above construction drawing review, construction design and other resource inputs, the project company should review costs estimates from the construction company for labor, materials, machinery and other inputs, in order to prepare a budget for the project company.

(4) Material preparation.

① Content of material preparation. If the contract stipulates that the project company is responsible for the procurement work, materiel preparation could be divided into the following categories. The first is building materials category: preparation of demand materials according to the construction schedule requirements, according to the material name, specifications, using time, material reserves quota and consumption quota. The second is preparation of the amount of components (accessories) category, according to the processing of pre-products, the name, specifications, quality and consumption. The third is construction installation of equipment category, preparing demand according to the construction plan, construction schedule, construction equipment type, quantity and move-in time. The last one is production equipment category, preparing demand according to the production process and layout of process equipment, equipment name, type, capacity, and number to make sure move-in time batch by batch and stage by stage and storage methods.

② Material preparation procedure. If project company is responsible for the procurement work, the first is preparing a demand plan about accessory, pre-products, construction machinery and technical equipment according to the construction budget, the divisional works or itemized works construction methods and construction schedule. Secondly, according to various materials demand planning, organizing supply of goods, determining processing, the supply locations and supply modes, and signing a material supply contract. Thirdly, based on a variety of materials demand planning and contracts, making transportation plans and transportation methods. Fourthly, according to requirements of the construction general layout, organizing materials planning move-in time, stacking and storage. Similarly, if the contractor undertakes acquisition tasks, the project owner should track the process of the acquisition of the contractor in order to meet the requirements of the owner.

(5) Site preparation.

① On site "four connections and one leveling". The large "four connections and one leveling" refers to jetty and receiving terminal, the small "four connections and one leveling" refers to each natural gas distribution stations, and valve chambers, "four connections" means roads, water and electric power and communication links, "one leveling" means the leveling construction site. Also according to the construction general layout and construction site, it needs to make on-site measurement control network, planning red line pile and engineering control coordinates and the level of pile measurement and location; to erect temporary facilities, including temporary walls and gates, living and office space, toilets, canteens, production facilities, site roads and drainage etc..

② The on-site storage and stacking of building materials and components. According to the requirements of various building materials, components, and the requirements of the general plan of construction, the construction company organizes the materials entering, acceptance, testing, stacking and storage.

③ The construction company organizes construction equipment move-in, installation and commissioning. According to the demand plan of machinery and technical equipment, and the requirements of the construction general layout, machinery and technical equipment will be placed in the specified location and stored in a warehouse, for all construction equipment must be inspected and tested before the start of construction.

④ Construction site preparation and installation of fire protection facilities under adverse weather windows. For the region where four seasons climate change a lot, shelter and protective measures should be taken in advance for building materials, machinery and equipment in winter and rainy season. Fire protection should be taken for fire flammable materials, and the contractor should arrange workers for the preservation of building materials. In southern provinces of China are mostly typhoon-affected areas, contingency plan against typhoon should be worked out, once

the arrival of the typhoon, appropriate measures should be taken.

7.2.3.3 Construction Organization

(1) Construction organization classification.

Considering project construction characteristics, technological process, resource utilization, plane or space layout, etc., the construction can be classified into 3 types as sequential construction, parallel construction, and flow process construction.

① Sequential construction. Sequential construction is a method to make the proposed project in construction objects decomposed into several construction processes, and complete each construction process according to the requirements of construction technology. After the completion of a construction object in the same order again to complete the next construction object, so on, order by analogy, until the completion of all construction objects. In Figure 7–3, for node method in the network diagram, technique (in next section 7.3.2 for details) said by a contractor wins the A, B, C 3 tenders pipeline laying tasks, each section is divided into 3 processes, but the implementation of contractor team every time can only bear one bid construction.

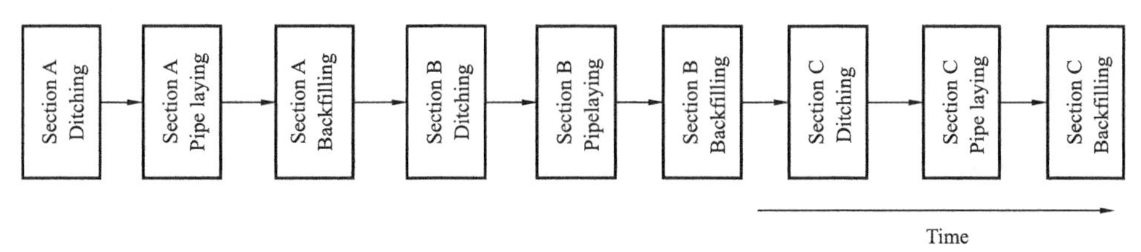

Figure 7–3 Pipeline sequential construction work diagram (node method)

Advantages: the organization and management to the construction site is simpler, the input of per unit time into the labor force, construction machinery, materials and other resources is less. It is conducive to resource supply and organization.

Weakness: it does not make full use of the working surface to carry on the construction, the construction period is long.If a team is set up by professional, the team can't continue to work, there is time to intermittent, labor and construction machinery and other resources can't be balanced. If a team completes all tasks, it's hard to achieve the specialized construction, which is not conducive to improve labor productivity and the quality of the project.

② Parallel construction. Parallel construction method is to organize several teams from the same labor organizations, at the same time and different spaces, to complete the construction objects according to the requirements of construction technology, which is usually used in the rush. Figure 7–4 shows that a contractor obtained 3 tender tasks of pipeline laying, the construction team can satisfy the working situation of 3 sections at the same time.

Advantages: it makes full use of the working surface in construction with short construction period.

Disadvantages too many professional construction teams, large personnel expenses, if there is no suitable work, then each professional team can't operate continuously at the same time,

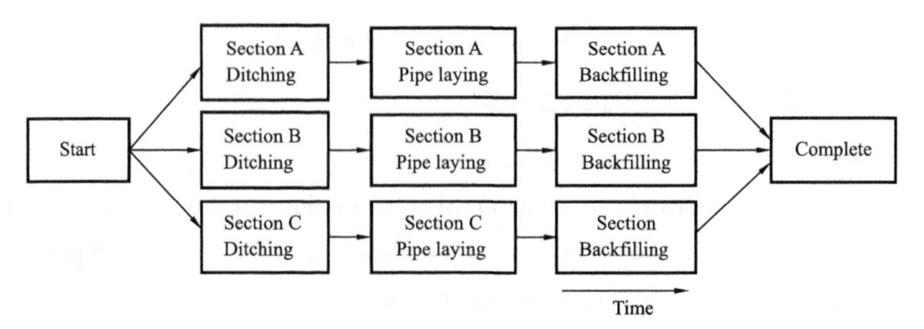

Figure 7–4 Pipeline parallel construction work diagram (node method)

and the resources such as labor and construction equipment can't be used in balance. If there are appropriate workload, per unit time into the labor force, construction machinery and material resources will increase exponentially, which is not conducive to the supply of resources and organizations. The construction site organization and management are more complex.

③ Flow process construction. Flow process construction is a method to make the proposed project in whole construction decomposed into several construction processes. That is divided into the same nature of portioned works, itemized works or process. At the same time, the proposed project on a flat surface is divided into a number of construction sections on the basis of the same amount of work, and the proposed project is divided into a number of construction layers in the vertical direction. In the completion of the first construction section of the construction task, under the same number of professional teams, equipment and materials, carrying out 1,2,3 section one by one until the last section of the construction. It is to ensure that the entire construction process of the proposed construction project in time and space, there is rhythm, continuous, balanced, until the completion of all construction tasks. Figure 7–5 shows that a contractor obtained 3 tender pipeline laying task, use the method of flow process construction, and the construction team can satisfy the working situation of 3 sections at the same time.

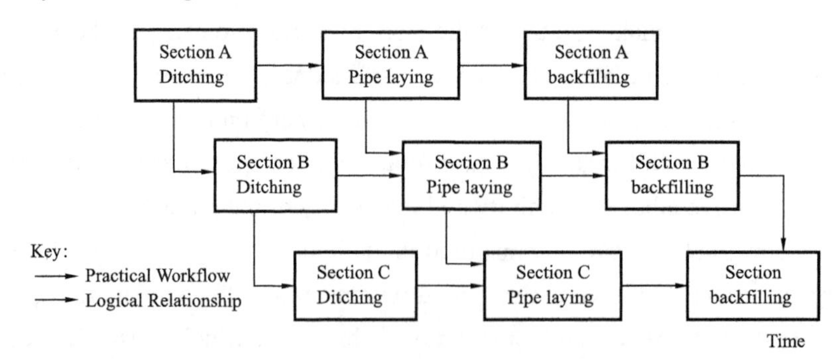

Figure 7–5 Pipeline flow process construction work diagram (node method)

(2) Flow process construction classification.

From the description above 3 construction methods, we can see that flow process construction is an advanced and scientific construction method. For the overall characteristics in the division process, timing and spatial arrangement, it is a frequently used method in the construction. Flow process construction can be subdivided into 4 types.

① The group flow process construction. Also known as large flow process construction, it is organized by a number of unit works among the flow process construction. Reflected in the project construction schedule, it is made of a sheet of project construction of general schedule.

② The unit works flow process construction is also known as integrated flow process construction, it is organized by a number of divisional works of the flow process construction in the interior of a unit works. On schedule sheet, it is expressed by several groups of the divisional works of the schedule indicator lines, which composed of a unit works of schedule plan.

③ The divisional works flow process construction. Divisional works flow process construction is also called the professional flow process construction. It is organized by with in the interior of a divisional works or between the itemized works of the flow process construction. On schedule sheet, it is expressed by a group of horizontal schedule indicator lines marked with construction section or working team numbers. Such as in pipeline construction, the divisional works is made up of 4 itemized works, trenching, pipe laying, welding and backfilling closely linked to each other in technology. The construction is divided into several tenders, organized by four professional teams, followed by continuously to complete their work in each section with the same construction process, which is called the divisional works flow process construction.

④ The itemized works flow process construction. Itemized works flow process construction is also called the detail flow process construction. It is a professional type of work within the organization of the flow process construction. On schedule sheet, it is expressed by horizontal schedule lines marked by a group of construction section or team numbers, for example, the working team of the pouring concrete of the storage tank is to finish the work of pouring concrete successively in the fan-shaped section of each partition.

(3) The definitions of the construction process.

① The definition of the construction process classification. It can be divided into 3 categories.

a.Prefabricated construction process. It refers to the construction process in order to improve the construction of product assembly, industrialization and mechanization and production capacity, no need to be in the construction site, such as preparation process of mortar, concrete, components, pre-products.

b. Transportation construction process. It refers to the construction process in order to transport construction materials, components, (semi-) products, pre products and equipment to the project site or warehouse operations.

c. Construction process. It refers to construction of the object in space, directly to form the final building products, such as jetty construction, storage tank construction, etc.. The construction process or professional team is usually represented by n.

The first two categories generally do not take up the space of construction object and will not affect the duration of the project, and generally do not reflect on the schedule. Only when they occupy the space of the construction object and affect the total duration of the project, which is included in the project construction schedule. The third category occupies the space of the construction object and affects the total project duration, which is the main content of the project construction schedule.

② The definition of working surface. It refers to the construction industry for some professional type of workers or some kind of machinery for the construction of the activity space. The size of working surface shows how many workers and machines can be arranged, which depends on the completed quantity and safety construction requirements in per unit time how many workers or machine in each work. Determined working surface is reasonable or not directly affect the productivity of professional teams. LNG project working surface parameters in each project require us summarizing information in the previous projects and providing the basis for new projects. Parameter representation usually is expressed as: m/person (set), m^2/person (set), m^3/person (set).

③ Construction section. It refers to the construction object in plane and in space divided into a number of labor amount roughly the same construction paragraphs, known as the construction section or assembly line section, the construction section of the number is expressed as m, which is one of the main parameters in assembly line construction. The purpose of assembly line construction is to make full use of the working surface, to avoid slowdown, and shorten the construction period as far as possible. For example, LNG pipeline construction is usually divided into several sections by different construction groups constructing at the same time.

④ Construction layers. It refers to the construction object in the vertical direction divided into a plurality of operating layers, for example, the process of building the storage tank can be divided into several circles for pouring concrete.

(4) The relevant time parameters.

The following are describing the relevant time parameters used in the calculation below, such as flow beat t_i, flow pace $K_{j,j+1}$, parallel lap time $C_{j,j+1}$, technical intermittent time $Z_{j,j+1}$ and organizations intermittent time $G_{j,j+1}$.

① Flow beat. It refers to the organization of assembly line construction, each professional team completes the task in the construction section for the working duration required, which is usually expressed in t_i. The construction process work duration is expressed in T. It is the basic parameter for assembly line construction. When the assembly line construction time is not equal, it is called no rhythm flow beat.

② Flow pace. It refers to the two adjacent construction process (or professional team) starting the construction one after another, the minimum interval time. Flow pace is generally used $K_{j,j+1}$, where j ($j=1,2,\cdots,n-1$) is the number for professional teams or construction process. Flow pace depends on the number of participating assembly line construction process. If the number of construction process is n, the total number of flow pace is n-1.

③ Parallel lap time. It refers to the organization of assembly line construction, in order to shorten the construction period under the condition of working surface allowed, if the previous professional team have completed the construction task or uncompleted construction tasks, but it could provide the working surface to a latter professional team, so that the latter one can enter the previous construction section in advance, both of them are working in parallel on the same construc-

tion section, the lap time is called parallel lap time or insertion time, usually expressed in $C_{j,j+1}$.

④ Technical intermittent time. It refers to the organization of assembly line construction, in addition to considering the flow pace between two adjacent teams, sometimes considering reasonable waiting time according to the technology properties of building materials or cast-in-situ components, such as setting time after concrete pouring. This waiting time is called technical intermittent time, usually expressed in $Z_{j,j+1}$.

⑤ Organizations intermittent time. It refers to the organization of assembly line construction, as the reasons of construction technology and construction organization, resulting in the increase in the flow pace. Such as construction workers, machinery relocation transfer time, examination and acceptance before pipeline backfilling soil and so on. Organizations intermittent time is usually represented in $G_{j,j+1}$.

7.2.4 LNG Project Work Package Resource input Management

7.2.4.1 Basic Work

(1) General concept.

To complete a job or task, we must ensure resource inputs. LNG project should have 5 resource inputs, they respectively are: human resources, time resources, equipment and material resources, financial resources and information resources (Figure 7-6). The work package plan is the basis for the project schedule. For the outsourcing of a project, the contractor should be responsible for the work package resource input management, but the relevant management departments of the project company must be followed. If you ignore the matching relationship between work package and resources, you can't complete the work package tasks.

The following will elaborate the matching problems between these 5 resources and work packages.

(2) The task directory preparation.

If the LNG project work breakdown structure directory (see Chapter 3) has been refined to perform specific tasks, you can omit this step work. If not, the project contractor who undertakes the specific task must do a detailed task to the preparation of the directory. Tasks that can be performed, also known as work packages, can be regarded as the supporting work of WBS or refinement. It must be included in the scope of the project, and task list should be accompanied by a detailed description of the task to ensure that the project team members can understand how to do it.

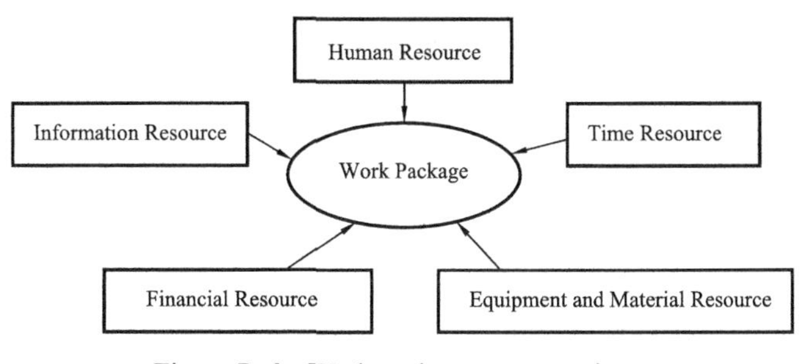

Figure 7-6 Work package resources input

(3) Preparation of task details.

For the bottom of a task to prepare a detailed description of the system, so that first-line managers and staff can operate conveniently. These words together into a book can form the LNG project task instruction, which should be included in the description of the mentioned work package resource allocation. Such instructions may lay the foundation for the use of the management process of other projects in the future. Detailed description should also include all assumptions and constraint conditions. Specific contents differ from each other due to specific tasks.

(4) Modifying WBS structure.

In the WBS to determine which tasks are necessary in the process, the project team will certainly be able to confirm that the particulars of what items were omitted or which project details description need to be modified or described more clearly. Any such changes shall be reflected in the WBS related documents (e.g., cost estimates), and the above changes are often more likely to occur when the project is involved in new or not validated technology.

7.2.4.2 Work Package and Human Resource Allocation

(1) Human resources of the project company.

Human resources involved here is just in terms of a work package, which is only a small part of the project company human resource management. On the current domestic LNG project, the project company is generally equipped with lean, personnel control more stringent, a lot of work to rely on the contractor, consulting firm to complete, but the relevant departments of Project company or individuals should participate in the coordination or audit work according to the engineering contracts between project company and contractors.

(2) Human resource allocation outside the project company.

① Contractor's human resource allocation. If a project is entrusted to the contractor, the contractor may have to spend a lot of manpower under the contract so as to complete the task within the prescribed time. Human resource inputs includes management, technicians and operating workers, and also should consider the increasing alternative staff in rush time.

② Supervisory company's human resources allocation. As a large-scale LNG project, supervisory company plays a significant role in it. The project company should make good use of human resources of the supervision company, and according to contract, learn the management and quality assurance experience in the work package from them.

③ Human resource management of consulting company. Also for key technologies and management issues, to carry out consulting activities, and give full play to the intellectual resources of the consulting experts, so that a work in the most understand the guidance of the best results delivered.

④ Responsibility matrix. Responsibility matrix is a method or tool to breakdown the work package, implement it to the associated units and individuals and make clear everyone' organization relationship, responsibility and status in the work package. From the perspective of the project company, the contractor is responsible for the work of the responsibility matrix of the work package, which is basis for project company to compile its responsibility matrix of the work package (Table 7-2).

Table 7-2 The Responsibility Matrix of A Sub Project

Division of labor / Task sequence NO.	A	B	C	D	E	F	G	H	I	J	K	L	…
I	√	√										√	
II			√		√								
III				√	√	√				√			
IV				√	√		√	√	√	√	√		
V				√	√		√	√	√	√	√		
……													

7.2.4.3 Work Package and Time Resource

(1) Schedule management basis.

Time resource here is also in terms of a work package. The work package schedule management is the foundation of the project time management. It faces the work package, namely, to complete the work package within the prescribed time, through the reasonable arrangement and schedule control. The schedule management is based on the following: the description of work package task, schedule management information for similar work packages, human resource allocation, equipment and materials sufficient degree and absolute time weather window, etc..

(2) Schedule management.

Based on above information, we should analyze work package tasks carefully, compile specific work processes and procedures, and make reasonable arrangement of human resources, equipment access, material preparation and vehicle arrangement. Considering coordination problems between working surface and other work package, we have to consider the climate window during absolute time, such as the dry season, rainy season, typhoon season or other unforeseen events, which could bring out the time delay and so on. Finally work out the schedule of work package, and base on it to control its schedule.

7.2.4.4 Work Package and Equipment Material Resource

(1) Construction equipment.

According to the specific work package, prepare for equipment in advance or deploy construction equipment on schedule, such as LNG pipeline engineering, equipment including trenching machines, welding machines, large-scale transport vehicles etc.. Equipment Integrity rate inspection must be carried out before entering, but also be ready to spare parts, etc..

(2) Equipment and materials.

Similarly, according to the work package tasks, it is prepared for related equipment and materials. Take work package of LNG gasification unit equipment installation as example, the ORV, SCV or IFV should be delivered to the site ahead of time. LNG jetty protective slope should be prepared the anti-wave preform, cement, steel, sand and other construction materials. Inner tank construction should be prepared with the Ni9% steel, perlite material etc..

7.2.4.5 Work Package and Capital Resource

Capital investment in work package is a small part of the whole project, and it generally does not need cash transactions, but reflected in equipment, materials. For the EPC contract, the owner and the contractors just talk about several total prices, not the specific work package cost; if the owner hires a consultant, it is to consider the occurrence of consulting fees. If the owner masters the cost of the single work package, it will help to calculate the exact cost of the entire project, must be highly valued from the beginning.

7.2.4.6 Work Package and Information Resources

Information resources are equally important, in terms of specific work package, it includes the following categories.

(1) Similar work package information.

Similar packages include resource allocation, work plan, procedure, program and so on, which are very good reference for the relevant work packages.

(2) Similar project information.

There are many similarities such as jetty construction information and terminal supporting engineering information, which can be used as a reference. The previous cost of equipment and material can be taken as a reference to the following project.

(3) Management information.

It includes previous work package management involved in project companies, contractors, supervisory management styles, methods, input levels, and experience, etc..

(4) Macro policy information.

It includes the change of the national and local policy on LNG project, such as energy-saving and emission-reduction. Sometimes a policy will promote or retard the advancement of the project.

7.2.4.7 Work Package and Resource Integration

(1) Summary of the work package investing resources.

Resource estimation is a comprehensive evaluation of human resources, time resources, equipment and material resources, capital resources and information resources for a specific work package. Estimation result is to identify and explain resource types and quantities required for each planning activity in detail. It can be summarized into a matrix Tablewith a work package and a variety of demand resources (Table 7-3). It is the basis Tablefor the summary of the total resources.

Table 7-3　Summary Tableof Work Package Resource Input

NO.	Task name	Human resource			Material and Equipment					Capital estimation	Time resource	Information resource
		Contractor	Supervisory Comany	owner	Machine name	Machine-team number	Mechanical quantities	Material name	Material quantities	(RMB)	Year Month Day~ Year Month Day	Similar Project information
1	I											
2	II											
3	III											
4	IV											
…	…											
Sum-mation												

(2) Resource calendar.

It is necessary to subdivide the resource requirements of the above work packages into each day, and to form a resource calendar of a work package, or do not use a specific resource of the working day. Work package resource calendars identify their respective holidays, as well as the time to use resources. The resource calendar also identifies the number of available during each available period.

(3) Project calendar.

When calculated from the theoretical LNG project respectively the earliest and latest start and end date, without considering resource constraints, and this is not the date calculated actual schedule, but said the duration required for the work package. Considering the resource constraints and other constraints of the specific work package of the project, the task is arranged in a specified time interval. The logical relationship between the preparation schedule can get the project schedule or sub project calendar, in the same way, the summary of each sub-project schedule can obtain the overall schedule of LNG project or overall project calendar.

7.3　LNG Project Task Sequencing

7.3.1　Logic Relationship Analysis of the LNG Project Task

7.3.1.1　General Concept of Task Sequencing

The sequencing process includes verifying and drawing up the correlation between work packages. The work packages must be scheduled correctly so as to make a feasible plan for the future. Sequencing can be performed by a computer (using computer software) or by hand scheduling. Small projects can be hand sorted, and the medium-sized projects in the early days (at this point the details of the project are poorly understood) are possible to be sorted by hand. However, for such a large LNG project, it is generally sorted by computer software. Specific to the LNG project task sequencing is based on the mentioned work breakdown structure, task list, and their instructions. It should be said that all the work must rely on the previous work. For the work packages on the same level, the relationship between them is very important. Some works have dependencies among them, while some can be performed simultaneously, in no particular order. We must take logical analysis among the work packages firstly. The following is manual sorting method, only to allow readers to understand the working process. The use of manual and computer sequencing are the two major means of project time management.

7.3.1.2 The Dependencies between Work Packages

Generally, sequencing should firstly analyze and determine the logical relationship between work packages especially the work packages on the same level. The same level is defined as the tasks at the same level in WBS structure diagram. In order to confirm the relationship of work packages, we should do enough analyses based on the logical relationship.

(1) Mandatory successively relationship.

The logical relationship between the work packages themselves or the technological requirements can't be changed, that is, the inherent dependence of the work package, also called the intrinsic correlation, which is usually caused by the objective conditions. For example, in gas pipeline sub-project, pipeline route survey in LNG project is the premise of pipeline design, and the pipeline permanent or temporary land acquisition is the premise for pipeline trenching. The design, site preparation and facilities procurement in jetty sub-project are the pre-work of the jetty construction, etc.. These are the successively logical relationships required between work packages and must be strictly enforced.

(2) Non mandatory successively relationship.

That is, this relationship is not the logical relationship between the works themselves, but rather the man-made organization. Because of the above mentioned five kinds of resource combination or match to the work of the organization caused by successively formed relationship, so it is also known as the specified correlation. Such as LNG receiving terminal sub-project equipment procurement, if the procurement team is powerful, it can purchase many types of equipment at the same time; if the power is limited, only can classify procurement, this procurement work has been artificially divided into successive relationship. Strictly speaking, this successive relationship is caused by the artificial organization. The specified correlation is defined by the project management team to determine the relevance, should be careful to use this correlation and fully to the statement. The selection of scheduling schemes is restricted by the recognition and use of

such correlations.

7.3.1.3 The Synchronous Relationship between Work Packages

Just as its name implies, synchronous relationship is non-working order between two or more jobs. Under the conditions of the working face, in order to speed up the schedule, the jobs can be arranged in parallel. The most prominent example is the pipeline project. It is generally divided into several tenders and awarded to several contractors, enabling them to carry out construction in the meantime. Furthermore, jetty, receiving terminal and pipeline projects are granted in 3 contractors, or a contractor whose technical strength and human resources are very powerful so that it can carry out 3 sub-projects in the meantime.

7.3.1.4 Work Package and External Correlation

External correlation refers to the relevance between this project task and external task. For example, to start overall LNG project, it needs to get the approval of the NDRC. Land acquisition of pipeline project should get the approval of relevant departments of local government. After obtaining land use certificates, pipeline trenching can be carried out. Jetty surrounding waters have to be approved by the State Oceanic Administration and then can use the waters. Therefore, in considering the duration of the project, uncontrollable time of external correlation should be considered.

7.3.2 Tools and Methods for LNG Project Task Sequencing

7.3.2.1 Gantt Chart Representation

Gantt chart method is one of the most commonly used sorting task representation method. As shown in Figure 7-7, the horizontal axis represents time and the vertical axis represents the construction process, each process is also divided into different working section, with the help of Gantt chart, represents the working relationship of tasks and duration. If all the tasks of a project are shown in a chart, you can get the total duration of the project. Seen the following Figure 7-7, there are respectively 3 construction processes A, B, C, where A is earlier than B, B is earlier than C. Each construction process is divided into 4 working sections, the precedence relationship is shown respectively by ①, ②, ③, ④ . If according to assembly line method to carry out the job, it can be seen from the Figurethat the project duration is 18 days.

Construction process	Construction days																	
	1	2	3	4	5	6	7	8	9	10	11	12	13	14	15	16	17	18
A		①			②			③			④							
B					①			②			③			④				
C								①			②			③			④	

Figure 7-7 Gantt chart tasks sequencing

7.3.2.2 Network Graph Representation

For network chart method, there are two forms: node method and arrow method.

(1) Activity-on-node (AON).

AON is also known as a single symbol network graph representation. As shown in Figure 7-8, the terminal node represents a task (represented by circles or boxes). Because each task has a unique job number, it is called a single symbol [Figure 7-8 (a)]. Different definitions of time may also be indicated, such as the earliest start time, the latest start time, the earliest completion time, the latest completion time, and so on. By means of arrow connection indicates the relationship between the sequences [Figure 7-8 (b)].

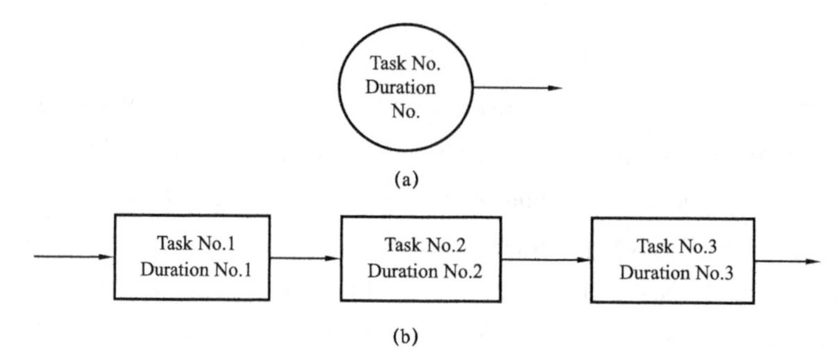

Figure 7-8 Node representation

① Duration representation. Usually node tasks have the following times.

a.The earliest start time: the earliest start time of the task node.

b. The latest start time: the latest start time of the task node.

c. The earliest finish time: the earliest time that the node task can be completed.

d. The latest finish time: the latest time that the node task must be completed.

② Dependencies between tasks. Node method also can represent 4 kinds of dependencies between two tasks (Figure 7-9).

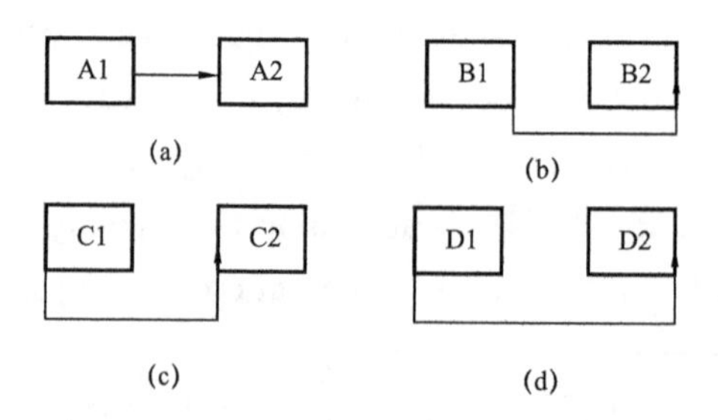

Figure 7-9 Node relation representation

a.End to start: the previous task must be finished before the next task is started[Figure 7-9(a)].

b.End to end: the previous task must be finished before the next task is finished[Figure 7-9(b)].

c.Start to start: the previous task must be started before the next task is started [Figure 7-9 (c)].

d.Start to end: the previous task must be started before the next task is finished[Figure 7-9(d)].

③ Definition of front closely tasks and back closely tasks. Relative to a task, the task that is located closely front of it is called the task's front closely task; the task that is located closely back of it is called the task's back closely task. In accordance with the task relationships that are described in these concepts, you can get a project's construction network diagram.

④ The basic principles to establish a network chart.

a. To draw network chart from left to right.

b. The task can begin only after all the related front closely tasks have been completed.

c. The arrow line in the network indicates the direction of the front and back flow, and the arrow lines can cross each other.

d. Each task can only have a unique identification code.

e. Task identification code must be larger than the front closely tasks.

f. Arrow lines can not appear loops.

g. If there are multiple start nodes, a common starting node should be selected to definitely show the beginning of the project, and similarly, the termination node is represented by a point.

⑤ Drawing project nodes relationship chart. Table 7-4 shows a pipeline valve chamber construction project. We gave numbers to each task A, B, C …… and gave specific tasks in the description section.

Table 7-4 Construction Project of a Pipeline Valve Chamber

Tasks	Description	Front closely task
A	Task book released	Non
B	Design	A
C	Construction plan	A
D	Resources allocation	A
E	Construction report	B、C、D
F	Expert review	E
G	Program approval	F
H	Equipment procurement	G
I	Construction	G
J	Equipment installation	G
K	Acceptance check	H、I、J
L	Put into use	K

According to Table 7-4, we can draw the construction of a pipeline valve room from the start of the project task book, followed by design, construction plan and allocation of resources. On the basis, the construction plan report is compiled, reviewed by experts and gotten the approval of construction programs, and then equipment procurement, construction and equipment installation can be carried out. The next is acceptance check, and finally put into use, the whole project is completed. Figure 7–10 is a node relation network diagram based on Table 7–4.

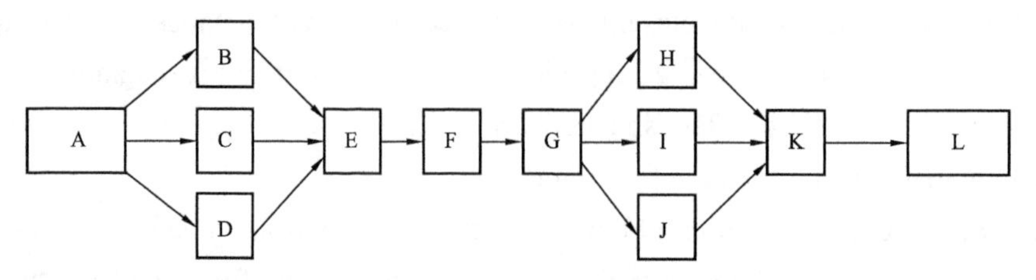

Figure 7–10 A pipeline valve chamber construction project tasks (node method) relationship diagram

(2) Activity-on-arrow (AOA).

AOA method, also known as double code network diagram representation, on the contrary with the above one, task is indicated by arrow; using nodes to represent the sequence (or called events), which means the end of the previous task and the beginning of the next task. Because of the arrow that describes a task process, at the both ends are described as two symbols, so it is called a double symbols Figure 7–11.

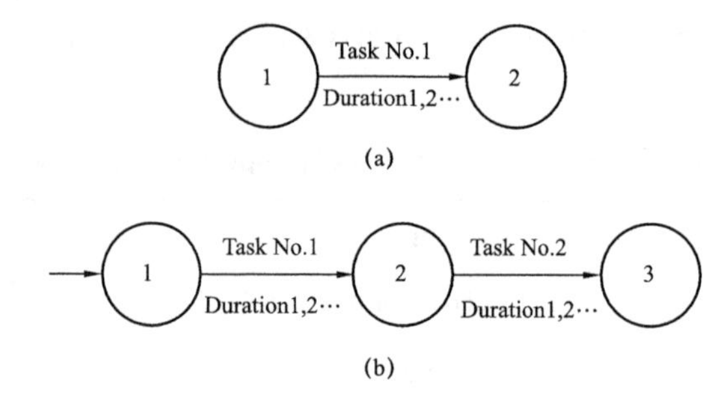

Figure 7–11 Arrow relation representation

① Duration representation. It can also be marked on the task number and duration.

② Virtual tasks. Tasks which need to take time to do down-to-earth, and are generally expressed by the solid arrow. while the virtual task is to Figurein convenience, it doesn't take time, just to show the relationship between the tasks, it is generally indicated by virtual arrow.

③ Arrow relationships plot method. Similarly, task relationships can be indicated by arrow method. Figure 7–12 shows this network chart notation.

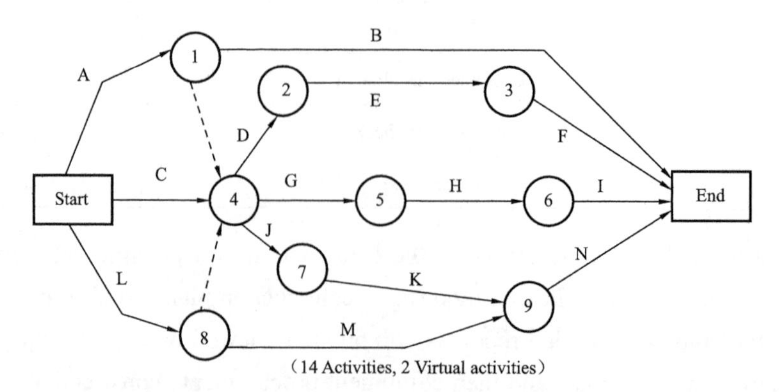

Figure 7–12 Relationship chart in arrow method

7.3.2.3 Planning Network Reference Model

In the preparation of LNG project activity network diagram, we can use the project network diagram completed at home and abroad or use standardized similar project network diagram, also we can use local networks and sub networks, which can accelerate the network diagram of the system. Generally, planning network reference model gathers a lot of experts' wisdom and experience. When the LNG project contains several identical or nearly identical contents, the sub network is particularly useful.

7.3.2.4 Task Sequencing Method Expression

There are 2 main methods to express assembly line construction progress chart: Gantt chart and network chart.

(1) Gantt chart.

Gantt chart method includes horizontal method and vertical method, and the following is describing horizontal method (Figure 7-13). Horizontal axis represents time of assembly line; vertical axis represents the construction process or number. m strips numbered horizontal lines that are numbered as ①, ②, ③, ④, ⑤ to represent different construction sections; n professional task teams that are numbered as A, B, C to represent the schedule arrangement of task team.

Itemized works No.	Construction schedule (d)																				
	1	2	3	4	5	6	7	8	9	10	11	12	13	14	15	16	17	18	19	20	21
A		①			②			③			④			⑤							
B		$K_{1,2}=3$			①			②			③			④			⑤				
C					$K_{2,3}=3$			①			②			③			④			①	

Figure 7-13 Assembly line construction sequencing—Gantt chart representation

Step: $K_{1,2}=K_{2,3}=3$ (d); construction section: $m=5$; professional teams: $n=3$; flow beat: $t_i=3$ (d); the total duration: $T=K_{1,2}+K_{2,3}+m\cdot t_i=3+3+5\times3=21$ (d).

(2) AOA method.

Network chart method includes transverse flow type, flow pace type and lap type, and the following only describes the flow pace network chart. As shown in Figure 7-14, solid arrows represent the real work, which is marked construction process. The circle marked with event number, under which is labeled with flow beat. Dotted arrows represent the imaginary task or work, and they show the relationship between tasks, no time-consuming. Flow pace is also represented by the solid arrows, with numbers and values marked under it.

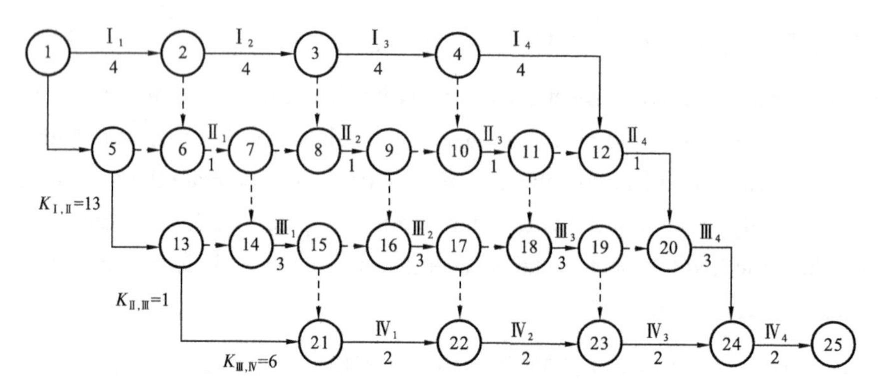

Figure 7–14 Flow process construction sequencing—network diagram representation

From the above Figurewe can see that the flow beats of the construction processes is different. The construction process is divided into Ⅰ, Ⅱ, Ⅲ, Ⅳ; flowing beats were $t_{Ⅰ}=4$, $t_{Ⅱ}=1$, $t_{Ⅲ}=3$, $t_{Ⅳ}=2$ time units; steps are separately as $K_{Ⅰ,Ⅱ}=13$, $K_{Ⅱ,Ⅲ}=1$, $K_{Ⅲ,Ⅳ}=6$; construction section: $m=4$; professional teams: $n=4$.

Total duration: $T=K_{Ⅰ,Ⅱ}+K_{Ⅱ,Ⅲ}+K_{Ⅲ,Ⅳ}+m\cdot t_4=13+1+6+4\times 2=28$ (time units).

7.4 The LNG Project Task Duration Estimation

Project duration estimation is an important work of the LNG project plan. The following will introduce the general LNG project time estimation under a certain resource condition matching time estimation, as well as time estimation tools and methods, etc..

7.4.1 LNG Project Task Time Estimation Basis, Tools and Methods

7.4.1.1 General Task Time Estimation Basis

Any work or task is completed under certain conditions. LNG project has no exception, and only in the case of the following conditions are clear can the whole task schedule be estimated accurately.

(1) The list of tasks.

The list of tasks is the foundation of schedule estimation, especially LNG project is supposed to have wide social influence, long industry chain, complex technical and huge investment. Firstly, the integrity of the task list is very important, if there is omission, it will definitely lead to inaccurate time estimates. Secondly, classifying similar kinds of work so that we can make a general arrangement for them, such as in the invitation of tender of the LNG jetty, LNG receiving terminal and pipeline design units, we can consider a unified bidding in the tender invitation and bid preparation, biding issue and bids evaluation process control; equipment or materials of same type is uniformly arranged in equipment procurement.

(2) Project constraints.

LNG project constraints also must be identified beforehand in time estimation. Constraints are divided into objective and subjective constraints.

① Objective constraints. Objective constraints are for project team, they cannot be controlled by project team. Mainly include the difficulty in looking for LNG resources supplier, the limitation that jetty natural conditions lead to workload, complexity of pipeline routine, material supply market change, users' tolerance to fuel price, tendency of national policy, and so on. For example, the current LNG resources are transmitting from buyer market to seller market. It will be more difficult to buy LNG resources in a cheaper price. It may take more time and efforts if we want to get the ideal LNG resources. Jetty natural conditions require more excavation quantities than filling construction, it will delay the jetty construction time. National policy encouraging the use of clean energy will help LNG project to be carried out, which can shorten the approval time of LNG project, and so on.

② Subjective constraints. For the project team itself, the project manager's ability, team members' work experience, division and collaboration, if the team has constructed LNG project before or not, if the other relevant resource is sufficient or not, and so on, the capable manager, good team spirit and sufficient resource will speed the progress, in time estimation, we could think less about the additional time. On the contrary, we need to think more.

(3) Resource matching.

Resource matching is the degree of fit of the type and amount of resources required for each job. Obviously, the completion of most of the work is connected with the assigned resources quantities quality. Any work package has a best resource matching problem. For example, the receiving terminal site preparation is subject to working surface constraints, such as the site area's accommodating staff (including contractors' workers, managers, supervisors, site project HSE management personnel, etc.), equipment operation, vehicle traffic, material stacking and handling. The time required for most tasks is determined by the relevant resources, but not the more the better in certain period, and combined with the nature of the work. If resources are poorly supplied, it will certainly affect the schedule to complete, and consequently delay the completion of the entire project. The time required for most tasks is related to the people capability and materials quality. For example, the same task takes a low-level technician one day, but takes a senior technician only half a day.

(4) Historical information.

Working files similar to LNG project, including work breakdown, worksheets, text description, summary reports, have a certain reference to the project; while the preliminary work information of the project, such as market information, report of LNG work scope, feasibility study report, are the basis of task time estimation. The historical information of the time required for various tasks possibly comes from the following situations.

① Project file. One or more of the companies that have been involved in the LNG project related to the project may retain records of the results of the previous project, which can be helpful to estimate duration if the records are very detailed. In many applications, individual team members may retain these records.

② Time estimation database for commercial use. Some past data are often valuable. They are very useful when the task time cannot be calculated by the actual work contents (such as concrete setting time, the time of the government agency's approval for certain types of applications).

③ Project team knowledge. Individual members of the project team may remember the previous task's actual or estimated number. Although this approach may be useful, its reliability is much lower than the archived file.

7.4.1.2 LNG Project Task Schedule Estimate's Tools and Methods

(1) Expert judgment.

It is often difficult to estimate the time needed for a person who has never experienced a LNG project, because there are many factors that can affect the accuracy of time estimation, such as the level of resources quality and labor productivity differences. But experts who have involved in LNG project will rely on the past data and information to make judgments. We can estimate with the help of experts' experience. If we can't find the right experts, the result is often unreliable with greater risks. In this case, we should fully consider the leeway while we are estimating the time.

(2) Analogy estimation.

Analogy estimation implies that a previous LNG project similar tasks' actual time is the basis of this project tasks' time estimates. We use this method in cases that we have to estimate a project's tasks time with very limited information and data. But we should face that the estimated time in this method has some error.

(3) Simulation.

Simulation uses different assumptions to calculate the appropriate time. The most common method is Monte Carlo method. In this method, it assumes the probability distribution of each task's time in order to calculate the probability distribution of the time required by entire project.

① Most likely duration. The most likely duration is the estimated duration in the case that resources allocated for the plan, resource productivity, the realistic possibility for the task, the dependence on other participants, and possible interruption time have been given.

② Optimistic duration. If the above conditions are formed the best combination, the estimated duration is positive duration.

③ Pessimistic duration. If the above conditions are formed the worst combination, the estimated duration is pessimistic duration. Usually the average of the above 3 durations can be used as the task's duration.

(4) Productivity estimation.

It is easy to know the productivity per unit time. The formulas are following:

$$Z=X\cdot y \tag{7-1}$$

$$T=Z/a \tag{7-2}$$

Where X——length of the pipeline, km;

Y——unit productivity, labor· d/km;

Z——total mon-hours, labor· d;

a——total personnel number, labor;

T——the pipe lay required time, d.

(5) Backup time.

Backup time is a given time that is more than the project scheduled time when there is a risk or uncertainty in a project. Sometimes it is called emergency time, reserved time or buffering time. For this large and complex LNG project, the arrangement of backup time is very important, because the LNG project is characterized by long duration, huge amount of work and involving widely, similar foreign projects for 2 to 3 years, and in such a long time, we may encounter many uncertain factors, so it is difficult to estimate the project duration. In order to have the initiative in time management, backup time must be considered on the basis of a percentage of task time that has been estimated, or given in a fixed length of time based on experts' opinion, or given quantitative time based on risk analysis. Since LNG project is composed by tens of thousands of specific tasks, be sure to consider successively relationship and parallel relationship among tasks, and finally consider the entire project's backup time. Backup time can be modified, extended or shortened, or even canceled according to project progress. Generally the entire LNG project's backup time is half a year.

7.4.2 LNG Project Task Time Estimation

7.4.2.1 Accumulative Sequence Dislocation Subtracting and Taking Large Difference Method

The method, also known as Paterkovsky method, is simple, accurate, and easy for beginners to master, especially used in the non-rhythm flow process construction.

(1) Basic steps.

① In turn accumulate the flow beat in its section of each construction process, and obtain the accumulating series of flow beat of each construction process.

② Put the flow beat of accumulating series dislocation in the adjacent construction process, after subtraction to get one difference sequence.

③ Take the maximum value of the series, which is the flow pace of both construction processes.

④ Project duration formula:

$$T_1 = \sum_{j=1}^{n1} K_{j,j+1} + \sum_{i=1}^{m} t_n \tag{7-3}$$

Where: t_n——The last professional construction team n flow beat in each construction section.

⑤ According to the meaning of this question, the resources affecting the duration can be calculated as follows:

$$T_2 = \sum Z_{j,j+1} + \sum G_{j,j+1} + \sum C_{j,j+1} \tag{7-4}$$

It is needed to explain that the resources factors affecting the work package or the sub project is vary widely, and to solve the practical problems, according to the different conditions the corresponding formula should be chosen.

⑥ Project duration, that is, the total duration of the project to consider the impact of resources:

$$T = T_1 + T_2 \tag{7-5}$$

$$T = \sum_{j=1}^{n1} K_{j,j+1} + \sum_{i=1}^{m} t_n + \sum Z_{j,j+1} + \sum G_{j,j+1} + \sum C_{j,j+1} \tag{7-6}$$

(2) Application example.

A LNG receiving terminal vaporizer engineering is drawn up for Ⅰ, Ⅱ, Ⅲ, Ⅳ, Ⅴ 5 construction processes. The construction is divided into 4 construction sections, and the flow beat of each section in the construction process is shown in Table 7-5. It is required to maintenance time for 6 days after the construction process Ⅱ completed; after construction process Ⅳ completed, the corresponding construction section has 2 days of preparation time. To be finished as soon as possible, construction process Ⅰ and Ⅱ are allowed to overlap one day. Please calculate the total duration.

Table 7-5 The Flow Beat of a Vaporizer Engineering (week)

Construction team	Construction process				
	Ⅰ	Ⅱ	Ⅲ	Ⅳ	Ⅴ
1	3	1	2	4	3
2	2	3	1	2	4
3	2	5	3	3	2
4	4	3	5	3	1

According to the above conditions and the above calculation steps:

① To calculate the accumulative sequence of flow beat:

Ⅰ: 3, 5, 7, 11

Ⅱ: 1, 4, 9, 12

Ⅲ: 2, 3, 6, 11

Ⅳ: 4, 6, 9, 12

Ⅴ: 3, 7, 9, 10

② To determine the flow pace:

$K_{\text{I},\text{II}}$:

$$\begin{array}{rl} & 3,\ 5,\ 7,\ 11 \\ -) & \quad 1,\ 4,\ 9,\ 12 \\ & 3,\ 4,\ 3,\ 2,-12 \end{array}$$

$\therefore K_{\text{I},\text{II}}=\max\{3,\ 4,\ 3,\ 2,\ -12\}=4$

$K_{\text{II},\text{III}}$:

$$\begin{array}{rl} & 1,\ 4,\ 9,\ 12 \\ -) & \quad 2,\ 3,\ 6,\ 11 \\ & 1,\ 2,\ 6,\ 6,-11 \end{array}$$

$\therefore K_{\text{II},\text{III}}=\max\{1,2,6,6,-11\}=6$

$K_{\text{III},\text{IV}}$:

$$\begin{array}{rl} & 2,\ 3,\ 6,\ 11 \\ -) & \quad 4,\ 6,\ 9,\ 12 \\ & 2,-1,\ 0,\ 2,-12 \end{array}$$

$\therefore K_{\text{III},\text{IV}}=\max\{2,-1,0,2,-12\}=2$

$K_{\text{IV},\text{V}}$:

$$\begin{array}{rl} & 4,\ 6,\ 9,\ 12 \\ -) & \quad 3,\ 7,\ 9,\ 10 \\ & 4,\ 3,\ 2,\ 3,-10 \end{array}$$

$\therefore K_{\text{IV},\text{V}}=\max\{4,3,2,3,-10\}=4$

③ Project duration

$$T_1=\sum_{j=1}^{n1}K_{j,j+1}+\sum_{i=1}^{m}t_n=4+6+2+4+3+4+2+1=26(\text{weeks})$$

④ According to the problems of resources affecting, the duration can be calculated as follows:

$$T_2=\sum Z_{j,j+1}+\sum G_{j,j+1}+\sum C_{j,j+1}=6/7+2/7-1/7=1(\text{weeks})$$

According to the conditions given by the question $Z_{\text{II},\text{III}}=6/7$ (weeks), $G_{\text{IV},\text{V}}=2/7$ (weeks), $C_{\text{I},\text{II}}=1/7$ (weeks).

⑤ Project duration, which takes into account the resource impact of the total duration of the program:

$$T=T_1+T_2$$

$$T=\sum_{j=1}^{n1}K_{j,j+1}+\sum_{i=1}^{m}t_n+\sum Z_{j,j+1}+\sum G_{j,j+1}-\sum C_{j,j+1}=27(\text{weeks})$$

7.4.2.2 Calculation of Task Process Time in Arrow Network Chart

Estimation task process time is one of the main contents of network planning. This method can be used to determine the time required to complete the total tasks, the estimation of a single task time is the basis of the project time plan.

(1) Network time base parameter calculation.

① Node time: one by one to calculate the earliest and latest start time of each node, and the total duration of the plan is obtained, which includes 2 kinds of time parameters.

② Operation time calculation: one by one to calculate the earliest and the latest start time of each process, the earliest and the latest finish time, which includes the calculation of 4 kinds of time parameters

③ Time difference (motorized time): The time difference is divided into total process time difference, process free time difference and process specific time difference.

(2) The relevant calculation formula introduction.

① Calculation of the earliest time (ET) of nodes. The earliest time of the node is the possible earliest start time of each process, which is equal to the end point of the node of the process may be the earliest time to complete.

The earliest time (ET_i) of Node i should start from the beginning of the network plan, calculate along the direction of the arrow line one by one, and in accordance with the following provisions of the symbolic representation:

If starting node i has not been assigned to the earliest time ET_i, assign it zero:

$$ET_i=0 \ (i=1) \tag{7-7}$$

When node j has only one arrow line, the earliest time ET_j is:

$$ET_j=ET_i+D_{i-j} \tag{7-8}$$

When node j has multiple arrow lines, the earliest time Et_j is:

$$ET_j=\max\{ET_i+D_{i-j}\} \tag{7-9}$$

Where: ET_i——the earliest time of start node i in task $i \sim j$;

ET_j——the earliest time of finished node j in task $i \sim j$;

D_{i-j}——the duration in task $i \sim j$, the same below.

② Calculation of the latest time (LT) of the node. The latest time of node is the latest finished time to the node all the process must be fully completed. It can also be regarded as the latest time of the node to complete all front closely process.

The latest time (LT_i) of Node i should start from the end of network plan, and calculate against the direction of the arrow line one by one. When part of the work is done by stages, the latest time of relevant node must be calculated reverse one by one from the stage completed nodes, and according to the following provisions for the symbolic representation:

The latest time LT_n of destination node n should be determined according to the network plan construction period T_p, namely:

$$LT_n=T_p \tag{7-10}$$

The latest time of other node i should be:

$$LT_i=\min\{LT_j-D_{i-j}\} \tag{7-11}$$

Where: LT_i—— the latest time of start node i in task $i \sim j$;

LT_j——the latest time of finished node j in task $i \sim j$.

③ Calculation of the procedure earliest start time (ES). Procedure earliest start time refers to the possible earliest start time of this procedure after the front closely procedures are all finished.

The earliest start time of procedure $i \sim j$ should be calculated one by one along the direction of the arrow lines, in accordance with the following provisions of the symbol represention:

Procedure $i \sim j$ takes start node i as the arrow tail node. When the ES_{i-j} is not assigned to the earliest time, artificial assignment is equal to zero, namely:

$$ES_{i-j}=0 \ (i=1) \tag{7-12}$$

When procedure $i \sim j$ has only one front closely procedure $h \sim i$, the earliest start time ES_{i-j} is:

$$ES_{i-j}=ES_{h-i}+D_{h-i} \tag{7-13}$$

When procedure $i \sim j$ has multiple front closely procedure, the earliest start time ES_{i-j} is:

$$ES_{i-j}=\max\{ES_{h-i}+D_{h-i}\} \tag{7-14}$$

Where: ES_{i-j}——The earliest start time of process $i \sim j$;

ES_{h-i}——Process $i \sim j$'s the earliest start time of front closely process $h \sim i$;

D_{h-i}——The duration of process $i \sim j$'s front closely process $h \sim i$, the same below.

④ Calculation of procedure earliest finished time (EF). Procedure earliest finished time refers to the possible earliest finished time of this procedure after the front closely procedures are finished. The earliest finished time EF_{i-j} of process $i \sim j$ should be calculated according to the following formula.

$$EF_{i-j}=ES_{i-j}+D_{i-j} \tag{7-15}$$

⑤ Calculation of procedure latest finished time (LF). The procedure latest time to complete the procedure shall not affect the entire task completed on time, the procedure must be completed at the latest time.

The latest finished time (LF_{i-j}) of procedure $i \sim j$ should start from the end of network plan, to be calculated reverse the direction of the arrow line one by one, in accordance with the following provisions of the symbolic representation.

Procedure latest finished time LF_{i-n} takes end node ($j=n$) as the arrow node. It should be determined according to the construction period T_p of network plan, namely:

$$LF_{i-n}=T_p \tag{7-16}$$

The latest finished time LF_{i-j} of other procedure $i \sim j$ should be calculated according to the following formula:

$$LF_{i-j}=\min\{LF_{j-k}-D_{j-k}\} \tag{7-17}$$

Where: LF_{j-k} —— The latest finished time of procedure ($i \sim j$)' s back closely procedure $j \sim k$;

D_{j-k} ——The duration of procedure ($i \sim j$)' s back closely procedure.

⑥ Calculation of procedure latest start time (LS). The procedure latest start time refer to the without affecting the entire task, the procedure must start at the latest time.

The latest start time LS_{i-j} of procedure $i \sim j$ should be calculated according to the following formula, namely:

$$LS_{i-j} = LF_{i-j} - D_{i-j} \tag{7-18}$$

⑦ Calculation of total time float (TF). In the network chart, the procedure can only be meaningful in the earliest start time and the latest finished time to complete the activity. During this time, in addition to meeting the requirements of the operation time of the procedure, it may have rich time. The rich time is the total time of the procedure can be flexible to master, called the total float of the process. The formula is as follows:

$$TF_{i-j} = LF_{i-j} - EF_{i-j} = LS_{i-j} - ES_{i-j} = LT_j - (ET_i + D_{i-j}) \tag{7-19}$$

Where: TF_{i-j} —— The total float of task $i \sim j$. Other symbols ditto.

⑧ Calculation of free float (FF). On the premise of not affect the earliest start time of its back closely task, free float refers to the maneuver time of the procedure. The process using the free float to change its start time or increase the duration does not affect the earliest start time of its back closely task. Calculation of procedure free float needs to consider the following 2 cases.

For tasks with back closely procedure, the free float equals to the minimum value of the back closely procedure's earliest start time minus this procedure's earliest finished time, namely:

$$FF_{i-j} = \min \{ ES_{j-k} - EF_{i-j} \} = \min \{ ES_{j-k} - ES_{i,j} - D_{i-j} \} \tag{7-20}$$

Where: FF_{i-j}—— The free float of procedure $i \sim j$. Other symbols ditto.

For tasks without back closely procedure, That is, the task that takes the network planning node as its completion node, the free float is equal to the difference of planning duration minus the earliest finished time, namely:

$$FF_{i-n} = T_p - EF_{i-j} = T_p - ES_{i-n} - D_{i-n} \tag{7-21}$$

Where: FF_{i-n}——The free float of task $i \sim n$ that takes the network terminal n as its finished node;

T_p——Network planning period;

EF_{i-n}——The earliest finished time of task $i \sim n$ that takes the network terminal n as its finished node;

ES_{i-n}——The earliest start time of task $i \sim n$ that takes the network terminal n as its finished node;

D_{i-n}——The duration of task $i \sim n$ that takes the network terminal n as its finished node.

(3) Calculation example.

The pipeline engineering is divided into trenching, pipe laying and backfilling. As seen in Figure 7–15, it is divided into 3 construction sections Ⅰ, Ⅱ, and Ⅲ. Duration is given in the figure. Please calculate the network chart time parameters of the engineering.

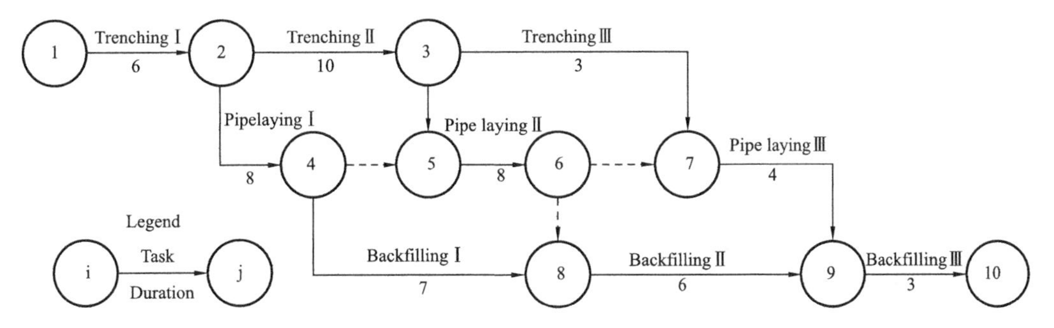

Figure 7–15 Pipeline engineering arrow network chart

① To calculate the earliest time ET_i of each node. Assuming $ET_1 = 0$, according to equation (7–8) and (7–9) we can get that:

$ET_2 = ET_1 + D_{1-2} = 0+6 = 6$

$ET_3 = ET_2 + D_{2-3} = 6+10 = 16$

$ET_4 = ET_2 + D_{2-4} = 6+8 = 14$

$$ET_5 = \max\begin{Bmatrix} ET_3 + D_{3-5} \\ ET_4 + D_{4-5} \end{Bmatrix} = \max\begin{Bmatrix} 16+0 \\ 14+0 \end{Bmatrix} = 16$$

$ET_6 = ET_5 + D_{5-6} = 16+8 = 24$

$$ET_7 = \max\begin{Bmatrix} ET_3 + D_{3-7} \\ ET_6 + D_{6-7} \end{Bmatrix} = \max\begin{Bmatrix} 16+3 \\ 24+0 \end{Bmatrix} = 24$$

$$ET_8 = \max\begin{Bmatrix} ET_6 + D_{6-8} \\ ET_4 + D_{4-8} \end{Bmatrix} = \max\begin{Bmatrix} 24+0 \\ 14+7 \end{Bmatrix} = 24$$

$$ET_9 = \max\begin{Bmatrix} ET_7 + D_{7-9} \\ ET_8 + D_{8-9} \end{Bmatrix} = \max\begin{Bmatrix} 24+4 \\ 24+6 \end{Bmatrix} = 30$$

$ET_{10} = ET_9 + D_{9-10} = 30+3 = 33$

② To calculate LT_i. Because the plan has no provision of the construction period, so $LT_{10} = ET_{10} = 33$, according to equation (7–10), (7–11) we can get that:

$LT_9 = LT_{10} - D_{9-10} = 33-3 = 30$

$LT_8 = LT_9 - D_{8-9} = 30-6 = 24$

$LT_7 = LT_9 - D_{7-9} = 30-4 = 26$

$$LT_6 = \min\begin{Bmatrix} LT_7 - D_{6-7} \\ LT_8 - D_{6-8} \end{Bmatrix} = \min\begin{Bmatrix} 26-0 \\ 24-0 \end{Bmatrix} = 24$$

$LT_5 = LT_6 - D_{5-6} = 24-8 = 16$

$$LT_4 = \min\begin{Bmatrix} LT_5 - D_{4-5} \\ LT_8 - D_{4-8} \end{Bmatrix} = \min\begin{Bmatrix} 16-0 \\ 24-7 \end{Bmatrix} = 16$$

$$LT_3 = \min\begin{Bmatrix} LT_7 - D_{3\text{-}7} \\ LT_5 - D_{3\text{-}5} \end{Bmatrix} = \min\begin{Bmatrix} 26-3 \\ 16-0 \end{Bmatrix} = 16$$

$$LT_2 = \min\begin{Bmatrix} LT_3 - D_{2\text{-}3} \\ LT_4 - D_{2\text{-}4} \end{Bmatrix} = \min\begin{Bmatrix} 16-10 \\ 16-8 \end{Bmatrix} = 6$$

$LT_1 = LT_2 - D_{1\text{-}2} = 6-6 = 0$

③ To calculate work time parameters $ES_{i\text{-}j}$, $EF_{i\text{-}j}$, $LF_{i\text{-}j}$ and $LS_{i\text{-}j}$. according to equation (7–12)~(7–18), we can get that:

Task1~2: $ES_{1\text{-}2} = ET_1 = 0$　　$EF_{1\text{-}2} = ES_{1\text{-}2}+D_{1\text{-}2} = 0+6 = 6$

$LF_{1\text{-}2} = LT_2 = 6$　　$LS_{1\text{-}2} = LF_{1\text{-}2}-D_{1\text{-}2} = 6-6 = 0$

Task 2~3: $ES_{2\text{-}3} = ET_2 = 6$　　$EF_{2\text{-}3} = ES_{2\text{-}3}+D_{2\text{-}3} = 6+10 = 16$

$LF_{2\text{-}3} = LT_3 = 16$　　$LS_{2\text{-}3} = LF_{2\text{-}3}-D_{2\text{-}3} = 16-10 = 6$

Task 2~4: $ES_{2\text{-}4} = ET_2 = 6$　　$EF_{2\text{-}4} = ES_{2\text{-}4}+D_{2\text{-}4} = 6+8 = 14$

$LF_{2\text{-}4} = LT_4 = 16$　　$LS_{2\text{-}4} = LF_{2\text{-}4}-D_{2\text{-}4} = 16-8 = 8$

Task 3~5: $ES_{3\text{-}5} = ET_3 = 16$　　$EF_{3\text{-}5} = ES_{3\text{-}5}+D_{3\text{-}5} = 16+0 = 16$

$LF_{3\text{-}5} = LT_5 = 16$　　$LS_{3\text{-}5} = LF_{3\text{-}5}-D_{3\text{-}5} = 16-0 = 16$

Task 3~7: $ES_{3\text{-}7} = ET_3 = 16$　　$EF_{3\text{-}7} = ES_{3\text{-}7}+D_{3\text{-}7} = 16+3 = 19$

$LF_{3\text{-}7} = LT_7 = 26$　　$LS_{3\text{-}7} = LF_{3\text{-}7}-D_{3\text{-}7} = 26-3 = 23$

Task 4~5: $ES_{4\text{-}5} = ET_4 = 14$　　$EF_{4\text{-}5} = ES_{4\text{-}5}+D_{4\text{-}5} = 14+0 = 14$

$LF_{4\text{-}5} = LT_5 = 16$　　$LS_{4\text{-}5} = LF_{4\text{-}5}-D_{4\text{-}5} = 16-0 = 16$

Task 4~8: $ES_{4\text{-}8} = ET_4 = 14$　　$EF_{4\text{-}8} = ES_{4\text{-}8}+D_{4\text{-}8} = 14+8 = 22$

$LF_{4\text{-}8} = LT_8 = 24$　　$LS_{4\text{-}8} = LF_{4\text{-}8}-D_{4\text{-}8} = 24-7 = 17$

Task 5~6: $ES_{5\text{-}6} = ET_5 = 16$　　$EF_{5\text{-}6} = ES_{5\text{-}6}+D_{5\text{-}6} = 16+8 = 24$

$LF_{5\text{-}6} = LT_6 = 24$　　$LS_{5\text{-}6} = LF_{5\text{-}6}-D_{5\text{-}6} = 24-8 = 16$

Task 6~7: $ES_{6\text{-}7} = ET_6 = 24$　　$EF_{6\text{-}7} = ES_{6\text{-}7}+D_{6\text{-}7} = 24+0 = 24$

$LF_{6\text{-}7} = LT_7 = 26$　　$LS_{6\text{-}7} = LF_{6\text{-}7}-D_{6\text{-}7} = 26-0 = 26$

Task 6~8: $ES_{6\text{-}8} = ET_6 = 24$　　$EF_{6\text{-}8} = ES_{6\text{-}8}+D_{6\text{-}8} = 24+0 = 24$

$LF_{6\text{-}8} = LT_8 = 24$　　$LS_{6\text{-}8} = LF_{6\text{-}8}-D_{6\text{-}8} = 24-0 = 24$

Task 7~9: $ES_{7\text{-}9} = ET_7 = 24$　　$EF_{7\text{-}9} = ES_{7\text{-}9}+D_{7\text{-}9} = 24+4 = 28$

$LF_{7\text{-}9} = LT_9 = 30$　　$LS_{7\text{-}9} = LF_{7\text{-}9}-D_{7\text{-}9} = 30-4 = 26$

Task 8~9: $ES_{8\text{-}9} = ET_8 = 24$　　$EF_{8\text{-}9} = ES_{8\text{-}9}+D_{8\text{-}9} = 24+6 = 30$

$LF_{8\text{-}9} = LT_9 = 30$　　$LS_{8\text{-}9} = LF_{8\text{-}9}-D_{8\text{-}9} = 30-6 = 24$

Task 9~10: $ES_{9\text{-}10} = ET_9 = 30$　　$EF_{9\text{-}10} = ES_{9\text{-}10}+D_{9\text{-}10} = 30+3 = 33$

$LF_{9\text{-}10} = LT_{10} = 33$　　$LS_{9\text{-}10} = LF_{9\text{-}10}-D_{9\text{-}10} = 33-3 = 30$

④ To calculate total floats $TF_{i\text{-}j}$ and free float $FF_{i\text{-}j}$. according to equation (7–19)~(7–21) we can get that:

Task 1~2: $TF_{1\text{-}2} = LS_{1\text{-}2} - ES_{1\text{-}2} = 0-0 = 0$, $FF_{1\text{-}2} = ET_2 - EF_{1\text{-}2} = 6-6 = 0$

Task 2~3: $TF_{2\text{-}3} = LS_{2\text{-}3} - ES_{2\text{-}3} = 6-6 = 0$, $FF_{2\text{-}3} = ET_3 - EF_{2\text{-}3} = 16-16 = 0$

Task 2~4: $TF_{2\text{-}4} = LS_{2\text{-}4} - ES_{2\text{-}4} = 8-6 = 2$, $FF_{2\text{-}4} = ET_4 - EF_{2\text{-}4} = 14-14 = 0$

Task 3～5 : $TF_{3-5} = LS_{3-5} - ES_{3-5} = 16-16 = 0$, $FF_{3-5} = ET_5 - EF_{3-5} = 16-16 = 0$

Task 3～7 : $TF_{3-7} = LS_{3-7} - ES_{3-7} = 23-16 = 7$, $FF_{3-7} = ET_7 - EF_{3-7} = 24-19 = 5$

Task 4～5 : $TF_{4-5} = LS_{4-5} - ES_{4-5} = 16-14 = 2$, $FF_{4-5} = ET_5 - EF_{4-5} = 16-14 = 2$

Task 4～8 : $TF_{4-8} = LS_{4-8} - ES_{4-8} = 17-14 = 3$, $FF_{4-8} = ET_8 - EF_{4-8} = 24-22 = 2$

Task 5～6 : $TF_{5-6} = LS_{5-6} - ES_{5-6} = 16-16 = 0$, $FF_{5-6} = ET_6 - EF_{5-6} = 24-24 = 0$

Task 6～7 : $TF_{6-7} = LS_{6-7} - ES_{6-7} = 26-24 = 2$, $FF_{6-7} = ET_7 - EF_{6-7} = 24-24 = 0$

Task 6～8 : $TF_{6-8} = LS_{6-8} - ES_{6-8} = 24-24 = 0$, $FF_{6-8} = ET_8 - EF_{6-8} = 24-24 = 0$

Task 7～9 : $TF_{7-9} = LS_{7-9} - ES_{7-9} = 26-24 = 2$, $FF_{7-9} = ET_9 - EF_{7-9} = 30-28 = 2$

Task 8～9 : $TF_{8-9} = LS_{8-9} - ES_{8-9} = 24-24 = 0$, $FF_{8-9} = ET_9 - EF_{8-9} = 30-30 = 0$

Task 9～10 : $TF_{9-10} = LS_{9-10} - ES_{9-10} = 30-30 = 0$, $FF_{9-10} = ET_{10} - EF_{9-10} = 33-33 = 0$

⑤ Use "Accumulative sequence dislocation subtracting and taking large difference method" to verify the total project duration: $T = 33$.

7.5 Preparation of LNG Project Schedule

7.5.1 Basis of Project Schedule Compilation

7.5.1.1 General Concept

Schedule planning is mainly to determine the start and end date of the whole project task, and the start and end time of the whole project is obtained by the time of the start and end date of thousands of work packages. If the start and end date of each work package is not accurate, or the corresponding resource calendar does not match, we can not work out an accurate LNG project schedule. Throughout the LNG project implementation process, the schedule preparation, time estimation, cost estimation and other processes are intertwined, these processes are repeated several times to finalize the project schedule.

7.5.1.2 Foundation of Previous Work

The above-mentioned LNG project decomposition, task definition, task sorting and individual task duration estimates are the foundations of the project schedule compilation. Especially similar LNG sub-project construction duration at home and abroad could be a good reference for the company to compile their schedule plan.

7.5.1.3 Resource Base Description

In terms of scheduling, it is necessary to know what resources, when and where to use them. For example, the above mentioned LNG project work package resource input management is the foundation of the LNG project resource base. We need to consider resource sharing and resource variable. In the resource base description, the level of detail of each resource is changeable. For example, a consulting project initial scheduling, we need only to know that there are 2 consultants for use within a certain period of time. However, in the same project final schedule must determine that a particular consulting personnel, and so on.

7.5.1.4 Resource Calendar Tableand Project Calendar Table

The resource calendar table gives the each resource and amount that can be used. When computing the earliest and latest start and end dates of the LNG project tasks respectively from the theory, without considering resource constraints, and this is not the date calculated actual schedule, but the length of time needed to continue. if the resource constraints and other constraints on the tasks in the project are considered, the schedule tasks within a prescribed time interval, it will be the real project schedule plan compilation process.

7.5.1.5 Constraint Conditions

(1) Mandatory date.

For example, some domestic project LNG resources in the early have been identified, suppliers will be scheduled for a certain date in the future to supply LNG. This means the whole LNG project must be completed in this specific date, if not completed, LNG resource of the receiving party will assume liability for breach of contract.

(2) Milestone events.

LNG project and the overall planning of local economic construction are closely linked. If the LNG user's project is incorporated into local government plan, the project milestone and user's project milestone will produce a strong correlations. The user's project and LNG project must be designed, constructed and operated synchronously. Thus, once the LNG project milestone events are identified, these dates are difficult to be changed.

7.5.1.6 Ahead and Lag

The amount of ahead and lag time should be reasonable. If it is used improperly, it will cause an unreasonable project schedule table. To precisely illustrate the relationship between tasks, the ahead and lag time of some logic relationships are needed. For example, there is a 2 weeks interval between purchasing and using an equipment, etc..

7.5.2 Schedule Compilation Methods and Tools

The following will describe the process of the schedule compilation and several common schedule compilation methods on the conditions of resource constraints.

7.5.2.1 LNG Project Schedule Levels

(1) General concept.

Once the national government authorities approved the launch of a LNG project next step work report, means that the project has "passes". In this case, the project office or the project company often has to propose a plan to match the LNG project initiator's goal plan and the preliminary milestone plan. If the Project Company takes the EPC contract, the Project Company and the contractors may jointly develope the initial milestone plan, and then start the preparation of the project schedule framework. In the following, we will describe the goal plan, milestone plan and schedule plan preparation, taking into account the project schedule planing levels(Figure 7-16), we would like to take this opportunity, but also introduce some concepts of bid plan and work package plan, prioritize each target, firstly ensure the achievement of the highest goal, and then achieve the second goal under the premise of without prejudice to the first goal as far as possible, in order to maximize the project overall achievement.

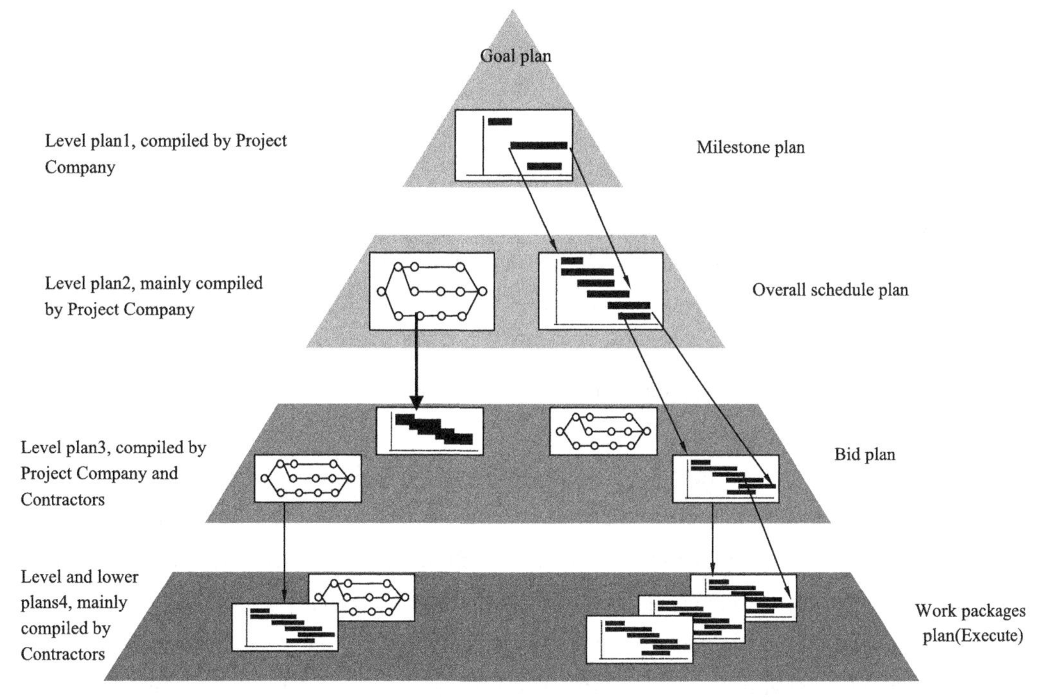

Figure 7-16 Project schedule management decomposition

Quoted from Shanghai Puhua science & technology development co., LTD, Xing Kejian, and modified.

(2) Goal plan.

Goal plan preparation is an important job that project company needs to propel the project after the national government approved the report of developing LNG project next step. Goal plan is compiled on the basis of the construction period of the sub projects, and is generally prepared by the project company control department. The Receiving Terminal (including storage tanks) takes 3.5～4 years. A $16\times10^4m^3$ storage tank takes 2～2.5 years, a 380 km pipeline takes 2～2.5 years construction period, considering the newly built LNG ships ($14.7\times10^4m^3$) will need 3 years. According to these construction periods, the project company can make a reasonable goal plan.

(3) Milestone plan.

Milestone plan is the start or finish time of several core sub-projects in project. Generally in the LNG project, milestone points are: jetty, receiving terminal field leveling start, tank start and finish, jetty and supporting engineering completion, the receiving terminal completion, pipeline completion, the LNG project fully completion, the first LNG ship unloading, trial production date, trial production end etc.. Milestone plans cover the segmentation implementation of the entire project and the deadline for completion from the large aspect. Milestone plans are usually compiled by sub-project contractor in accordance with the project company's target plan. Then the 2 companies will make decision together after discussing. Once the milestone plan is completed, it will be the goal that the whole teams strive for, and can't be arbitrarily changed.

(4) Overall schedule plan.

Overall schedule plan includes the main part of milestone plan and each bid section plan, usually not include the executing plan. Overall schedule plan demands to reflect the arrangement contents of the entire plan. While making overall schedule plan, we can divided it into 2 steps. The first step, at the initial stage, the project company or contractor compiles a framework of overall schedule plan in accordance with milestone plan; The second step, after the completion of the sub plan and then combine with the former, as far as possible not to change the entire LNG project under the premise of full completion of the appropriate adjustments, before and after the echo, comprehensive consideration, and gradually improve. This stage will be finished after the discussion by the project company, supervisory company and contractors.

(5) Bid section plan.

It is mentioned above that the LNG project consists of 3 big sub-projects. According to the operation of the project, can be contracted to a construction company. Sub-projects of large quantity of constructions can also be called for bids in each section, divided into several secondary sub-projects and contracted to several construction companies. Such as LNG pipeline sub-project, it is divided into several bid sections in common. The contractors compile their own schedule plan and implement it separately, which will speed up the whole sub-project ahead of schedule. For example, Guangdong Dapeng LNG pipeline project is divided into 6 sections. Of course, the coordination workload in each bid section will be enlarged during the project, for the project company responsible person and project schedule control personnel need to cooperate with the contractor closely. Plan compilation is a repeated coordination process for many times, we should consider it from partial to whole and then whole to partial.

(6) Work package plan.

As the name suggests, the work package plan is also called the execution plan, is the most basic, can carry on the plan of the operation, if the plan is thicker, but cannot achieve the operational ability, also cannot be called the execution plan. LNG project execution plan may have hundreds or even thousands of items. When compiling the execution plan, it needs the contractor to consider his own resource for schedule arrangement, including human resources, equipment, capacity, and so on. It also needs the contractor to consult with relative sub-project control department of project company until achieve the agreeable execution plan.

7.5.2.2 LNG Project Schedule Compilation Procedure

The preparation processe of the project schedule is different that is based on whether the project company has a own contracting team. At present, CNOOC conducted a number of liquefied natural gas projects, was to take out issuing engineering contracting project, adopt general contractor EPC and also sub contractors, but the project company always put a lot of effort to compile the project schedule before the start of the project. Generally speaking, the project company is the first to compile a goal plan (Figure 7-17), and then the preparation of milestone plan; general contractors according to the owner's goal plan and milestone plan, prepare the schedule from the bottom to the top. First of all, start from the schedule level 4. The project company control department and the engineering department and the project team and the contractor, supervision company for a lot of repeated consultation and coordination, to confirm the schedule level 4, set up a construction schedule level 4 and weekly construction schedule goal level 4; Based on the schedule level 4 to prepare the schedule level 3, and the schedule level 2. Finally, they are matched and coordinated with the owner's milestone plan.

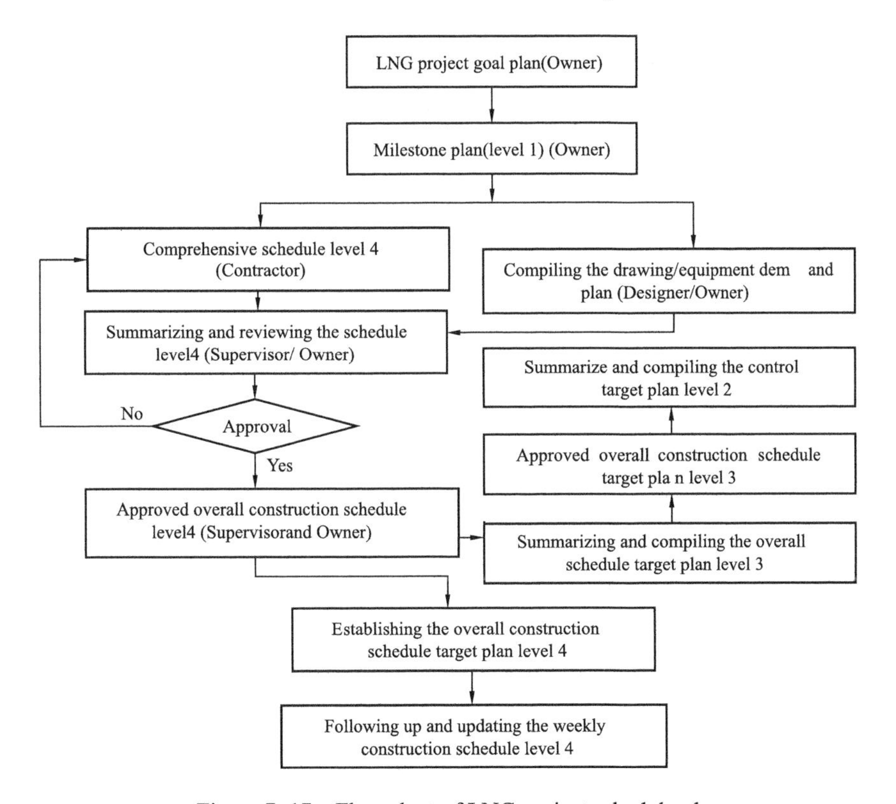

Figure 7-17 Flow chart of LNG project schedule plan

7.5.2.3 Compilation of LNG Project Milestone Plan

(1) Preparation significance.

① Milestone plan is based on some key milestones start or finish time in LNG construction projects. It is a strategic plan or schedule framework of construction projects, and it sets out achievable intermediate results in a construction project. Such as ceiling of a storage tank is finished, a section of pipeline laying is completed etc..

② Milestone plan is based on the LNG project construction project to achieve the ultimate goal must be experienced in the work process, and the major and key work sequence is determined. Each milestone represents a critical event, and shows its time limits must be completed. Such as all construction projects must be completed before joint commissioning.

③ Milestone plan is very suiTablefor the group engineerings such as the longer duration and complex LNG projects. Milestone is a part that cannot be ignored in project management, and it is a major event in a project. In the project, milestone is a time point, and it usually refers to the completion of a deliverable.

④ The preparation of milestone plans for the project objectives and scope of management is very important, which assists in the scope of the audit, to provide guidance to the project implementation, is the project leader to guide and check the work of the road map.

⑤ The establishment of the milestone plan is the collective wisdom of the key managers and key project partakers of the project company. Generally a special meeting is held to discuss and formulate it. The way through all group members to make it is much better than that the project chief manager formulates the milestone plan alone and forcing the project team to run it. It can make milestone plan to get a wider range of support.

(2) The specific steps of the establishment of the milestone plan.

① To collectively discuss all possible milestones. All the members record these ideas on the flip chart, through the brainstorm to select the final milestone.

② To audit the alternative milestones. In alternative milestones, some of them are a part of other milestones; some are activities that cannot be regarded as milestones. But these activities can help us to understand some milestones clearly. In particular, to determine milestones with inclusion relationships.

③ To test on the path of each result. It needs to figure out the results path, list each milestone and compare them with the appropriate adjustments and changes in the order of their occurrence.

④ To show the logical relationship among the milestones in lines. The lines which show the logical relationship among the milestones are started from the final results of the projects, and to draw the logical relationship with a backward way. This step may urge to reconsider the definition of milestones, or to add a new milestone, merge milestones, even to change the definition of the results path.

⑤ To determine the final milestone plan, it needs to require all the participants unanimously to approve the final milestone and reach a consensus and offer it for the important partakers to check and examine, then post it in the project management office, so that everyone can grasp.

The above are the common steps in the preparation of milestones, but because of the uniqueness of the project, we should not rigidly adhere to the form in practice, and can be used flexibly.

(3) Representation method.

Usually there are 2 ways to show the milestone plan—tabular representation and diagram representation. Please see the next section.

7.5.2.4 Critical Path Method (CPM)

(1) Definitions.

① Critical path. Starting from the beginning of the network diagram, a series of nodes along the direction of the arrow to the end of a series of nodes, which must have one of the longest total duration and determine the shortest finish time of the entire project. This path is called the critical path. The time of the path is also known as the total duration. Tasks on the critical path are called critical tasks, there is no time to reserve on the critical path. The characteristics of the critical path are as follows.

a.The duration of the activities of the critical path determines the duration of the project, and the duration of the critical path is added to the duration of the project.

b.Any activity on the critical path is the key activity. And any activity delay will lead to the entire project completion time delay.

c.The critical path is the longest path from the starting point to the end in the project. So if you want to shorten the time limit for a project, you must make the improvement on the critical path. On the contrary, if the critical path time is prolonged, the entire project completed period will be extended.

d.The time consuming of the critical path is the shortest amount of time to complete the project.

e.Activity on the critical path is the smallest total float activity.

② Non critical path. From the starting point on the network, along the direction of the arrow until the end, the other entire path in addition to the critical path is called non critical path. All the time of those paths are less than the total duration. They all have the maneuver time and time difference, every task on the non critical path is called non critical task, and all of them have the time reserves.

③ The relationship between the critical path and non critical path. The relationship between the critical path and non critical path is not invariable. They can transform to each other under certain conditions. If the time of the critical path is shorten by the improvement of the measures and technologies in the organization, the critical path will demote to non critical path, and some non critical path will upgrade to critical path.

(2) Selection of critical path.

As shown by Figure 7-18, we set up from the start point of A, the time is assumed to be zero, the next node number B, C, D, ……, the number in the box for the task, we can list to find out the duration of each path.

From Table 7-6, A → D → E → F → H → I as the critical path. The critical path method is used to estimate the time required for the entire LNG project task, and the result is to get the earliest or latest start and end time of the project. CPM method can also calculate the time difference of each activity and the total time of the whole project.

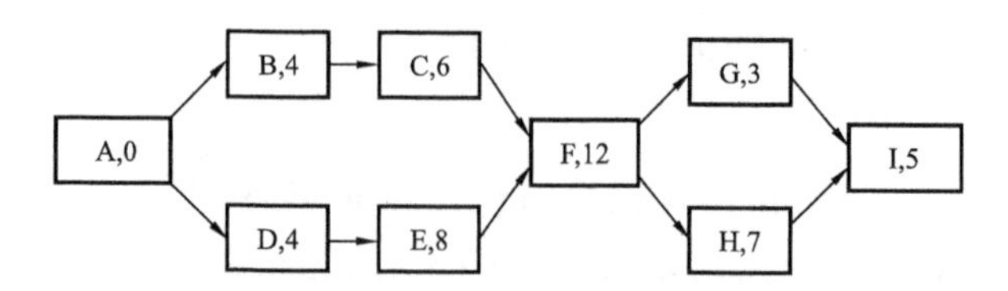

Figure 7-18 The node representation of the project network

Table 7-6 Path Time and Critical Path Analysis

No.	Path	Time	Critical path or not
1	A→B→C→F→G→I	30	No
2	A→B→C→F→H→I	34	No
3	A→D→E→F→G→I	32	No
4	A→D→E→F→H→I	36	Yes

7.5.2.5 Project Management Software

Project management software is widely used to help the project scheduling, such as P3e/c software, microsoft project management software and other project management software. These softwares can automatically carry out the mathematical calculation and resource adjustment, which can be considered and selected for many schemes quickly. These software can also be used to point and display the results of planning.

7.5.3 Schedule Compilation Results

7.5.3.1 LNG Project Schedule Compilation Procedures

(1) Compiling each task network plan and schedule plan.

After the contractors obtained the engineering contracting tasks, according to the owner's target plan, initial milestone plan, engineering technical standards and their technology and economic strength, they will break each work in each level, identify the bottom task and build the task schedule from down to up. The bottom level works include schedule network diagram and schedule planning. It is basis for the upper level planning [Figure 7-19 (a)].

(2) Compiling bid section plan.

Based on each network diagram and schedule plan, we should sort the same level's tasks, comprehensively consider resource calendar, and also special concern critical path and prepare the bid plan [Figure 7-19 (b)].

(3) Compiling sub project schedule.

Based on each section of the schedule planning, the owner should strictly examine the progress of the bid schedule. Especially in each section is awarded to various contractors, the owners should analyze each situation of resource input and the reasonableness of schedule plan, with the contractor together to develop sub project schedule [Figure 7-19 (c)].

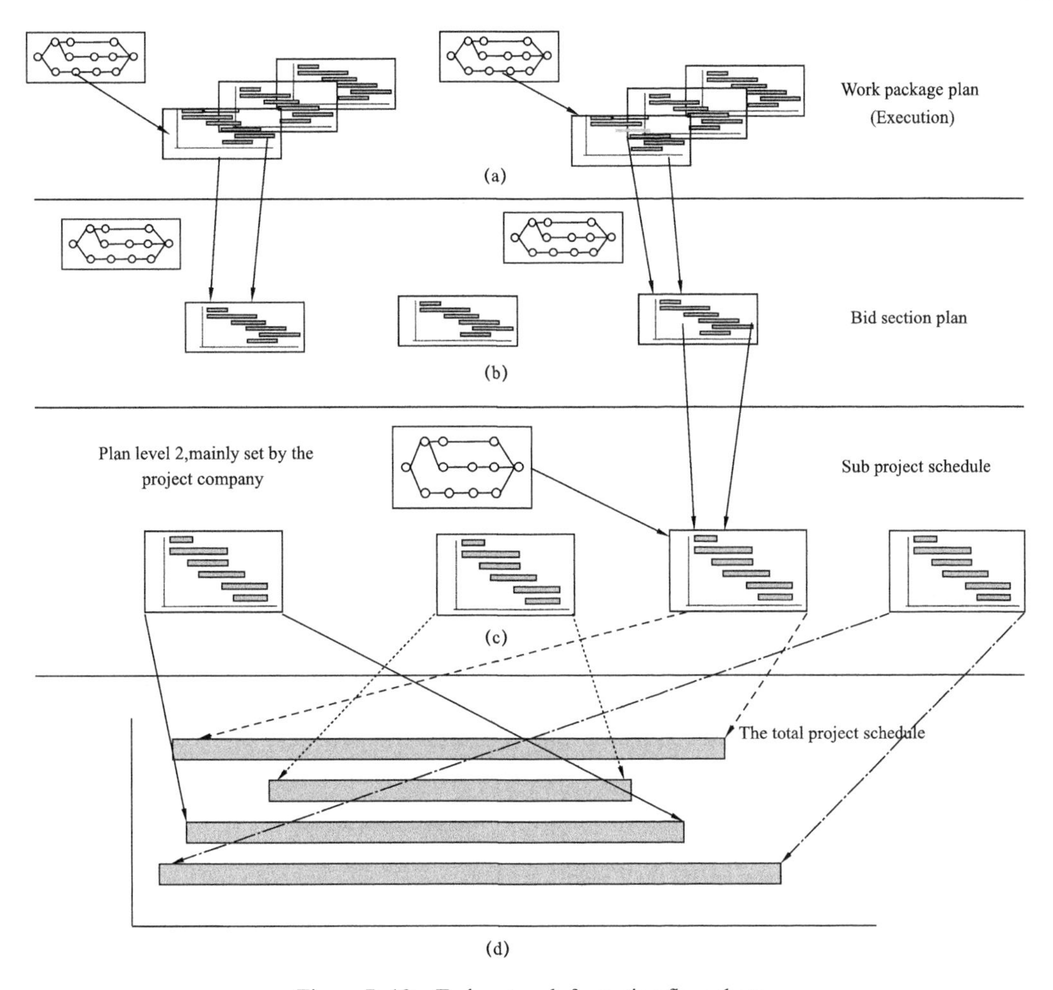

Figure 7–19 Task network formation flow chart

(4) Compiling the overall project schedule.

The final step is to prepare the overall schedule of the LNG project plan. In addition to considering the sub project schedule factors, but also considering the owner's resources input, supervisory company's resources input and the government's policy support and other factors [Figure 7–19 (d)].

7.5.3.2 The Expression Forms of LNG Project Schedule

Using a graphical form to indicate that the LNG project schedule is the result of this stage, here are a few common charts and tables.

(1) Gantt chart.

Table 7–7 is picked from the Shanghai LNG receiving terminal project schedule level 2, which is a typical Gantt chart to express the critical path plan. The Tablerepresents the start and end time of the important work on the critical path. It is detailed by the milestone control plan, and is also the basis for the compilation of the other zoning plans and stages control plans.

Table 7-7 Shanghai LNG Project Control Schedule

Number	Task name	Start time	Finish time	Notes on chart
1	1. Pre work and project management	20/9/04	7/7/09	
2	Previous work	20/9/04	1/4/07	
3	Feasibility study and evaluation report	20/9/04	12/8/05	
4	Project application report	1/6/05	12/8/05	
5	LNG resource transportation principle agreement	16/6/05	19/10/05	
6	Market sales principle agreement	16/6/05	17/10/05	
7	LNG resources procurement contract	12/2/05	1/4/07	
8	Project approval	8/12/06	8/12/06	12-8
9	Project management	1/1/06	7/7/09	
10	Market sales contract	1/1/06	7/3/07	
11	Tentative sale price mechanism	2/6/08	8/2/09	
12	Signing construction period insurance contract	13/6/06	11/2/07	
13	Signing operation insurance contract	10/12/08	7/7/09	
14	Financing contract	29/8/06	16/3/07	
15	LNG transport contract	23/10/06	30/12/08	
16	Equipment/materials imported duty-free approval	4/4/07	31/3/08	
17	LNG import duty-free approval	25/5/07	11/6/08	
18	2. Project implementation	25/3/05	7/7/09	
19	Land formation	25/3/05	31/12/08	
20	Preliminary design, construction drawing design tender	25/3/05	13/6/06	
21	Land formation and construction	22/8/05	29/5/07	
22	EPC construction site handover	20/12/06	6/6/07	
23	Contact dike completion	26/12/06	26/12/06	12-26
24	Land top construction	31/8/08	31/12/08	
25	Reserving Terminal sub project	3/5/05	10/4/09	
26	FEED design	3/5/05	7/10/05	
27	EPC bidding	8/10/05	8/8/06	
28	EPC signed and effective	8/8/06	8/8/06	8-8
29	Supervisor bidding	20/7/06	17/10/06	
30	Terminal preliminary design and approval	16/10/06	29/12/06	
31	Design	9/8/06	8/12/08	
32	FEED examined	9/8/06	6/11/06	
33	Basic design	9/8/06	5/4/07	
34	Detailed design	6/4/07	8/12/08	
35	procurement	20/11/06	29/11/08	
36	Tank procurement	20/11/06	20/11/08	
37	Terminal procurement	11/12/06	29/11/08	
38	CFRE equipment procurement	9/4/07	15/11/08	
39	Construction	11/1/07	10/4/09	
40	Tank construction site preparation	21/1/07	28/3/07	
41	The first tank construction and installation	28/3/07	11/12/08	
42	The second tank construction and installation	16/4/07	30/12/08	
43	The third tank construction and installation	17/5/07	12/3/09	
44	First tank hydrostatic test	4/8/08	18/9/08	
45	Second tank hydrostatic test	28/8/08	8/10/08	
46	Third tank hydrostatic test	15/10/08	24/11/08	
47	First tank pre-trial operation	25/12/08	5/2/09	
48	Second tank pre-trial operation	9/2/09	6/3/09	
49	Third tank pre-trial operation	13/3/09	10/4/09	
50	Mechanical completion of first tank	9/2/09	9/2/09	2-9

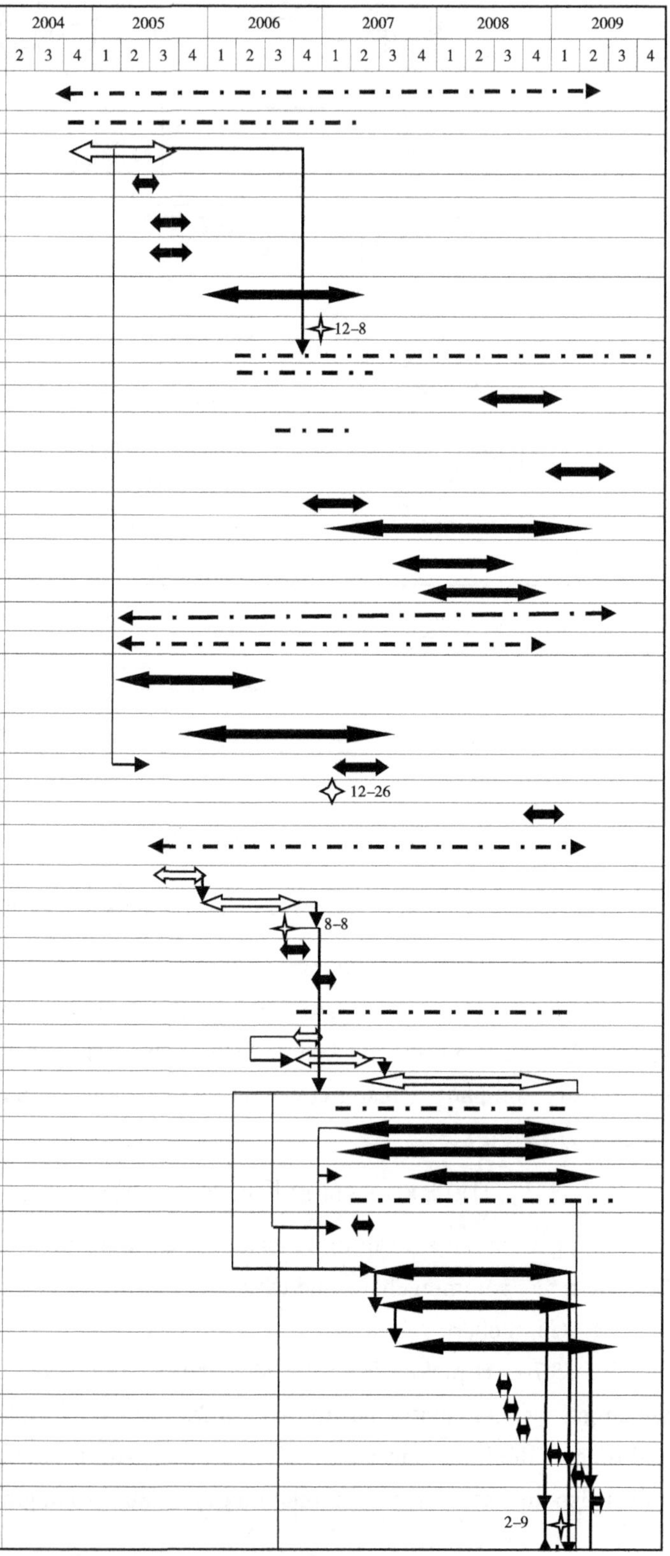

Timeline columns (Gantt chart): 2004 (quarters 2, 3, 4); 2005 (1, 2, 3, 4); 2006 (1, 2, 3, 4); 2007 (1, 2, 3, 4); 2008 (1, 2, 3, 4); 2009 (1, 2, 3, 4)

Continued

Num-ber	Task name	Start time	Finish time	2004	2005	2006	2007	2008	2009
				2 3 4	1 2 3 4	1 2 3 4	1 2 3 4	1 2 3 4	1 2 3 4
51	Mechanical completion of second tank	9/3/09	9/3/09					3–9	
52	Mechanical completion of third tank	10/4/09	10/4/09						4–10
53	Terminal (non-tank) construction	11/1/07	8/2/09						
54	BOG condenser installation	25/4/08	30/6/08						
55	Torch installation	30/8/08	8/9/08						
56	IFV installation	4/8/08	19/9/08						
57	Intake seawater lift pump installation	23/8/08	28/10/08						
58	BOG compressor installation	15/8/08	31/10/08						
59	Unloading arm installation	28/5/08	24/12/08						
60	Terminal (non-tank) mechanical completion	9/2/09	9/2/09						2–9
61	District 9000 eng. bidding/construction	15/2/08	26/12/08						
62	District 9000 eng. bidding	15/2/08	15/4/08						
63	Bidding documents preparing	15/2/08	6/3/08						
64	Issue/receive bidding	7/3/08	27/3/08						
65	Bidding and Contract negotiation	28/3/08	6/4/08						
66	Awarding contract	15/4/08	15/4/08					4–15	
67	District 9000 eng. construction	16/4/08	26/12/08						
68	Off-site construction	29/8/08	30/1/09						
69	Water, electricity, road, communication agreement	29/8/08	29/8/08					8–29	
70	Power supply construction	20/9/08	17/1/09						
71	Water supply construction	22/9/08	20/11/08						
72	Road construction	30/9/08	27/1/09						
73	Communications construction	1/1/09	30/1/09						
74	Gas pipeline	7/4/06	30/1/09						
75	Submarine pipeline engineering	7/4/06	30/1/09						
76	Sea pipe research	7/4/06	31/5/07						
77	Sea pipe basic design	16/10/06	12/7/07						
78	Bidding of pipe materials	15/3/07	11/4/07						
79	Signing contract of sea pipe material	13/7/07	13/7/07				7–13		
80	Manufacture and delivery of sea pipe material	14/7/07	5/7/08						
81	Signing contract of land pipe material	10/9/07	10/9/07				9–10		
82	Manufacture and delivery of land pipe material	11/9/07	14/7/08						
83	Pipe coating negotiation	28/7/07	29/12/07						
84	Pipe coating construction	30/12/07	7/7/08						
85	Third party check the bidding	18/8/07	12/10/07						
86	Bidding of detailed design and pipe laying	19/3/07	29/12/07						
87	Detail design for sea pipe	31/12/07	15/10/08						
88	Optical fiber cable, cable compensation agreement signed	18/7/08	18/7/08					7–18	
89	Fisheries compensation agreement signed	8/9/08	8/9/08					9–8	
90	Getting the permission of through the levees	30/9/08	30/9/08					9–30	
91	Submarine pipeline construction permit, marine work permit for processing	25/8/08	25/8/08					8–25	
92	Operation in Nanhuizui	1/7/08	30/1/09						
93	Operation in Ximentang	1/7/08	24/1/09						
94	Sea pipe laying	18/9/08	29/12/08						
95	Ditching	10/10/08	8/12/08						
96	Ball through, pressure test, drainage and dry	5/1/09	30/1/09						
97	Land pipe and the end of the gas station	23/5/07	11/1/09						
98	Preliminary design and report	23/5/07	9/10/07						
99	Preliminary design and approval	1/7/08	1/7/08					7–1	
100	The permit of planning construction land and project	1/7/08	29/8/08						
101	Construction drawing design	13/2/08	18/11/08						
102	Pressure regulating equipment acquisition	12/11/07	15/11/08						
103	Permanent land acquisition agreement of the end station and valve room	16/5/08	23/8/08						

Continued

Num-ber	Task name	Start time	Finish time	2004–2009 (quarters)
104	Signing the temporary land borrow agreement of pipe construction	30/5/08	30/5/08	5-30
105	Supervisor bidding	4/7/08	20/8/08	
106	Bidding the construction of pipe and final station	3/6/08	15/8/08	
107	Construction of final station	16/8/08	7/1/09	
108	Pre-test run of final station	2/1/09	10/1/09	
109	Mechanical completion of final station	11/1/09	11/1/09	1-11
110	Land pipe construction	16/8/08	6/12/08	
111	Ball through, pressure test, drainage and dry	6/12/08	30/12/08	
112	Mechanical completion of land pipe	11/1/09	11/1/09	1-11
113	Port engineering	11/11/05	29/12/08	
114	jetty engineering	11/11/05	15/10/08	
115	Preliminary design, construction drawing design tender	11/11/05	7/4/06	
116	Preliminary design	8/4/06	15/12/06	
117	Detailed design	1/12/06	15/5/07	
118	Equipment procurement	4/6/06	15/9/08	
119	Supervisor bidding	12/10/06	30/3/07	
120	Construction biddi ng	17/12/06	25/3/07	
121	LNG jetty construction	8/4/07	30/9/08	
122	EPC construction site handover	15/12/07	15/4/08	
123	Working boat wharf construction	8/4/07	29/8/08	
124	Equipment procurement	1/10/08	15/10/08	
125	Waterway engineering	5/9/06	29/12/08	
126	Preliminary design	5/9/06	15/12/06	
127	Detailed design	1/12/06	10/12/07	
128	Signing implementation agreement of waterway widening engineering	29/3/08	29/3/08	3-29
129	Waterway widening	28/2/08	31/10/08	
130	Handling temporary Anchorage procedure	16/5/08	31/8/08	
131	Signing the sweeping agreement	30/9/08	30/9/08	9-30
132	Sweeping	1/10/08	29/11/08	
133	Signing the agreement for manufacturing and installation of basin front beacon	31/8/08	31/8/08	8-31
134	Manufacturing and installation of basin front beacon	1/9/08	29/12/08	
135	Production preparation	27/9/08	7/7/09	
136	Ensuring the production management system	4/2/08	23/4/08	
137	Personnel preparing and training	27/9/07	7/2/09	
138	Preparing the handbook of preliminary run beforehand/ preliminary run/production	20/11/07	7/2/09	
139	LNG quantity and Quality Inspection Commission	20/11/07	13/11/08	
140	Open ports agreement	2/1/08	7/2/09	
141	LNG ship berthing and departing simulation test	31/3/08	30/6/08	
142	Forensics and acceptance for Pressure vessel, pressure pipe and safety valve etc.	25/4/08	10/4/09	
143	Establish of HSE system in operating period	24/4/08	7/2/09	
144	Life logistics support	23/7/08	7/2/09	
145	Singing LNG berthing auxiliary ship lease agreement	10/11/08	7/2/09	
146	Handling production permits	9/11/08	7/2/09	
147	Handling safety production permits	5/12/08	7/2/09	
148	Preparation production material	9/1/09	7/7/09	
149	The first LNG ship arriving	8/2/09	8/2/09	2-8
150	Project commissioning	8/2/09	8/4/09	
151	Terminal performance test and reliability test	9/4/09	7/7/09	
152	3.Production	8/7/09	8/7/09	7-8
	Legend	Task:	Key task:	Milestone: Abstract:

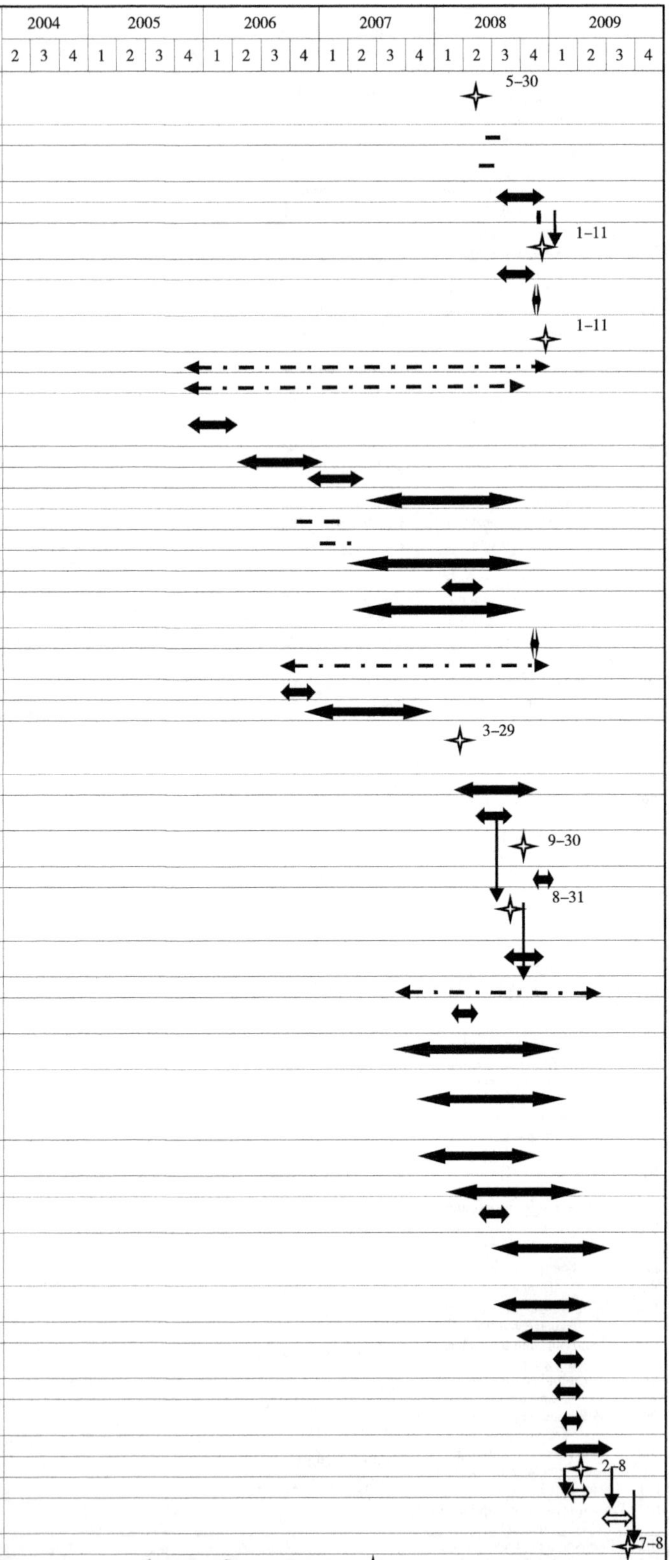

(2) Other representation.

It can also use the summary Tablemethod and milestone chart.

① Tabular representation. Table 7–8 and Table 7–9 respectively list Chinese Guangdong Dapeng LNG project pipeline and terminal milestone, also give planning time and actual completion time, and show in the progress of the sub project construction, the time changes due to a variety of reasons. The characteristic of this representation is that it can clearly list each important milestone and easily compare planning time and actual completion time. But it cannot directly express each sub project's duration, and cannot clearly show the logical relationship.

Table 7–8 Guangdong Dapeng LNG Project Pipeline Milestone

Category	Project milestone	Planning time (D/M/Y)	Actual completion time (D/M/Y)
1	Award of CPPE Contract	28/01/04	28/01/04
2	Preliminary Design Approved	29/09/04	29/09/04
3	Award of Pearl River Crossing Contract	20/09/04	12/11/04
4	Award of HDD Contract	20/09/04	22/12/04
5	Steel Pipe 1st Delivery	15/10/04	25/10/04
6	Commence Trunkline Construction	03/01/05`	29/01/05
7	Commence Pearl River Tunnel Construction	28/10/04	01/12/04
8	Completion of Trunkline Engineering (except SCADA)	31/12/04	21/02/2006
9	Completion of Land Acquisition	31/12/04	20/04/2006
10	Steel Pipe Final Delivery	30/06/05	
11	Trunk line Mechanical Completion (Early Finish)	31/12/05	
12	Trunk line Mechanical Completion (Late Finish)	03/01/06	By the end of July 2006
13	Commence Trunkline hydro test and Pre-commissioning	03/01/06	30/3/2006
14	Complete Trunkline Hydro test and Pre-commissioning	31/03/06	undeterminded

Table 7–9 Guangdong Dapeng LNG Project Receiving Terminal Milestone

Category	Project milestone	Planning time (D/M/Y)	Actual completion time (D/M/Y)
1	Award of EPC Contract	30/05/03	30/05/03
2	BOG Compressor Order Placed	22/04/04	22/04/04
3	Camps and Facilities Complete	30/05/04	30/05/04

Continued

Category	Project milestone	Planning time (D/M/Y)	Actual completion time (D/M/Y)
4	Tank One Commence Construction (Piles)	20/03/04	20/03/04
5	Marine Area Usage Certification	01/05/04	09/06/04
7	Construction Land Usage Scheme Certification	30/09/04	03/11/04
8	Approval Project Prelimilary Design	01/06/04	08/06/04
6	Tank detail Engineering Design Complete	22/02/05	20/12/05
9	All Earth & Civil Complete	12/12/04	By the end of July, 2006
10	Completion of Detail Engineering Design (Except as-built &Comm)	25/03/05	31/12/2005
11	Interim complex Certificate for the Building Sect.	23/11/05	undeterminded
12	All the Process and Plant & Facilities Installed	07/12/05	undeterminded
13	All Utilities Instslled	15/09/05	14/03/2006
14	Complete of Operation and Maintenance Manuals	13/12/05	30/03/2006
15	Start Training Program for Company's Personnel	01/01/06	05/01/2006
16	Tank One Complete Construction	17/04/06	13/04/2006
17	Commence Commissioning Tank One	21/04/06	20/04/2006
18	Commissioning Complete Tank One	27/04/06	02/05/2006
19	Ready for Start-Up/Cool down of Tank One Compl.	30/04/06	06/06/2006
20	First Gas Sendout	15/06/06	26/06/06
21	Complete As-built Documentation	30/06/06	By the end of Oct 2006
22	Interim complex Certificate for the LNG Term. Sect.	30/06/06	On or before 15/08/2006
23	Interim complex Certificate for the 2nd Tank Sect.	15/07/06	On or before 30/08/2006
24	Issued Reliability Run Certificate	15/01/07	On or before 30/08/2007

② Block diagram method. Figure 7–20 is the block diagram of Fujian LNG project milestone. The characteristic of this method is that it can clearly show the logical relationship between each milestone, but the block diagram of the impact, cannot do too much description. Schedule management plan refers to how to manage the change of the schedule. According to actual needs, schedule management could make very detailed or just a framework, it could be expressed by normal forms or informal forms. It is a part of the whole project plan.

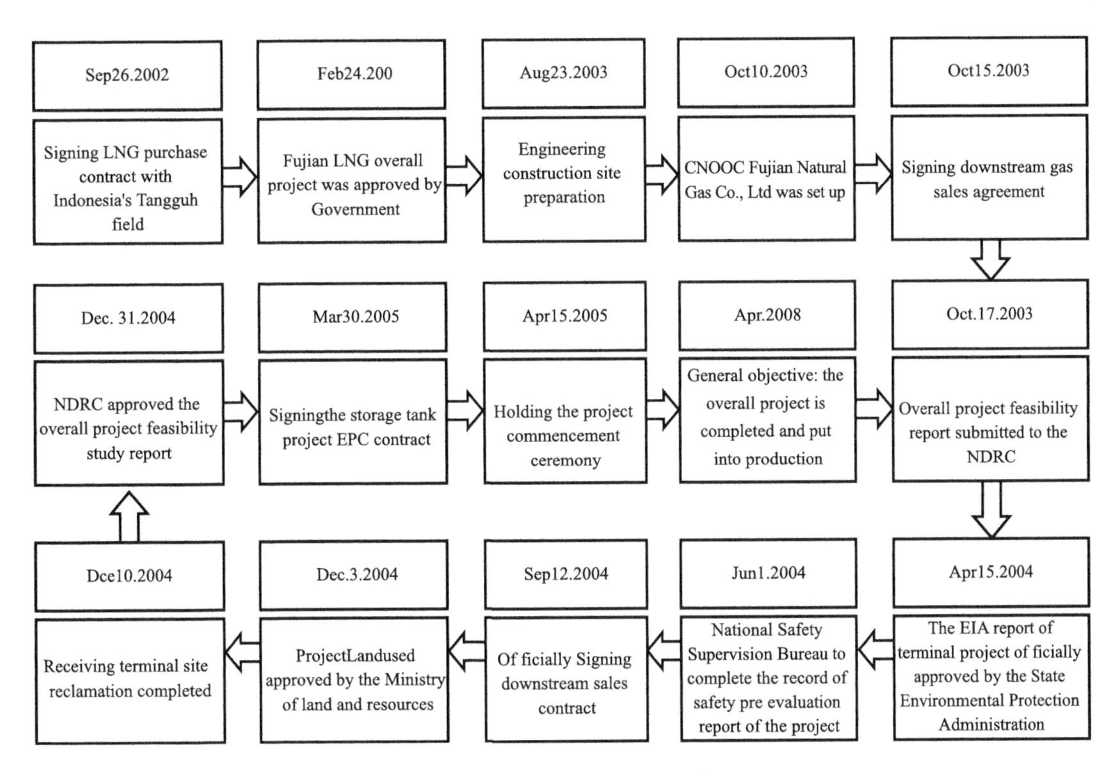

Figure 7–20 Fujian LNG project milestones

7.6 LNG Project Schedule Control

7.6.1 LNG project schedule control basis

7.6.1.1 Project Schedule Plan

The work package schedule is the basis of the higher level engineering project schedule. Similarly, LNG receiving terminal, tank, jetty, pipeline, supporting project sub-project schedule is part of the overall project plan. It provides the basis of the measure and the schedule implementation report. To carry out schedule control, we must take the schedule plan of each level as the basis.

7.6.1.2 Implementation Reports

(1) EPC contracting.

Despite the use of EPC contracting, the project company needs to deeply know the information of the progress of the project. General contractor needs to report to the owner the situation of the work finished or not finished on time. If the general contractor also takes subsequent sub projects, it needs to propose the measures that will be taken and the matters cooperated with the owner in the condition of ensuring the quality of the project. These have become the basis of the schedule controlling of the project company.

(2) Owner management.

If the project company manages the engineering construction by themselves, it still needs to strictly control the implementation of the LNG project plan. Generally it is controlled by the Control Department to send someone responsible for each of the contractor's schedule, to compile the project planning and control process in advance (Figure 7-21).

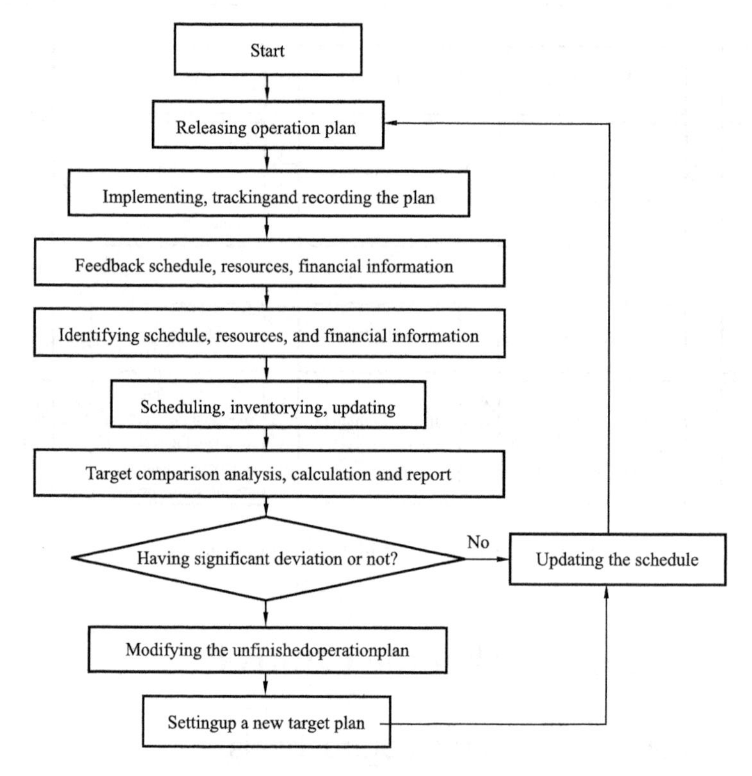

Figure 7-21 LNG project schedule change approval flow chart

7.6.1.3 Application for Schedule Change

General contractor or contractor must prepare a written report if they want to change the schedule. The report needs to clearly present the external or internal factors causing changes in progress. And it also needs to report the taken measures and variety of options after adjusting, and recommend the perfect solution. The owner must analyze these factors carefully because the consequence of these specific changes will accelerate or extend the schedule.

7.6.2 Methods and Tools for Schedule Control

7.6.2.1 Project Schedule Change Management Process

Any schedule changes will have an impact on the advance of the entire project, but in the face of various conditions and factors change, the project company (owners) should objectively and rationally treat the changes, both from the national government, or from LNG resources, both from domestic and foreign suppliers, or from the contractor, as long as it is reasonable, the owner needs to take into account of schedule change, but must be done strictly in accordance with the process. If the project uses foreign contracting, generally in the early time to sign the contract, the consideration of project changing should be written into it. One is the progress of the change control procedures; two is allowed to schedule deviation. The following is a schedule changes process (Figure 7–22).

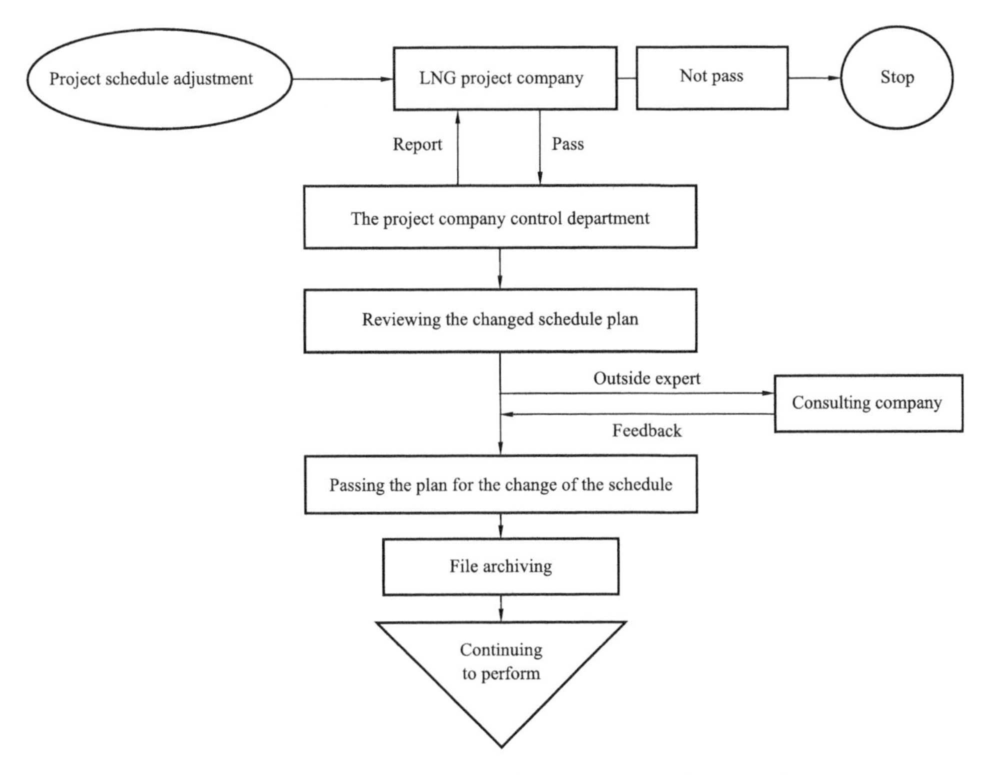

Figure 7–22 LNG project schedule change approval process chart

7.6.2.2 Method for Measuring the Schedule Change

(1) General concept.

Method for measuring the schedule change can be used to evaluate the size of the difference between the actual and planning schedule progress, ordinary at this point do not wish to have any differences. Usually it is reflected in the project implementation report, which is linked to the change of budget control index, and it is the important document in schedule control changes, which will be elaborate in the LNG project cost management. One of the most important functions of schedule control is to correct the schedule deviation. For example, in the non-critical activities, a larger time delay may only produce a smaller effect on the project, but in a critical activity, the small delay needs to be corrected immediately, etc.. The following introduces 2 common diagrammatic methods of schedule control.

(2) The Gantt chart that reflects expected and actual schedule.

This chart is put the project schedule prepared and the actual operation schedule at a certain point into the same Gantt chart, so that project managers can compare with them at a glance (Figure 7-23). There are several time concepts in the diagram, the planning time is the estimated finished time of project changes; the actual completion time is the task completion time after resource input, it has 3 possibilities when compared with planning time, that is advance, postpone or completed on time; jet lag is the flexible time of planning time; the remaining time is extra control time in the project schedule; the expected time is the corrected excepted start time and the end time.

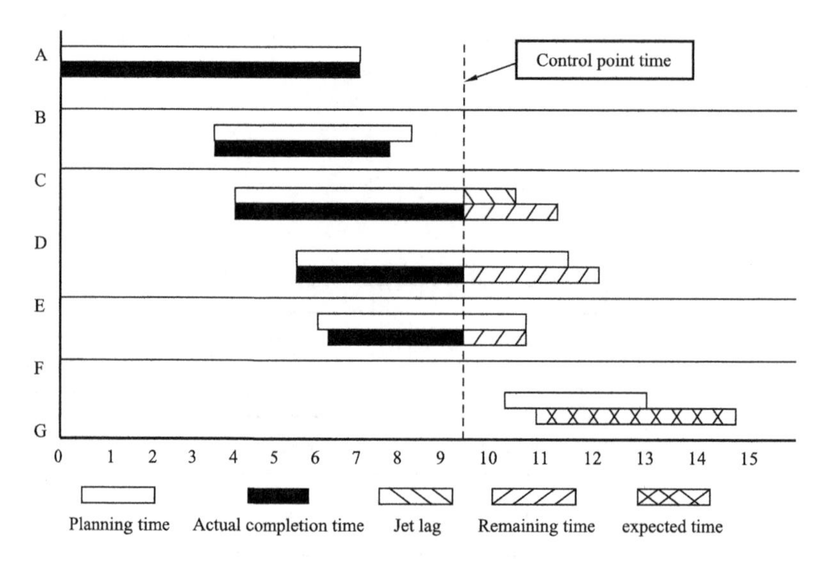

Figure 7–23 Schedule tracking gantt chart

(3) The baseline Figurethat reflects schedule and cost.

This Figure puts the relationship between the project schedule and the subsequent budget into the same figure, and the milestone points also can be put into the figure. In Figure 7–24, the median line (linear) is the schedule cost control target; the overlapping curve on it is reasonable control line of the schedule cost. The curve above the linear is the failure control line of the schedule cost; the overlapping arc is the smoothing curve. The curve and overlapping curve are the ideal schedule cost control lines. The project company can establish reasonable schedule cost baseline according to own actual situation, and then draw different schedule cost curve in different stages of project, compare with the baseline in order to determine the cost of the project schedule control.

7.6.2.3 LNG Project Milestones Plan Control

LNG project is composed by several milestones of sub projects. And the final milestone is to provide the customers with gas on time. Generally speaking, the milestone plan is the core of the project schedule control, and catching the milestone plan control is the key to guarantee the final goal of the project (Figure 7–24).

(1) The milestone tracking chart.

The milestone tracking chart is a simple and practical method to control milestone plan. The tracking chart shows the implementation and changes of milestone with vertical and horizontal

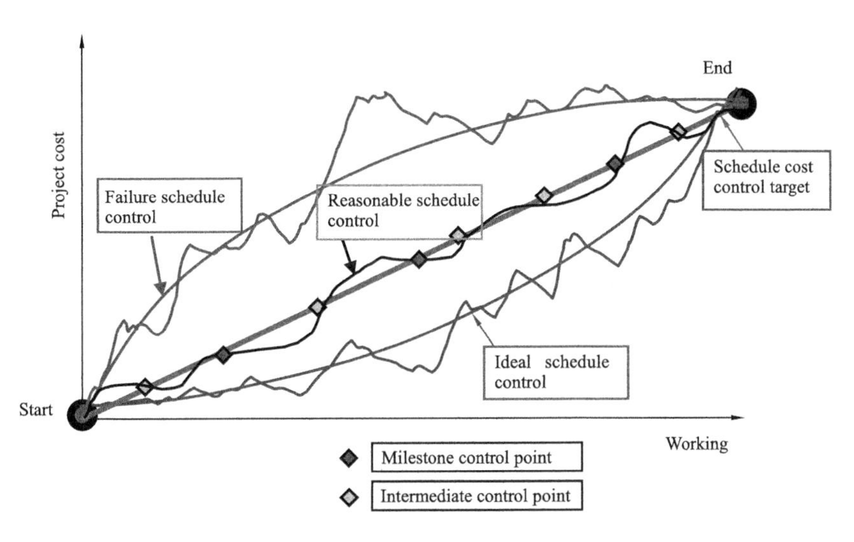

Figure 7–24 LNG project schedule and budget relationship chart

time axis and a variety of simple diagrams. Figure 7–25 is the actual completed tracking chart based on the part of Guangdong Dapeng LNG pipeline milestone plan, the solid octagons are scheduled milestone dates, and the hollow octagons are actual milestone dates. The fold(straight) line between them is actual time track based on weekly, and short-term is predicted milestone. This Figuregives the plan and control department a good reminder. For example, the planning signing date of Award of Pearl River Crossing Contract and large and medium-sized rivers crossing contract was September 20, 2004, but because of some reasons, the date delayed to November 12 and December 22. This Figurereflects many fold lines, and the owner should alert and take measures to deal with it.

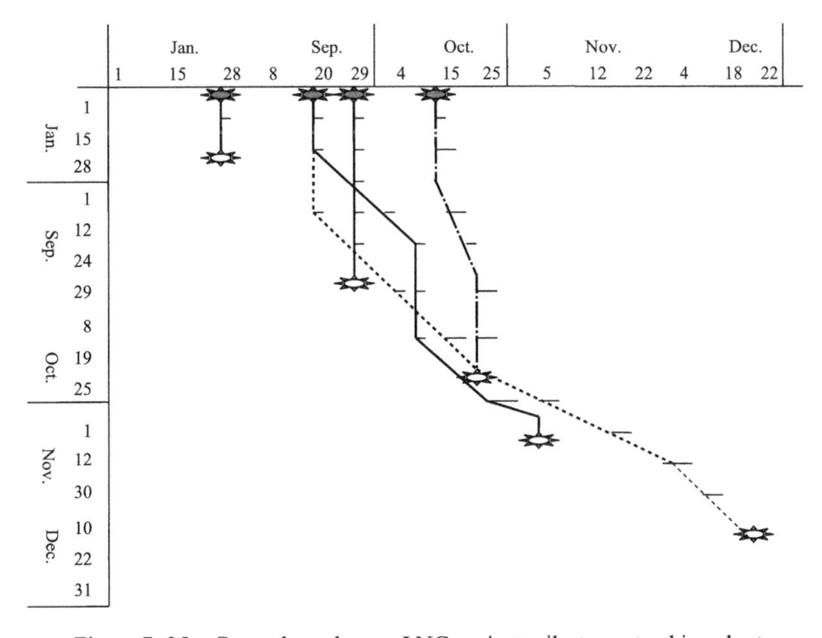

Figure 7–25 Guangdong dapeng LNG project milestones tracking chart

Table 7-10 Part of Guangdong Dapeng LNG Pipeline Sub-roject Milestone Plan

Number	Milestone	Plan time	Actual time
1	Award of CPPE Contract	28/01/2004	28/01/2004
2	Preliminary Design Approved	29/09/2004	29/09/2004
3	Award of Pearl River Crossing Contract	20/09/2004	12/11/2004
4	Award of HDD Contract	20/09/2004	22/12/2004
5	Steel Pipe 1st Delivery	15/10/2004	25/10/2004

(2) Milestone schedule control chart.

The chart is put the expected to complete the time points (baseline) and the actual completion time points of each milestone are plotted into one chart (Figure 7-26) in order to compare directly. And it also can decompose one milestone into several control point then put them in one chart.

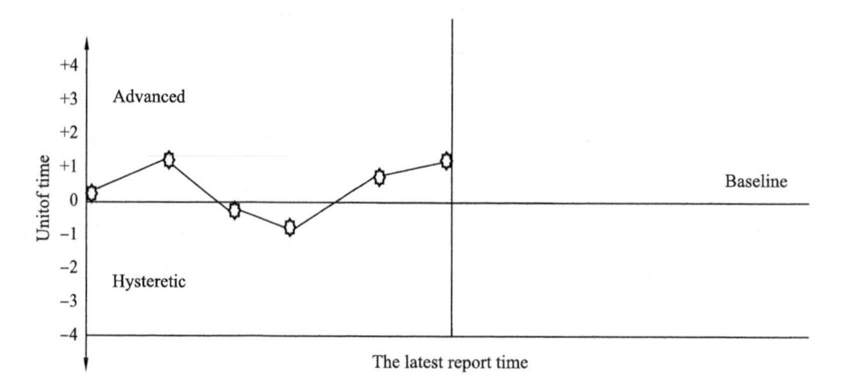

Figure 7–26 Milestone schedule control chart

(3) The milestone plan change procedure.

The program is shown in Figure 7–27. Milestone plan change is based on the previous milestone plan. In order to guarantee the ultimate goal of LNG project, it must be based on the original milestone plan, starting with the final milestone, the project end point, to reverse for a new milestone, the main purpose is because of the delay of the project milestone would lead to the ultimate goal of the LNG project cannot be completed on time. It would change the milestone plan between the deferred milestone and final milestone, so that it can catch up with the time gap, to ensure the realization of the ultimate goal—to supply gas to the downstream users on time.

7.6.2.4 Critical Path Change

The critical path is generally not time reserves, but because in the case of taking organizational measures or new technology conditions, it also can shorten the total time, which reflects the critical path changed. It has 2 meanings, one is that the critical path time is shortened, but it is still the critical path, another is the critical path demotes to a non-critical path.After E process

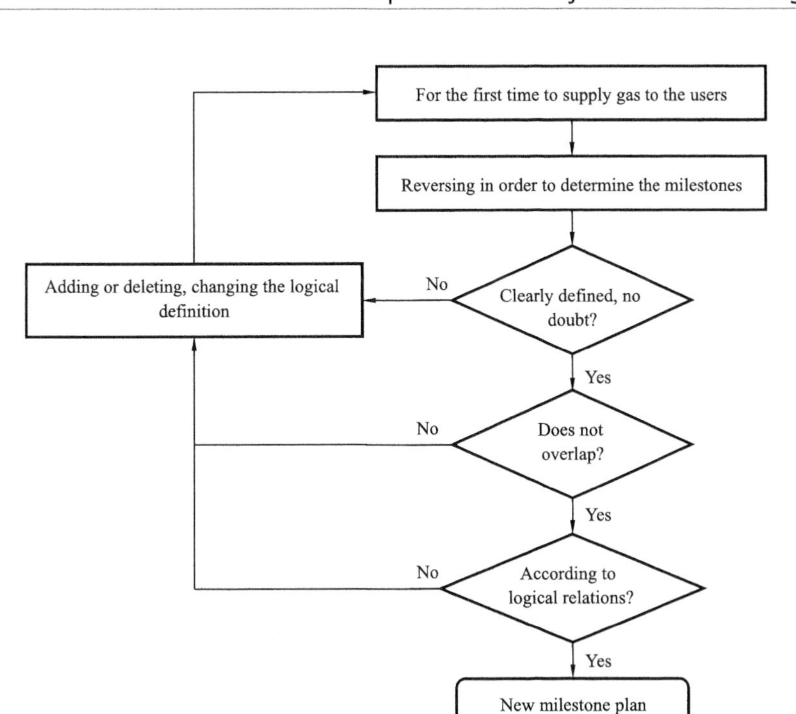

Figure 7-27 LNG project milestone schedule control

adjusted in the critical path for E1 and E2 2 processes, time respectively by 4 unit and 5 unit, but can be done at the same time in the D process, so the longest time changes from the original 8 unit to 5 unit (Figure 7-28). The final critical path turned into A—B—C—F—H—I (Table 7-11).

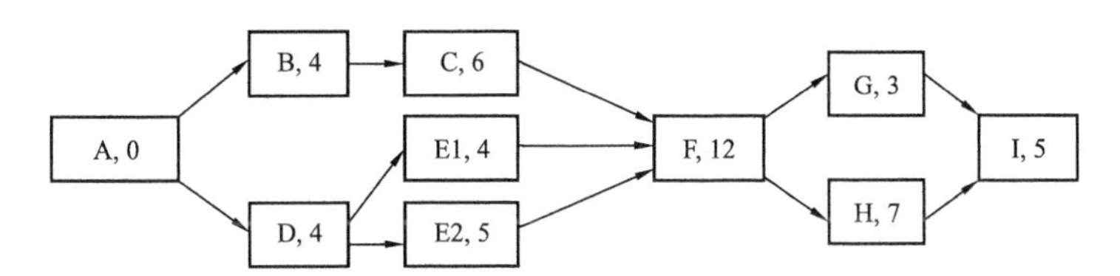

Figure 7-28 A project node method relation graph

Table 7-11 Path Time and Critical Path Analysis

No.	Path	Time	Critical path or not
1	A → B → C → F → G → I	30	Not
2	A → B → C → F → H → I	34	Yes
3	A → D → E1 → F → G → I	28	Not
4	A → D → E1 → F → H → I	32	Not
5	A → D → E2 → F → H → I	33	Not
6	A → D → E2 → F → G → I	29	Not

7.6.2.5 Time Compression Method

(1) General concept.

Time compression refers to the critical path optimization, combined with the cost, resource, time, activity of a feasible schedule factors to adjust to the project as a whole, until a critical path time can't be re compression so far, the best time to schedule obtained.

Time compression is a mathematical analysis method. The approach is to find ways to shorten the project schedule without changing the scope of the project (for example, to meet the specified date or to meet other planning objectives).

(2) Catching up method.

The catching up method is due to catch up the schedule, the owner requires the contractor to increase the investment of resources in work package, LNG project company agreed to add an incremental cost than the corresponding cost of the original schedule because of rush. There is a way to weight the success and defeat relation between the cost and schedule to determine how to get the maximum amount of time compression with the minimum incremental cost. The application of catching up method still needs to analyze the increasing or decreasing cost of the whole project caused by the implementation of the program, because the increasing cost of this work package may not cause the increasing of the total cost.

(3) Parallel construction method.

As mentioned earlier, parallel construction method is a way to reduce the project construction time, but it must consider 2 prerequisites, first one is the permission of the engineering operation; another is the possibility of the increasing resources. If these conditions are met, the implementation plan can be worked out. But it must be considered in parallel construction and lead to rework and increasing risk.

7.6.2.6 Other Measures

(1) Resource adjustment approach.

Mathematical analysis is usually to produce an initial schedule, and the implementation of this plan may need more resources than actually we have, or it needs the resources to have a big change (it brings difficulties to management). Resource adjustment approach (such as assigned the scarce resources to the critical path first) can make a schedule in the resource constrained. The completion time calculated by resource adjustment trying is usually longer than initial schedule. Preparing the schedule plan with the computer optimization software is a trying method based on the resource constrained.

Resource constrained scheduling is a special case of resource adjustment. It is only limited by the amount of available resources.

(2) Alternative Plan.

Very few projects are accurately planned. The predictable changes need to be re estimated, re modified activities, or analysis of a variety of scheduling plans.

(3) Project management software.

Project management software can compare the plan date and actual date, and it can predict the effects due to the schedule change. This software is a useful tool to control the schedule.

7.6.3 Schedule Control Results

7.6.3.1 Schedule Update

Schedule update refers to the implementation of the plan to adjust the plan (Table 7-12). It is necessary to inform the relevant parties about the planning update results. The schedule is sometimes needed to adjust the other plans of the project, such as resource planning, procurement planning, cost planning, financing plan, and so on. In some cases, the schedule delay is so serious that it is necessary to propose a new benchmark schedule.

The results after the schedule update are:

Table 7-12 Schedule Update Result Summary

Level 1	
Updated project integration plan	Updated project scope control plan
Updated project milestone plan	Updated project overall schedule control plan
Level 2	
Updated project schedule control plan	Updated plan network model
Updated task list	Updated the plan of tenders
Updated project calendar	Updated repository
Updated design control plan	Updated procurement control plan
Updated construction control plan	Updated test run control plan
Level 3	
Updated work package plan	Updated critical path chart
Updated planning establishment process	Updated resource calendar
Updated backup time hypothesis	Updated productivity calculation formula
Updated simulation technology	Updated similar estimation method
Updated work package resource assigned table	
Chart	
Updated milestone chart	Updated block chart
Updated crossing chart (Gantt chart)	Updated network chart
Updated flow chart	Updated referenced chart
Updated milestone tracking chart	Updated milestone time control chart

7.6.3.2 Corrective Measures

It refers to take corrective action to make schedule in line with the project plan. In time management, corrective action is taken to accelerate the activity in order to ensure that the activity can be completed on time or as far as possible to reduce the delay time.

(1) The drive effect evaluation.

No matter which method is adopted, the parallel operation method, or the use of new technology to change the critical path, after effect of these implementation methods are evaluated and methods can improve work efficiency many times, to quantify the concept to illustrate the problem. It needs to sum up experiences and lessons, to provide reference for the future similar work.

(2) Expert consultation.

Sometimes an expert with experience in some construction organization and technology, the project company may make recommendations to the contractor, by consulting the experts, to improve construction management and personnel allocation, get twice the result with half the effort, similarly, afterwards to summarize.

(3) Work interface coordination.

Sometimes because of several contractors' construction in the same space, the work often delays and cannot play their role. The Control and Engineering Departments of the project company must play a role of the owner, supervise and coordinate each contractors work time and work space, and take the time-sharing, segmentation and batches ways to make every contractor play a role in the project schedule.

7.6.3.3 Lessons and Experiences

The reasons of the schedule differences, the reasons of the corrective measures and other aspects of the lessons and experiences should be recorded to become the historical data and information for this project and other project in the future.

(1) Assumptions.

The assumptions on the construction conditions may deviate at the beginning because of the cognition limitations at that stage. The project company should summarize carefully and find out the real changed reasons of the original assumptions to judge the assumptions more accurately in the future.

(2) Constraint condition.

The constraint conditions of construction will also change with conditions change. For example, the human resources of the project company are inadequate in the early days, and it limits the owners' initiative to the project management. Later with the strengthening of the human resources, the improvement original constraint conditions would be conductive to the promotion of the project.

(3) Organizational matching.

LNG project usually has early project team, project office and project company three stages. The project leader, especially the project general manager must grasp the period of transition, change the structure of the organization in time. Especially in commissioning adjustment phase, it needs to supplement production operators immediately and take strict training to adapt to the changes in the direction of the work.

(4) Management strategy.

In terms of the project company, project management can be divided into direct and indirect management. In terms of the management control level, direct management work refers to the project to set up specialized agencies, and arrange specific person to management. Indirect management refers to the project company through the external power to manage. Therefore, management strategies should be based on the number of human resources of the project company, especially the quality of human resources of the project company to choose the universal, self-oriented, auxiliary or relying on the type of management.

Chapter 8 LNG Project Quality Management

8.1 Concept of LNG Project Quality Management

8.1.1 Relevant Definitions

8.1.1.1 Definition of Quality

International Standard Organization (ISO) 9000 series of standards in the definition of quality is: a set of inherent characteristics to meet the requirements of the degree. Also known as "quality is the ability to reflect the entity (product, process or activity, etc.) to meet the needs of a clear and implied capacity of the total."

For a project, quality is reflected in the process, service, and final product to meet the level of written specifications and implied requirements. In terms of the LNG project, the quality is measured by the operation process and the results of the project. The process reflects the completion of the task and the intermediate product, and the results reflect in the construction of the receiving terminal, storage tank, pipeline and supporting sub-projects, and also reflect the project provides the gas to meet the quality requirements and quality transportation service. In particular, the definition of quality should include the followings.

(1) Quality refers to the object.

The specific things that carry the quality attribute not only refer to the quality of the solid products, but also the quality of the work and the process of the activity. In this way, the so-called "entity" refers to products, processes (services) and tasks (work). Such as the construction of LNG sub projects must meet the quality requirements, while the contract terms, operation procedures, etc. must be in line with the pre-established index requirements.

(2) The meaning of quality itself.

The quality itself is the sum of the ability and the inherent characteristic of the "entity" to meet the needs of the users. It is made up of a set of characteristics, and the quality can be quantitative and qualitative. Such as LNG sub projects should meet the requirements of the standard specification, meet the LNG receiving terminal, storage, gasification and transmission functions; and natural gas delivered to the users should meet the physical and chemical analysis indicators of LNG products stipulated in the contract.

(3) Quality express.

Different "entity" is different from the substance of the quality, it by their own standards, technical contract in the form of express also is extended to must comply with the laws, regulations and rules of the industry. In terms of the LNG project, it must be in accordance with the

LNG project design, construction and service standards, for a wide range of social influence of LNG project, must comply with the health, safety and environment protection laws and regulations of our country.

(4) The relativity of the quality.

Different countries and regions, due to the different natural environment conditions, the degree of technological development, consumption level and different customs, the users of the "entity" may also be different, which is mainly determined by the different needs of different customers. From this point of view, the quality also has the characteristic of relativity.

(5) Progressive quality.

The quality index that reflects the characteristics of "entity" is invariable in a certain period. However, with the improvement of people's demand, the development of science and technology, and the progress of society, the quality of products is constantly changing and the general quality is from low to high, this is the continuous improvement of the quality of new products and new products continue to emerge power. LNG project is the same, and the revision of the specifications and standards is a long-term task.

8.1.1.2 Definition of Project Quality Management

(1) Definition of general quality management.

International organization for standardization of project quality management is defined as: quality management is to determine the quality guidelines, objectives and responsibilities, and in the quality system through such as quality planning, quality control and quality improvement, so that all management activities that enables the quality to be achieved.

Project quality management is to meet customer needs and social responsibility, through a series of coordination and control activities and processes, to achieve the established quality indicators. Project quality management must consider 3 aspects of the project operation, the service and the product of the project, the project quality management is suitable for all projects, the quality of the project product is vary by project type, considering the above all aspects, that is the complete project quality management.

(2) Definition of LNG project quality management.

LNG project quality management is to carry out quality planning, quality assurance, quality audit, quality improvement in the project quality system through the formulation of quality policies, objectives and responsibilities. All the management activities of LNG receiving terminal, jetty and gas transmission trank line constructed and natural gas and services provided to users meet the requirements of specification. It should include two aspects: soft task quality management and hard task quality management.

① Definition of soft task quality management for LNG project.

The soft task quality management of LNG project is to achieve the expected results and objectives through the pre formulation of the standards, requirements and evaluation conditions.

② Definition of hard task quality management for LNG project.

Quality management of hard tasks for LNG projects is through the project company(owners) and the contractors, the supervision company and the project related units, in accordance with the existing national laws and regulations, LNG technical standards and specifications, design blueprint and documents required to coordinate and control the project construction process, to meet the established quality standards and specifications.

8.1.2 The Significance and Characteristics of LNG Project Quality Management

8.1.2.1 The Significance of Quality Management

(1) Quality management is the key to the success of a project.

As LNG project owner, quality management is one of the key points in project management. Whether it is soft or hard task, only every work and steps to ensure quality, it can achieve the project quality objectives. In the process of project operation, it must be in accordance with the quality standards for monitoring at any time, if the work does not reach the quality requirements, it will not be taken into the next step process and procedures, so that everyone, everything, anytime to control the quality, only in this way, to ensure the success of the entire project.

(2) Quality management is the prerequisite for the project to obtain benefits.

Quality is the core of the project. The project benefit comes directly from the quality of the project. LNG project can achieve the desired economic index and social benefits, in addition to the project put into operation successfully, the more importantly is running smoothly throughout the project production operation periods. Generally LNG project ensures the production period of 25 years or more. Only quality management is in place, it can guarantee the gas supply to the downstream users need, and get profit from the payment by the end users. If there is no quality guarantee, physical chain will appear rupture, inevitably lead to capital chain broken, so as not to reap the benefits from the LNG project.

(3) Quality management is the basis of the project expansion.

In view of the practical operation of LNG project in China, the market cultivation and development of natural gas demand has been increasing dramatically, which depends on China's current economic development demand and the change of energy structure. The first period after the completion of the project, the second phase of the project will be started, such as Guangdong Dapeng, Fujian, and Shanghai LNG project in construction the phase 1, at the same time, the phase 2 plan has also been considered. So, the phase 1 of the project quality management will provide the foundation for the phase 2 project and new project.

8.1.2.2 Characteristics of Quality Management

(1) Soft task quality control is more difficult.

① The more difficulties in early user market implementation. The degree of implementation of the end users of the LNG project is an important factor in determining the size and the invest-

ment project, in order to achieve the requirements of this goal, LNG project sponsor and LNG project company will devote more energy and financial resources, in-depth research to the target market, on this basis, focus on the implementation of specific projects for large gas uses, but also have a more accurate prediction of the potential market. User market estimate is too small to reach the size of the economy, at the same time, the user market estimate is too large that may brings difficulties for future sales.

② The more contract types. LNG project industry chain involves a variety of contract types (see Chapter 15), including the LNG purchase contract, LNG ocean shipping contract, equipment procurement contract, supervision contract, financing insurance contract, professional management consulting contract, natural gas sales contract, etc.. Contracts involving a wide range of professional, technical complexity, it will provide a great deal of difficulties for the specific contract negotiations, such as to ensure the interests of the project company, so that the future of the contracts do not bring risks, and to meet the requirements of the project company, it requires project company to recruit and cultivate good negotiators, in order to cope with various types of contract negotiation and implementation.

③ The more project management links. LNG project management involves project pre-feasibility study, feasibility study, FEED design, preliminary design, detailed design, EPC tender, the choice of contractors, supervision company, the sub-project construction stage of the schedule, cost, quality and HSE control, each sub-project completion acceptance and the entire project acceptance, the first ship delivery, storage, LNG gasification, pipeline, gas distribution. It needs project company to invest a lot of manpower, material and financial resources, but also has each link management procedures and the measures. If a quality link appears to be broken, the LNG project will be affected.

(2) Hard task quality control interfaces are vast.

① Single project quality control points are more. In terms of single items of project, we need to strictly control points with design, materials, machinery, contractors, construction organization, construction technology, construction method, operating procedures and methods, equipment installation, equipment debugging, health, safety, environmental management and so on. It can ensure the quality of the whole individual project without the delivery of the non- qualified intermediate products to the next process.

② Multi project portfolio interface. The LNG project consists of receiving stations, terminals, storage tanks, pipelines and ancillary works. Each sub project has its own characteristics. It must be managed according to itself design, construction and acceptance standard management. In addition, these sub projects also form a whole, to complete the transfer of qualified natural gas to the user function. As a result, among them the hardware combination to complete, the quality of individual sub project meets the requirements. Only the first step in the quality control and management and the interface between single items of projects as a seamless connection meet the quality requirements. The ultimate goal of the whole LNG project engineering quality can be achieved.

③ LNG project is not repeatable. One of the features of project is unrepeatable. So is LNG project. It unlike in industrial production, with fixed forming production line, mature management process, formed a complete set of production equipment and stable production conditions, relatively easy to quality control. Project one-time feature requires us to control every link, ranging from design, materials, equipment purchase, construction, acceptance and commissioning. The operation of LNG project and every environment is different. The quality management condition is also different. The specific reference method is few. It must develop the line with its own quality management strategy according to the respective project environment. Once there is a quality problem, rework to remedy will be costly.

8.1.3 LNG Project Quality Management Process

As LNG project schedule and cost management, quality management is also an important content of project management. The above three constitute the traditional sense of the project three controls.

8.1.3.1 Three Components of Quality Management Process

Figure 8–1 shows the LNG project quality management flow chart. It includes the quality management system, quality management methods and measures, and quality control. The figure shows that, firstly, LNG project quality management system includes the frame of overall quality management, the LNG project company internal organization system and standard system of three parts. The middle box is the repeated crossing process of the quality management procedure, which includes the quality planning and quality assurance of the two parts, in response to the quality problems prior to the preparation of response measures. Quality audit is the quality of the pre-and post appraisal of the deployment and procedures, the work can be fed back to the upper two layers of work, and further amendments to it. Quality control is the activities throughout the project life cycle.

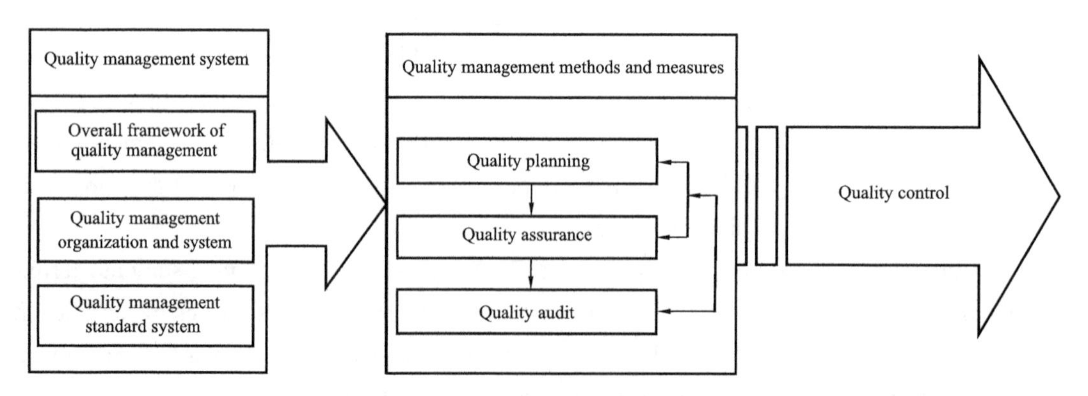

Figure 8–1 LNG project quality management flow chart

8.1.3.2 The Relationship between the Three Components of the Process

(1) The quality management system.

As a whole, the LNG project involves the government departments, survey and design units, engineering contractors, suppliers, supervision companies, users and LNG project com-

pany itself. The above units constitute the overall project quality assurance integrated ability, each endows with its quality responsibility. From the LNG project itself, the project company is the main body of quality management, according to the company's quality management scope, objectives and guidelines, the quality management organization and LNG project of the international, domestic technology and management standards and specification system must be established. Only the organization is set up, in order to clarify the quality management responsibilities of each department of LNG project company, and to distinguish the quality management level and division of labor. From the process of LNG project design, construction, acceptance, it is necessary to consider both the use of foreign LNG standards and norms which is currently lack in China, and also consider the use of domestic mature jetty, pipeline, public works standards and norms. Only the establishment of quality management system, in order to lay the foundation for the LNG project quality management.

(2) Quality management methods and measures.

As shown in Figure 8-1, according to their own organization of technology and management capabilities and the characteristics of the LNG project, LNG project company should formulate quality management methods and measures, including quality planning, quality assurance and quality audit. The general manager of LNG project company is the first responsible person of quality management, at the same time should give full play to the role of project company up and down on the quality management, also together with the directly involved in the supervision company, contractors, suppliers and project related organizations and units, develop quality management objectives, principles and control process; also consider investment in quality management and cost control, and so on.

(3) Quality control.

Quality control is the content of dynamic management, it should be based on mathematical statistics method, the quality of statistical analysis, the results will provide the basis for quality control and improvement. In addition to the normal process of quality management, also according to the changes in the internal and external factors of the project, the quality of LNG project should be adjusted and controlled. So that the LNG project has always been under the effective quality control operation, and quality control information should be timely feedback to the project company quality management institutions and individuals, in order to carry on the revision of the quality of the static management part.

Above three parts of the quality management are interdependent, tightly linked, forming a whole. The continuous improvement of quality management will also be included in the process of quality control, which includes accumulated experience as reference quality management model for the new LNG projects.

8.2 LNG Project Quality Management Systems

Quality management system is an organic whole which is composed of organization, responsibility, procedure, activity, ability and resources to ensure the quality of products, processes or services to meet the requirements or potential requirements.

8.2.1 LNG Project Quality Policies and Objectives

8.2.1.1 LNG Project Quality Policies

Quality policies are "the organization's top management official release of the group overall quality objective and direction". For LNG project company, quality policies should reflect the company's quality concept, quality awareness and quality pursue. It is the company's code of conduct, and should reflect the user's hope and commitment to customers. Quality policies generally are authorized and promulgated by general manager of the LNG project company. Such as the quality policy proposed by one LNG project company is: project construction quality is the first, the project quality control is under the leadership of general manager quality objectives management, and quality management depends on everyone, every position and every process.

8.2.1.2 LNG Project Quality Objectives

Quality objectives are the directions of the quality pursuit, which are the concrete manifestation of the implementation of the quality policies, and belong to the quality policies, echoed with the schedule target and cost target. Quality objectives are generally decomposed into project company, contractors, supervision companies, and even at the each level of the various departments, for the implementation, inspection and assessment.

Fujian LNG Project Company put forward the overall objective of quality management: creating quality engineering, and striving to project into a leading domestic and international first-class project. Specific objectives are "to grasp the project safety, quality, schedule, cost of the coordination among the four major controls, supervise and control all the activities that affect quality in the project implementation process, so that the quality of the factors under controlled state, in order to prevent, reduce, eliminate the existence of quality problems, striving for LNG project to obtain national engineering construction 'Luban Award' ".

One project company puts forward to meet the requirements of relevant laws and regulations in the engineering, procurement, construction, production. It should keep safe, environment protection, energy-saving and realizing its function. High construction quality is achieved. The number of the quality accident is zero.

8.2.2 LNG Project Quality Management Organization Structure and Responsibilities

8.2.2.1 LNG Project Quality Management Organization Structure

LNG project quality management is not only a matter of the project company, but all the units and individuals involved in the project directly or indirectly in the joint responsibility. LNG project company is the main body of quality management, the general manager of the project company is the first person responsible for quality control. It should also include the related government departments at all levels, building designers, contractors, suppliers, supervision company and users. Above is for the overall LNG project quality management framework map (Figure 8–2).

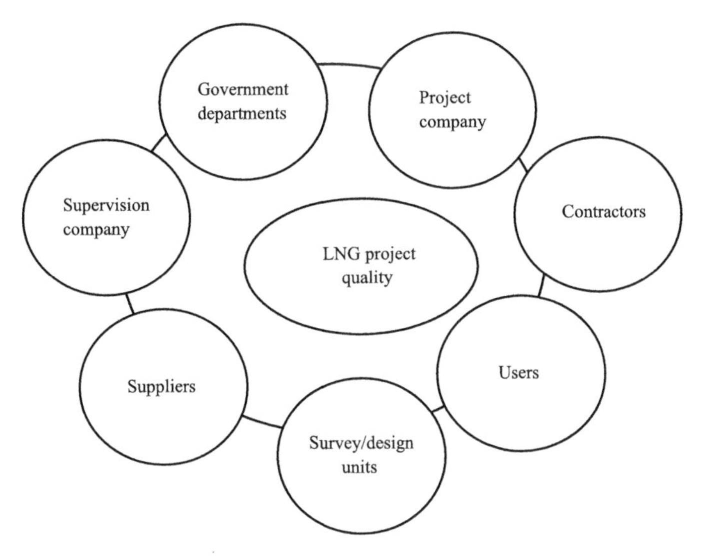

Figure 8–2 LNG project quality management framework

8.2.2.2 Related Departments or Units of Quality Responsibilities and Obligations

(1) LNG project company.

① The project company shall award the project to the construction contractor with the corresponding qualification grade. In general, the construction engineering contracting units shall not dismember the project to several subcontractors.

② The project company shall, in accordance with the law, bid for the construction projects of design, construction, supervision, and purchase of the important equipment and materials (except in the procurement of EPC contract), and choose a qualified supplier.

③ The project company must provide the original information relating to the construction project to the relevant units in the investigation, design, construction and supervision. The original data must be true, accurate and complete.

④ The project company shall not force the contractor to bid on the price below the cost of the project, and shall not compress the reasonable period of time. The company shall not express or imply that the design unit or the construction unit violates the mandatory standards for the construction, and reduces the quality of construction projects.

⑤ The project company shall submit the construction plan to the people's government at or above the county level, or other relevant departments to review. Specific measures for the review of construction drawings design documents are formulated by the construction administrative department in conjunction with other relevant departments under the State Council. Construction drawing design documents are not allowed to be used without examination and approval.

⑥ LNG project belongs to national key construction projects. The project company shall entrust with the corresponding quality level of the project supervision units to carry out supervision, also can entrust with the corresponding qualification level and without subordinate relations or other interest of the project design units for supervision.

⑦ Project company should handle the procedures of engineering quality supervision in accordance with the relevant regulations of the state before receiving the construction permit or work-start reports.

⑧ In accordance with the stipulations of the contract, if the project company procures the construction materials, components and equipment and should guarantee those products complied with the design documents and contract requirements. Project company may not express or imply construction units using unqualified building materials, components and equipment.

⑨ The renovation project involving the change of building body and load-bearing structure, the project company shall entrust the original design unit or a design unit with the corresponding qualification grade to put forward the design scheme. No design, no construction.

⑩ After receiving the construction project completion report, the project company shall organize design, construction contractors, project supervision and other relevant units to carry on the inspection and acceptance.

⑪ The project company shall support and cooperate with the supervision and inspection of the people's governments at or above the county level, and shall not refuse or obstruct the construction project quality supervision and inspection personnel shall perform their duties according to law. After the construction project is approved, the project can be delivered to use.

⑫ Project company shall comply strictly the relevant state stipulations on the management of archives, collect and sort out each link of the construction project documents, establish and improve the construction project file. After the construction completion inspection and acceptance, the company should transfer project files to the construction administrative department or other relevant departments in time.

(2) Government departments at all levels.

① The State shall implement the system of quality supervision and management of construction projects. Construction administrative departments of the local people's governments at or above the county level shall supervise and manage the quality of construction projects within the administrative area of this administrative region. The quality of state key construction projects shall be implemented by the Department of Construction Administration under the State Council or the entrusted agencies in the implementation of a unified quality supervision and management.

② The competent department of construction administration under the State Council shall strengthen the supervision and inspection of the implementation of laws, regulations and mandatory standards on the quality of construction projects. The administrative departments of construction and other relevant departments of the local people's governments at or above the county level shall strengthen the supervision and inspection of the implementation of laws, regulations and mandatory standards on the quality of construction projects.

③ Development Planning Department under the State Council shall organize the supervision and inspection of state-funded major construction project, in accordance with the provisions of the State Council's responsibilities.

④ Construction quality supervision and management may be entrusted by the construction administrative department or other relevant departments of the construction project quality supervision entrusted to specific implementation. Institutions engaged in professional construction project quality supervision must be in accordance with relevant state regulations by the relevant departments of the State Council or the provinces, autonomous regions and municipalities directly under the central government examination. After passing the examination, the institutions can do the quality supervision.

⑤ Construction Administrative Department of the people's government at or above the county level and other relevant departments which fulfill their duties of supervision and inspection shall have the right to take the following measures:

a. To ask the inspected units to provide documents and materials on the quality of engineering;

b. To enter the construction site for inspection of the units inspected;

c. When find the problems that affect the quality of the project, and shall be ordered to make corrections.

⑥ If construction administrative department or other relevant departments find that the construction units in the process of completion inspection and acceptance are in violation of relevant state construction engineering quality management rules, the project company shall be ordered to stop using and reorganizing the completion inspection and acceptance.

⑦ Water, power, gas supply, Public Security Fire Department and other departments or units shall not express or implied by the construction units, to buy their designated production and supply units of construction materials, construction parts and equipment.

⑧ Construction administrative departments should strongly support the relevant units and individuals to report, accuse and complaint the construction of the quality of the accident, the quality defects. Special major quality accident investigation procedure shall be handled in accordance with the provisions of the State Council.

(3) Survey and design units.

① Survey, design units engaged in the construction project shall obtain corresponding levels of qualification certificate according to law, and contract projects within the scope as allowed by their level of qualification. There is prohibited for survey, design units beyond the permitted scope of their level of qualification or contract projects in the name of other survey, design units. There is prohibited for survey, design units to allow other units or individuals to contract projects in the name of the unit. Survey and design units shall not subcontract or illcgal subcontracting contracts by the project.

② Survey, design units must survey, design in accordance with the project construction mandatory standards. It is responsible for the quality of the survey and design. Registered architects, registered structural engineers and other practitioners should sign in the design document, responsible for the design document.

③ Geological, surveying, hydrologic survey results provided by survey units must be truthful and accurate.

④ Design units shall be based on the survey results of the document for construction engineering design.

⑤ Design documents shall be in accordance with the requirements of the design depth of the state, indicating the use of project contract period.

⑥ Construction materials, components and equipment in the design document should be indicated the type, model, performance and other technical indicators. Its quality requirement must meet the requirements of international and national standards. In addition to the special requirements of LNG project of building materials, special equipment and process line endures, the design unit shall not specify materials and equipment.

(4) Contractors.

The contractor is an entity that provides services to the LNG project through contractual relationships.

① The Contractor shall obtain the qualification certificate of the corresponding grade in accordance with the law and contract works within the scope of its qualification grade. The contractor is prohibited from exceeding the business scope permitted by the qualification grade of the entity or in the name of other units. The contractor is prohibited from allowing other units or individuals to undertake projects in the name of this unit. Without the consent of the project company (owners), shall not transfer or illegal subcontracting.

② The international contractor shall have the corresponding qualifications and undertake the excellent performance of the LNG project, according to the relevant provisions of the relevant ministries, when they undertake LNG project construction in our country, they should set up a joint venture company with domestic contractor together, the international contractors can be used as EPC general contractor, responsible for project quality.

③ The contractor shall be responsible for the construction quality of the project. The contractor shall establish the quality responsibility system and implement the project manager, technical director and the responsible person of the construction management. If the construction works to implement the general contracting mechanism, the general contractor shall be responsible for the quality of all construction projects; construction engineering survey, design, construction, and equipment procurement of one or more of the implementation of the general contract, the general contractor shall be responsible for the quality of its construction projects or procurement of equipment.

④ If the general contractor in accordance with the construction project subcontract to other units, the subcontractors shall be responsible for the quality to the general contractor in accordance with the provisions of the subcontract. The general contractor and subcontractors shall jointly and severally liable for the quality of the subcontract works.

⑤ If international EPC general contractor is responsible for the procurement of materials and equipment. It shall be in accordance with the contract to the owner, and provide the supplier information and quality assurance files, etc..

⑥ The contractor must be in accordance with the project design drawings and construction technical standards, not allowed to modify the engineering design, not cut corners. During the construction process, the contractor finds the design documents and drawings are in error, and should timely make comments and suggestions.

⑦ The contractor must inspect the construction materials, components, equipment and the commodity concrete in accordance with technical standards, design requirements and construction contract, which shall not be used without inspection or unqualified. The inspection shall have a written records and personnel signature.

⑧ The contractor must establish and improve the construction quality inspection system, strict process management, ready to take concealed engineering quality inspection and record. Before concealed work, the constructor shall inform the project company and construction engineering quality supervision institutions to take engineering test, and it can be concealed after test qualified.

⑨ The construction personnel should take on-site samples of the structure safety and specimen block, and the relevant materials under the supervision of the contractor or the construction supervision company, and these samples need to be sent to the company with appropriate level of qualification for quality testing.

⑩ When the construction engineering quality problems found or the construction project acceptance unqualified, the contractor shall be responsible for repairing.

⑪ The contractor shall establish and improve the education and training system to strengthen the staff education training. Personnel without education training or test unqualified may not be permitted to work at its position.

(5) Suppliers.

Suppliers are an entity that provides materials, equipment, instruments, meters, and spare parts for the LNG project.

① The suppliers shall meet the qualification of LNG project company supplier, and provide materials and facilities within the scope of its qualification. Suppliers are forbidden to provide materials and facilities beyond its qualification, and suppliers shall not transfer its supply business.

② The suppliers shall provide all kinds of materials and facilities conforming to quality standards in accordance with the contractual requirements, and overseas suppliers shall apply China's export right in advance to ensure material facilities export customs clearance.

③ Within the prescribed period of time in accordance with the contractual requirements, suppliers shall provide the quality and quantity of materials and facilities, provide China customs inspection of the declaration materials and procedures, and assist the project company related department to check and accept.

④ When the inspection found that the material facilities have quality problems or other problems, the suppliers shall cooperate with the LNG project company and the Chinese government departments for investigation and evidence collection.

⑤ When quality problems of materials and facilities found and verified, the suppliers shall make timely remedial measures, in the specified time to submit an alternative, and in accordance with the terms of the contract to compensate for the loss caused by the project company.

(6) Engineering supervision unit.

① The construction supervision company shall obtain corresponding levels of qualification certificate, and take engineering supervision business within the scope as allowed by their level of qualification. The construction supervision company is prohibited to undertake the engineering supervision business beyond the scope of this unit level of qualification permission or in the name of other construction supervision company. The construction supervision company is prohibited to allow other company or individuals to undertake project supervision business in its own name. The construction supervision company shall not transfer the supervision business.

② If the construction supervision company and supervision of engineering construction contractor, and construction materials, fittings and equipment supply units have subordinate relationship or other interests, the construction supervision company should not undertake the project construction supervision business.

③ The construction supervision company shall be on behalf of the project company for construction quality supervision, and take the construction quality supervision responsibility, according to the laws, regulations and relevant technical standards and design documents and construction project contract.

④ The construction supervision company shall send corresponding qualifications general supervision engineer and supervision engineers in the construction site. Without supervision engineer's signature, building materials, components and equipment shall not be used in engineering or installation, and the contractor is not allowed to carry out the next process of construction. Without the signature of the chief supervision engineer, the project company shall not pay project funds, and shall not conduct the final acceptance.

⑤ Supervision engineer shall take side watch, patrol and parallel test as forms to implement the construction engineering supervision in accordance with the requirements of the project supervision specification.

(7) Users.

① According to the contract of supply and demand, the users need adjust their matching project based on the arrangement of LNG project schedule to synchronous completion and acceptance with the LNG project, including the quality and safety evaluation, and preparation for participating in synchronization commissioning of LNG project.

② Users should cooperate with experts organized by LNG project company to conduct a comprehensive assessment to the project commissioning conditions, correct the existing problems timely rectification, and strive to achieve synchronization of upstream and downstream of

the commissioning conditions.

③ Users need sign the contract with LNG project company involving trial production and production period of gas, gas quality, gas pressure and gas price, in order to ascertain responsibilities, rights and obligations of both parties.

④ During the commissioning, users need keep the communication with LNG project company and timely feedback on both sides. Users also need summarize the gas output and input procedures and ways from the commissioning process and practice, in order to accumulate experience for formal production.

8.2.3 Quality Management Organization Structure and System

8.2.3.1 LNG Project Quality Management Organization Chart and Responsibility Matrix

(1) LNG quality management organization.

As shown in the Figure 8-3, Fujian LNG project company quality management has its own characteristics. The company set up a special quality management committee and quality management staff-expert group. In charge with quality was the engineering department, for the overall coordination of LNG project quality; the corresponding receiving terminal, jetty, pipeline sub-projects, respectively, three sub-project teams were set up to take responsibility for quality management, and equipped with the corresponding project quality control managers, the sub project of the receiving terminal is divided into the secondary storage tank engineering and other engineering, and the individual supervision and quality control managers were arranged, pipeline and terminal were also equipped with the supervision and quality control manager, the subcontract work was also equipped with quality control manager. The model reflects under the

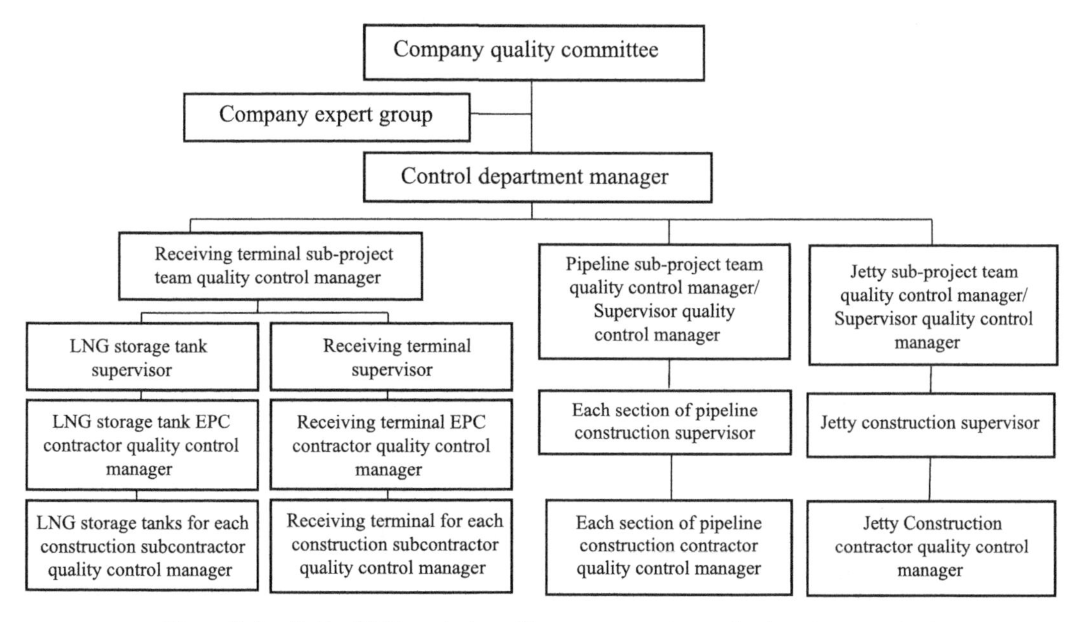

Figure 8-3 Fujian LNG project quality management organization structure chart

leadership of quality management committee, the project general manager was responsible for the overall quality system, engineering department was the lead unit of quality management, the project teams and supervision units were responsible for the quality of the engineering.

(2) Responsibility matrix.

Quality responsibility matrix reflects responsibilities of quality management in each stage, each link, division of labor of the project, here takes the quality management of Shanghai LNG project company quality responsibility matrix as an example (Table 8-1).

8.2.3.2 LNG Project Company Key Personnel Quality Management Responsibilities and Obligations

(1) General manager.

① The first responsible person for the quality management of the project construction.

② To preside over the development and approval of the company's quality policies, objectives and strategies.

③ Be responsible for the organization of the company to set up the rational allocation of quality and responsibility, the implementation of quality responsibility.

④ To lead the establishment and improvement of quality management rules and regulations, and provide necessary and sufficient resources for the implementation of quality policies, effective operation and continuous improvement of quality management.

(2) Deputy general manager.

① To assist general manager to make the company quality policies, objectives and strategies, and promote the implementation of quality policies and objectives.

② To take the responsibilities for quality management work of the in charge departments, and preside over the development of quality objectives and related management documents in charge of the Department.

③ To take the responsibilities for the assessment of the quality management work of the in charge department.

④ To assist the general manager for the company's quality management organization, promotion and examination.

⑤ To assist the general manager to adhere to the effective implementation and continuous improve the quality management.

⑥ Fully responsible for the company's quality management work when the general manager absence.

(3) The chief engineer.

① Engineering project technical person in charge.

② To provide the design basis and control the quality of the key data.

③ To organize the technical cooperation and control the quality of the technical documents.

④ To assist deputy general manager in charge of the company's technical quality management.

Table 8-1 Shanghai LNG Project Company Quality Management Responsibility Matrix

Management Content	Responsible Department									
	Technical Department	Engineering Department	QHSE Department	Marketing Department	Planning and Control Department	Resources and Transportation Department	Business Department	Coordination Department	Finance Department	Personnel and Administration Department
Quality Policy, Objectives and Strategies	○	○	●	○	○	○	○	○	○	○
Quality Control Plan	●	●	○	△	△	△	●	△	△	△
Documents Quality Control	○	○	●	△	△	△	○	△	△	○
Corrective Actions for Unqualified	○	○	●	△	△	△	○	△	△	△
Quality Event /Accident Investigation and Treatment	○	○	●	△	△	△	○	△	△	△
Survey and Design Quality Control	●	○	○	△	△	△	△	△	△	△
Quality Control of Acquisition and Acceptance	○	○	○	△	△	○	●	△	△	△
Manufacturing and Construction Quality Control	○	●	○	△	△	△	○	△	△	△
Pre Commissioning and Commissioning	○	●	○	△	△	△	○	△	△	△
Trial Production / Transfer of project	○	●	○	△	△	△	○	△	△	△
Project Quality Management Assessment	○	○	●	△	△	△	○	△	△	△
Remarks: "●" main responsibility department, "○" secondary responsibility department, "△" relevant support departments										

(4) The chief accountant.

To ensure the quality management of the company's capital investment, the cost of quality management will be included in the annual budget, and supervision of the management and implementation of related expenses.

8.2.3.3 LNG Project Company Department's Quality Management Responsibilities and Obligations

The following introduction is Shanghai LNG project company quality management responsibilities and obligations of the various functional departments.

(1) QHSE department.

① The Department is the lead unit of quality management, which organizes and coordinates quality management work of all departments.

② To prepare the quality management requirements in tender and contract documents.

③ To audit requirements for quality management in the process of design, construction, acquisition, etc..

④ To coordinate design report of the relevant government departments, construction for examination and approval relating to the quality documents with responsibility department.

⑤ To participate in the review of the quality assurance and quality control of contractors during the bid invitation, and participate in the qualification examination and evaluation of the suppliers.

⑥ Be responsible for the compilation and supervision of the implementation of project quality management regulations.

⑦ To organize engineering quality acceptance, and supervises serious unqualified rectification.

⑧ To organize the investigation and handling of major quality events and organizes verification.

⑨ To participate in examination and verification to the requirements of the quality with the project progress payment and change.

⑩ To organize the regular and irregular inspection for the quality management in construction site.

⑪ To organize the assessment of project quality management.

(2) Planning and control department.

① To ensure project bidding, procurement documents for engineering contractors and suppliers of schedule and cost requirements to meet the quality management requirements.

② When the schedule, cost and quality objectives are in contradiction, it should be taken a comprehensive consideration for schedule, cost, and quality, in order to meet the requirements of project quality management, the schedule and cost should be adjusted.

(3) Technical department.

① To organize "design basis" data collection, collation, archiving and providing outward,

ensure that the original information is true, accurate, complete, etc..

② To organize technical document encoding, clear technical documents, records of the style and format.

③ To organize the preparation of the technical parts of the tender.

④ To organize the work to compile the technical requirements of the formulation of project bidding, procurement document for the contractors and the suppliers, and organize technical bid evaluation work.

⑤ To organize the company to set up the quality control plan for the investigation and design process of the project, and approve the work plan of the contractor.

⑥ To participate in major projects of key process inspection and test, the witness of the witness point, etc..

⑦ To organize project survey and design review and acceptance.

⑧ To strengthen the reliability and accuracy of technical documentation for technical quality control of the project.

⑨ To report to the relevant government departments for examination and approval in design documents related to the quality, and obtain the approval.

(4) Engineering Department.

① Be responsible for the quality, schedule, cost and safety of daily control management of the project construction site.

② To establish and improve the company project construction management system.

③ Be responsible for manufacture, construction, installation, testing and commissioning in project construction site.

④ To set up the construction quality assurance engineer post, responsible for organizing the preparation of project construction quality control plan, and organizing professional engineers for project construction quality control management.

⑤ To organize and compile the control requirements of the project contractors and the supervision company in the bid invitation documents.

⑥ To organize for reviewing, approval of the construction organization design, construction stage technical documents, construction quality control procedures, inspection / test plan and quality assurance documents, etc., to implement effective control of project construction management.

⑦ To examine and approve supervision plan and other supervision of files submitted by supervision company.

⑧ To organize itemized works and portioned works and quality acceptance and supervise the unqualified rectification.

⑨ To organize survey and treatment about project construction general quality accident.

⑩ According to the drawings, technical specification and the requirements of the contract, to review the supplier product quality inspection reports, and participate in key equipment, material, supervision and inspection work.

⑪ To report to the relevant government departments about documents related to the quality for examination and approval in construction, and obtain the approval.

(5) Business Department.

① To prepare acquisition strategy, acquisition and acceptance quality control plan, and make the project procurement meet the requirements of quality and function.

② To follow up and manage the procurement and acceptance activities, ensure the suppliers to submit the drawings, documents and quality certificates on time, and is responsible to transfer the relevant certificates to the QHSE department for the record.

③ To organize relevant departments and personnel for inspection and ensure that the material, structure and equipment purchased by the project company (OFE) in accordance with the design documents and contract requirements.

④ To organize the key equipment, materials supervision and manufacture.

⑤ To arrange the supplier to the on-site technical service work.

(6) Resource and Transportation Department.

① To implement company quality policy and control of LNG import quality.

② To organize customs clearance and commodity inspection of the first LNG ship.

(7) Marketing Department.

The department collects and feedbacks the user's technical requirements, including providing explanation of LNG quality parameters, the grid connected pressure, peak shaving capacity and calorific value adjustment.

(8) Financial Department.

The department appropriates the company quality management expenses according to the plan.

(9) Coordination Department.

The department is responsible for the daily communication, opinion collection and processing of the relevant departments of the state and local government, and the approval of documents.

(10) The personnel and administration department.

① Be responsible for the company's archives management, formulate relevant regulations, and put forward the requirements of quality management for administrative document.

② Be responsible for the company's human resources allocation, choose a professional training, with sufficient experience, and competent personnel to participate in project management.

③ To organize company staff quality management training.

(11) Company quality committee.

If the project company sets up quality management committee, it can refer to the following duties and obligations.

① To approve the company's quality policy, and guide principle, strategy.

② Overall management of project quality, so as to achieve the project quality objectives.

③ To carry on the inspection, guidance and supervision of quality management work of the

project team and department.

④ To solve the major quality problems.

8.2.3.4 LNG Project Quality Management System and Management Process

(1) LNG project quality management system.

Guangdong Dapeng LNG project company established a quality management system (Figure 8–4), which was divided into three levels. The first level, the files as ISO 9000 series standards of quality management was established, such as quality management regulations, manuals, for the project company in accordance with the quality policy and objectives. The second level, the files were the project company and the contractor as quality management procedures and plans, which described the implementation of quality system elements involved in the activities of the various all participant units. The third level, the files were described as detailed assignments of all contractors, including operation instruction, technological process, records.

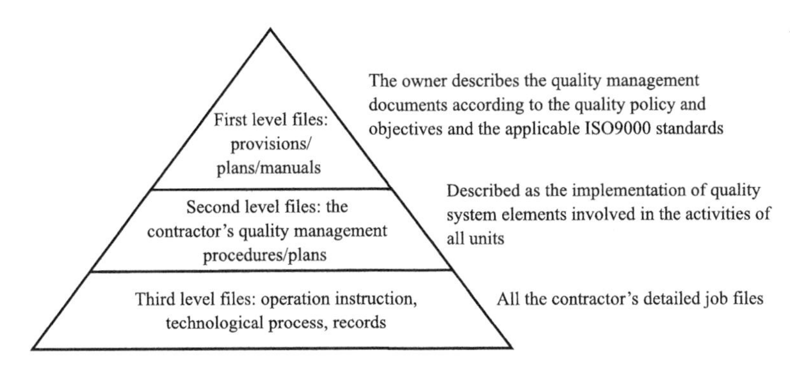

Figure 8–4 Quality management system

(2) LNG project quality management process.

Flow chart method is also used to express the LNG project quality management, including the project quality management of the owners, supervision, the general contractor, sub contractor's quality system of mutual contact, such as Guangdong Dapeng LNG project not only in quality organization quality control system (Figure 8–5), but in the implementation of quality control system (Figure 8–6) with its characteristics, which can provide a reference for other LNG projects.

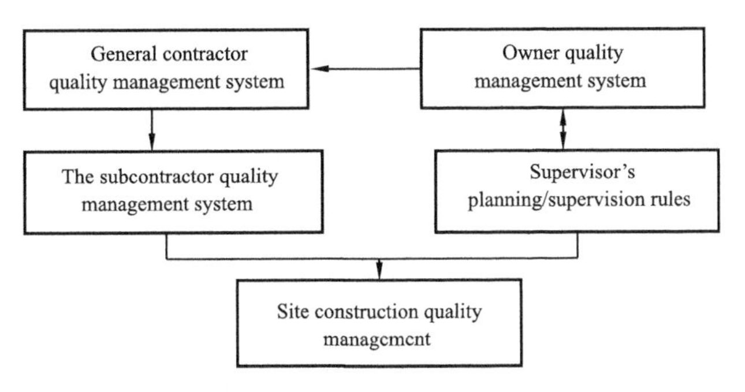

Figure 8–5 Project quality system control framework

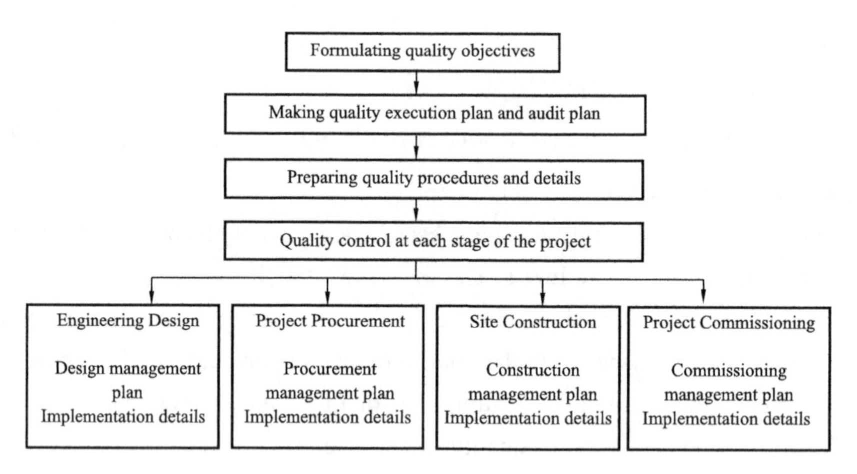

Figure 8–6 Project quality control execution system

8.2.3.5 LNG Project Quality Management Laws, Regulations and Standard System

(1) The national laws and regulations.

In China, LNG project must have a set of quality management system of laws, regulations and standards, which include two aspects. One is our country's laws and regulations, and it must comply with the mandatory terms; another is standard and specification. Because our country has not yet set up a complete system of LNG standard, in this respect, it must use more foreign standards and a complete set of LNG project involved in other domestic standards which constitutes the LNG project quality management requirements of the file system.

As long as the construction of the project in China, it must be carried out in the framework of the relevant laws and regulations of our country, and compulsory execution. According to the contents involved in the LNG import project, the national and ministries and commissions laws and regulations need to be followed, including the health, safety, environmental protection, engineering construction quality specifications, regulations. The frame work is shown in the Figure 8–7.

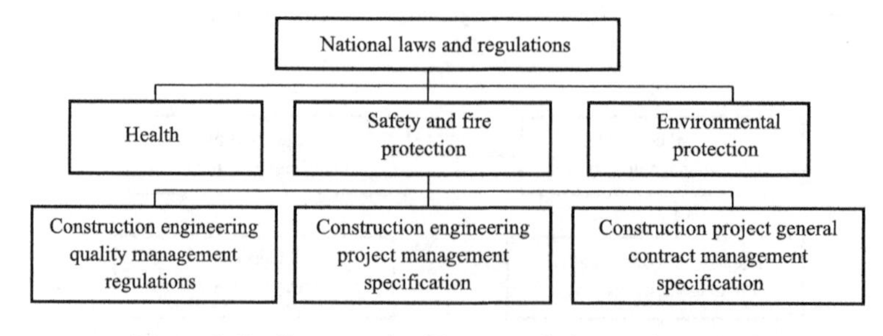

Figure 8–7 Framework of laws regulations and standards

(2) International standards and domestic standards.

Since CNOOC introduced LNG as a strategic goal, it began to consider the necessity of formulating the LNG industry national, industry and enterprise standard system. In order to promote the cause, the principles adopted in the standards are as follows. it is necessary to combine the advanced foreign LNG standard with domestic engineering construction standard. If there are domestic and applicable standards, domestic standards can be used; and if no domestic standards, the foreign advanced standard will be directly referenced. Through the company's actual work of our first pilot project——Guangdong Dapeng LNG project, at home and abroad for LNG standards and related jetty, terminal, tank, pipeline and the utility on the domestic standards have been researched, and now forms the LNG project (Figure 8-8) with a standard system framework, a total of more than 1140 standards, which lay the standard foundation for other construction of LNG project.

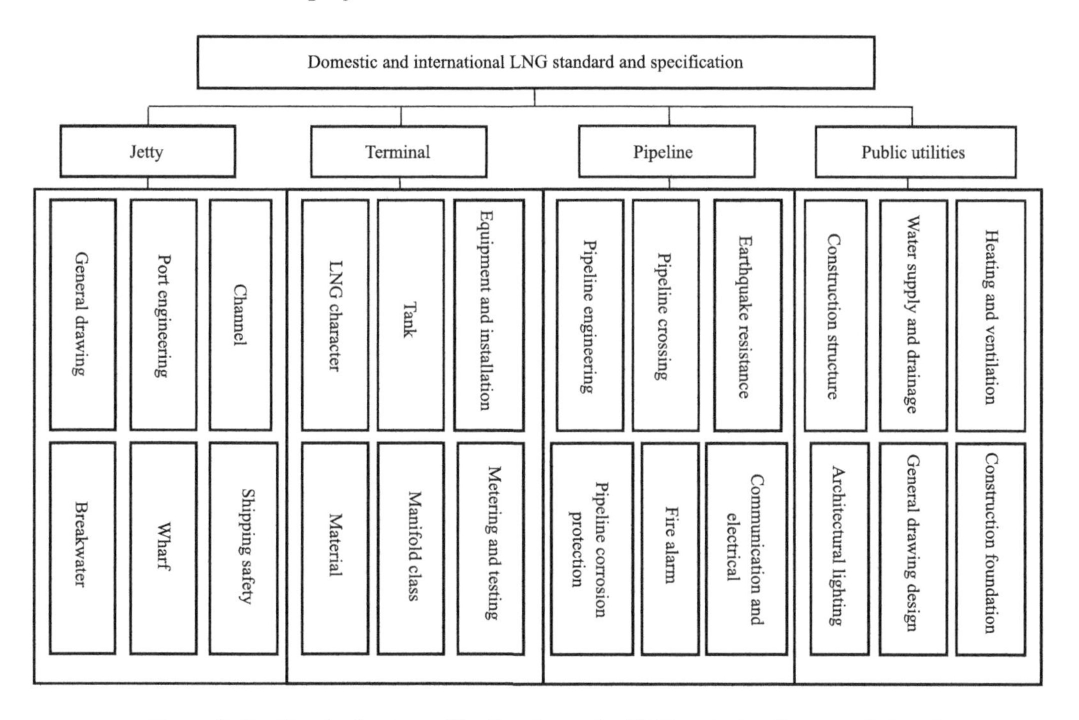

Figure 8-8 Standard and specification frame for LNG projects at home and abroad

8.3 LNG Project Quality Planning

8.3.1 Environmental Analysis of LNG Project Quality Management at Home and Abroad

8.3.1.1 Present Situation of International LNG Standardization

(1) International quality standard system.

International Organization for Standardization (ISO) Quality Management and Quality Assurance Technology Committee, on the basis of relevant quality management standards in various countries, promulgated and revised the quality standards respectively in 1986, 1994 and 2000, and provided a unified platform of quality management and quality assurance for all countries. On the basis of this, ISO carried on the thorough research to the project quality management, released the international project management quality standard——ISO10006, which cited a large number of ISO9000 series standard. China's Standardization Committee has adopted a series of international quality standards in 2000, and as the national standard GB/T 19000 series released. At present, ISO has been updated ISO9000-2000 version to the ISO9000-2008 version of the quality standard series, which is a scientific summary of the practical experience of quality management in many developed countries in the world, and also is the crystallization of the experience and wisdom of human quality management with the versatility and guidance.

(2) The applicability of the international quality standard system.

① The ISO9000-2008 series and the updated series of standards are applicable to all product categories, different organization sizes and types.

② The standard system focuses on the effectiveness of the quality management system, continuous improvement and coordination, and emphasizes the process management.

③ In the standard system, environmental management has been included in quality management system, which is in line with the current international popular environment friendly, and coordinates development trend.

④ For China's LNG project company beginning to get involved in the LNG industry, if not fully established its own quality management system standard, it needs to formulate the quality management standards according to their actual needs, so as to meet the needs of users and the purpose of project management.

8.3.1.2 The Contribution of Various Countries to the LNG Standard System

(1) Summary.

Stones from other hills may serve to polish jade. With the emergence of LNG industry in Europe, the United States and Japan, the standards have been brewed and released. In order to create the LNG industry in our country, using the LNG standards of foreign industrial countries, in-depth study of the national LNG standard system, is the best way to accelerate the LNG industry in China. Standard types include: basic general standards, product standards, project design, construction standards, material and equipment standards, management standards, safety standards and environmental standards. The existing LNG standards in various countries have laid a foundation for the establishment of a unified LNG international standard.

(2) European LNG standardization construction.

In 1959, 5000m^3 LNG shipments in "Methane Pioneer" modified by general cargo ship, was taken from lake Charles, Louisiana in the United States to Canvey Island in England, which meaned that United Kingdom had become the LNG consumers. After 40 years, the British Standardization Committee prepared a standard "Low temperature vertical flat bottomed cylindrical storage tank", involving in the design, construction, installation and operation of the LNG important equipment, and the design and construction of the prestressed and reinforced concrete of LNG tanks and its foundation. At the same time, Germany had also created its own tank stan-

dard. European Committee for Standardization, the continent's leader, gathered the power of the experts and scholars from Austria, Belgium, Denmark, Finland, France, Germany, Greece, Iceland, Ireland, Italy, Luxembourg, the Netherlands, Norway, Portugal, Spain, Switzerland, Sweden and Britain, on the basis of the above countries' LNG standards, prepared "The general characteristics of liquefied natural gas" and LNG equipment and installation, low temperature pipe, flange, ship shore connection, loading and unloading arm test and design, fire prevention and other aspects of the standard, which has played an important role to guide the development of LNG industry in Europe.

(3) The United States LNG standardization construction.

The United States is the first country in the LNG industry. In many years of production practice, it givs full play to the strength of the professional association of technical experts, such as the National Fire Protection Association(NFPA), the American Petroleum Association(ASME), and other professional associations. The overall has formed a series of LNG standards, especially in LNG production and storage, large welded low pressure tank design and construction, LNG vehicle fuel products, insulation materials, pipeline transportation, fire prevention, safety. In particular, the federal government has legislation in LNG security.

(4) Japanese LNG standardization construction.

By the end of 2015, Japan has more than 30 LNG receiving terminals, about 42% of the world's total number of receiving terminals, and becomes the world's largest LNG importer, accounting for 39% of the world's total imports in 2009. It has accumulated a large number of LNG operational data, and laid a very good foundation for the preparation of their own standards.

Japan Gas Association, the Japanese Industrial Standard (JIS), Japan Electric Power Association and the Japan High Pressure Safety Gas Association, respectively, LNG standardization construction has contributed to the formation of their own series of standards, and some of which are also used in South Korea and Taiwan region of China, such as LNG underground storage recommended standards, LNG receiving terminal, the satellite station equipment recommended standards, gas production equipment and safety regulations recommended standards, gas production equipment design recommended standards, general industry in the use of combustible gas electrically charged equipment, electrically charged equipment explosion-proof protection recommended standards, LNG receiving terminal regular inspection standards, regular inspection manual, the general specification of pressure vessels, liquefied gas equipment standards, and so on.

8.3.1.3 International Road of LNG Standard

(1) The necessity of establishing the international LNG standard system.

① The need for international cooperation. In recent years, LNG industry and LNG market are active. LNG production is expected to increase from 250 million tons per year in 2010 to 390 million tons per year in 2020, and 150 to 200 billion dollars (at current prices) will be invested to build new gasification plants and liquefaction plants. This kind of ultra large multi project portfolio, by a company or unit is very difficult to complete, the need for multinational resource owners, banks, investors, equipment manufacturers and design engineering companies to understand each other and work together to make a contribution to this large-scale project. In this way, we need to recognize the standard platform which is a top priority.

② The New industry needs. LNG industry involves resources and consumer countries, such as China, India and other LNG consumer countries. They need to build LNG receiving gasification facilities and devices. Nigeria, Iran and other countries need to build natural gas liquefaction facilities and devices. These new industry countries are often lack of supporting LNG standard system. In order to develop their own new industries, it is a shortcut to draw support from the standards and norms of LNG industry mature countries. From this point of view, LNG international standards approved by the countries will not only be conducive to the import of equipment and materials, promote the improvement of domestic localization, avoid technical barriers and technology monopoly, but also accelerate the new industrial countries to master and upgrade technology.

③ The need for new technology. At the technical level, new equipment, new liquefied process and new ideas (LNG, modular design of the floating gasification system) will be or have been designed by the engineering design enterprises. For those countries that have developed the LNG industry as a new industry, the use of these new technologies requires the appropriate standards and norms to confirm and identify these technologies to achieve the cost reduction and guarantee the safety and reliability of the project.

(2) Establish international LNG standards working group.

For the above reasons, establishing the LNG international standardization working group is imperative, which will promote international LNG standards forward so that the LNG standard will form a system as soon as possible. For this purpose, in October 2006, LNG standardization working group was established at meeting ISO193TC67. The three major oil companies of China were all represented. The main tasks of the working group were as follows:

① To evaluate the existing LNG industry standards and norms fixed demand of each country;

② To evaluation of the applicability of the existing LNG international standards and systems in the LNG industry;

③ To determine the existing standards and norms of the owner and the LNG working group for the framework;

④ To determine the priority order of the standard establishment.

We believe that with the organization and coordination of LNG Standards Technical Committee, the international LNG standard system will be gradually improved.

8.3.1.4 China's LNG Standards towards Internationalization

(1) LNG standardization organization in China.

The Guangdong Dapeng LNG project has been put into operation, which marks the rise of

China's LNG industry. In China, CNOOC first set foot in the LNG industry, paid more attention to LNG standards at the beginning, had led the research and preparation of LNG standards. According to the spirit of the document of the National Standard Committee Comprehensive [2008] No. 125, CNOOC was director unit of national LNG Professional Standards Sub Technical Committee formally established in September 2008 (Figure 8-9). This is the major event related to the three major oil companies and staff engaged in LNG standardization. The establishment of the Sub Technical Committee will help the work under the direct leadership of the National Standardization Management Committee and the National Oil and Gas Standardization Technical Committee, accelerate the pace of LNG standards amendments, and improve China's LNG standard system. It is helpful to contact with similar international standardization organizations, learn from the experience and practice of the LNG standard amendments of international organizations, and give full play to the industrial experience of advanced LNG industrial countries and connect the tracks as soon as possible with international standards. It is also contributes to the domestic all engaged in the integration of LNG industry enterprises, research institutes, universities and other relevant units of the scientific research and technical force to play to their strengths, and contribute ideas and exert efforts together for China's LNG industry. Under the guidance of the national and industry LNG standard system, we can draw up the new, high level, high quality LNG standards as soon as possible, so as to meet the needs of LNG industry market and production.

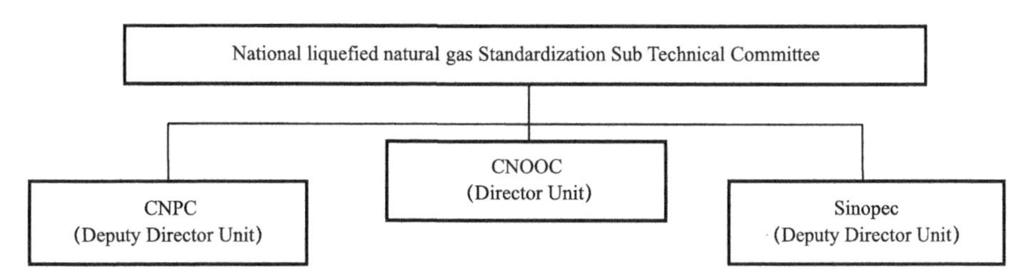

Figure 8-9 China liquefied natural gas standardization organization structure

(2) The work progress of LNG Professional Standards Sub Technical Committee.

① National and industrial standard system table.

LNG Professional Standards Sub Technical Committee and the previous working group have systematically studied foreign LNG professional standards system from 2000, which is carried out in three stages: foreign standards survey, collection, aggregation; the recognition and translation of core standards; research, screen, contrast and forming reports. Based on the above work, the "LNG professional national and industrial standards system table" (Table 8-2) is compiled. The table highlights the analysis of measurement and design, construction and operation of core standards, focusing on the products and health safety and environment (HSE) standards that are lack in domestic. Till 2015,40 national, industry standards has been completed, and 30 national and industry standards will be gradually completed in the future.

Table 8-2 LNG Professional National and Industrial Standards System Table (2016 Edition)

Prepared to complete the standards				
No.	Category	Standard number	Standard name	Int. st. encoding
1	General basis	GB/T 19204—2003	General characteristic of liquefied natural gas	EN1160 : 1997
2		SY/T 6936—2013	Liquefied natural gas glossary	
3	Design	SY/T 6807—2010	Guideline to programming project application report for LNG projects	
4		SY/T 6935—2013	Specification for preliminary design of liquefied natural gas receiving terminal	
5		JTS 165—5—2009	Code for design of liquefied natural cas port and jetty	
6		SY/T 6986.1—2014	Installation and equipment for liquefied natural gas——Design and testing of marine transfer systems. Part 1 : Design and testing of transfer arms	BS EN1474-1 : 2008
7		GB/T 22724—2008	Installation and equipment for liquefied natural gas——Design of onshore installations	EN1473 : 1997
8	Analysis Measurement	GB/T 20603—2006	Refrigerated light hydrocarbon fluids——Sampling of liquefied natural gas——Continuous method	ISO8943 : 1991
9		GB/T 24962—2010	Refrigerated hydrocarbon liquids——Static measurement——Calculation procedure	ISO 6578 : 1991
10		GB/T 24964—2010	Refrigerated light hydrocarbon fluids——Liquefied natural gas——Procedure for custody transfer on board ship	ISO 13398 : 1997
11		GB/T 21068—2007	Standard specification for LNG density calculation models	ASTM D4784-93
12		GB/T 24959—2010	Refrigerated light hydrocarbon fluids——Measurement of temperature in tanks containing liquefied gases——Resistance thermometers and thermocouples	ISO 8310 : 1991
13		GB/T 24957—2010	Refrigerated light hydrocarbon fluids——Calibration of membrane tanks and independent prismatic tanks in ships——Physical measurement	ISO 8311 : 1989
14		GB/T 24960—2010	Refrigerated light hydrocarbon fluids—Measurement of liquid levels in tanks containing liquefied gases—Electrical capacitance gauges	ISO 8309 : 1991

Continued

Prepared to complete the standards				
No.	Category	Standard number	Standard name	Int. st. encoding
15	Analysis Measurement	GB/T 24961—2010	Refrigerated light hydrocarbon fluids—Measurement of liquid levels in tanks containing liquefied gases—Float-type level gauges	ISO 19504 : 1993
16		GB/T 24958.1—2010	Refrigerated light hydrocarbon fluids—Calibration of spherical tanks in ships. Part 1 : Stereo photogrammetry	ISO 9091-1 : 1990
17		GB/T 24958.2—2010	Calibration of spherical tanks for refrigerated light-hydrocarbon fluids in ships—Part 2 : Triangulation method	ISO 9091-2 : 1992
18	Construction Installation Operation	GB/T 26978.1—2011	Design and manufacture of site built, vertical, cylindrical, flat-bottomed steel tanks for the storage of refrigerated, liquefied gases with operating temperatures between 0~-165℃. Part 1 : General	EN14620-1 : 2006
19		GB/T 26978.2—2011	Design and manufacture of site built, vertical, cylindrical, flat-bottomed steel tanks for the storage of refrigerated, liquefied gases with operating temperatures between 0~-165℃. Part 2 : Metallic components	EN14620-2 : 2006
20		GB/T 26978.3—2011	Design and manufacture of site built, vertical, cylindrical, flat-bottomed steel tanks for the storage of refrigerated, liquefied gases with operating temperatures between 0~-165℃. Part 3 : Concrete components	EN14620-3 : 2006
21		GB/T 26978.4—2011	Design and manufacture of site built, vertical, cylindrical, flat-bottomed steel tanks for the storage of refrigerated, liquefied gases with operating temperatures between 0~-165℃. Part 4 : Insulation components	EN14620-4 : 2006
22		GB/T 26978.5—2011	Design and manufacture of site built, vertical, cylindrical, flat-bottomed steel tanks for the storage of refrigerated, liquefied gases with operating temperatures between 0~-165℃. Part 5 : Testing, drying, purging and cool-down	EN14620-5 : 2006
23		GB/T 24963—2010	Installation and equipment for liquefied natural gas—Ship to shore interface	EN 1532 : 1997

Continued

Prepared to complete the standards				
No.	Category	Standard number	Standard name	Int. st. encoding
24	Construction Installation Operation	GB/T 20368—2006	Production, storage and handling of liquefied natural gas (LIG)	NFPA59 : 2003
25		SY/T6711—2014	Technical code for liquefied natural gas receiving terminal	
26		SY/T 6928—2012	Operation regulation for liquefied natural gas receiving terminal	
27		SY/T 6929—2012	Operation regulation for liquefied natural gas jetty	
28		GB/T 26980—2011	Liquefied natural gas (LNG) vehicular fueling systems code	NFPA52-2006
29		JB/T 4780—2002	Tank containers for liquefied natural gas	
30		GB/T 25986—2010	Filling Device of Natural Gas Vehicles	
31		GB/T 20734—2006	Mounting requirements for liquefied natural gas vehicle special equipment	
32		SY/T 6934—2013	Operating regulation for liquefied natural gas (LNG) vehicle fueling station	
33		QC/T 755—2006	Technology requirements of special equipment for LNG vehicle	
34		JJG 1114—2015	Verification regulation of liquefied natural gas filling machine	
35		JJF 1524—2015	Program of pattern evaluation of liquefied natural gas dispensers	
36		DB11/ 1093—2014	Safety technology requirements of container skid-mounted refueling device for liquefied natural gas vehicle	
37		SY/T 10029—2004	Recommended practice for planning, designing, and constructing floating production system	
38		CCS	Specification for entry level and construction of liquefied natural gas filling pontoon at water	
39		SY/T 6933.1—2013	Technical standard for natural gas liquefaction factory design construction and operation. Part 1 : Design construction	
40		SY/T 6933.2—2014	Technical standard for natural gas liquefaction factory design construction and operation. Part 2 : Operation	

Continued

Prepared to complete the standards				
No.	Category	Standard number	Standard name	Int. st. encoding
Drawing up and planning standards				
1	General basis	GB	Liquified Natural Gas	
2	Design	SY/T	Installation and equipment for liquefied natural gas——Design and testing of marine transfer systems. Part 2 : Design and testing of transfer hoses	BS EN 1474-2-2008
3		SY/T	Installation and equipment for liquefied natural gas——Design and testing of marine transfer systems. Part 3 : Ofshore transfer system	BS EN 1474-3-2008
4		GB	Specification for liquefied natural gas station (onshore installations) fire protection design	
5		GB	Design specification for liquefied natural gas pipeline	
6		GB	Design guide for floating liquefied natural gas devices	
7		GB	Specification for ship shore liquefied natural gas filling station design	
8		GB	Code for design of offshore fire protection of the liquefied natural gas receiving terminal	
9		SY/T	Guidelines for land formation and civil engineering technology of liquefied natural gas receiving terminal	
10	Analysis Measurement	GB	Refrigerated liquid hydrocarbon liquid level measurement of the liquefied gas container microwave liquid level gauge	ISO 13869
11		GB	Refrigerated light hydrocarbon fluids general requirements of automatic tank gauge. Part 1 : The boats carrying liquefied gas	ISO 18132-1 : 2006
12		GB	Refrigerated light hydrocarbon fluids general requirements of automatic tank gauge. Part 2 : Frozen type of onshore tank level gauge	ISO 18132-2 : 2006
13		GB	Measurement of total sulfur in liquefied natural gas (LNG) ultraviolet fluorescence method	
14	Construction Installation Operation	GB	Guidelines for implementation of risk assessment in the design of onshore liquefied natural gas facilities (including the ship / shore interface)	

Continued

Prepared to complete the standards				
No.	Category	Standard number	Standard name	Int. st. encoding
15	Construction Installation Operation	GB	Specification for anticorrosion and thermal insulation technology for cryogenic storage tanks	
16		SY/T	Specification for design, construction and acceptance of insulation for cryogenic piping and components	
17		SY/T	Acceptance specification for natural gas liquefaction plant	
18		SY/T	Guidelines for ship to ship transport of liquefied natural gas	
19		GB	Shipping procedure for LNG receiving terminal	
20		SY/T	Specification for liquefied natural gas station static equipment installation and inspection	
21		SY/T	Specification for liquefied natural gas station dynamic equipment installation and inspection	
22		SY/T	Code for design and construction of concrete structures for liquefied natural gas storage tanks	
23		SY/T	Specification for nitrogen pre - cooling technology for liquefied natural gas pipeline	
24		SY/T	Technical specification for small - sized satellite station of liquefied natural gas	
25		GB	Acceptance specification for small - sized satellite station project of liquefied natural gas	
26		GB	Technical specification for receiving terminal on water for liquefied natural gas	
27		GB	Acceptance specification for receiving terminal on water for liquefied natural gas	
28		SY/T	Specification for design and construction of membrane tanks for liquefied natural gas	
29		SY/T	Specification for construction and acceptance of large scale full containment tank of liquefied natural gas	
30		SY/T	Specification for design and construction of membrane tanks for liquefied natural gas	
31		GB	Cryogenic energy utilization	

② LNG enterprise standard system table.

When it comes to the detailed aspects of technology, management and work, especially related to the operation of LNG project quality management procedures and specific technical requirements, can only rely on the actual situation of enterprises to strengthen the enterprise standardization construction to complete. According to the needs of enterprise management, from the standardization of the institutional settings and working funds, CNOOC has given enough attention. At the beginning, LNG receiving terminal location setting, target market research, feasibility studies, application reports, engineering design, project acceptance and post-assessments, were lack of standards. For this reason, CNOOC drew up its LNG enterprise standards system table (Table 8-3). It was hoped to have each link from the project established to the project completion to have standards to check, in order to meet the needs of the development of the industry. In general, the structure of the enterprise standard system should be consistent with the structure of enterprise management. Modification and improvement of enterprise standard system is a very complex system engineering, involving a wide range of work. Although we have compiled and released 15 corporate standards, there is a lot of work to do. It cannot be achieved overnight, so it needs to draw up and improve gradually.

Table 8-3 LNG Enterprise Standard System

No.	Category	Standard name	Status
1	Early study stage	LNG project opportunity research (subject) report preparation guide	To be prepared
2		Economic work requirements for construction projects	Promulgated
3	Pre-feasibility study/ feasibility study stage	Guideline to programming a site selection report for LNG projects	Promulgated
4		Guidelines for LNG project pre-feasibility study report preparation	To be prepared
5		Guidelines for LNG project feasibility study report preparation——general report	Promulgated
6		Guidelines for LNG project feasibility study report preparation——jetty engineering report	Promulgated
7		Guidelines for LNG project feasibility study report preparation——terminal engineering report	Promulgated
8		Guidelines for LNG project feasibility study report preparation——pipeline engineering report	Promulgated
9		Guidelines for LNG project feasibility study report preparation——market report	Promulgated
10		Guidelines for LNG project feasibility study report preparation—— LNG resource and transportation report preparation	Promulgated
11		Guideline on compiling safety assessment prior to start report of liquefied natural gas project. Part1: Receiving terminal	Promulgated
12		Guideline on compiling safety assessment prior to start report of liquefied natural gas project. Part 2: LNG jetty	Promulgated

Continued

No.	Category	Standard name	Status
13		Guideline on compiling safety assessment prior to start report of liquefied natural gas project. Part 3 : Gas pipeline	Promulgated
14		LNG project investment estimation index——feasibility study stage	Promulgated
15		Guidelines for LNG project feasibility study of investment estimate preparation	Promulgated
16		Guidelines for other LNG project engineering cost estimate preparation	Promulgated
17		LNG project economic evaluation methods and parameters	To be prepared
18	Design stage	Liquefied natural gas terminal general layout	To be prepared
19		LNG project investment estimate index——preliminary design stage	To be prepared
20	Construction stage	Guidelines for LNG project management manual preparation	To be prepared
21		LNG project bill of quantities valuation model	To be prepared
22	Acceptance stage	Inspection and acceptance procedures for liquefied natural gas project. Part 1 : Receiving terminal engineering	Promulgated
23		Inspection and acceptance procedure for liquefied natural gas project. Part 2 : Port engineering	Promulgated
24		Inspection and acceptance procedures for liquefied natural gas project. Part 3 : Gas pipeline engineering	Promulgated
25		Guidelines for liquefied natural gas (LNG) project evaluation after the completion acceptance report preparation	To be prepared

8.3.2 The Preparation of LNG Project Quality Planning

LNG project quality planning refers to the use of which standards to regulate the product, process and engineering, and determine who, when to meet these quality standards. In terms of engineering, it is mainly for the quality policy, quality objectives, procedures and corresponding standards, to implement standard quality plan, control of process management, realize the goal of the project quality. Planning is one of the results of the quality plan.

8.3.2.1 The Scope of Quality Management

The quality scope is determined by the scope of the project. LNG project involves from upstream LNG resources to downstream users, involving LNG ocean transportation, LNG terminal receiving, storage, gasification, land natural gas pipeline transportation, gas distribution and downstream utilization of the other aspects.

(1) LNG project pre-feasibility study / feasibility study.

Due to different national conditions, in the LNG project site selection study and to the approval of the competent department of the state, it is not possible to make foreign standards as reference. Therefore, we compiled the series of the "Guidelines for LNG project feasibility study report preparation" "Economic parameters and index series standard" and "Guideline to programming project application report for LNG projects" . Now we are embarking on a compilation of enterprise standards: "Guidelines for LNG project site comparison report preparation", and "Guidelines for LNG project pre-feasibility study report preparation" and so on, which are the standards for the preparation of the report and the content of the research, all of those are the guide for the work of this period.

(2) LNG resources.

LNG resources is the basis of the project. How to buy qualified LNG resources is the key, generally by the resource buyers and sellers through negotiations to finalize. We first compiled the national standard "General characteristic of liquefied natural gas" . One of the recommended LNG range index, physical and chemical indicators or similar to the specification of natural gas components index and the use of safe, it is the most basic standard for negotiations. We are working on the preparation of the product standard for "liquefied natural gas" .

(3) LNG marine transport.

It is determined by the terms of the contract signed by the LNG project company and ocean transportation management company. Their respective rights and obligations must be defined. In addition to a single project of transport (point to point) company, multi-project transportation companies have certain restrictive provisions, to ensure the objective of the transportation task to be completed. To this end, we have adopted the guiding international standards ISO "Installation and equipment for liquefied natural gas——ship to shore interface" .

(4) Design.

In the design stage, we have completed the receiving terminal, storage tanks, jetty, low temperature pipeline, key equipment, device and other design standards. We are working on preparing the "Liquefied natural gas terminal general layout" "Specification for preliminary design of liquefied natural gas receiving terminal", and "LNG project investment estimate index——preliminary design stage", with service to international, national and industry standards, to promote the LNG project design work more solid, more perfect.

(5) Engineering construction safety.

LNG project management is an integrated management processes that involves in project schedule, budget, quality and HSE. Foreign project management experience and practice cannot be copied completely, and must be summarized and improved on the basis of our own projects practice, to be used in the future LNG project in our country. According to Guangdong Dapeng, Fujian LNG project practice, we have released "Technical code for liquefied natural gas receiving terminal" "Operation regulation for liquefied natural gas receiving terminal" and are preparing for the compilation of "Guidelines for LNG project management manual preparation", which will provide a reference for LNG project quality management.

(6) LNG specialized jetty.

In addition to comply with general technical standard of jetty, specialized research is made for LNG jetty design, construction, the waterway, the traffic density, water depth, wind flow condition, turning round, the size of ocean LNG transportation ships, safe distance, norms and standards at home and abroad. The State Department of Transportation issued guidance significance to the standard "Code for Design of Liquefied Natural Gas Port and Jetty". We have compiled the "Operation regulation for liquefied natural gas jetty", and "Shipping procedure for LNG receiving terminal".

(7) Storage tank.

LNG tank is the core engineering of projects, which can solve the contradiction LNG storage space between shipping arrangements and spare gas consumption. Before this, our country has never been involved in the project, so lack of such standards. Because LNG is cryogenic and flammable, tank support structure and insulation material have their special requirements, and generally adopt international standards as national series standards, such as "EN 14620-1-2006: Design and manufacture of site built, vertical, cylindrical, flat-bottomed steel tanks for the storage of refrigerated, liquefied gases with operating temperatures between 0~-165℃", and "Installation and equipment for liquefied natural gas" series standards, which provides us rare engineering design, construction and acceptance specification to grasp the LNG project core engineering.

(8) Gasification.

LNG gasification is a prerequisite to transport natural gas. The design is according to the needs of each project, the maximum daily gasification capacity, and equipped with the gasification devices. Is the use of ORV? Or is the use of IFV? In order to ensure the daily, monthly, annual peak gas demand, the general use of international standards to buy, but also consider the design and construction of reference to international standards.

(9) Measurement.

In the trial production process, LNG measurement needs to be geared to international standards. In addition to concerns about LNG volume or weight measurement methods, we need to pay particular attention to energy metering of LNG. Because LNG components are different, and the main factor that truly reflects the price is the calorific value energy, while energy metering is used widely currently in international condition, so we use the international series of measurement standards (Table 8-2), in order to be better in international cooperation and exchanges.

(10) Gas transmission and distribution stations.

It is a necessary part of the natural gas distribution. The transmission and distribution station design and construction can be carried out in accordance with national standards.

(11) Onshore gas pipeline.

It is an important bridge to ensure the supply of natural gas. Domestic long-distance transportation pipeline design and construction standards have been very mature.

(12) Overall acceptance.

The individual sub-project acceptance is the basis of overall LNG project. The joint acceptance of the whole LNG project and the downstream supporting user's project is the true sense. Through research and operation of the project, LNG project acceptance and assessment after LNG project completion and inspection have no foreign standards as a reference. As a result,

based on the practice of Guangdong Dapeng and Fujian LNG project, we complete the series of standards of "LNG project acceptance procedures", and are preparing to compile the "LNG project post assessment procedures", and provide the methods and tools for project management.

(13) Non-pipeline transportation.

Non-pipeline transportation is used to solve inadequate pipeline capacity, or for some reason that needs to adopt non-pipeline transportation, including tankers and tank vessels. The mode of transportation has appeared in domestic for many years. We can refer to domestic and international standards, design to build the loading station, and select transport and transport safety.

(14) LNG utilization.

LNG cold energy utilization, LNG satellite station, gas stations, have been formed in the domestic industry. It has been prepared to complete the "Operating regulation for liquefied natural gas (LNG) vehicle fueling station", and "Filling device of natural gas vehicles", and is organizing the "Technical specification for small - sized satellite station of liquefied natural gas" "Acceptance specification for small - sized satellite station project of liquefied natural gas" and "Liquefied natural gas utilization" .

8.3.2.2 Project Quality Control Interfaces

LNG project is a complex and large group of projects. For example, Guangdong Dapeng LNG project is divided into 25 unit works, 376 divisional works, and 3033 itemized works. Because of the large number of engineering, sometimes the situation is occurring that several construction groups work in one construction site, so as to appear that the interface for the engineering quality is not clear, by late quality problem occurres which causes wrangling and evading. To this end, Guangdong Dapeng LNG project through groping for a start on the establishment of the project quality control interface (Table 8–4), effectively solved this kind of problem.

8.3.2.3 Project Quality Management Plan

Project quality management plan is to make an arrangement involving what the quality standard of project should be achieved and how to achieve the project quality standards.

(1) Compiling principles of quality management plan.

① Targeted quality policy. Referring to the ISO9001-2000 standards, it is necessary to compile the overall quality management plan of the LNG project which is in line with the project quality policy.

② Focused quality control units. Overall quality management plan focuses on the control management of the contractors /suppliers during project construction.

③ Quality requirements of effective survey and design. Engineering survey and design is the basis of the quality of the project. The management departments should compile the survey and design quality management plan and requirements.

④ Strict procurement quality control procedures. Equipment and material procurement is another important content of quality control. The management departments shall establish procurement management plans and procedures.

Table 8-4 Project Quality Control Interface

No.	Control project names or activities	Design units	Manufacturing units	Detection units	Construction units	Design supervision units	Manufacturing supervision	Construction supervision	Construction project division	Supervision headquarters	Project management department
1	To create a national quality project activities	●	●	●	●	●	●	●	●	●	◇
2	Project division	—	—	—	—	—	—	—	—	■	★
3	Regulations for completion data management	—	—	—	—	—	—	—	—	■	★
4	Survey and design quality inspection acceptance plan	●	—	—	—	■	—	—	—	★	⊙
5	Survey process quality inspection acceptance	○	—	—	—	☆	—	—	—	△	△
6	Survey results acceptance	○	—	—	—	●	—	—	—	●	☆
7	Preliminary design process quality inspection acceptance	○	—	—	—	☆	—	—	—	△	△
8	Pipeline route comparison results confirmation	○	—	—	—	◇	—	—	●	●	☆
9	Confirmation of process flow	○	—	—	—	◇	—	—	—	●	☆
10	Supporting system results confirmation	○	—	—	—	◇	—	—	◎	●	☆
11	Preliminary design results Acceptance	○	—	—	—	●	—	—	—	●	☆
12	Materials, equipment technical specification confirmation	○	—	—	—	⊙	—	—	—	⊙	☆
13	Construction drawing design process quality inspection acceptance	○	—	—	—	☆	—	—	—	△	△
14	Construction drawing results acceptance	○	—	—	—	●	—	—	—	●	☆
15	Materials in factory acceptance	—	○	—	—	—	☆	—	—	△	△

Continued

No.	Control project names or activities	Design units	Manufacturing units	Detection units	Construction units	Design supervision units	Manufacturing supervision	Construction supervision	Construction project division	Supervision headquarters	Project management department
16	Manufacturing process examination and acceptance	—	●	—	—	—	☆	—	—	△	△
17	Material, equipment factory inspection (supervision and construction projects)	—	○	—	—	—	☆	—	—	△	△
18	Materials and equipment in-site acceptance	—	—	—	○	—	—	☆	△	△	△
19	Construction quality inspection acceptance plan	●	—	—	●	—	—	■	⊙	★	⊙
20	Construction process quality inspection acceptance	◎	—	—	○	—	—	☆	●	△	△
21	Operation group job completion acceptance	◎	—	—	○	—	—	☆	●	△	△
22	Section completion handover acceptance	●	—	●	●	—	—	●	●	●	◇
23	Completion pre-acceptance preparation	●	—	●	●	—	—	●	●	●	◇
24	Supervision headquarters special quality inspection	●	●	●	●	●	●	●	◎	◇	◎
25	Design supervision special quality inspection	●	—	—	—	◇	—	—	—	◎	◎
26	Construction supervision special quality inspection	●	—	—	●	—	—	◇	◎	◎	◎
27	General quality problem processing	●	●	◎	●	—	—	◇	★	△	△
28	Major quality problem processing	●	●	◎	●	—	—	●	●	◇	★
29	Survey design quality evaluation report	—	—	—	—	○	—	—	—	△	△
30	Manufacturing quality evaluation report	—	—	—	—	—	○	—	—	△	△
31	Construction quality evaluation report	—	—	—	—	—	—	○	—	△	△

Remarks: "●" Attend, "○" Report, "△" Keep on record, "◎" Invite, "◇" Organize, "★" Approval, "☆" Acceptance, "⊙" Examine, "■" Chief editor, "—" No action.

⑤ Quality acceptance specification. Acceptance of hardware and software are the final links of engineering quality control. The project quality acceptance requirements should be compiled from the start, and be improved continuously through practice.

⑥ File form. Total quality management department is responsible for the audit, the company leader is responsible for approval, and the file is issued to the relevant department in the form of controlled documents.

(2) Content of the quality management plan.

① Enforcement of laws and standards. The design and construction of LNG project in addition to strictly implement our government's relevant laws, regulations, rules, and national mandatory technical standards, in principle, the main works are mainly based on international standards, foreign advanced standards and recognized authority standards, and according to the relevant laws and regulations of the government and the international standards, the national industry standards, norms for inspection.

② Contractor and subcontractor. According to the project company's personnel and the specific situation of the project, select sub-project engineering as the external contractor and subcontractor to ensure engineering quality.

(a) Guangdong Dapeng LNG project is a pilot project. BP was introduced as a partner with the main management of the construction of the receiving terminal. The receiving terminal and jetty were EPC contracted by the international contractor, so the technical risk was small, and the owner's coordinate interface was small. And it can meet the requirements of SPA effective conditions. The owner is easier to control the cost and risk. It is more in line with the requirements of the pilot project.

(b) Fujian LNG receiving terminal sub-project was adopt in the way of EPC contractor containing the subcontract works. Tank engineering was separately bid in advance, in order to meet the needs of the project critical path. For domestic familiar engineering construction (e.g., jetty, pipeline), it can give full play to the advantage of domestic to bid separately. Owners had certain control ability of every single sub project. As a result, it increased the work coordination interface and risk, but could fully exercise the team.

(c) Procurement. According to the equipment, materials, especially the key equipment and materials, which are involved in the sub project, the procurement forms should be stipulated in the contract. For example, the contractor is responsible for procurement operations, the project company is responsible for approval and payment (CFRE); the contractor is responsible for the whole process of procurement (CFE); or the project company is responsible for the procurement by their own (OFE); whether it is necessary to entrust a third-party inspection agency for quality control.

(d) Adoption of new technology and trial production. According to the project company facilities environment condition and the importance of safety, we should determine whether to adopt the trial products. Such as Shanghai LNG project, the new product and technology is not used in principle, unless it is approved by an expert review or inspection test, and is approved by government authorities or the recognized authority.

(e) Selection of the contractor/supplier. Firstly, through qualification examination to exclude contractors / suppliers without similar engineering performance, then we adopt the way of bidding and finally confirm the qualified contractors/suppliers after a rigorous evaluation process. Directly related to the quality of the project materials, equipment suppliers and engineering contractor, we should implement effective quality assurance system, comply with the corresponding requirements of ISO9000 series standards and certified qualified authority. We should actively cooperate with the government department's quality supervision, to ensure all aspects including occupational health, fire protection, safety and environmental certificates to be handled in time.

(f) Process supervision and inspection. In addition to the project company's own test, company entrusts supervision or a third-party inspection agency to carry out project supervision or inspection for the design, procurement, construction, commissioning, and production.

(g) Nonconforming products control. Through the inspection, testing and statistics to identify the presence of nonconforming products and propose targeted prevention and improvement measures.

(h) Quality audits. We should confirm the scope and content of the quality audit, which includes the system, procedures, process and methods audit. Through audit, we can achieve continuous improvement.

(3) The implementation, supervision inspection and modification of the quality management plan.

① In carrying out various quality activities, all departments should manage in strict accordance with the contents stipulated in the quality management plan, and timely couple the plan execution situation and existing problems back to the responsible department.

② The quality management department should organize relevant departments to review or validate the quality management plan implementation, and supervise the implementation and coordination of the plan.

③ In the implementation process of the quality management plan, when the objective factors changes and needs to be modified by the original preparation department, as appropriate, to change, after approval to re-issue.

8.3.2.4 Document Quality Management

(1) Project file content.

It is formed in the process of project construction, at least consisting of standards, norms, recording proofs, drawings, reports, regulatory documents with conservation value. According to the filing system, archive the data, including paper documents, electronic documents, kind, audio and video files and other.

(2) Document management principles.

① The principle of centralized and unified management of documents. Quality management department is responsible for document quality control. Part-time archivists of every department are responsible for implementation.

② Technology Department is responsible for offering style and format that are applicable to technical documentation and records, and to submit to Quality Management Department for summarizing.

③ Procurement and engineering contract is by a complete and accurate description of the procurement project or to provide engineering services, in a unified style and format of the document. Business Department is responsible for the business documents and records required to report Quality Management Department for summarizing.

④ Quality Management Department is responsible for referring to the relevant construction projects to archive requirements, acceptance specification and recording proofs, summarizing company responsible departments of documents and records requirements, compiling project document management requirements, and confirming unified company file format, content and coding.

⑤ Each department should strictly enforce the company's relevant documents management regulations in the release of documents and records.

⑥ Each department should promptly transfer to company's Archives Management Department when received important documents. The department should select, classify and file according to archiving requirements.

⑦ Specific document identification, storage, protection, retrieval, retention time and disposition are performed in accordance with "Company File Management Approach".

⑧ Based on the engineering project implementation progress, Engineering Department should supervise, urge and carry out the collection of documents with project construction synchronously. According to project stage, promptly hand over various types of files that should be archived to file management department. Accomplish files acceptance in project completion acceptance.

(3) File management structure.

① Relevant departments dynamically manage the documents, preserve one original file (paper media), and establish the appropriate classification folders.

② Engineering technical department will receive the design results, technical information and other documents, sort according to unified data, stamp with approval seal and issue to the relevant departments of the company, contractors, supervision and government quality supervision departments.

③ Business Department should compile, review, approve, publish and archive the procurement and construction contract files according to normal programming.

④ The contractor shall provide the equipment, materials, works or services to match the quality, technical, contractual and government documents, as well as any other additional records and certificates.

(4) Document sorting principle.

Following the rules of the formation of documents, it needs to maintain the organic link between the documents, so that the project file can be complete, system, and accurate. The organic connection file include: the source, content, time, languages and professional, file types, project, stage.

(5) File archiving.

After the completion of the project, the completion of the contract and the relevant documents is required to be archived by the engineering, and technical departments organize the collation, fill in the file directory. The Quality Management Department organizes the relevant personnel in accordance with the requirements of the state and the company file requirements for review, after the inspection and acceptance, and Archive Management Department archives the documents.

8.4 LNG Project Quality Assurances

8.4.1 The Concept and Basis of LNG Project Quality Assurances

8.4.1.1 The Concept of Project Quality Assurances

Quality assurance means that all planned and systematic activities to ensure the operation of LNG projects, construction of facilities and natural gas products provided to users to meet the quality requirements.

Quality assurance and quality continuous improvement is a progressive relationship. The former is to achieve initial quality indicators, and the latter is more emphasis on improving the original basis. Quality continuous improvement is designed to provide more benefits to the organization and its customers, throughout the improvement activities undertaken in the organization to bring significant effectiveness and efficiency.

8.4.1.2 The Basis of Project Quality Assurance

(1) Project quality plan.

Quality management plan shows how to implement LNG project quality management policy and objectives, and to take specific methods and procedures. On this basis, quality assurance measures are put forward.

(2) Measurement results of project quality.

The discovery of any quality problems must be based on the results of measurement, experiment and evaluation, so we should establish the quality system in the process of project operation, and provide basic data for quality improvement.

(3) Work description of project quality.

Whether LNG project is either hard task or soft task, there must be work guidelines and operating manuals prepared in advance in order to guide its expanded work. Through the implementation, we can find what is applicable and feasible, and what needs to be improved.

(4) Implementation of change results.

When the quality problem is found, it needs to propose to change the plan immediately, including the changes in management process, working procedure, methods, detection method, measurement frequency, and the analysis method.

8.4.2 LNG Project Quality Assurance Work Contents

Project quality assurance work mainly has the following aspects: the system, the resources, the process and the cost assurance. It also includes the quality improvement activities of the continuous development and the full control of the project change.

8.4.2.1 System Assurance

System assurance includes the effectiveness of the entire project's quality management framework, project company's internal quality management system and laws / regulations / standards / specifications of file system.

(1) The effectiveness of the project quality management framework.

① Components of the project quality management framework. LNG project is an energy based industry, the local government usually pay more attention to it, and it will be listed as key project. All levels of government departments are responsible for quality management. Project company should take the initiative to promote and introduce it, so that government departments can understand the project contents. The local government shall decompose the responsibilities to specific departments which include quality supervision departments, engineering construction department, port and shipping departments, and match with LNG project quality management organization structure. To establish normal communication channels, and to clear the project quality responsibilities is the primary task of the project quality assurance.

② To establish normal engineering quality reporting system. After conneeting with the quality supervision with local governments, the Quality Management Department of the project company shall take the initiative to communicate with the relevant departments to establish a normal quality reporting system, and regularly or not regularly accept the relevant departments of the project quality inspection and supervision.

(2) Internal quality management system of project company.

① To maintain the integrity of the project company's internal quality management system. Generally, the project company's internal quality management system is established before the project start. The initial system may differ from the actual quality management. Moreover, with the progress of the project, the original system does not adapt to the place, and even the emergence of missing some contents, which requires the project company to adjust and strengthen internal quality management system in accordance with the situation charging timely, to make it meet the requirements of the developing situation.

② To avoid the absence of quality management links and interfaces. Project company is the subject of quality management. General manager of the project is the first responsible person of quality management. Quality Control Department is the operating entity of quality management. In terms of LNG project quality management, it involves the seven units and departments mentioned in Figure. 8-2. To maintain the quality management of these units is not dislocation, but also to guarantee the quality management of the project company's internal interfaces is clear, and do not appear quality management of gray and fuzzy zone.

(3) Laws / regulations / standards / specifications and other file system.

① To implement national laws and regulations without deviation. Implementation of LNG project is related to national energy strategy and planning, laws and regulations of environmental protection and public safety, and the project such as regulations and norms of project construction of jetty, storage tanks and pipelines. The project company's quality management departments should collect laws and regulations of national, ministries and local government seriously before the project starts, and unswervingly implement during the project implementation process.

② To keep the connection between foreign and domestic standards. Project company should investigate the domestic existing engineering design and construction standards of jetty, storage tanks and pipeline, collect and organize the foreign LNG standards that have not been formulated in domestic, choose foreign standards which are suitable for China with domestic technology and market demand, especially LNG health safety and environmental standards, and form a complete project standards system to ensure the smooth progress of the project engineering design, construction and acceptance.

③ To formulate enterprise standards timely. Enterprise standards are usefully supplements to the international standards, national standards and industrial standards. In accordance with the practice of a few CNOOC LNG projects, the establishment of enterprise standards is crucial for project to promote. Because of its simple procedure, strong pertinence, flexible and easy to modify, enterprise standards become an essential element of project management. Enterprise standards are focus on the management standards and guidelines in the early project phase and later acceptance.

8.4.2.2 Resources Assurance

To ensure the quality of the project, the amount of investment and qualified resources are the first. In terms of the LNG project, the resources are mainly five kinds of human, financial, material, information and time. The following will mainly explain the quality requirements of the invested resources.

(1) Human resources.

① High quality leaders, decision makers, organizers and commanders. To ensure the high quality of LNG projects, the primary factor is the need for a number of high-quality project sponsors, director level of the project company, the leadership of the project company, the design and construction contractor and senior decision-makers of supervision company that directly participate in the project. High quality refers to both the strategic vision of the LNG project and the macro-control of the direction of investment; it includes both the strategy capabilities of overall operation and the decisive decisions when facing a major crisis. It specifically refers to the selection and decision that are taken by LNG project initiator when facing project investment opportunities, major decisions that are taken by project company's board member during the project operation, the management and control that are taken by general manager of the project company and its leadership during the various stages of project, and the overall grasp and command that are taken by the executives of contractors and supervision company on the properties and processes of the project.

② Pragmatic middle commanders. To ensure high quality of the LNG project, in addition to the high level of command and decision making, the middle-level commanders cannot be ignored. This mainly refers to the managers of the various departments LNG project company, design / construction contractors involved in the project and supervision company's middle leadership. They should understand both technology and management, and possess both strong executive power and pragmatic professional spirit. At every stage or process of project operation, they command and lead the team, always put work and project quality in an important position, and implement the quality policy, objectives, strategies and plans truly.

③ Hard working staff. To ensure LNG project quality, genuine performers and operators are the key factor. It includes the project company's general technical and management personnel, design and construction contractors that are involved in the project and front-line workers and staff of supervision company. The project company's technical and management personnel should be proficient in business and technology, and also with a spirit of ownership, play a role in communication, coordination and lubricants during the operation of the project. Contractor team's general management staff and workers need to have not only technical quality, management quality, service attitude and reputation in their own work, but also a certificate of special qualification jobs, technical jobs, working at height. The general supervisors of supervision companies need to hold not only post qualification certificates, but also a strong sense of responsibility, and it is necessary to be familiar with state laws and regulations and the requirements of the standard specifications.

④ Other intellectual introducing. LNG project is a new industry that is introduced into China. We are unfamiliar with both in technical and management. Therefore, in accordance with the overall level of the project company, we should introduce external intelligence, including special technical consultants or LNG resources negotiating consultants.

(2) Funds.

Project funds on schedule and in full place are the basic conditions for the project schedule, quality assurance and the production on time. From the source and composition of the project investment, there are the following two aspects.

① Shareholders' own funds. This is the first batch of funds invested in preliminary study of the project and the start of the project. It is mainly invested by the project sponsors. The sponsors that have good economic benefit and funds accumulation is key factor. For example, the upstream industry of CNOOC maintains rapid growth of economic benefit year after year; its own financial strength is strong, which lays the foundation for LNG project investment.

② Bank loans. To obtain bank loans, the first condition is that the project company's investors have a good reputation to repay, and the second is a good investment project. LNG projects are the basic energy projects supported by state. It can improve China's energy structure, protect the environment, and promote local economic growth and social harmony. Currently, LNG project financing has become a hot investment project that major investment banks eager to participate.

(3) Equipment and materials.

LNG project involves jetty, receiving terminal, storage tanks, and gasification devices. A lot of materials rely on imports to solve, such as nickel 9 steel and perlite in storage tank, as well as cryogenic pumps and valves, and so on. To ensure the quality requirements, it needs for the following measures.

① Test standards of materials and equipment. According to the characteristics of its purposes, it needs to clearly define the suppliers or the manufacturers to provide performance standards and quality standards that meet the requirements of owners in the procurement contract, so that the owners or a third-party can take inspection.

② Test methods of materials and equipment. It includes written examination, visual inspection, physical inspection, and nondestructive inspection. According to the specific requirements of materials and equipment, it also needs to determine whether to take the spot inspections, all inspections or overall test machine inspections.

③ Special equipment inspections. Some special equipment also requires manufacturers, owners and contractors in the construction site to check out of the box. Manufacturers provide installation, debugging and commissioning.

(4) Time.

Every project is completed within a certain range of time and space. Time assurance includes the following.

① Normal project schedule arrangement. LNG project generally includes several stages: early project general market research, feasibility studies, LNG resources implementation, project approval, project engineering design, construction, acceptance and trial production. These stages are interlocking and insurmountable. If the market is not implemented, we cannot determine the scale of the project. If the resources are not implemented, the project would not be approved by the state. The project company formation and personnel applications should be carried out with the progress of the project. Special attention should be paid to the "four connections and one leveling" project started, in order to avoid the project personnel idling and engineering pause.

② Climate window selection. Specific LNG project implementation should combine with the local climate windows. Particularly, the coastal area of China is the region with much typhoon, rain and high temperature in summer. It should combine with the jetty and receiving terminal's location, avoid the influence caused by typhoon and flow and waves, and reduce the construction downtime and damage caused by climate window.

(5) Information.

① To establish project information communication channels. Project information mainly includes technical information and management information. In the project implementation process, in addition to the project company itself, it also needs to ensure other participants, such as contractors, supervision companies, suppliers, users, government departments' right to know, and timely transfer project information.

② Information sharing. At present, it has entered information society period. In order to deliver LNG project information timely, we can take multi-channels and means, such as the normal monthly report,, annual report, e-mail, LNG project company website, and the regularly or irregularly reporting meeting, inspections, special assessment and other forms, to feedback and share information, and avoid one-way information flow.

8.4.2.3 Process Assurance

(1) LNG project scope composition between the various links connection.

LNG project scope, involving many aspects of the upstream and downstream, from the project decision-making, user's market, LNG resources, feasibility studies, FEED design, preliminary design, detailed design, project construction, project acceptance. Every aspect of the work is rich in contents, the challenge is strong, and the link between them have certain logic relation and impassable. For example, only in the case of market fixed in advance, can be signed with buyer of the purchase and sale of LNG resources contract. Project company should according to the project scope, reasonably arrange the time and manpower, so that to make the front and rear links orderly.

(2) Process control between hard and soft tasks.

As mentioned above, LNG projects include both invisible soft tasks and tangible hard tasks. Each hard task connects and coordinates with each other through the soft tasks that are between them. Such as pipeline construction, it must involve the implement of organizations agencies and formulating rules and regulations, and also the process of each operation, such as cleaning up—trench excavation—welding and inspection—site anticorrosion—pipe laying—trench backfilling—landforms recovery. Finally, it needs to compile the acceptance manual and hold experts assessment meeting, and so on.

(3) Coordination among sub-projects construction.

From the view of project construction, there are tanks, receiving terminal, pipeline, jetty and public sub-projects. Among the sub-projects, there are the space-time distributions in construction area, the connection of process technology, as well as the entire system's integrity coordination. Good coordination among the sub-projects construction is the key to success.

(4) Classification and combination of sub-projects.

For each sub-project, it is composed by each unit works. To do a good job of the combination of these unit works, according to the time and space and process technology, carry out the resources investment, and seize the process control. It is no doubt that all of those can ensure the sub-projects' quality.

(5) Working procedure operation.

The daily work of staff is the basic work of engineering construction, and it is the quality

assurance of the source, as the basis for quality process management is to have each kind of work operating procedure, to guide the workers and technical personnel to operate, according to the operating procedure, in order to lay the foundation of quality building.

8.4.2.4 Quality Costs

(1) Definition of quality costs.

Quality costs refer to the cost of all the work carried out in the LNG project construction and service to reached quality, as well as the gas quality requirements, including the costs incurred in maintaining the quality of the existing provisions and the costs incurred in taking remedial measures when the quality standards have not reached the level.

(2) Input of quality costs.

LNG project fund investments have many classified subjects. In order to ensure project quality for quality management, control and continuous quality improvement and other work, it must be put into a special cost, and then quality costs become the guarantee to ensure project quality. LNG project Quality Management Department should firstly compile the quality cost estimation; secondly ensure the special money to be used in special area, to allow the costs of quality to generate greater economic benefits. It is an important part of the project management costing. Investment of quality costs will generate long-term economic benefits.

8.4.3 Methods and Tools of Project Quality Assurance

8.4.3.1 Cost / benefit analysis

It is also called the quality economic analysis, which needs to consider the economics of the project quality when formulating the project quality plan.

Quality management of any project needs two work aspects. One is the quality assurance work, and the second is the quality inspection and remedial work. The former produces the project quality assurance cost, and the later produces the project quality inspection and correction cost. Project quality assurance is the same as quality plans compilation, and must firstly consider the balance of the cost / benefit. Adopting cost / income method is to arrange these two quality costs of the project reasonably, so that the total costs of project quality are relatively low. The difference between LNG projects and other projects providing products is that unqualified products can be replaced by standard products to compensate. While the LNG project must control the construction process to meet the quality requirements of all sub-projects, as well as to provide safe, reliable, and standard natural gas. Once the LNG resources supply contract is signed, it should be strictly enforced for 20 to 25 years.

8.4.3.2 Quality Benchmark Method

It refers to the use of other LNG projects in the actual or planned project quality management results or plans, as the project quality target, through the comparison of the results of the project to develop a quality plan and guarantee method. Such as Guangdong Dapeng LNG project and Fujian LNG project, quality plan and guarantee measures can provide benchmark for other LNG project quality management.

8.4.3.3 Quality Audit

Quality audit, especially the quality audit independent of the project's takeholders, the goal is to identify that the LNG project is the use of low efficiency, low efficiency of policies, processes and procedures, by the project Quality Management Department in accordance with the requirements of the general manager, to be used in the review of application for change, corrective action plan approval, demonstration of defect and remedy measures.

8.4.3.4 Experimental Design Method

The use of experimental design information is an analysis technique based on statistic. It helps to identify what affects the project quality largest in variety of variables, and then identify the key factors of project quality to guide the project quality plans and ensure the compilation of measures.

8.5 LNG Project Quality Controls

8.5.1 The Concept and Basis of LNG Project Quality Controls

8.5.1.1 The Concept of Project Quality Controls

Project quality control refers to the process supervision and management in order to ensure that the natural gas, process or service provided by the LNG project meets the prescribed quality requirements, which includes advance control, in-event control and ex-post control.

8.5.1.2 The Principles of Project Quality Control

Based on the LNG project quality objectives and requirements, the principles of quality control includes the supervision, inspection, statistics, analysis of the quality of project process, to find the deviation and feedback timely, take corrective measures so that the results achieve the objectives, and summarize and improve continuously to make the entire LNG project quality meet standards.

(1) To establish quality standard system.

If there is no standard of hard tasks, either no requirement of soft tasks, you cannot carry out quality control on LNG sub-projects, and cannot judge on the advantages and disadvantages of the process. Therefore, the establishment of standards system is the basis for project quality control.

(2) To execute testing procedures.

If the establishment of the standard system is the basis, the implementation of testing procedures is the means. Through testing, statistics, analysis, find the quality problems and timely feedback to the relevant departments and personnel, and provide information for correction.

(3) To correct quality deviations.

Through comparing test results or statistical analysis results with standards and requirements, we should take measures timely when find problems. After changing processes, work procedures, materials, work parts, and so on, we can finally meet the quality indicators.

(4) To summarize and improve continuously.

Meeting the quality requirements by practice is only the preliminary results. We are required summarizing the quality tests and corrective process timely, so that on one side to meet the needs of this project, promote actively and expand the quality assurance results, on the other side to gain experience for new projects and lay the foundation for the continuous quality improvement.

8.5.2 Methods and Tools of LNG Project Quality Control

8.5.2.1 Common Methods and Tools

(1) Pareto chart method.

Pareto chart is known as Permutation chart (Figure 8-10), by which we can find the factors affecting the primary and secondary of quality. Pareto chart is composed of two vertical axes, a horizontal axis, several rectangles and a summation curve. Histogram is plotted according to the order of occurrence frequency of the statistical factors. We should pay special attention to those main factors affecting quality and propose corrective measures. In most cases, 80% of the quality problems are caused by 20% of the causes.

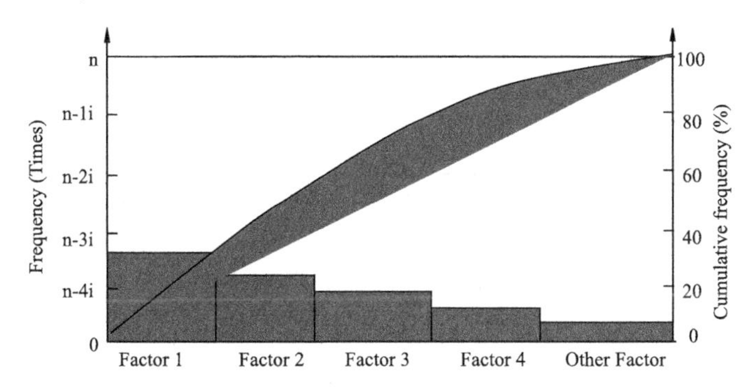

Figure 8-10 Pareto chart

(2) Control chart method.

Basic pattern of control chart is shown in Figure 8-11. There are generally three lines. The top line is called upper control line that is represented by UCL; the line at the bottom is called lower control line that is represented by LCL; the line in the middle is called center line that is represented by CL; at the top and bottom also set the requirement upper limit (UL) and requirement lower limit (LL). In quality control sampling process, the measured data points are marked on the map. If all the points are within the control limits and no abnormal, it indicates that the quality meets the requirements; if a points outside the upper control line or the lower control line, it indicates that the quality is abnormal, and we need to take measures to remedy.

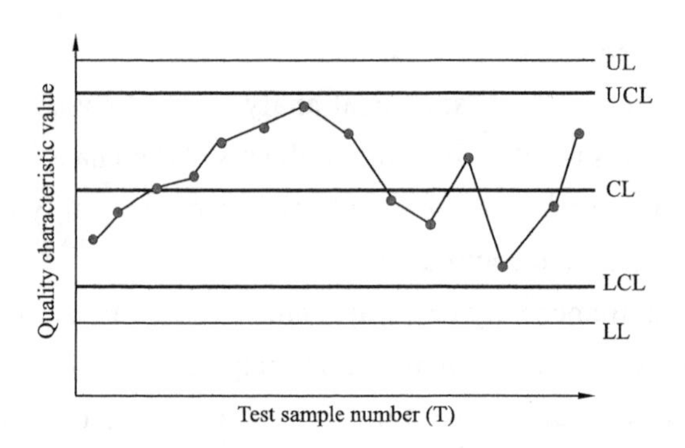

Figure 8-11 Quality control diagram

(3) Flowchart method.

Flowchart is used for process analysis to identify the causes of quality problems, the project process parts in which the control project quality problems occur, and the development and forming processes of the problems. Guangdong Dapeng LNG project pipeline network project quality management adopted the following quality control and management processes (Figure 8-12).

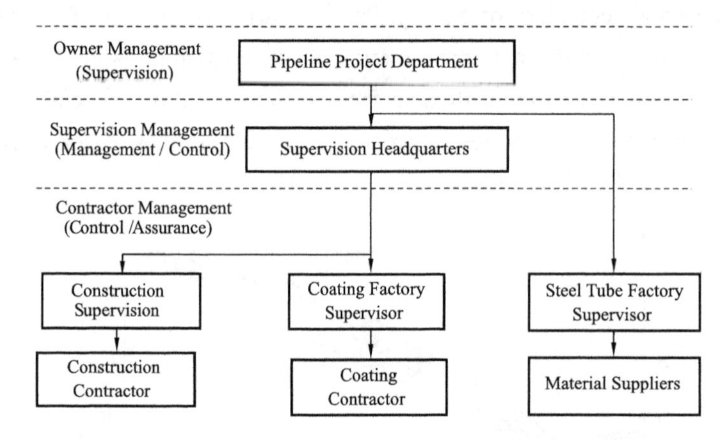

Figure 8-12 Quality control and management

(4) Trend analysis.

Trend chart is a line graph plotted in the form of dots according to the order of the data occurs to discover the history and law of quality deviation. Trend analysis uses mathematical tools to predict future quality changes in accordance with the historical results.

(5) Scatter point analysis.

Scatter plot is used to analyze whether there is a causal relationship between two variables. Using the rectangular coordinate system, it generally shows the reasons as abscissa, and results as ordinate. From the point of view of the scatter plot, the two variables are positive correlation, negative correlation, similar positive correlation, similar negative correlation, nonlinear correlation, irrelevant relation and so on. Therefore, we can take proper measures to solve the quality problems.

(6) Causality diagram analysis.

Causality diagram is called fishbone chart, which is composed by a number of branches. Branches are divided into large branches, middle branches, small branches and thin branches.

They represent different levels of reasons respectively. Take Guangdong Dapeng pipe welding quality causality diagram (Figure 8-13) as example.

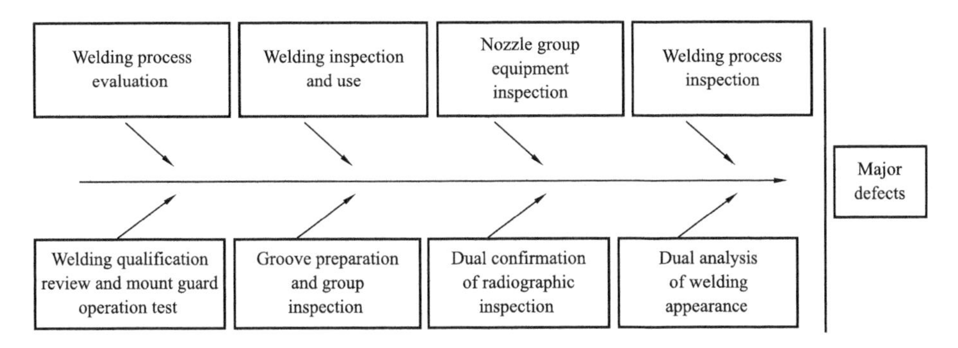

Figure 8-13 Quality control causality diagram

(7) Histogram analysis.

The histogram is a more common chart analysis. According to the appearing frequency of certain properties, the statistical data are shown by means cylinders of different heights in the rectangular coordinate system. By analyzing the distribution of these cylinders, it may be the normal type, positive skewness type, negative skewness type, insolated island type, diauxie type, zigzag type. Take different measures according to the quality requirements.

The above 7 methods are referred to as 7 quality tools.

8.5.2.2 Other Methods and Tools

(1) Checklist method.

According to the check items, prepare the check list in advance for inspection. Checklist is a unique structured quality control method in project quality control.

(2) Quality Inspection method.

Quality inspection refers to the quality control used to ensure work results to be consistent with quality requirements, such as measurement, inspection and test.

(3) Statistical sampling method.

It means selecting a certain number of samples to check to infer the overall quality, and acquire the quality information and carry out quality control.

8.5.3 Process Quality Control of LNG Project

8.5.3.1 Survey and Design Quality Control

(1) Contract signature.

Generally, engineering and control management departments draws up the "design basis", technical specification or contract according to the applicable LNG engineering design of laws, regulations, standards and norms. The survey and design results content, process control requirements should be made clear in work scope. Through the bidding process, the departments determine the survey and design project contractor, make survey and design management procedure and clear among the sub-project interface, and are responsible for organizing and processing the conflict of different sub-projects due to the survey and design issues. The departments should supervise contractor to rationally allocate qualified survey and design personnel in accordance with contract requirements. Major sub-projects survey and design issues shall be discussed and determined by the Quality Committee, expert group and the contractor.

(2) Tracking process.

Engineering and control management departments are responsible for organizing the implementation of the survey and design quality control plan, the survey and design process for tracking management, allocating professionals to supervise the entire engineering survey and design process, and timely feeding back the work report to the relevant departments of the company.

(3) Results review.

Survey and design results shall be reviewed in accordance with the relevant national and industry technical regulations and specification or by experts organized by government departments, or through entrusted third party (including supervision) when necessary. The contents of the review of the previous survey and design of the project include project feasibility study, special subject research. Preliminary design technical documents and drawings review includes: whether engineering technical solutions meet the design depth requirements, the predetermined quality standards at the project decision-making stage. Construction design review includes equipment, facilities, buildings, pipelines and other engineering object size, layout, material selection, structure, interrelation, construction and installation quality requirements of the detailed drawings and instructions, to ensure the use of functions to meet the quality objectives and level. Without examination or failing to pass the examination, the construction drawing design documents are not allowed to use.

8.5.3.2 Quality Control of Procurement Plan, Delivery and Acceptance

According to the requirements, there are three acquisition ways: contractor furnished reimbursable equipment (CFRE), contractor furnished equipment (CFE) and owner furnished equipment (OFE).

(1) The purchase plan.

Company Purchase Department or Business Department is responsible for drawing up the quality control plan of purchase and acceptance. From CFRE to CFE, the contractor / supplier submits and implements the procurement and acceptance of the quality control procedures, and the department is responsible for organizing the relevant departments to carry out the plan approval. But by the OFE, the department organizes relevant departments to prepare procurement and acceptance of quality control procedures, and makes clear rules for technical parameters, expediting, acceptance, transportation of equipment and materials, in order to ensure the quality of goods purchased in line with the requirements.

(2) Supplier qualification requirements.

The suppliers are only from those who have the ability to meet the contract, technical and

quality requirements, after approval procedures, and the general manager of the company approved the "supplier list" in the selection of equipment and materials. For equipment and materials with high importance level, in the contract or purchase order signed by the company before the responsible departments organize relevant departments to evaluate the qualification pre-selected suppliers, and conduct due diligence on the supply of plant facilities or production process and quality inspection procedures when necessary. The findings should form a report as part of the evaluation of the contractor / supplier.

(3) Expediting.

Engineering Department formulates the equipment and materials expediting plan. Purchase Department organizes the implementation and regularly prepares the status report to inform the relevant departments. The report shall at least indicates the date of delivery of the contract, the latest delivery date of the supplier's commitment, and the status of the current equipment (including the manufacturing, transportation and customs clearance status), site need date, actual arrival date.

(4) Inspection before delivery.

For important equipment and materials, the company should assign the inspection team to witness the test before leaving the factory, if necessary, to invite third party expert assistance. Purchase Department is generally responsible for inspection organization and coordination. The inspection team shall, in accordance with the specific requirements of the purchase package, consist of the appropriate personnel authorized by the company. The inspection team shall prepare the acceptance report after the completion of the inspection work, which shall be signed by all the members of the inspection team and filed for the management requirements.

(5) Acceptance.

For the foreign procurement of important equipment and materials which belong to the national compulsory inspection items, the company's responsible departments organize commodity inspection applications, coordination and other work after entering the customs. For OFE, the company's responsible department should be responsible for customs clearance and organize relevant departments and supervision company for acceptance, if necessary, entrust the third party independent agency for inspection. For CFRE, the contractor is responsible for customs clearance, and the company's responsible department organizes relevant departments and supervision company for acceptance, if necessary, entrust the third party independent agency for inspection. For CFE, the contractor according to the arrival of notice, in advance informs the company's responsible department, based to the equipment and materials procurement of the important degree to decide whether to participate in the acceptance. If it needs to participate in the inspection, the company's responsible department organizes other departments for inspection, when necessary, entrust third party independent inspection agency for inspection.

Inspection should be carefully checked. The receipt of the goods are consistent with the order requirements, inspection and verification of equipment, and materials shall be carried out in accordance with the requirements of the contract. The inspection report shall be submitted after the completion of inspection. In the acceptance of any discrepancies with the purchase order, such as the number of inadequate, the appearance of damage, according to the relevant provisions of the deal, the company's responsibility department should prepare a special report submitted to the company. For important equipment and materials manufacturing process, the company's responsible departments shall organize the relevant departments or commissioned third party inspection agency to witness the inspection process.

(6) LNG component quality assurance.

Company Resources and Transportation Department centralizedly manages for LNG procurement and transportation plan/scheduling, by the department in accordance with the user group of LNG component requirements, signs LNG resource purchase and sale contract and transport contract with resource supplier and transportation partner. When necessary, take third party to do the LNG component analysis, in order to ensure the quality requirements of LNG.

(7) Archiving.

Company's responsible department organizes the relevant departments for acceptance of purchase equipment and materials quality records, factory certificate and related documents, and archives according to the project company archives management regulations.

8.5.3.3 Manufacture and Construction Quality Control

(1) General practice.

Manufacturing and construction are the content of the implementation stage, and they are also the most important process of project quality. The implementation stage of the quality control according to the assessment and acceptance scope planning, work breakdown and quality planning for every single step, and then according to the results of the inspection process quality of each individual (such as the quality level of individual processes into unqualified, qualified, good, excellent grades, 4 grades) summary statistics, results in the formation of quality the procedure (qualified rate or excellent rate), and so on, and ultimately to form all the project quality inspection results.

(2) Contractor construction organization management.

① The contractor shall design the construction organization at the beginning of all manufacturing, construction and installation work, and the contents include at least:

(a) The work scope description;

(b) Project management organization;

(c) Management responsibilities;

(d) The quality objectives, receive and reject standards;

(e) Manufacturing, construction or installation method (should be consistent with the provisions of the technical and quality requirements);

(f) Job order, work plan;

(g) Human—histogram;

(h) The use of machinery and equipment (including test equipment), certificates and technical performance;

(i) Any special working environment requirements;

(j) Test requirements;

(k) The special construction process;

(l) Identification requirements;

(m) Handling, storage and transportation.

② Before the start of the construction project, the Engineering Department of Project Company shall organize contractor and supervision company for technical clarification; Quality Planning Department shall prepare manufacturing engineering and construction quality control plan in advance and supervise contractor in accordance with the approved construction organization design documents and drawings of construction work.

③ Construction drawings, technical specifications and other technical documents shall obtain technical management departments for approval; the contractor's project construction organization design, quality control plan, the relevant procedures shall obtain engineering department approval. The above documents require the contractor according to characteristics of the project preparation, to ensure the effective control of project management.

④ The quality management department carries on the control to the construction process by the principle of daily quality management, supervision and inspection, the inspection of the third party and the contractor's self-inspection. It also organizes site representative of engineering department and supervision company to conduct a monthly on-site overall quality check, supervises the company's responsible department, supervision company and construction unit for quality management situation, participates in the key process of key engineering, the implementation of quality assurance measures to supervise the implementation of the project, and promotes the implementation of quality assurance measures.

(3) Inspection.

The professional qualification certificate personnel in accordance with the basic design, specifications, adopted specification requirements, the use of suitable equipment, measuring tools, in the appropriate environmental conditions to carry out inspection. The hold points specified in inspection documents, without the consent by the quality supervision department, can't be carried out down next process. The provisions of the certification and inspection institutions hold points, prior to the written approval of the certification inspection agency, also cannot continue.

Inspection results should be formed records to demonstrate the required test projects completed, and by the authorized qualified release personnel to determine whether to accept or not; after inspection, the rework, repair and replacement, should be inspected in accordance with the initial inspection requirements, and review criteria can't be lower than the initial inspection standard; general inspection work is composed of the contractor's self-inspection and supervision company of the sampling inspection. Key and important work inspection consists of contractor self-test, supervision company re-inspection and the professional staff of the sampling.

(4) Special construction technology.

① Special construction process shall be subject to the professional training of qualified personnel, in accordance with the approved procedures, the use of suitable equipment, in a suitable environment to carry out inspection. Engineering Technology Department organizes relevant departments for examination and approval of implementation of procedures, equipment and personnel, special construction process in accordance with the applicable codes, standards, specifications and industrial practices, including at least as follows:

(a) The welding process and welding procedure assessment procedures;

(b) Welding repair procedures and repair inspection procedures;

(c) The welder training and acquisition certificate process;

(d) Welding equipment, welding rod oven and heat preservation cylinder control;

(e) Non destructive testing (NDT) procedure;

(f) The welder and NDT personnel qualification certificate;

(g) Design and calculation of concrete mix ratio;

(h) Concrete precast, pouring construction procedures;

(i) The piling operation procedure;

(j) The heat treatment program (if any);

(k) Flame deformation correction program.

② The contractor shall establish procedures, equipment and personnel qualification records. These records shall be submitted to the relevant departments of the project company. The Engineering Technical Department shall organize the relevant professional personnel to inspect the quality control of the special construction process.

(5) Identification.

① All materials, parts and components, including prefabricated parts, should always be controlled according to the following requirements:

(a) With the specification of identification and traceability requirements;

(b) It can quickly and accurately find the applicable drawings, specifications, quality records or other relevant documents;

(c) Only if it has been determined to be qualified, can it be used and installed.

② The identification method used can be directly in kind for marking (such as printing, label, permanent marking material identification), or using a traceable record card (such as cargo cards, labels). Pay attention to the direct in kind for marking:

(a) The identification method, position and nature of the object can't be harmful to the function or quality of the object;

(b) The logo should be clear;

(c) When the subsequent work may cause identification wear or blurred, it should be used to identify the transfer.

Engineering and Technical Departments through regular and irregular sampling, supervises the management of all kinds of identification.

(6) Handling, storage and transportation.

Purchase Department shall make provisions for packaging materials, transportation packaging, environmental conditions and handling equipment while signing the purchase contract. Materials can only be issued by authorized personnel, if materials with restricted storage period, when in the extended and not let out or use; materials, parts and components during storage, transportation should have packaging. The packing size should meet the requirements for the safe transport. On the package, it shall be marked by the location of the goods. Engineering and Technology Department supervises the construction site handling, storage and transportation management through regular and irregular sampling.

(7) Test equipment.

The Engineering Department and Control Department shall manage all test equipment of the contractor as required by the following requirements:

① Under the appropriate environmental conditions, qualified organization or personnel checks the test equipment with appropriate detection method using known accuracy to confirm whether it meets the requirements or not.

② All test equipment should be in accordance with the provisions of periodic maintenance and calibration by qualified institutions or personnel.

③ All test equipment shall be stored and kept under appropriate conditions.

④ Each equipment calibration or verification as requirements shall be used only as unique and permanent identification number.

⑤ The calibrator on the calibration device shall be protected safely, and no-unauthorized personnel contact is strictly forbidden.

When some problems of checkout equipment are found, the Engineering Department organizes qualified organization or personnel to evaluate the inspection and test records of the equipment, and to confirm whether the items are valid or not.

(8) Design change.

Design change is also applied to the design unit of the formal issue of the design document to be modified. The proposed changes shall be subject to the supervision company for review and get its signature and review opinion, and then submitted to the Engineering Department for approval (if necessary, government departments review and approve). When the change is related to the design basis and design principle, the Engineering Department should report the results of the assessment approved by the company's engineering and technical director and then inform the design unit. Change should be designed by the design unit of professional design personnel, specify the reasons for the design, the original design document number and name, to explain the specific details of the design change, to fulfill the design of the internal review procedures.

Design changes need to get the Engineering Department to receive, sort, classify and affix the special approval seal, and then issued, without permission. Design changes shall not be issued to the general contractor to implement, if the design changes are likely to affect the government department in charge of approval of the project, it should get the prior consent of the government department.

(9) Construction technology verification.

If the construction unit has doubt or need to put forward design alternatives, it can be proposed that the approved application for construction technology and approved by the design unit. The approved application shall be drafted by the general contractor's professional person in charge, explained the need for approved content, reasons, the number and name of design documents, signed by the general contractor project manager, through the project company Engineering Technology Department transfers to design units. The Engineering Department should supervise and urge the design unit in time to review and report, add the necessary design instructions or drawings. After receiving response from the design unit, the Engineering Department should organizes supervision company and company's relative departments to review the response, and put forward the opinion of examination and approval. If the design unit's response refers to the design changes, it should mention the necessity and operability of design changes, and through the relevant departments for approval (if necessary, government departments review and approve) and then reply to the general contractor. If the design unit does not involve the design changes, the approved application will be submitted directly to the general contractor. At the same time, report it to project company Engineering and Technical Departments for record.

(10) Examples of quality control.

Guangdong Dapeng project quality management system in strict accordance with the ISO9000 "LNG Quality Management Handbook" "Pipeline Project Quality Management Measures" "LNG owners Focus on the Quality of Links" and a total of 13 copies of quality control documents, issued a notice of about 50, timely and accurately handled the large quality problems, and resolutely put an end to possible quality problems. The following is the example of pipe quality control.

① Pipeline welding quality: the whole line of welding quality firstly pass rate of 98.86%, corrosion prevention and repair of a pass rate of 99.95% (Table 8–5).

② Civil engineering quality: through the division and section of the station, valve, hydraulic protection and other civil works of the comprehensive inspection and strict control of the material, all civil engineering quality are in line with the requirements of the standard.

③ The quality of the pipeline cleaning pressure test: the whole pipeline cleaning pressure test is in accordance with the standard, the pressure test is one time success.

④ The Pearl River crossing shield engineering: one time success.

8.5.3.4 Trial Production Operation

(1) General quality requirements for trial production.

The company sets up production management organization in the later stage of engineering construction, which is responsible for the production preparation and production management, in order to meet the needs of trial production operation and production, to ensure the project construction and production of convergence. The contractor shall, in accordance with the control

Table 8-5 Pipeline Welding Quality Statistics

Tenders / Construction Units	Total Welding Interface Number	Nondestructive Testing						Cutting Number	Repaired Mouth		
		Ultrasonic Testing (UT)			Radiographic Testing (RT)						
		Testing Number	Unqualified Number	Primary Qualification Rate (%)	Testing Number	Unqualified Number	Primary Qualification Rate (%)		Repaired Mouth Number	Qualified Number	Primary Qualification Rate (%)
1/A	17196	1260	34	97.30	16244	169	98.96	19	17196	17196	100.00
2/B	6087	2202	19	99.14	3986	74	98.14	12	6087	6087	100.00
3/C	9579	2965	28	99.06	6687	66	99.01	4	9499	9499	100.00
4/D	10056	2752	6	99.78	7404	31	99.58	0	10056	10052	99.96
5/E	9451	2400	5	99.79	7152	138	98.07	12	9451	9430	99.78
5/E	7609	1168	11	99.06	6545	46	99.30	2	7597	7594	99.96
5/E	6314	0	0		6314	51	99.19	0	6314	6314	100.00
6/F	11089	2109	19	99.10	9083	124	98.63	7	11078	11078	100.00
Total	107472	14856	122	99.18	63415	699	98.90	56	77278	77250	99.96

points sets in the project schedule, complete the construction on schedule, and the engineering quality shall comply with relevant standards, regulations and commissioning requirements. Commissioning should meet the requirements of the conditions. The commissioning file should include all the contents of the design document. The results should be approved by the higher authorities to check the qualified, to reach the required standards. After the approval of the commissioning plan, it start feeding test. When an accident or failure occurs in the process of debugging, it should immediately identify the causes, and take measures to be excluded, otherwise not allowed to continue debugging. Environmental protection engineering should be synchronized with the production device debugging. The treatment of waste water and waste gas discharged by each stage shall meet the emission standards stipulated by the state. The principle of "Safety first, prevention first" should be implemented, and the safety work should run through the whole process.

(2) Overall plan of trial production operation.

Prior to the installation of production equipment, the contractor shall submit the overall plan of the trial production operation to Engineering Department and the organization of production management for review. After approval by the general manager of the company, submit to higher authorities for approval. The overall plan shall be carried out, which shall at least include the following contents:

① General;

② The trial production organization;

③ personnel, goods and materials preparation and preparation planning of rules and regulations;

④ Preparation of various specific trial production operation plan;

⑤ External conditions for equilibrium and staging prediction of water, electricity, gas, raw material and external output;

⑥ The trial production plan and the overall plan chart;

⑦ HSE management plan, emergency plan, safety assessment report format;

⑧ The existing problems and solutions.

(3) Pre-commissioning includes at least the following system:

① The pipeline system and internal processing equipment;

② The electrical system debugging;

③ Instrument system debugging;

④ Single machine debugging.

After the commissioning and qualifiing, the relevant units sign and confirm it in the prescribed form.

(4) Debugging.

The following conditions should be available before commissioning:

① The contractor's project handover certificate has been signed;

② Project contractor has prepared the test plan, and approved by the competent authorities;

③ Production management organization has been established, the responsibility system has been clear, management personnel, operation and maintenance personnel through the examination qualified, has held a certificate in place;

④ The post responsibility system as the center of the rules and regulations, process regulations, safety regulations, electrical and mechanical instrument maintenance procedures, analytical procedures and post operation method and commissioning plan have been implemented;

⑤ Production management organization staff have accepted the security, fire education, production command; management personnel, operation and maintenance personnel by the examination of qualified, has received a safe operation certificate;

⑥ Machine, electric, instrument has been normal working; maintenance management system has been established;

⑦ Machine, electric and instrument maintenance team, during debugging by the contractor and production management organization composed on duty, has in place;

⑧ Automatic analysis instrument, laboratory analysis instrument has been tested and qualified; routine analysis of the standard solution has been prepared; analysis personnel have been in place;

⑨ Communication facilities within the production command, scheduling system and device are already open;

⑩ Safety, first aid, fire-fighting facilities have been prepared, and has confirmed that the sensitive and reliable and consistent with the provisions of the relevant safety;

⑪ Commissioning spare parts, tools, test instruments, repair materials have been prepared, and the normal management system has been established;

⑫ Debugging resources have been implemented; the first ship LNG has confirmed the arrival of the goods at the time, and can meet the needs of debugging;

⑬ Water, electricity, gas, have been able to ensure continuous and stable supply; the accident motor, uninterrupted power source, instrument automatic control system and protection system have been able for normal operation;

⑭ Lubricating oil consumption goods have been in accordance with the provisions of the design documents and specifications of the number of complete debugging programs, and to ensure continuous and stable supply;

⑮ Machinery, pipe insulation and anticorrosion work has been completed;

⑯ All roads, lighting can meet the needs of debugging;

⑰ The three waste treatment devices have been built; pre-commissioning is qualified, and has the condition to use;

⑱ Plant life and health facilities, security measures have been able to meet the needs of the commissioning work;

⑲ All kinds of measuring instruments have been calibrated and are in the period of validity;

⑳ Downstream markets are already available for receiving of natural gas.

(5) Trial production operation and operation adjustment.

After the competent department of higher authorities organizes the inspection and approval, trial production run and operation adjustment can be carry out, and timely eliminate the defects exposed in the test. All stages of the trial production and operation adjustment are qualified. The company and the contractor signed a certificate, to create conditions for normal production.

8.5.3.5 Project Handover Management

(1) Handover.

The contractor shall work with the item company to transfer the project after the commissioning of the equipment is qualified. The Engineering Department organizes the production management organization and the contractor joint inspection, inspection qualified inspection certificate issued by the project company confirmed the project. In the inspection process, it may find that equipment and systems do not meet the standard requirements or affect the safe operation of the required rectification. The Engineering Department shall organize the contractor, in accordance with the terms and conditions of the contract, to resolve them promptly.

(2) Internal handover.

Transfer the work between the Engineering Department and Production Management Organization, including the transfer of engineering facilities and the transfer of the relevant documents.

① The main contents of the transfer of engineering facilities include:

(a) The jetty, receiving terminal and gas pipeline engineering facilities;

(b) The equipment, spare parts, special tools and instruments;

(c) The remaining engineering bill of material and residual materials.

② Transfer of test papers. Major relevant documents required to be transferred include:

(a) The technical documents such as the outline of debugging, completion drawing;

(b) The inspection report and the certificate of quality;

(c) The operation / repair regulations;

(d) Emergency plan, guarantee measures.

8.5.3.6 Quality Event / Accident Investigation and Handling

(1) The general quality event / accident.

When the treatment measures are for rework or repair, inspection and testing shall be carried out again after the completion. When necessary, test items or frequency, to prove rework or repair quality. For general unqualified, the responsible departments organize relevant professionals to unqualified review, analyze the reasons, focus on the following contents.

① To prevent the unqualified again.

② To determine and approve the method and the degree of deviation of the relevant provisions.

③ Statistics due to not fully comply with the approved practices, procedures, or the relevant provisions or the relevant requirements to cause the unqualified.

(2) The major quality accident.

In the construction process, when major quality events / accidents happen, the relevant

departments should require contractor to suspend the construction, and report to the company's Quality Department.

① When major quality event / accident occurres in equipment supervision made process, supervision company should promptly issue instructions to suspend the manufacturing, and report to the Purchase Department, and the Quality Department.

② When major quality event / accident occurres in design process, Technical Department shall timely report to the Quality Department. When necessary, organize experts to review.

③ Contractor submits the written report content: quality event / accident survey (accident time, place, reason for the preliminary, economic losses), cause analysis, corrective action and the responsible units, and persons responsible for handling submissions.

④ Company related responsible departments organize survey, hold the accident analysis to determine the event / nature of the accident, put forward opinions on the handling of the report to the superior departments in charge and relevant departments and units responsible for the accident work.

⑤ Contractor protocols processing program to the company's responsible department for approval and implementation, and continues to work through the examination of the self-inspection and supervision, to meet the requirements.

⑥ On the quality accident hidden, false, or to delay the report, the responsible departments will be investigated.

⑦ For quality serious failure, the company's Quality Department organizes relevant departments to review and deal with, and submit a report to the company in a timely manner.

⑧ The responsible departments of the company supervise the contractor / supplier to implement the unqualified processing and validation.

⑨ Contractor shall compile non-conforming and corrective action report, record the unqualified the nature, scope, review the treatment measures and corrective action verification results and submit to Quality Department for the record. If the corrective action is related to the requirements of the certification inspection agency, the corrective procedure shall be approved by the certification and inspection agency before implementation.

8.5.3.7 Warranty Period Quality Management and Project Quality Management Assessment

(1) Warranty period quality management.

After the project is transferred, the Production Management Organization has become a site management organization, and the Engineering Department organizes the relevant departments to assist the Production Management Organization to complete the following main management contents:

① To handle due to quality problems caused by design flaws;

② To handle quality problems due to equipment supplier responsibility;

③ To assist the Production Management Organization to deal with quality problems caused by operation;

④ To organize and arrange contractors and equipment suppliers in the warranty period, to solve existing quality problems, especially in the repair work;

⑤ One month before the expiration of the warranty period, the Engineering Department shall submit the application of quality guarantee expiration date to the company, and the Production Management Organization shall organize the final production facilities inspection and approval application.

After the end of the warranty period, the Production Management Organization is responsible for device maintenance, repair and operation.

(2) Project quality management assessment.

The quality management assessment of the project is generally carried out one month after the completion of a single project. After the completion of the project, the Quality Management Department summarizes the process of quality control management, on the basis of reference to the assessment of the quality of the government departments. The relevant departments to make a reward and punishment recommendation. The vice general manager of the company in charge of quality, who organizes to prepare the individual project quality assessment report and submits to the company for approval. The general manager has the right to decide the special award for quality management. Responsible for the occurrence of quality incidents / accidents, it will be based on the severity of the problem to give the appropriate punishment, the specific penalties determined by the company's leadership. Do not report on the quality of incident / accident cheat, false, omission or intentional delay, in addition to instruct to make a supplementary report, the responsible departments and the person need to be given corresponding punishment.

8.5.4 LNG Project Quality Control Results

8.5.4.1 Improving Project Quality Management System

Through the whole process of LNG project quality management, it can be used to guide the LNG project quality management system, whether the system is complete and effective or not, or whether the modified system is perfect or not.

(1) Quality standards.

Only in the case of standards and norms, can it make the project hard tasks and soft tasks have the goal. Through a project of complete operation and put into production, it is an effective inspection of the prior use of standards and norms. LNG project comes from the use of international standards. Now by the efforts of LNG industry, national and industry standards have been prepared. CNOOC has also compiled a series of enterprise standards, which are the supplement to the national and industry standards, and will be conducive to the development of China's LNG industry.

(2) Process and regulations.

Similar to the standard system, there must be a set of complete quality management process-

es and regulations. And through the LNG operation of the project, find deficiencies of processes and regulations. Through practice, modify, practice again and modify again, to refine these processes and regulations which will be the project company valuable assets.

(3) Quality management organization system.

As mentioned above, Guangdong Dapeng and Fujian LNG projects, their quality management system is not the same. According to the project company organization, contractors, supervision company of quality management system and project contract and form and the different ways of making choices, through practice of several restructuring and modification, the companies increase the allocation of staff, and have their own characteristics of organizational structure. The organization structure of quality management will still have referential significance to other new projects.

8.5.4.2 Project Quality Control Measures

Quality control measures are based on the quality assurance process, the implementation of the organization's quality standards and processes for re-evaluation and analysis.

(1) The quality control plan.

On the basis of quality control plan, it will find the shortage of the original plan, through practice to amend, in order to achieve better quality control.

(2) The numerical amount of allocated resources.

Control quality requires resources, such as human, financial, material, time and information. Through the quality control, it will find out which resource input is not enough, which is surplus, and which is appropriate, in order to find the best balance point of the quality control work.

(3) The corrective and preventive measures.

Inevitably, this and that quality problems appear in the process of LNG project operation. Through practice, at any time to summarize effective corrective and preventive measures, such as pipeline welding quality is closely related to welder' technology, in order to prevent the emergence of welding quality problems. At the beginning, welder must be taken quality inspection and on-site welding examination, and the unqualified person must be reject from the posts.

8.5.4.3 Selection of Effective Quality Control Methods and Tools

(1) The best methods and tools.

Above introduced ten kinds of quality control tools, the methods and tools explored in practice are the most effective, so that both can save labor costs, but also to seize the core of quality control. The statistical classification should be taken for the best method in each checking item, and recorded for the follow-up project to provide reference.

(2) The best combination method.

Some quality inspection work often requires a combination of methods and tools, in order to carry out mutual verification, such as the evaluation of the construction team work quality is the highest, can take a full set of quality inspection and causality diagram method. The first is to find from the statistics of the whole set of the highest value, the lowest value and average value, the second is through causal analysis to find that the main factors affecting the quality.

(3) Comprehensive analysis.

After a number of methods and tools are examined, a set of data is obtained. Data processing and analysis is the way to find the cause of quality. Is it necessary to use the probability statistics, or to carry out the regression curve? Is it the computer to carry on the complex computation, or trend analysis? Only after the comprehensive analysis, can get the best analysis method.

Chapter 9 LNG Project HSE Management

9.1 General Concepts of the LNG Project HSE Management

9.1.1 Relevant Definitions

9.1.1.1 Overview of HSE Management

HSE is short for "Health, Safety and Environment". HSE management is commonly referred to health, safety and environmental management. Just as its name implies, it refers to the country, enterprises, social organizations even families and individuals on the health, safety and environment management and attention. At present, China's government attaches great importance to the HSE management, and regards the HSE management not only as a basic national policy, but also the basic conditions to maintain social stability and unity, promotes sustainable and healthy development. So the local government, enterprises and units will see HSE management as the focus of their daily work to catch.

In recent years, HSE management system in the international petroleum and natural gas industry in practice tend to mature. It sets the national peer management experience, highlighting prevention, leadership commitment, full participation, continuous improvement of the scientific management thought. Petroleum and natural gas industry realizes the modern management is the permit to the international market.

9.1.1.2 Definition of LNG Project HSE Management

LNG project HSE management lasts throughout the life circle of whole project. Project companies and the direct and indirect participation units should carry out the evaluation on environment, health and safety during different stages, including pre-research, pre-feasibility study, feasibility study, project design, construction, acceptance and trial production, in particular, the HSE plan, organization, implementation, control and improvement of the safety pre-evaluation, construction safety, contractor's safety, acceptance, and emergency safety process management.

9.1.2 The Significance and Characteristics of LNG Project HSE Management

9.1.2.1 The Significance of the LNG Project HSE Management

(1) The special significance of LNG project HSE management.

The previous project management books mostly emphasize the nine knowledge system, but rarely put HSE management from risk management to peel out. We will specially introduce

HSE management in this book, and use one chapter to elaborate the HSE management in LNG project, for the following reasons. Firstly, the LNG project has a long industrial chain, involved in many industries, and is closely related to social groups and natural environment, so the HSE management is indispensable. Secondly, mixed with air, LNG will turn into flammable substances after gasification, easy to burn and explode, and strengthening the HSE management is the basic guarantee for project acceptance and commissioning work. Thirdly, HSE management is as important as cost, schedule and quality management in the LNG project, called as the four controls of project management, and is one of the important content of the company project management of the LNG project. Fourthly, influenced by CNOOC enterprise culture, the author recognizes the importance of HSE management. Therefore, keeping the HSE management into an independent chapter is a feature of the book, and is also to supplement and perfect the theory of project management.

(2) The extensive sociality of the project needs to strengthen the HSE management.

LNG project belongs to the basic energy industry. LNG or gasified gas users including gas-fired power plants, industrial fuel industry, chemical industry, raw materials enterprises, city residents, vehicle fuel filling users, involving a broad social beneficial to the people's livelihood, and whether Guangdong Dapeng or Fujian LNG project, or Shanghai LNG project, benefits the local population of tens of millions. Naturally, HSE management of this project becomes more evident and important than the general project. For project companies, especially initiators, should always take the HSE management as a reflection of social responsibility.

(3) An environmental-friendly project is one of the goals of the project company.

LNG project is different from other construction project, that is, not exists in isolation, but involves many companies and factories from upstream to downstream of a complete industrial chain, at the same time, the LNG project covers large area, with long pipeline path, and having a lot of distribution stations and pressure stations. To carry out the LNG project, the first thing to consider is the occupation of land, coastline and sea area resources to build LNG receiving terminal and jetty, also the land of long distance pipeline route should be considered. Its users should also consider supporting the construction or transformation of the station and the occupation of land resources, etc. There is no doubt that the construction of the stations and the pipeline, are both involved with the environment-friendly and coordination problems. To build, not at the expense of the environment, and to develop, more not to take the environment damage as the prerequisite, HSE management can effectively prevent the damage to the environment, in order to ensure the friendship and coordination between the LNG project and environment.

(4) The extension of HSE culture of controlling shareholder.

Project company HSE management principles, ideas and policies are mostly influenced by the controlling shareholder's company. CNOOC, as the shareholders of China's first LNG project company, through the long-term cooperation with foreign companies, undertakes the upstream of oil and gas exploration and production, which has accumulated rich experience in the HSE management of upstream, and has formed a most important part of its own enterprise cul-

ture. HSE management principles, ideas and policies, as a part of the CNOOC strategic planning and management aim, naturally want to be reflected on their downstream industries; meanwhile, CNOOC cooperats with BP in Guangdong Dapeng project, and has learned the HSE management experience from BP. CNOOC's Fujian LNG project has carried out the CNOOC's HSE principles, ideas and policies, and has made certain experience. This chapter is a summary of the above LNG project HSE management. Starting from the LNG projects, summing up the HSE management model is an extension of the HSE culture of the controlling shareholder.

9.1.2.2 The Characteristics of the LNG Project HSE Management

(1) HSE management system is the foundation of project success.

Like the quality management system, establishment and perfection of HSE management system is the foundation of the LNG project success. From the perspective of China's existing LNG project operation, such as the Fujian and Zhejiang LNG project, at the beginning of the project company setting up, they had established HSE management system, and constantly improved it. From the human resources, capital investment, organization and management system, HSE management makes a comprehensive deployment. HSE management is always running through the project site selection, FEED design, preliminary design, detailed design, engineering construction and production acceptance. Such as Fujian LNG project, it created an excellent record of zero accident of the project HSE, which is the good example of other LNG project.

(2) Forming synergy is the premise of LNG project HSE management.

As a national key energy infrastructure project, LNG projects involve all aspects of the upstream, midstream and downstream, in addition to the project company HSE management itself, but we must accept the inspection and supervision of the state and local governments in HSE, to avoid major accidents. It is not only the national requirements, but also the local government hopes. Investment holding parties, such as CNOOC, will put HSE management into an important part of their enterprise culture, and focus on the midstream and downstream of the project HSE system management, in particular, strengthen the LNG project HSE management macro control. LNG project company also requires surveyors, designers, suppliers, construction contractors, in accordance with the project company's HSE management philosophy and procedures, in order to achieve the same level of project company HSE management. The project located in the community, so it is also needs to accept the community HSE management, integrated into the scope of its harmonious society; at the same time it should accept the supervision of the media, and insist on openness and transparency through the performance of HSE management, to gain the positive and objective publicity. Only the above parties to form a joint force in HSE management, can they achieve the goal of LNG project HSE.

(3) Taking HSE three parties into account is a means of project to reach the standard.

LNG project is different from the general product project or software project, these projects are only responsible for the quality of the product itself, no need for special management or less attention to HSE. The LNG project involves both terminal site and pipeline environmental issues. The project also involves the design, construction, inspection safety issues. LNG product is necessary to consider the health problems, and also should pay attention to safety and quality standards of natural gas after gasification. we should face on not only safety performance issues of materials and equipments, but also pay attention to the control, operation safety and health issues. Therefore, during the project management process, we must integrate the health, safety and environmental management, and can't neglect either, which is not only the need of the project, but also the need of the society.

(4) Staged HSE management is the way to accomplish the project objectives.

For all the stages of the LNG project, at the different stages of health, safety, environment, the key points of concern is not the same. In the early stage, we mainly focus on the environment protection, concerning jetties, terminal sites and pipelines, as well as optimized use of land, coastline and sea area. That is to say, we put priority on protecting environment and saving land resources. In the stage of designing, what we mainly consider is choosing materials and facilities, and determining safe distance. In the stage of construction, the safety of contractors and front-line staff are the most important. In the stage of receiving and production test, we should consider health, safety and environment as a whole, and validate whether the standards are reached.

9.1.3 LNG Projects and HSE Management Chain

As is shown in Figure 9-1, viewing in the whole aspect of industrial chain, LNG industry involves natural gas production, liquefaction, maritime transport, LNG receiving, land pipeline transportation and consumption of end users. The producers and operators in different parts of the chain are responsible for themselves. If LNG is imported from abroad, the Chinese LNG project company is responsible for the LNG receiving terminal and land pipeline transportation of these two aspects, and the users are responsible for their respective HSE management.

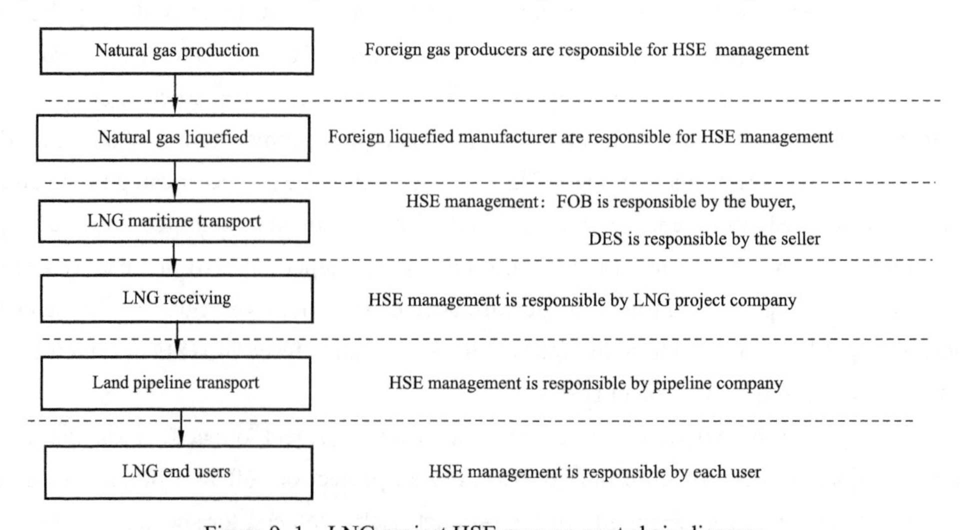

Figure 9-1 LNG project HSE management chain diagram

9.1.4 The Relationship between LNG Project HSE Management and International Organization HSE Management

9.1.4.1 In Line with International HSE Management Concepts and Systems

(1) Application of occupational safety and health system in China.

International Labor Organization (ILO) has been committed to the Global Labor Occupational Safety and Health, ILO held a meeting of experts in April 2001, revised and unanimously adopted the "Occupational Safety and Health Management System (OSHMS) Technical Guidelines". OSHMS is another important standardized management system, which is followed by ISO9000 (Quality Management System) and ISO14000 (Environmental Management System). It is also known as the management method of post industrialization era. In May 2001, representatives of the Chinese government, Trade unions and Entrepreneurs Association in Kuala Lumpur, participated in the promotion of ILO in the Asia-Pacific region to promote the application of the OSHMS guidelines for regional meetings. The Chinese government submitted bilateral technical cooperation proposals to the International Labor Bureau in the field. Since then, OSHMS, as a scientific management mode and system, has been popularized and applied in Chinese country. OSHMS is the advantage and market access pass that the enterprise wins in the domestic and international market competition. As the first company to enter the LNG industry, CNOOC, in the upstream oil and gas field with the experience of years of cooperation with the international petroleum companies, has been incorporated the OSHMS system into the CNOOC's enterprise culture. The LNG project downstream industries also have been explored and promoted the OSHMS system.

(2) Basic contents of OSHMS.

OSHMS system is a part of enterprise management system, which is to achieve the goal of occupational safety and health policy, and to ensure that this policy can be effectively implemented. It is combined with the enterprise's overall management function, and it is a dynamic, self-regulating and improving system, which involves all activities of the enterprise's safety and health. OSHMS's general requirement is to establish and maintain a professional safety system, and promote enterprise's continuous improvement of occupational safety performance, and it complies with occupational safety and health laws, regulations and other requirements, to ensure the safety and health of employees. Occupation health and safety management system, including the principle, idea, policy, organization, planning, implementation, evaluation and improvement measures of eight elements, and these elements are with constant circulation, continuous improvement. Its core content is the identification, evaluation and control of risk factors.

(3) ISO14000 environment management system.

ISO14000 environmental management system standards are established by the ISO/TC207 (International Environmental Management Technical Committee), which is an international standard for environmental management system, including a lot of focus on environmental management issues within the field of international environmental management systems, environmental auditing, environmental mark, life cycle analysis. Its purpose is to guide various types of organizations (business, company) to obtain the correct environmental behavior (not including the test method standards for pollutants, contaminants and sewage limit standards and product standards, etc.). This standard is not only suitable for manufacturing and processing industries, but also for the construction, transport, waste management, maintenance and consulting services. The standards are set aside 100 standard numbers, from ISO14000 to ISO14100, divided into seven series. ISO14000 series of standards are also reflected in the theme of whole life-cycle, which requires companies to control factors that affect the environment during the whole process of product design, production, use, disposal and recycling. Based on "Environmental Policy", ISO14000 reflects the idea of life cycle. TC207 sets up a Life Cycle Assessment Technology Committee, to evaluate the size of the environmental impact of the product at each stage of production, so that enables enterprises to analyze and improve their products.

(4) The relationship between ISO9000 and ISO14000.

For many of the requirements of the enterprise, ISO9000 quality system certification standards and ISO14000 environmental management system standards are universal, and the two sets of standards can be used together. Many enterprises and companies in the world have passed the ISO9000 series standard certification. These enterprises or companies can put in through the ISO9000 system certification experience when applied to environmental management certification. The new version of the ISO9000 family standards are more embodied with the principles of the use of the two sets of standards, so that the ISO9000 family standards and ISO14000 series of standards are more closely linked.

(5) The features and objectives of ISO14000.

ISO14000 series of standards are to promote the improvement of the global environmental quality. It is through a set of environmental management framework documents to strengthen the enterprises' environmental awareness, management capacity and security measures, so as to achieve the purpose of improving the quality of the environment. At present, they are the standards that enterprises adopt voluntarily, and it is the conscious behavior of enterprises. In our country, a third party independently verifies whether the products produced by enterprises meet the requirements. The goal of ISO14000 is through the establishment to meet the requirements of national environmental protection laws, regulations of the international standard. ISO14000 series standards in the global scope, improve the global environment quality, promote world trade, and the ultimate goal is eliminating trade barriers. LNG project company should strive to obtain ISO14000 certification.

(6) The Chinese government recognizes the conventions and recommendations of a number of international organizations.

In October 22, 1994, adopted at the tenth meeting of the Standing Committee of the Eighth National People's Congress, The Chinese government has formally approved "The Convention on the Safety of the Safe Use of Chemicals" in the 170 conventions of the International Labor Organization; in October 27, 2001, adopted at the twenty-fourth meeting of the Standing Committee

of the Ninth National People's Congress, the Chinese government formally approved "The Construction Industry Safety and Health Convention". The international Labor Organization conventions and recommendations have a significant impact on our country: "Benzene Convention, 1971", "Occupational Disease Conventions and Recommendations, 1974", "Working Environment (Air Pollution, Noise and Vibration) Recommendation, 1977", "List of Occupational Diseases (revised at 2010)", "Occupational Safety and Health Recommendation, 1981", "Occupational Health Services Recommendation, 1985", "Chemicals Convention, 1990", "Prevention of Major Industrial Accidents Convention, 1993".

9.1.4.2 International HSE Management Systems and Conventions Will Have a Positive Impact on LNG Projects

China's LNG project has just started. The project HSE management is still lack of mature system and experience. No doubt, actively adopting internationall-recognized systems, conventions and proposals, including the OSHMS system and ISO14000 will have a positive impact on the LNG project

(1) To adopt directly.

OSHMS and ISO14000 are proved to be a mature HSE system. The project company can boldly learn and directly use in the construction of HSE system, especially in the HSE planning and implementation, evaluation, rectification, contractor management, procurement management, prevention and control measures, emergency prevention, emergency preparedness and response.

(2) To adopt after modification.

The above mentioned "Code of Practice for the Prevention of Major Industrial Accidents, 1991", "Working Environment (Air Pollution, Noise and Vibration) Recommendation, 1977", "Occupational Safety and Health Recommendation, 1981" can serve as a good text. Combined with the LNG project, project company can reduce the research and exploration time and formulate regulations.

(3) The reference for formulating.

The implementation of the international conventions and recommendations are the results of millions of management and technical experts on decades of practice. We have reasons to believe that, in the LNG Project Company HSE management work, learning its principles, ideas, policies, procedures, terms, will play a positive role in the preparation of company management procedures and regulations, and the establishment of HSE management system.

9.1.5 LNG Project HSE Management Process

LNG project HSE management and time management, cost management, quality management together, constitute the four controls of project management.

9.1.5.1 Three Components of the HSE Management Process

Figure 9-2 shows the flow chart of LNG HSE project management. It can be seen from the figure, the first is the LNG project HSE management system, which is composed of three parts: HSE management mode, LNG internal organization and institutional system. The middle box is the HSE stage management parts, mainly aiming at different construction stages of the LNG project. The HSE management measures prepared in advance, whose stages include early stage (including pre-feasibility and feasibility study stage), design stage, construction stage and acceptance stage of commissioning / trial operation. HSE audit and continuous improvement is to carry out the dynamic management process of the HSE ex-ante evaluation and ex-post identification. Through the audit of HSE principles, ideas, policies, goals, systems, procedures, methods and measures, company should find out the problem and shortage, then carry on the further revision. HSE audit and continuous improvement is the activities throughout the project life cycle.

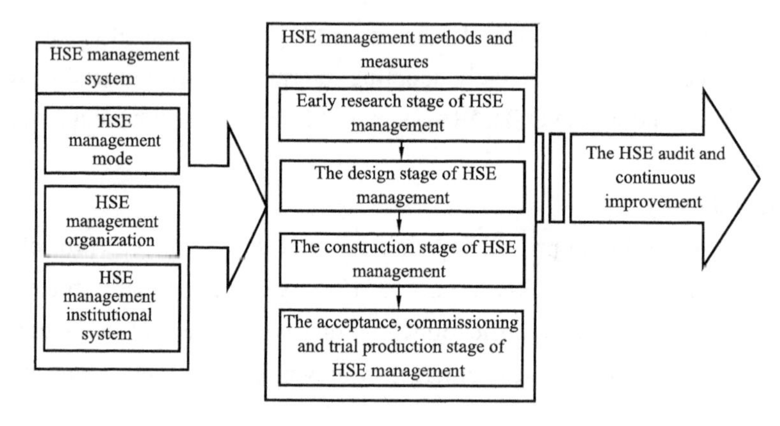

Figure 9–2 LNG project HSE management flowchart

9.1.5.2 The Relationship between Three Components of the Process

(1) The relationship between HSE management system and methods / measures.

Like quality management, LNG project HSE management involves units, including government departments, construction contractors, supervision companies, community, the relevant units (including survey / design units, suppliers), media and LNG project company itself. The above units constitute the whole synergy of the project HSE management, and each possesses its HSE responsibility. From the LNG project itself, the project company is the main body of HSE management for each project stage. HSE management organization and the HSE regime system must be set up, including the choice of international and domestic HSE technology and management standards and normal system. Only when organizations have been implemented, in order to make clear the HSE management responsibility of each department of the project company in different stages of the project life cycle, can it distinguish the HSE management level and working interface among the owner, supervisor and contractor in the project construction. HSE management system is the foundation, methods, measures and means.

(2) The relationship between HSE management methods / measures and the audit and continuous improvement.

As shown in Figure 9–2, according to the organization's technology and management ability, and the characteristics of LNG project, combined with the LNG project management stages, the LNG project company formulate the corresponding HSE management methods and measures, which including the prophase research stage, design stage, construction stage and commissioning / trial operation stage. At the same time, in the HSE management, the project company should give full play to each individual in the HSE management, and give full play to the role of directly involved in construction contractors, supervision companies, communities, related units and also give full play to the role of indirectly involved in the government and the media organizations. In the process of project operation, HSE audit and continuous improvement analyze and reflect on HSE guidelines, ideas, policies, objectives, process, system, standards and so on to see those which need to improve, and which need to enrich and repeal, to ensure the integration of static and dynamic management of HSE.

(3) The relationship among the HSE audit and continuous improvement and the former two.

HSE audit and continuous improvement is the content of dynamic management. It is necessary to use the tracking, investigation, statistical methods to analyze the HSE problem. The results will provide the basis for HSE control and improvement. In addition to the normal process of HSE management, according to the change of internal and external environmental factors of the project, we should adjust and control the HSE management of LNG project timely, in order to make the LNG project work under the effective HSE control operation. The HSE audit information should be timely feedback to the HSE management institutions and individuals of the project company that revise HSE static management parts. The continuous improvement includes the HSE organization system and process improvement.

The three parts of the HSE management process are dependent on each other, acting in cooperation, forming a whole. Continuous improvement of HSE management is the last action of HSE management, including unceasingly enhancing the HSE project management level, and including the accumulation of experience to provide some HSE management mode for the new LNG project.

9.2 LNG Project HSE Management Systems

9.2.1 LNG Project HSE Management Mode, Organization and Responsibilities

9.2.1.1 Summary

HSE management system is to ensure that the project construction, commissioning and production work smoothly throughout the project life cycle and also to meet the requirements of HSE laws and regulations, which generally include the establishment of HSE management mode, internal HSE management organization system, standards, norms and regulations and relative laws and regulations of these three parts. HSE management mode is the guarantee of LNG project HSE management and control. LNG project company is the main body of the HSE management and control, and its internal functional departments of the division of HSE responsibilities and even the specific responsibilities of each employee, will be decisive for the project company HSE objectives. The HSE laws, regulations and standards are basis for the measurement of HSE management. HSE rules, regulations and procedures are developed by project company as self-constraint documents, which has the characteristics of self-control, revision and flexible. The three parties work together to make contributions to realize the LNG project's HSE goal.

9.2.1.2 LNG Project HSE Management Mode

The main body of LNG HSE management is project company. At the same time, design departments, construction contractors, supervision companies, and suppliers also take charge for HSE responsibilities. The project company should influence them in order to ensure the implementation of HSE management system, and accept the supervision of government, community and media all levels. The following is the LNG project HSE management mode (Figure 9-3).

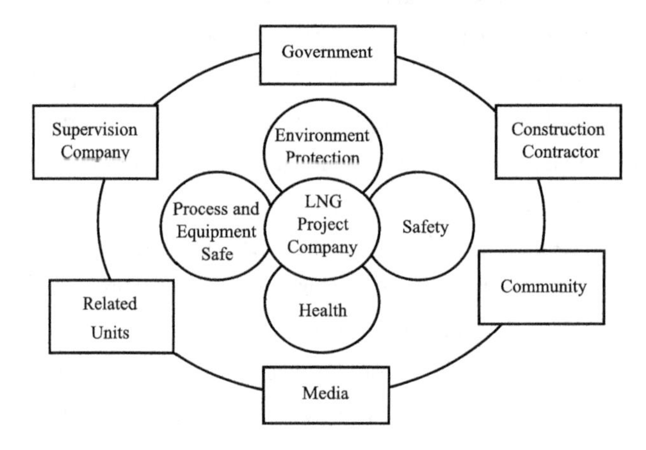

Figure 9-3 LNG project HSE management mode

9.2.1.3 The HSE Responsibilities and Obligations of Involving Departments or Units

(1) LNG project company.

① First of all, establish correct HSE policies, ideas, HSE management system, including HSE management mode, HSE organization and standard system.

② Initiatively build up the management channel with all level government departments. In the project site selection stage, construction stage and acceptance stage, accept the supervision and inspection of the environmental protection deportment, safety deportment, land resources deportment, cultural relics deportment, military sector, and get guidance and help from government departments at all levels.

③ Put HSE management guidelines, ideas and policies into the engineering construction survey, design and construction, and remind contractors, designers, supervision company and suppliers to execute HSE management.

④ Conduct a serious HSE argumentation in the project site selection, pipeline routing, the survey and design of gas transmission substation, so that the construction works and the environment maintain the best coordination.

⑤ Rigidly manage and control contractors in the construction process of sub-projects. When several sub-projects are constructing at the same time, HSE management interfaces should coordinate different contractors, in order to insure no dead angle and leak hole, and urge contractors to train and supervise front-line staff for safe production and work.

⑥ Cooperate with local community extensively and propagandize HSE of the project. Let them know the difficulty of HSE management in the process of construction, so community people will not disturb the construction. Project company should set up harmonious relationship with local people in the process of site selection, demolition and construction.

⑦ Pay attention to the industry comments of supervision units in bidding inviting stage, and pay attention to the evaluation of supervision units in the process of project construction management.

⑧ Integrate HSE management into daily management during construction stage, and execute HSE daily meeting, weekly analysis and monthly inspection system.

⑨ Pay attention to the HSE incident and accident prevention. When HSE accident happenes, immediately start the emergency response procedures, to prevent the further expansion of the accident, to carry on the accident investigations, analyze the reasons and draw a lesson.

⑩ Pay attention to establish the reward and punishment mechanism of HSE management in the construction stage, and reward the units and individuals who have contribution to the HSE management, including material rewards, then a good atmosphere of awarding contributors and punishing offenders would be formed.

⑪ Insure the HSE overall acceptance in the phase of the project completion and the commissioning, and organize the designers, construction contractors, engineering supervisors and national authorities to carry on the field inspection and acceptance. Normal commercial operation should be started up after acceptance and rectification.

⑫ Collect HSE documents and establish HSE files timely by the files management regulation, then transfer to project files management department after project completing acceptance.

(2) Government departments at all levels.

① Safety Production Supervision and Management Department. The country implements construction safety management system. Local people's governments at or above the county level of Construction and Administrative Departments and Safety Production Departments should implement supervision and management of the project. Project company should initiatively accepts the Production Safety Department to review safety pre-evaluation report of the project, to supervise and inspect in the process of project construction and to audit and rectify in the acceptance phase.

② Environmental Protection Department. The compctent department of environmental protection under the State Council shall strengthen the supervision and inspection on the implementation of environmental laws, regulations and mandatory standards of the state key construction projects. The Environmental Protection Department and other relevant departments of the local people's governments at or above the county level shall strengthen the implementation of mandatory supervision and inspection on the water and soil conservation and restoration of vegetation.

③ Health Department. The National Health Department shall, in accordance with the duties prescribed by the State Council, carry out occupational health supervision and inspection on major construction projects, and control the impact of noise on the surrounding community, to prevent the emergence of infectious diseases on the front-line staff.

④ Cultural Relics Management Department. Heritage Management Department should strictly enforce the protection of the heritage of the land which is occupied by the project, to dispose the change of the location of the relics in accordance with the regulations, and ensure the protection of the local ancient cultural heritage and the site.

⑤ Military Management Department. The Military Management Sector should coordinate with the project company to avoid military zone is occupied, and provide help for national key projects in their power.

(3) Construction contractor.

① To build up HSE management organization and regulations according to owner's HSE management requirements, in particular the rules and operation manual, train and educate the front-line staff for HSE management, and build up HSE inspection system to detect HSE incidents, and deal with them timely.

② International contractors should build up HSE management system for LNG projects according to our government regulations, and implement the international popular HSE management measures and methods.

③ Be responsible for the HSE construction, find problem and communicate timely. When HSE incident and accident occurs, contractor should try to minimize the casualties and investigate the cause of it, and accumulate experience for avoiding later accidents.

④ Subcontractors are responsible for the subcontracting engineer HSE management of the general contractor, and general contractor and subcontractors are responsible for subcontracting engineer HSE management together.

⑤ To organize the safety examination, discovery and eliminate the hidden dangers timely in the running process.

⑥ To provide necessary labor protection supplies for staff, and ensure the right use of them, contractor can't use materials, equipments, devices, protective equipment and safety testing instruments which are not comply with national and industry standards.

(4) Engineering supervision unit.

① To put HSE management as one of the important contents of the supervision, allocate the HSE supervision personnel, and establish the HSE records.

② Engineering supervision unit and construction contractor HSE management personnel should communicate effectively, discover and deal with HSE problem timely, in order to avoid HSE accident expand or spread.

③ To strictly control the HSE standards and indexes of building materials, building components, fittings and equipment, find the problems which are not comply with HSE management, then report to owners timely, change and replace them.

④ Check up the safety technology measures or special construction scheme.

⑤ When hidden dangers of safety accidents are discovered in the inspection process, ask construction units to change or stop the construction in accordance with the relevant regulations by written instruction promptly.

⑥ Supervise and inspect the construction safety production by laws, regulations and mandatory standards or the commission contracts.

⑦ Main contents of safety supervision in construction preparation stage are as follows.

(a) To prepare project HSE management plan, clear the scope of safety supervision, content, work procedures and system measures, as well as personnel with plans and responsibilities, in accordance with the mandatory standards for engineering construction, GB50319 "The Code of Construction Project Management" and the relevant industry standards.

(b) Draw up HSE supervision implementation details, which contain the methods, measures, control points of safety supervision, and inspection program of safety technology measures for construction units.

(c) Check whether the safety technology measures in the construction design process, the divisional and itemized works of special construction scheme with large risk, meet the mandatory standards of engineering construction or not.

(d) Inspect construction units in engineering items on the safety of production rules and regulations, and the establishment of safety supervision agencies, and full-time safety management personnel with the situation. Supervise and urge the construction unit to check the safety production rules and regulations system of the subcontractors.

(e) Check whether the construction qualification and safety production license are valid or not.

(f) Check whether the project managers and full time safety production managers have the legal qualifications or not, and whether is consistent with the tender documents.

(g) Check whether the special operation qualifications of personnel is legally valid or not.

(h) Check and review the emergency rescue plans and safety protection measure expanse using plans.

⑧ Main work contents of safety supervision in the construction stage are as follows.

(a) In accordance with the construction organization design of the safety technology measures and special construction plan, supervise the construction units, organize the construction, and stop illegal construction work.

(b) Inspect and check more dangerous operating conditions in the construction process regularly.

(c) Check the acceptance procedures of safety and erection facilities in the construction site, for example, cranes, overall enhancing scaffolding, templates and so on.

(d) Check whether the safety signs and protection measures are comply with the mandatory standards in the construction sites, and check the safety production expanse.

(e) Supervise the construction units to examine safety by themselves, and check the results, attend the safety production special inspections of the construction units.

(5) Relative units.

① Investors. Investors should provide guidance and assistance to the HSE management of the LNG project company, and allow the project company to innovate, highlight the characteristics of the LNG project, and develop its own LNG management culture, under the unified framework of the investor's HSE management policies, ideas and policies.

② Suppliers. Suppliers should provide all kinds of materials and facilities which meet HSE standards according to contract requirements. Foreign suppliers should provide equipment and materials which are proven by foreign LNG projects and comply with LNG projects HSE requirements.

③ Visitors. Visitors should be strictly in accordance with safety education tips, accompanied by HSE management personnel, wear safety helmet and clothing according to the provisions of the line site visit, and shall not enter explicitly prohibited places in the process of construction projects.

(6) Community government.

① Carry out the propaganda of the LNG project to the community masses, introduce the risk and harm that may appear in the construction phase of the project, and avoid local people straying into the construction site, which may bring unnecessary harm.

② Issue the local regulations on the construction of key LNG projects, and provide the necessary road, water supply, electricity and communication for the project construction.

③ Carry out the persuasion and interpretation work to local people and units for the demolition of houses, factories and public facilities in accordance with the requirements of the project, to form a good social atmosphere for national key construction projects.

(7) Media.

① In order to promote the local economy development, media should report national key LNG projects objectively, and propagandize the advantages of LNG projects, for example, environment protection, low price and so on.

② When the HSE incident or accident happenes, media should realisticly and practically report the initiative with the LNG project company, and carry out the masses of psychological counseling and comfort, to avoid causing unnecessary panic among the local people.

9.2.2 LNG Project Company Interior HSE Management Organization Structure and Responsibilities

9.2.2.1 LNG Project HSE Management Circle and Organization Chart

(1) HSE management circle.

HSE management should be based on the HSE organization and division of responsibilities of the project company, and there is no fixed pattern or routine. The following is the LNG project company HSE management responsibilities of the size (Figure 9–4), introducing from the inside to the outside.

① Core circle. The general manager is primarily responsible for project HSE, and the core members of the project company, like the deputy general manager, the chief engineer also should be responsible for project HSE.

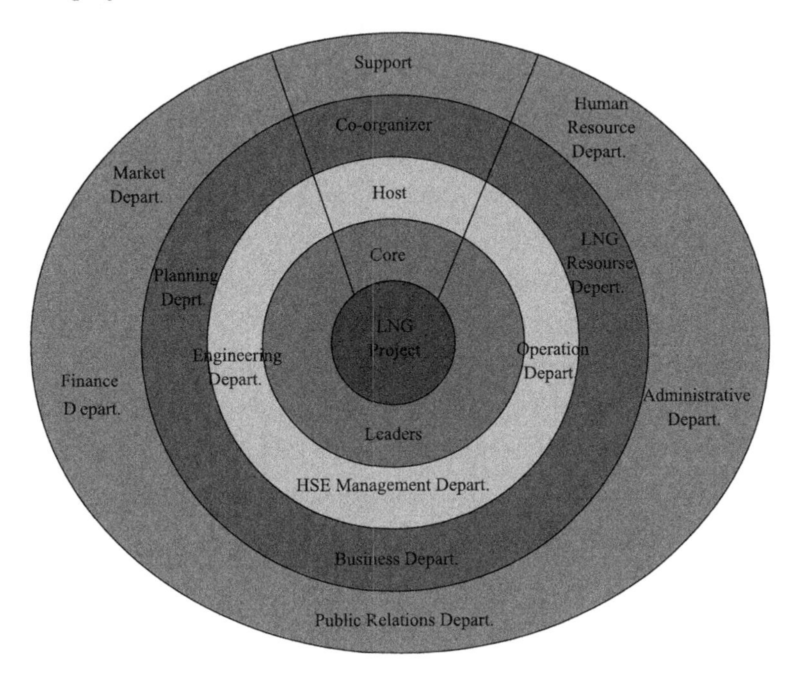

Figure 9–4 HSE management circle

② Host circle. Engineering and Operation Departments are the main responsible units for the project HSE, and HSE Department is responsible for supervision. Engineering and Operation Departments should negotiate and deal with contractors, supervision units and suppliers about HSE contents directly, but HSE Department should communicate with relative government departments, community and media besides supervision.

③ Co-organizer circle. On the one hand, it is responsible for the HSE. On the ofter hand, it provides HSE professional gatekeeper to the competent department, including the LNG project company's LNG Resources Department, Planning Department, and Business Departments.

④ Support circle. Support circle provides all kinds of support about HSE managements. For example, Finance Department provides funds, Public Relation Department communicates with local government, Market Department provides market operation, Human Resource Department provides HSE recruitment and training, and Administrative Department provides data archiving management.

From the HSE management circle graph, it can be seen that the LNG HSE project management reflects the general manager is the first responsible person, other members of thc core are responsible for duties themselves, department in charge plays the leading role, and the whole company participates the HSE management.

(2) LNG project company HSE management agency.

LNG project company HSE management is related to department settings, personnel arrangements. There is no unified model. Figure 9–5 is a combination of a project company in the storage tank for the EPC implementation of subcontractors and pipeline sectionalization bidding, and also contains pipeline shielding and crossing engineering. The case shows an organization chart of HSE management of the project company. We should emphasize that no matter what the form of EPC contract is, contractor is a main body for LNG project company HSE management, because may be several sub-contractors are working at the same time in the process of construction, and HSE interface's coordination work is very huge among every construction units. And then, in the latter period of project construction and trial operation period, HSE management preparation should be studied in advance in the operation period by assigned person. It should be said that the HSE management work is conducted throughout the entire project stages. From the management method, it can be divided into organization management and document management, program management, legal standards and regulations system management and site management. From project company and contract firms, it can be divided into internal HSE management and contractor HSE management. From the project life cycle, and it can be divided into pre-research stage, design stage, construction stage, acceptance stage HSE management and production trial running HSE management.

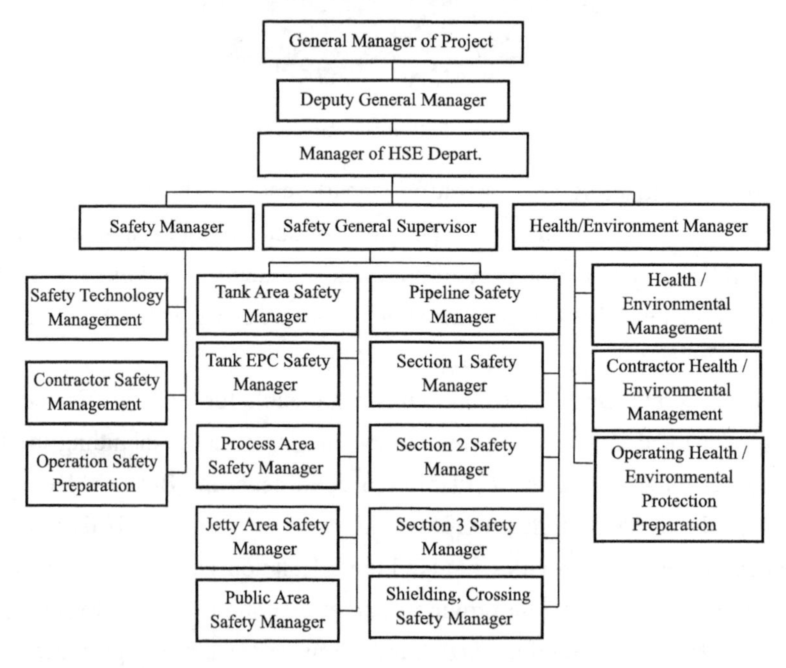

Figure 9–5 A LNG project HSE management organization structure chart

9.2.2.2 LNG Project Company Main Personnel in HSE Management Duties and Obligations

Following introduction describes the general LNG project company main personnel's HSE

management duties and obligations, and then readers can grasp the scope as a whole, providing a framework for project companies to develop their own HSE management duties and obligations.

(1) General Manager.

① The first person is responsible for the project company HSE management.

② Implement HSE laws, regulations and standards, preside over the development and approval of the company's HSE ideas and policies.

③ Be responsible for setting HSE organization and allocate responsibilities in company.

④ Establish and improve HSE management system as a leader role, and provide necessary and sufficient resources for HSE management's effective operation and continuous improvement.

⑤ Examine and approve HSE management objectives and program.

⑥ Be responsible for examining and approving company property and personal accident insurance plan, HSE performance and incentive policy.

⑦ Preside over the development of HSE continuous improvement program, and maintain management system suitability and effectiveness.

(2) Deputy General Manager.

① Assist the general manager in the development of the HSE concept and policies, and promote the implementation of the HSE management objectives.

② Be responsible for HSE management of daily major work, and preside and draw up HSE objectives and related documents.

③ Assist the general manager to improve the HSE organization and management, and promote the HSE performance assessment.

④ Assist general manager to implement and continuously improve HSE management.

⑤ Be responsible for HSE management, when general manager is absent.

(3) Chief Engineer.

① Be responsible for engineering project HSE management technology and system implementation.

② Be responsible for coordination with the technical person in charge of the contractor for HSE management.

③ Organize technical documentation for the audit of HSE management.

④ Assist deputy general manager in the technical HSE management.

(4) Chief Accountant.

To ensure the company's HSE management of capital investment, HSE management costs will be included in the annual budget, and the implementation of the relevant expenditure management and implementation will be supervised.

9.2.2.3 LNG Project Company Department's HSE Management Responsibilities and Obligations

(1) HSE Department.

① The leading department is responsible for organizing and coordinating HSE management work among various departments.

② Organize HSE management system for establishment, operation process supervision and control.

③ Compile the HSE management requirements in the tender and contract documents.

④ Audit HSE management requirements of various functional departments in the design, construction, procurement and other processes.

⑤ Cooperate with the responsible departments to report the design, construction and other documents relating to the government for approval.

⑥ Participate in the review of contractor's HSE commitment and HSE control level during the period of bid invitation, and participate in the review and comment on the supplier's HSE.

⑦ Be responsible for the compilation and supervision of the implementation of the project HSE management regulations.

⑧ Organize HSE acceptance of unit works, and supervise the HSE work which is not qualified for rectification.

⑨ Organize the investigation and handling of major HSE events / incidents, and organize verification.

⑩ Participate in project HSE payment, change and other requirements for HSE audit.

⑪ Organize HSE management regular and irregular inspection in construction sites.

⑫ Organize HSE management performance examination.

(2) Engineering and Operation Department.

① Be responsible for the daily control of HSE in the project construction site.

② Establish HSE daily management system in project construction.

③ Be responsible for HSE engineering management in the process of manufacture, construction, installation, testing and trial operation.

④ Supervise construction unit to establish HSE engineer post, and be responsible for organizing the writing of HSE control plan and HSE construction control management by professional engineers.

⑤ Participate in the preparation of HSE control requirements in project bidding documents for engineering contractors and supervision company.

⑥ Organize to review and approve the construction organization design, construction stage technical documents, construction HSE technology control procedures, inspection / test plan and HSE guarantee documents, and implement effective HSE technology control for project construction.

⑦ Participate in the examination and approval of the supervision of the HSE planning and other supervision work documents submitted by supervision company.

⑧ Participate in organizing HSE inspection of portioned and itemized works, and supervise unqualified rectification.

⑨ Participate in the project construction of general HSE event / accident investigation and treatment.

(3) Plan and Control Department.

① According to the project schedule, carry out the decomposition of project tasks, prepare the work plan, sum balance, and guarantee the implementation of the HSE safety work during the same period.

② Be responsible for the control, inspection and reporting of the implementation process of the project HSE.

③ In project bidding and procurement documents, audit HSE management of schedule requirements for engineering contractors and suppliers.

④ Be responsible for the HSE cost budget preparation, responsible for execution, inspection, supervision and rectification of the HSE costs in contractor's construction.

⑤ When HSE, schedule and cost objectives are in conflict, it should comprehensively consider the HSE, schedule, cost and quality, in order to meet the requirements of engineering HSE and quality management, then adjust the schedule plan and cost.

(4) Business Department.

① Participate in the preparation of the HSE plan in the project contract, and provide the HSE approval authority for approval.

② Participate in the preparation of procurement strategy, procurement and acceptance of HSE control plan, so that the project procurement to meet the HSE, quality, functional requirements.

③ Participate in HSE tracking managements in procurement and acceptance activities, and make sure suppliers submit drawings, documents and related HSE certificates on time.

④ Participate in HSE inspection of relevant departments and personnel, and make sure the project company's own procurement materials, components, fittings and equipment comply with the requirements of design documents and contracts.

(5) LNG Resources and Transportation Department.

① Organize HSE problem implementations in LNG resources and transportation program, and arrange transport ships. Draw safe LNG procurement and transportation dispatch plan, and supervise the implementation.

② Implement project HSE policy, and execute LNG import of HSE management.

③ Supervise the quality of LNG import.

(6) Market Department.

① Provide users with LNG or natural gas technical requirements, including parameters descriptions, safety network pressure and heat value adjustment.

② Fully consider and analyze the HSE risk caused by marketing activities in the formulation of marketing plans and strategies.

③ Be responsible for providing safety recommendations for LNG or natural gas users.

(7) Finance Department.

Allocate HSE management fees by plans, and supervise whether the fees are used reasonably.

(8) Public Relationship Department.

Be responsible for the company and national, local government departments of the HSE daily contact, communication, opinion collection and processing, and file approval.

(9) Human Resource and Administration Department.

① Be responsible for allocating human resources, select person who has accepted professional HSE training, with sufficient experience, and participate in project HSE management.

② Coordinate and organize the company staff for HSE management training.

③ Be responsible for HSE documents management, draw relevant regulations, and propose HSE documents management requirements.

9.2.3 HSE Management Levels

LNG project HSE management is a complex system engineering. It is a part of the company culture and requires high attention from leaders, advanced theoretical guidance, matching management system as support, don't go about several tasks at a time. Especially a new LNG company is build up, management levels should be made sure before the system set up. The following chart introduces the management levels (Figure 9-6).

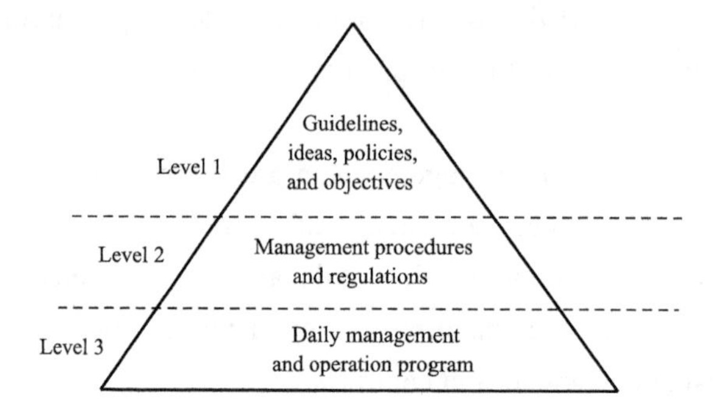

Figure 9-6 Project company HSE management levels

9.2.3.1 HSE Management Level 1

HSE management system includes institution system. The highest level of the system are the guidelines, ideas, policies and objectives, which reflects the project company's macro management ideas and strategies.

(1) HSE guidelines.

HSE guidelines are the statement of intent and principles for enterprises to manage their HSE. Usually, they are expressed as the form of slogans, easy to remember and execute.

(2) HSE ideas.

HSE ideas reflect the enterprise culture of company, and for society, people, company and employee conduct guidelines. What are the ideas? Simply say, it is the method of doing things by company. HSE ideas are the important part of enterprise culture, which reflect HSE social responsibility and obligation, and also reflect the degree of importance of the project company and the level of investment in HSE. If the leaders of project company have a high sense of responsibility, they should emphasize HSE management very much, and implement it to works. HSE ideas are drawn by project company, which are the accumulation of leader experience and enterprise culture. HSE ideas are improved gradually with practice, and are continuous improvement process.

(3) HSE policies.

Policies are restricted by HSE ideas. HSE policies play the role of guideline and criterion for management rule and regulation, and are mandatory. Similarly, the project company's HSE policies are not the same, and there is no fixed pattern.

(4) Objectives.

HSE objectives must be set up by project company, and the objectives can be achieved by efforts usually in one year or many years. When HSE objectives are set up, the period is one year usually. HSE objectives can be higher than the goals of national or local HSE regulations. The specific contents are drawn by the requirements of company conditions, management strategies and management levels.

9.2.3.2 HSE Management Level 2

(1) Management procedures.

Management procedures are the part of the second level contents. Procedures are necessary in doing things, and are the contents of basic management. HSE management procedures are the most emergent task for HSE management department. For example, like the Zhejiang LNG project, in the early days of the project, company is set up, and HSE system is formulated immediately, so that HSE management work can be done according to the procedures.

(2) Management regulations.

Management regulations are defined by the LNG project scope, which are the daily management, engineering design, construction, acceptance, production and other aspects of the HSE management of the text description. When investors have systematic and standardized methods in HSE management, project company can refer to the HSE management ideas and experiences. Project company should set up special HSE department, draw HSE management regulations, and plan gradually according to project schedule and emphasis. When investors don't have HSE management system or enough experiences, Project company can invite foreign and domestic consultation company to draw HSE management system in writing plans. What needs to be emphasized is that every LNG project company must put the HSE problems mentioned on a very important position. Our government pays more attention to the problems on the management, and HSE must be mandatory. This is also a big move to enhance the credibility of the company, serve the community, and create a better life for each project company.

9.2.3.3 HSE Management Level 3

(1) HSE daily management.

Daily management includes HSE department staffs' conventional work process and work contents. Generally speaking, each stage's work is different according to the progress of the project. In the early stage of the project, the main consideration is the national security laws and regulations of the LNG project, the development of the company's HSE project ideas and policies, and then prepares for future HSE management regulations. The environment of LNG terminal/jetty should be appraised in the pre-feasibility and feasibility study stages, and health/occupational illness should be appraised in the reports. There are special chapters in the HSE management. In the project construction period, according to the specific provisions of the management, project company HSE department and the supervision company carry out supervision for the construction project of the health, safety and environment. At the same time supervise the contractor HSE management work. In the project acceptance period, according to the engineering design, it needs to carry out the HSE inspection, find out the content of the rectification and conduct the rectification. In the initial stage, check the project safety production, including the HSE acceptance.

(2) Operation program.

Operation program is the process and step of a specific work, and each work should be operated according to the way of protecting personal safety, environment and nature in harmony. Particularly in pre-production period, operation manual should be drawn gradually according to the specific production operation contents, in order to find out a set of completed operation program for formal production.

9.2.4 HSE Management Guidelines, Ideas, Policies and Objectives

9.2.4.1 HSE Management Guidelines

HSE guidelines are the central theme of HSE management system. In order to effectively implement the management system, there must be correct policy guidance. Recommended HSE management guidelines are as following.

Safety first, prevention first, to prevent major accidents.

People-oriented, caring for life, to ensure the health and safety of employees.

Science based, following the rules, to provide quality service products.

Energy saving, pollution prevention, protection of ecological environment.

Full participation, continuous improvement, and striving for first-class enterprise performance.

9.2.4.2 HSE Management Ideas

The formation of the company's HSE ideas requires the top leadership to promote the establishment and shaping of the positive work, which needs to keep in practice constantly and root deeply in the hearts of all employees. Mr. Marvin Bower, who has served as chairman of Mckinsey, elaborates the ideas of company, and he believes the ideas are a series of rules or guidelines, and the ideas are built up gradually by several trials and leadership of leaders. And then they become the expected behavior. For HSE management ideas of LNG project company, ideas include the following aspects.

(1) HSE management is the basic guarantee and requirement of the development of the company.

HSE is an important part of the company's project construction, production and management activities. With reference to the management mode of the international similar companies, the company can quickly improve the management level of the company's health and safety, and regard it as an important part of enterprise culture and the core competitiveness of enterprise.

(2) HSE is far beyond the scope of the economy.

HSE is not only economic issue, but also social issue. For LNG project company, the LNG project itself and supporting user projects involving multi-industry, multi-user and urban residents. Managing HSE affairs is not only from the economic considerations, but also as a social responsibility.

(3) HSE management exists throughout all aspects of project.

The final responsibility of HSE management should be implemented in the whole process of the project operation, and all levels of management should be committed to the HSE management and integration into each job and each post. Who is in charge, who is responsible for. It needs to reach the pursuit of the company's conscious management, self-constraint.

(4) HSE should persist in internal and external supervision.

LNG project company should adhere to the principle "internal audit is an important means of self-assessment and monitoring". The main way is to improve and learn advanced management experience, combined with internal and external supervision, so that the HSE management is becoming more and more perfect.

(5) HSE target management.

Project company should set goals, strive to practice HSE management, firmly believe that only through the "implementation" of the efforts can achieve the established objectives. Make efforts to pursue outstanding HSE performance. HSE management not only should be complied with national and local laws and regulations, but also should play an active role in improving management levels.

(6) Staff is one of the most valuable assets of the company.

Staff is the first-class asset of company, and is non-renewable. Company should be people-oriented and care for staff. The pursuit of harmless target should become conscious action of LNG project company.

(7) The close relationship between material / energy and environmental protection.

As the product of the LNG project, natural gas itself is a high quality, efficient and clean energy. The project company in the construction stage should do as much as possible to use efficient, clean and harmless materials and energy, to achieve comprehensive protection of environment and resources.

(8) HSE management of contractor is a part of company management.

The project company should pay attention to contractors' HSE management, and appreciate contractors' HSE management culture, at the same time learn the advanced management methods from contractors. Integrate the EPC contractors', material suppliers', consulting service providers' HSE management into company management.

(9) HSE management is a continuously improving process.

HSE management is not a short-term behavior, but requires continuous improvement and advance. The principle of continuous improvement should be persisted in the whole project periods. Good HSE records should be maintained. HSE performance should be improved gradually by system management methods. There is no best, only better.

9.2.4.3 HSE Management Policies

All activities of LNG project company should comply with the relative laws and regulations, and industrial standards. Personnel safety, health, environmental protection and company's property are the company's starting point and destination. Through the systematic management, health, safety and environmental protection performance are increased year by year. Under the guidance of LNG project company and management framework, company's HSE policies are determined, and the contents are as follows.

(1) To establish HSE management system gradually.

Health, safety and environment protection management system should be established, and personnel behavior should be uniformed, in order to realize the goal of HSE and execute "continuous improvement plan".

(2) Everyone to participate in HSE activity.

In order to enhance the awareness and quality of employees, project company should propagandize HSE policies to employees, and provide training and rewarding or punishment for them. Project company should encourage employees to study scientific technology and use HSE scientific products.

(3) To provide safe work condition for employees.

Project company should provide safe workplace, reliable facilities and equipment, and standard labor protection supplies for employees, and encourage employees to maintain workplace safe and healthy, and create good working condition, and create the harmony between providing clean energy and protecting natural environment.

(4) To deal with new and updated equipment for safety and environmental protection.

On the use of facilities, equipment, the new and updated equipment should be carried out HSE review and regular inspection. The project's total health, safety and environmental management should be reviewed regularly.

(5) To ensure the emergency plan with the times.

Project company should amend the emergency plan in time, establish emergency resources pool, regularly organize the emergency exercises and minimize casualties, pollution accident.

(6) To take risk identification as the basis of HSE management.

Risk identification, risk analysis and safety evaluation should be implemented in design, construction and production process, and implemented into each of the rectification program.

(7) To strengthen contractor's HSE management.

The contractor is required to establish a safety and environmental management system, to implement the industry standard, and to carry out HSE management according to the company's HSE policy, to train employees continuously.

(8) Responsible for social HSE management.

Project company should establish good relationship with government agencies, social communities and public, cooperate with HSE problems closely, and make contribution for better community environment.

9.2.4.4 HSE Management Objectives

(1) The general concept.

① The HSE objectives should not only take full account of the requirements of laws and regulations, but also the major risk sources, important environmental factors, and the technical solutions in the project cycle, then carry out targets decomposition, consistent with the company's commitment to continuously improve HSE performance.

② The annual objectives are proposed by the HSE department and approved by the general manager of the company.

③ Each department and contractor should develop the project HSE overall goal and indexes according to HSE requirements, and submit to HSE department.

(2) Objectives setting.

Annual objectives as an important part of the management of the project company, the implementation of one vote veto, generally contains the following indicators.

① Personnel death accident index.

② Occupational injuries and occupational disease control index.

③ Environmental pollution index.

④ The single accident direct economic loss control index.

⑤ The performance of other HSE management measure index.

At present, the LNG project is managed by project company, and all kinds of contractors undertake specific research, design, construction work. Therefore, in the formulation of the LNG annual objectives, the project company first sets its annual objectives, then project company urges the contractors to propose their own annual targets in the project, so that the HSE management interface is clear, easy to make HSE performance evaluation.

9.2.5 LNG Project HSE Laws, Regulations and Standard Systems and HSE Management Procedures and Regulations

9.2.5.1 National HSE Laws and Regulations, International and Domestic HSE Standards

(1) National Laws and Regulations.

Within the territory of China, the construction must be carried out in the framework of relevant laws and regulations of our country, and it is a mandatory implementation. According to the content of the LNG project introduced, it needs to follow the national and ministerial level laws and regulations, including health, safety and environmental protection three aspects.

(2) Domestic and international HSE standards.

HSE standard specification includes international, national / industry, enterprise three levels. Because our country is establishing the LNG standard system, some aspects need to draw lessons from foreign standards. If there are domestic HSE standards, they should be used firstly, and if there are no domestic HSE standards, the foreign advanced standards should be used. Through the practical work of China's first LNG pilot project, Guangdong Dapeng Project, domestic and foreign LNG standards and jetty, receiving terminal, storage tank, pipeline and utilities on HSE standards were investigated and the LNG project HSE standard system framework was formed.

9.2.5.2 HSE Management Procedures and Regulations

(1) HSE management procedures.

HSE management procedures should be established in the early time of project, improved to be perfect constantly. HSE management procedures include HSE risk identification and management procedures, emergency control procedures, monitoring procedures, special HSE analysis procedures, on-site HSE procedures and the contractor management procedures (Figure 9–7).

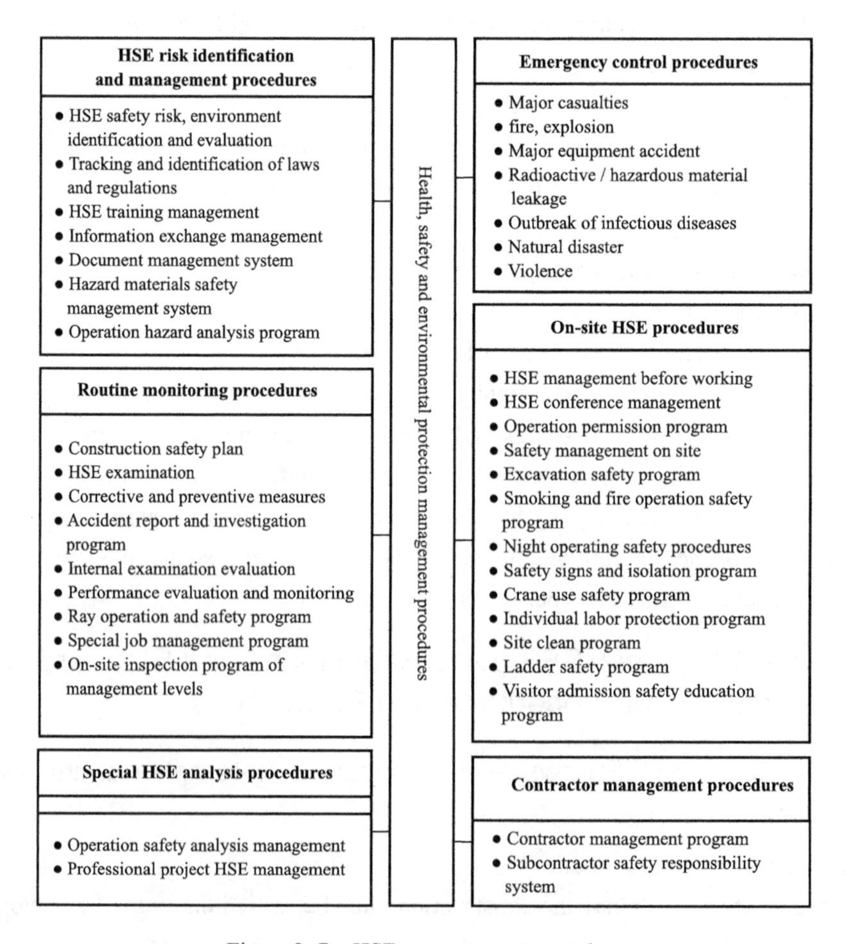

Figure 9–7 HSE management procedures

(2) HSE management regulations.

If the health and safety management procedure is the major pillar of the management of HSE, then the management regulation is another major pillar of HSE management. From the existing LNG project company HSE management regulations, they can be divided into three categories (Figure 9–8), that is, health management, safety management and environmental protection regulations. They are introduced as following.

Health management regulations	Environment management regulations
• Site management regulation for cleaning operation • Site medical and health regulation • Alcohol and drug regulation • Hazardous material handling requirements	• Project construction environment control • Waste management regulation • Noise and water pollution control • Gas emissions and dust control requirements

Daily safety management regulations	Safety management regulations	Special workplace management regulations
• Regulations on safety signs and protective railings • Safety lighting and night operation safety management regulations • Provisions on the administration of personal protective equipment • Fire safety management regulations • Vehicle management regulations • Holiday safety management regulations • Office safety management regulations • Safety management regulations for staff travel • Operation environment gas detection and management regulations • Lightning and static electricity safety protection • Excavation operation safety management • Lifting operation safety management regulations • Safety requirements for electrical work • Scaffold safety management regulations • General requirements of construction machinery, equipment and vehicles • Ship safety management • Security and guard system • Manual tool safety operation regulations • Safety management regulations for machine protection cover • Regulations for welding and cutting safety management		• Listing lock security regulations • Radioactive rays operation safety • Stop or supply power operation management • Hoisting crane rigging safety management regulations • Aerial work operation safety management regulations • Safety management for blasting operation • Regulations for temporary use of electrical safety • Chemicals management regulations • Safety management for thermal engineering • Inspection regulations of hand and portable electric tools (equipment) leakage protection • Regulations on the safety management of confined or enclosed space operations • Safety management regulations for industrial compressed gas cylinders • Safety requirements for pressure test • Debug safety management regulations • Safety management regulations for diving operations • Regulations on safety management explosives and blasting safety • Operating license system • Equipment certificate management • Compressed air management

Figure 9–8 HSE management regulations

9.3 HSE Management in LNG Project Pre research Stage

LNG project pre-research includes project market investigation, selection of receiving terminal site (shoreline, port), pre-feasibility study, feasibility study, project application report preparation and pipeline route selection. HSE management in this period is mainly to carry out safety pre-evaluation, environment evaluation and occupation hazard evaluation. It is necessary to select an appropriate analysis and evaluation methods, and the results are embodied in the text report.

9.3.1 HSE Analysis Methods

HSE analysis methods are the tools for analyzing and evaluating the risk of LNG project.

The characteristics, principles, objectives, application conditions, and work load of each method are not the same, and they have own advantages and disadvantages. The following are only a few of the main, commonly used HSE analysis methods, which are introduced one by one.

9.3.1.1 HSE Check List Method

HSE check list (HCL) is a kind of most basic, simple, widely used systematic HSE analysis method in LNG project. At present, the HSE check list is not only used to find a variety of potential accidents in the system, but also to the inspection items and systems to carry out assigned rating.

HSE check lists are formulated by person who is familiar with process, mechanical equipment and operation, and is rich in HSE technology and management experience. Prior to detailed analysis and full discussion of analysis object, list the inspection units and parts, inspection items and requirements, and develop the supposed HSE rating score standards and evaluation system. When analyzing and inspecting the system, compare the HSE check list item by item and give a score, then evaluate HSE levels, as following Table 9–1 to Table 9–5. When HSE check lists are used in looking for defects or problems in design, maintaining, environment and management and so on, the evaluation contents and procedures can be omitted.

9.3.1.2 Operation Condition Hazard Analysis Method (Graham–Kinney Method)

Operation condition hazard analysis method is a simple semi-quantitative analysis method, which is brought forward by K.J. Graham and G.F. Kinney. They think the risk factors of effecting operation conditions are L (probability of an accident), E (frequency extent of exposed to dangerous) and C (consequences of accidents). D represents the risk score, Using the product of these three factors, $D = L \times E \times C$, which is used to analyze the risk of operating conditions. The greater the D value, the greater the risk of operating conditions.

(1) Analysis procedures.

① Based on the working condition of analogy, the expert group is composed of people who are familiar with the working condition.

② Expert gives scores for L, E and C according to regulations and standards, and the average value is calculated as L, E, C. Finally the risk level of operating conditions is analyzed by using the risk score D.

Because the expert scoring method is used to analyze the results, the accuracy of the analysis results will be affected by the experts' experience and the ability to judge.

(2) Scoring standards.

① The possibility of an accident L.

The probability of an accident L qualitatively expresses the probability of the accident. The probability of inevitable accident is 1, the corresponding value is 10. The probability of absolutely no accident is 0, but the production operation does not exist none accident occurrence absolutely, so the provisions of the fact that the no possibility of an accident, the corresponding value of 0.1. Other scores of corresponding situation are as following (Table 9–1), according to the regulations above.

Table 9-1 Accident Probability Score (L)

Score value	Probability
10	Inevitable accident
6	Quite possible
3	Possible, but not usual
1	Absolute suddenness, less possible
0.5	Assume, very impossible
0.2	Very impossible
0.1	Actually impossible

② Frequency extent of exposed to dangerous environment E.

The longer time of exposed to dangerous environment, the greater likelihood of harm and dangerous. The score of continuously present to hazardous environment is 10, very rarely exposed to hazardous environments with a score of 0.5, none exposed to hazardous environment is 0, but it is no practical meaning. Scoring standards are as following at Table 9-2.

Table 9-2 Frequency Extent of Exposed to Dangerous Environment (E)

Score value	Frequency
10	Continuous exposure
6	Exposure in working hours everyday
3	Once every week or sometimes
2	Once every month
1	Sometimes a year
0.5	Very vare exposure

③ Consequences of accidents C.

Because the scope of injury extent is very large due to accidents, the score of minor injury is 1, and the score of more than 10 people died is 100. The score standards are as Table 9-3. The interpolation method can be used in scoring according to the severe extent after accidents.

Table 9-3 Consequences of Accidents (C)

Score value	Consequence
100	More than 10 people die
40	Many people die but less than to
15	One people die
7	Severe injury
3	Injury
1	Minor injury, need ambulance

④ Standard of division of risk.

The score of less than 20 is lowly dangerous, less than the risk of riding to work daily by experience. The score of 70～160 is obviously dangerous, and corrective measure should be adopted. The score of 160～320 is highly dangerous, and immediate correction is needed. The score of more than 320 is extremely dangerous, and operation should be stopped and changed immediately. Risk classification standard is as Table 9–4.

Table 9–4 Risk Classification Standard

Risk score (D)	Risk degree
≥320	Extremely dangerous, stop and change operation immediately
160～320	Highly dangerous, need immediate correction
70～160	Obviously dangerous, need correction
20～70	Relatively dangerous, need attention
<20	Slightly dangerous, acceptable

9.3.1.3 American Dow Chemical Company (DOW) Fire Explosion Index Analysis Method

The fire explosion index analysis is based on the past statistics, the material's potential energy and the existing safety measures. In the process of the system, the use of material, equipment, volume and other data is a gradually calculating process.

It is based on the element material coefficient MF, process conditions (general process risk coefficient F_1 and special process risk F_2). Through a series of coefficient calculation (unit fire explosion index F&EI, influence area, damage coefficient DF calculation) to determine the unit fire explosion risk level (basic maximum possible property damage BMPPD and actual maximum possible property damage AMPPD, maximum possible loss of working days MPDO and business interruption BI). It is mainly for the analysis of the production, storage, handling of flammable and explosive, chemical and reactive substances in the chemical process and other related processes (usually used in the LNG project).

This method is used in a large number of charts and tables, so that the analysis can be made simple and clear, but there are some difficulties in the development of safety indicators.

(1) Analysis program.

After each accident, the thermal radiation intensity and pressure are calculated, and the influence of every possible accident to the surrounding personnel and property is obtained. The parameters are put forward for the further measures, and the evaluation process is shown in Figure 9–9.

(2) Basic contents analysis.

① Data preparation.

Engineering design and accurate device (production unit) design, process flow diagram, installation cost table, associated equipment replacement cost data, Dow analysis method and detailed procedure, calculation formulas, tables and appendices should be prepared.

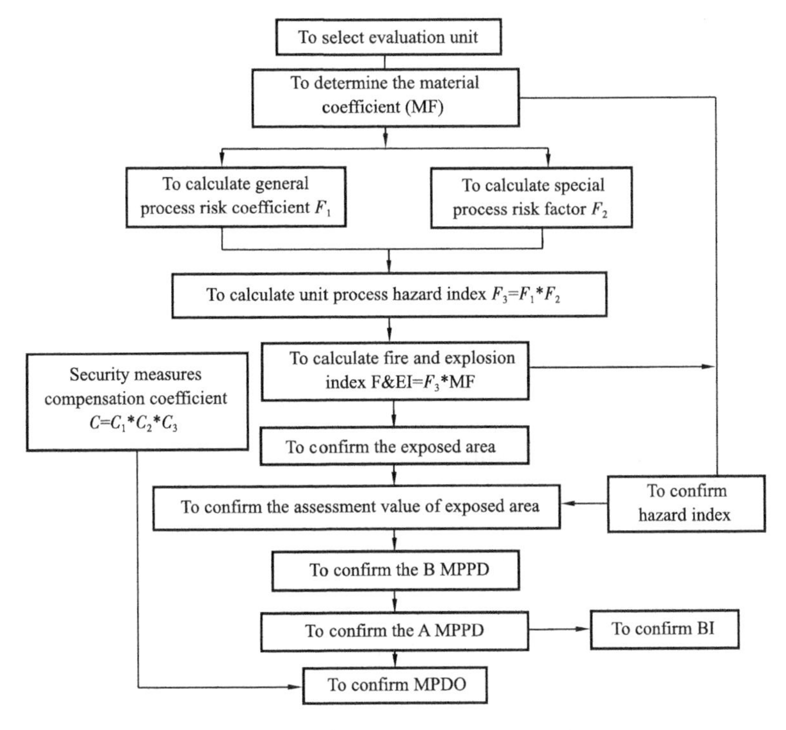

Figure 9-9 DOW chemical company analysis program

② To confirm the process unit.

③ To calculate the material factor MF of each process unit.

④ To calculate general process risk factor F_1 and special process risk factor F_2 (equal to the basic coefficient and the sum of each correction factor).

⑤ Calculation.

(a) Unit process risk factor $F_3 = F_1 \times F_2$.

(b) Fire explosion index F&EI = $F_3 \times$MF.

According to the fire explosion index, evaluate the risk level of the unit, see Table 9-5.

Table 9-5 Risk Degree Level

Fire explosion index	Risk degree
1~60	Low
61~96	Light
97~127	Medium
128~158	High
>158	Serious

⑥ To calculate the basic maximum possible loss of property (BMPPD) in the exposed area.

(a) Determination of exposed areas.

$$R = \text{F\&EI} \times \delta \tag{9-1}$$

δ——the correction coefficient;

F&EI——fire and explosion index;

R——exposed area radius, m.

$$S = \pi R^2 \tag{9-2}$$

S——exposed area, m^2.

(b) By F_3, MF can detect the unit damage coefficient.

(c) Property value in exposed areas, A.

$$A = \text{original investment} \times \xi \times \text{price coefficient} \tag{9-3}$$

ξ——the correction coefficient;

A——property value in exposed areas, million dollars.

(d) Maximum possible property damage (BMPPD).

$$\text{BMPPD} = A \times DF \tag{9-4}$$

⑦ To calculate the actual maximum possible property damage (AMPPD).

(a) Compensation coefficient of safety measures, C.

$$C = C_1 \times C_2 \times C_3 \tag{9-5}$$

C_1——Process control measure compensation coefficient;

C_2——Isolation measure compensation coefficient;

C_3——Fire protection measure compensation coefficient.

(b) Actual maximum possible property damage (AMPPD). Unit: million dollars.

$$\text{AMPPD} = \text{BMPPD} \times \text{C} \tag{9-6}$$

⑧ Maximum possible damage of working days (MPDO).

MPDO may be derived from the actual maximum possible damage of AMPPD.

⑨ Business interruption.

$$\text{BI} = (\text{MPDO}/30) \times \text{VPM} \times \phi \tag{9-7}$$

BI——business interruption, million dollars;

VPM——Daily output value;

ϕ—— Unfixed costs and profits.

⑩ Compare AMPPD and MPDO with analysis index (safety index). Safety compensation measure should be adopted if they don't meet the requirement, then calculate AMPPD and MPDO. Compare and improve them until they meet the requirement, and then improve until it meets the requirements.

9.3.2 Safety Pre-evaluation of Project

LNG project safety pre-evaluation work is generally employed with the grade A qualification of project safety evaluation and consulting agency to complete.

9.3.2.1 Evaluation purpose and significance

(1) Safety issues in engineering construction period.

① Through the LNG project construction process of the main security risk analysis, distinguish differences between the LNG project construction period specific safety risk factors and the general project safety risk factors.

② According to the relevant LNG engineering construction safety laws and regulations, standards, evaluate whether the construction program is in compliance with safety regulations and requirements. Through comparison and selection of safety construction plan during construction period, put forward the best safety construction plan.

③ To evaluate the potential accident consequences in the construction period.

④ To put forward safety countermeasures for the construction period of the project based on the analysis and evaluation results.

⑤ To provide basis for safety inspection during construction period.

(2) Safety issues in project completion.

① Analyze safety characteristics of project completion, confirm the major dangerous distribution, and analyze the major risk factors in LNG loading and unloading, storage and gasification in the production process.

② According to domestic and foreign relevant laws, design regulations and standards, evaluate the construction safety program, including site selection, general layout, production technology and equipment, facilities.

③ To simulate and evaluate the consequences of LNG or natural gas leakage, diffusion, fire and explosion accidents.

④ To propose safety measures according to analysis and evaluation results.

⑤ To provide safety design, management, supervision and approval basis.

(3) Significance of evaluation.

By carrying out the project safety pre-evaluation, implementing the principle of "safety first, prevention first", promote the engineering safety facilities and the main project design, construction, production and use. Simultaneously, it is helpful for guaranteeing the labor safety in construction process, and reducing the risk of engineering accidents. Safety pre-evaluation report provides foundation for safety design, construction management and supervision.

9.3.2.2 Evaluation Procedure

Safety pre-evaluation procedures shall be implemented in accordance with the requirements of "Safety Evaluation Rules for the Construction Project of Dangerous Chemicals". Specifically, there are four stages: pre-preparation, safety evaluation, exchange of views with the construction unit, the preparation of safety evaluation report. And the main work flow is shown in figure 9-10.

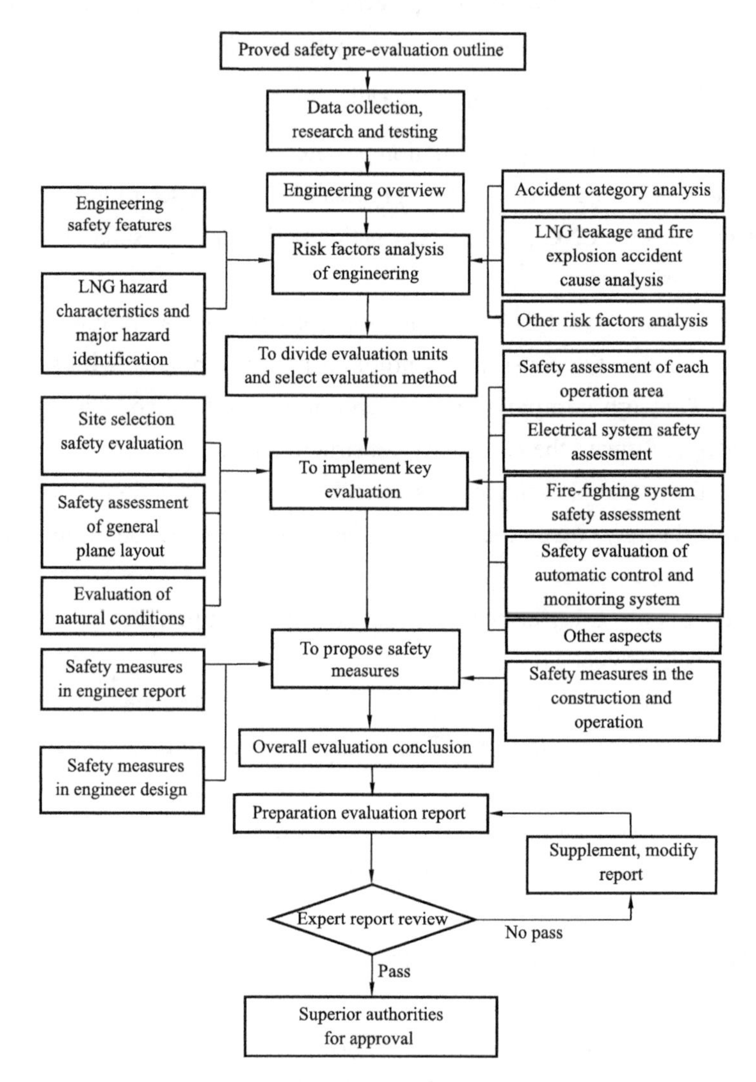

Figure 9-10 LNG project safety pre-evaluation procedures

9.3.2.3 Basis and Scope of Safety Pre-evaluation

(1) Basis for safety pre-evaluation.

1) Documents relating to laws, regulations, rules, regulations, standards, etc.;

2) Engineering feasibility study report;

3) Engineering safety pre-evaluation commission contract;

4) Other relevant engineering materials provided by construction units.

(2) Safety pre evaluation scope.

Safety pre-evaluation is consistent with geography scope and engineer scope of LNG project. Geography scope includes the waters and the road area involved in the course of the route, the port and the pipeline and the gas transmission substation. Engineer scope includes LNG transport ship, jetty, receiving terminal, gas pipelines, gas distribution station, LNG shipping, unloading, storage, gasification, pressurized external transportation and all other processes and related activities.

9.3.2.4 Contents of Safety Pre-evaluation Report

At present, in accordance with the national examination and approval units and procedures, LNG project safety pre-evaluation report concludes three parts: receiving terminal, jetty and pipeline. The report chapters are as following.

(1) Receiving terminal.

Table 9-6 Contents of Safety Pre-evaluation Report of LNG Project Receiving Terminal

Chapter	Content	
1	**Preface**	
2	**Instruction** 2.1 Evaluation purposes 2.2 Evaluation foundation	2.3 Evaluation scope 2.4 Evaluation program
3	**General Situation of Construction Project** 3.1 Introduction of project 3.2 Geographical location and relevant natural conditions 3.2.1 Geographical location 3.2.2 Relevant natural conditions 3.2.3 Surrounding environment 3.3 Engineering construction program 3.3.1 Construction scale	3.3.2 General layout 3.3.3 Process and equipment facilities 3.3.4 Power supply and lighting 3.3.5 Control and monitoring system 3.3.6 Fire-fighting facilities 3.3.7 Auxiliary production system and equipment facilities 3.3.8 Organization management and staffing
4	**Risk and Harmful Factors and Degrees in Construction Project** 4.1 Physical and chemical performance index 4.2 Dangerous and harmful factors 4.2.1 Characteristics of LNG 4.2.2 The natural gas characteristics of easy to spread, flammable, explosive, choking 4.3 Dangerous and harmful degrees 4.3.1 Evaluation unit division	4.3.2 Determination of safety assessment method 4.3.3 Inherent risk degree analysis 4.3.4 The consequences and causes of the accident cases occurred during the operation of the receiving terminal project 4.3.5 Risk degree analysis 4.3.6 Impact analysis of natural conditions
5	**Safety Conditions in Receiving Terminal** 5.1 External circumstances 5.1.1 Casualties scope of explosion, fire, suffocation, according to calculation 5.1.2 Distance between surrounding area and storage facilities (composed by the process devices and LNG storage quantity, major dangerous source)	5.2 Safety condition analysis of receiving terminal 5.2.1 The influence of receiving terminal to the surrounding units of production, business activities or residents living 5.2.2 The influence of surrounding units production, business activities and residents living to receiving terminal

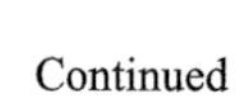

Continued

Chapter	Content
6	**Analysis of the Safety Production Conditions of the Receiving Terminal Project** 6.1 Safety assessment of total plane layout 6.1.1 General plane layout 6.1.2 Vertical layout 6.2 Safety evaluation of storage zone 6.2.1 Accident cause analysis in storage tank zone operation 6.2.2 Qualitative evaluation of process scheme in storage tank zone 6.2.3 Qualitative evaluation of equipment and facility in storage tank zone 6.2.4 Simulation evaluation of leakage, diffusion, fire and explosion accidents in storage tank zone 6.3 Safety evaluation of process zone 6.3.1 Accident cause analysis in process zone operation 6.3.2 Qualitative evaluation of process scheme in process zone 6.3.3 Qualitative evaluation of equipment and facility one by one in process zone 6.3.4 Simulation evaluation of leakage, diffusion, fire and explosion accidents in process zone 6.4 Safety evaluation of loading and unloading process and equipment, facilities in jetty zone 6.4.1 Accident cause analysis in jetty zone operation 6.4.2 Qualitative evaluation of loading and unloading process scheme in jetty zone 6.4.3 Qualitative evaluation of equipment and facility from the aspects of LNG loading arms in jetty zone 6.4.4 Simulation evaluation of leakage, diffusion, fire and explosion accidents in jetty zone 6.5 Safety evaluation of tank car loading zone 6.5.1 Accident cause analysis in tank car loading process zone operation 6.5.2 Qualitative evaluation of process scheme in tank car loading zone 6.5.3 Qualitative evaluation of equipment and facility from the aspects of tank car loading in jetty zone 6.5.4 Simulation evaluation of leakage, diffusion, fire and explosion accidents in tank car loading zone 6.6 Safety evaluation of pipeline and components 6.6.1 Pipe material selection 6.6.2 Pipe arrangement and connection 6.6.3 Pipe frame design 6.6.4 Pipeline components 6.7 Safety evaluation of automatic control system 6.7.1 The validity and reliability of automatic control 6.7.2 Quality of operating personnel 6.7.3 Maintenance of control system 6.7.4 Safety protection system 6.8 Safety evaluation of electrical system 6.8.1 Power supply 6.8.2 High and low voltage cable laying method 6.8.3 Protection and control of electrical equipment 6.8.4 Main equipment selection 6.8.5 Lighting 6.8.6 Lightning protection 6.8.7 Static electricity 6.9 Safety evaluation of fire-fighting system 6.9.1 Reliability of fire-fighting design 6.9.2 Corrosion protection of pipeline and pump 6.9.3 Configuration and layout of fire extinguishing system 6.9.4 Selection of fire extinguishing device 6.9.5 Selection of material and fire extinguishing agent 6.9.6 Maintenance of fire-fighting system 6.10 Safety evaluation of utilities 6.10.1 Drainage system 6.10.2 Construction ventilation engineering 6.10.3 Heating ventilation and air conditioning engineering 6.10.4 Telecommunication system engineering 6.11 Safety evaluation of other aspects 6.11.1 Flare system 6.11.2 Seawater supply system 6.11.3 Compressed air and nitrogen gas station 6.11.4 Central laboratory 6.11.5 Maintenance room 6.11.6 Warehouse 6.11.7 Impact of external personnel to enter 6.11.8 Quality of internal staff 6.11.9 Management system 6.11.10 Individual protective equipment 6.11.11 Personnel rescue and escape 6.11.12 Engineering is far away from the mainland, island construction personnel rescue and escape

Continued

Chapter	Content	
7	**Safety Measures** 7.1 Safety measures in the engineering feasibility study report 7.2 Safety measures complement in the engineering design 7.2.1 Safety measures proposed from terminal site and general plane layout 7.2.2 Safety measures proposed from loading, unloading and storage process, equipment and facilities, fire-fighting system, emergency rescue equipment and facilities 7.3 Safety management measures	7.3.1 Safety measures proposed from construction installation process and production operation 7.3.2 Safety measures proposed from major hazard source management 7.3.3 Safety measures proposed from ensuring safe operation 7.4 Major accidents emergency plan framework 7.4.1 Safety production accidents emergency plan system 7.4.2 Major accidents emergency plan framework
8	**Evaluation Conclusion and Recommendation** 8.1 It is pointed out that the safety problems of LNG receiving station should be paid attention to in the design, construction and operation 8.2 Whether the receiving station project is completed in accordance with the relevant national laws and regulations, rules, standards, safety requirements	8.3 The safety evaluation conclusions of the location of the receiving terminal, the general layout, the selection of process equipment and the auxiliary production facilities 8.4 Summary of the main dangerous and harmful factors in the receiving terminal project, and major risk and harmful factors in prevention

(2) Jetty.

Table 9-7 LNG Project Jetty Safety Pre-evaluation Report

Chapter	Content	
1	**Preface**	
2	**Instruction** 2.1 Evaluation purposes 2.2 Evaluation foundation	2.3 Evaluation scope 2.4 Evaluation program
3	**General Situation of Construction Project** 3.1 Introduction of project 3.2 Geographical location and relevant natural conditions 3.2.1 Geographical location 3.2.2 Relevant natural conditions 3.2.3 Surrounding environments 3.3 Engineering construction program 3.3.1 Construction scale 3.3.2 Cargo throughput and flow 3.3.3 Ship design	3.3.4 General plane layout 3.3.5 Unloading process and equipment 3.3.6 Power supply and lighting 3.3.7 Control and monitoring system 3.3.8 Telecommunication 3.3.9 Water supply and drainage system 3.3.10 Fire-fighting facilities 3.3.11 Jetty operation standard and working days 3.3.12 Organization management and staffing

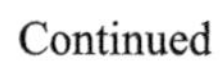
Continued

Chapter	Content	
4	**Analysis of engineering risk and harmful factors** 4.1 Safety characteristics 4.1.1 LNG has easy evaporating, easy flow, easy to produce static, low temperature, fast phase change and other characteristics 4.1.2 Natural gas is easy to spread, flammable, explosive, choking 4.2 Identification of major hazards 4.3 Analysis of engineering accident category 4.3.1 LNG or natural gas leakage, diffusion, fire, explosion accidents 4.3.2 Personnel of asphyxia, hypothermia frostbite, mechanical injury, object striking, drowning, electric shock, falling and other personal injury accidents 4.3.3 LNG transport vessel, cryogenic vessel, piping and other equipment damage accidents 4.4 LNG/natural gas leakage accident cause analysis	4.4.1 Human unsafe behavior 4.4.2 Quality defects or failures of equipment and facilities 4.4.3 Influence of external factors 4.5 LNG/ natural gas fire and explosion accident cause analysis 4.5.1 Static electricity 4.5.2 Non-explosion proof electrical equipment 4.5.3 Sources of ignition such as thermal operation 4.6 Other risk factors analysis 4.6.1 Personal poison and suffocation 4.6.2 Cryogenic frostbite 4.6.3 Mechanical injury 4.6.4 Object striking 4.6.5 Drowning 4.6.6 Electric shock 4.6.7 Falling and other personal injury accidents
5	**Evaluation unit division and evaluation method selection** 5.1 Evaluation unit division	5.2 Pre-evaluation method selection and introduction 5.2.1 Pre-evaluation method selection 5.2.2 Introduction of safety evaluation method
6	**Qualitative and quantitative evaluation** 6.1 Safety evaluation of engineering construction program 6.1.1 LNG jetty site 6.1.2 LNG jetty general plane layout 6.1.3 Unloading process and equipment, facilities 6.1.4 Fire-fighting system 6.1.5 Related supporting engineering 6.1.6 Ship berthing and unberthing operation safety 6.1.7 Influence of natural conditions 6.1.8 Influence of other factors 6.2 Safety evaluation of leakage and diffusion risk 6.2.1 LNG leakage accident of the jetty	6.2.2 Quantitative evaluation of LNG leakage and diffusion using approved software 6.2.3 Determination of vapor cloud hazard and harmful scope and key preventive measure 6.3 Risk assessment of fire and explosion 6.3.1 The fire and explosion hazard of LNG ship was calculated by the method of Dow chemical fire and explosion hazard index in LNG unloading process unit 6.3.2 To set up the LNG leakage accident of the jetty pipeline, using the approved software to simulate the pool fire and flash fire accident
7	**Safety measures** 7.1 Safety measures in the engineering feasibility study report 7.2 Safety measures complement in the engineering design 7.2.1 Safety measures proposed from jetty site and general plane layout	7.2.2 Safety measures proposed from unloading process, equipment and facilities, fire-fighting system and relevant engineering. 7.3 Safety management measure 7.3.1 Safety measures proposed from construction installation process and production operation

Continued

Chapter	Content	
7	7.3.2 Safety measures proposed from major hazard source management 7.3.3 Safety measures proposed from ensuring safe operation of LNG jetty 7.4 Major accident emergency plan framework	7.4.1 Safety production accidents emergency plan system 7.4.2 Major accidents emergency plan framework
8	**Evaluation conclusion and recommendation** 8.1 Safety issues need to be focused on the jetty design, construction and operation 8.2 Summary of major risks, harmful factors 8.3 From the jetty location, general layout, unloading process, whether it meets the requirements of the given evaluation conclusion	8.4 Whether the safety facilities and safety measures proposed in the report of the feasibility study of the jetty meet with the relevant safety regulations and standards. After adopting the evaluation report to put forward the safety measures, whether they meet the safety requirements of production

(3) Pipeline.

Table 9-8 LNG Project Pipeline Safety Pre-evaluation Report

Chapter	Content	
1	**Preface**	
2	**Instruction** 2.1 Evaluation purposes 2.2 Evaluation foundation	2.3 Evaluation scope 2.4 Evaluation program
3	**General Situation of Construction Project** 3.1 Introduction of project 3.2 Pipeline overview 3.2.1 Geographical location 3.2.2 Gas transmission scale 3.2.3 Pipeline components 3.2.4 Administrative divisions along the pipeline 3.2.5 Gas transportation process 3.2.6 Piping trend charts and corresponding tables 3.3 Pipeline engineering 3.3.1 Pipeline trend 3.3.2 Natural condition along the pipeline 3.3.3 Social environment 3.3.4 Pipe material and size 3.3.5 Crossing engineering 3.3.6 Pipeline corrosion protection 3.3.7 Cut off valve room 3.4 Process station 3.4.1 Station setting	3.4.2 Process 3.5 Power, automatic control and communication 3.5.1 Power supply and distribution program 3.5.2 Lightning protection 3.5.3 Prevention of static electricity 3.5.4 Automatic control system 3.5.5 Telecommunication system 3.6 Water supply, drainage and fire fighting 3.6.1 Water supply 3.6.2 Drainage 3.6.3 Fire fighting 3.7 Main engineering quantity 3.7.1 Pipeline engineering 3.7.2 Cathodic protection 3.7.3 Sub-transmission station engineering 3.7.4 Automatic control system 3.7.5 Telecommunication system 3.7.6 Power supply system 3.7.7 Thermal engineering 3.7.8 Construction 3.7.9 Crossing project

Continued

Chapter	Content	
3	3.7.10 Auxiliary engineering 3.7.11 Pipeline corrosion protection 3.7.12 Station lists specification 3.8 Pipeline extension 3.8.1 transport capacity check 3.8.2 Safety	3.8.3 Health 3.8.4 Fire fighting 3.8.5 Safety management 3.9 Road engineering 3.10 Organization and personnel
4	**Analysis of hazard and harmful factors** 4.1 Engineering safety features 4.1.1 Flooding along the pipeline 4.1.2 Landslide 4.1.3 Railway 4.1.4 Village of dense population 4.1.5 Sensitive military facilities and other external conditions 4.1.6 Surrounding circumstance of main valve chamber 4.2 Analysis of natural gas physical and chemical properties and hazard, harmful factors 4.2.1 Natural gas components and basic properties	4.2.2 Natural gas has easy diffusion, flammable, explosive, suffocation and other characteristics 4.3 Analysis of hazard factors in process 4.3.1 Analysis of hazard factors in station process, mainly including pipeline rupture and equipment failure in station 4.3.2 Analysis of hazard factors in pipeline gas transmission process, mainly including pipeline leakage and fire, explosion 4.3.3 Other risk factors analysis, the main natural conditions, social environment
5	**Evaluation units division and method selection** 5.1 Qualitative evaluation method of onshore pipeline 5.1.1 Safety check method 5.1.2 Pipeline risk assessment method 5.2 Quantitative evaluation method selection of onshore pipeline 5.2.1 Selecting Dow chemical fire evaluation method	5.2.2 Explosion hazard index evaluation method 5.2.3 Approved software 5.3 Qualitative evaluation method selection of offshore pipeline 5.3.1 Preliminary hazard analysis method 5.3.2 Safety check method 5.4 Approved software
6	**Qualitative and quantitative evaluation** 6.1 Onshore gas transmission pipeline 6.1.1 Qualitative evaluation of gas transmission station 6.1.2 Accident simulation of gas station gas leakage 6.1.3 Qualitative evaluation of gas transmission pipeline 6.1.4 Gas leakage accident simulation of gas pipeline 6.1.5 Evaluating leakage accidents, risk and probability of gas transmission pipeline using fault tree analysis method, expert evaluation and statistical analysis method 6.2 Offshore gas transmission pipeline	6.2.1 Qualitative evaluation of offshore gas transmission pipeline design from the effects of surrounding environment and natural disaster on offshore pipeline design and future plan 6.2.2 Qualitative evaluation of pipeline laying phase from the aspects of pipe rupture, frame damage, structural damage of vessel, ship capsizing, losing anchor, cable breakage, collision, pipeline shift, freedom suspending, pipe corrosion and landing hazard 6.2.3 Qualitative evaluation of operation stage from the effects of waterways, fishery activity, anchoring, shipwreck, dredging operation on the offshore pipeline

Continued

Chapter	Content	
7	**Safety measures** 7.1 Safety measures in the engineering feasibility study report 7.2 Safety measures complement in the engineering design 7.2.1 pipeline direction 7.2.2 Pipeline laying 7.2.3 Pipeline corrosion protection 7.2.4 Line pipe 7.2.5 Cut-off valve chamber 7.2.6 pipeline logo 7.2.7 Utilities 7.2.8 Natural disasters protection 7.3 Safety measures in the construction installation process 7.3.1 Comprehensive measures	7.3.2 Security measures in bad geological conditions 7.4 Safety management measure 7.4.1 Comprehensive safety measures 7.4.2 Security measures proposed from major hazard source management 7.4.3 Safety measures proposed from ensuring safe operation of natural gas transmission pipeline 7.5 Major accidents emergency plan framework 7.5.1 Safety production accidents emergency plan system 7.5.2 Major accidents emergency plan framework
8	**Evaluation conclusion and recommendation** 8.1 Summary of major risks, harmful factors existing in natural gas transmission pipeline 8.2 Evaluation conclusion of gas transmission pipelines complied with standards 8.2.1 pipeline direction 8.2.2 Pipeline laying 8.2.3 Crossing and shield engineering 8.2.4 Pipe material selection 8.2.5 Pipe thickness design 8.2.6 Pipeline corrosion protection	8.2.7 Station engineering 8.2.8 Cut-off valve chamber setting 8.2.9 Auxiliary production facilities 8.3 General evaluation conclusion of safety production requirements complies with national relevant laws, regulations, rules, specifications and standards 8.4 Safety problems needs highly paid attention to in gas transmission engineer design, construction and production

9.3.3 Pre-evaluation of Project Occupational Disease Hazard

The Pre-evaluation work of LNG project occupational disease hazard should be done by occupational health evaluation agency with relevant qualification.

9.3.3.1 Evaluation Purposes and Significance

(1) Analysis of occupational disease risk factors in LNG project.

According to the occupational health characteristics of LNG project, the main occupational disease hazard factor's kinds, distribution and damage degrees can be analyzed and identified in LNG loading and unloading, storage, gasification, pressurized transmission and other production process, and prevention measures of occupational disease can be proposed, and labor health can be protected, according to the occupational health characteristics of LNG project.

(2) Providing scientific foundation for the feasibility study report.

According to occupational health laws, regulations, norms, and standards, it needs to determine the occupational disease hazard categories of construction projects, evaluate project feasibility study report consistent with occupational disease prevention and control requirements, in order to provide scientific basis for health administrative departments to review the LNG overall project receiving terminal and gas pipeline feasibility study report.

(3) Providing foundation for the preliminary design of the project.

It also needs to provide design foundation of occupational health relevant for the preliminary design of LNG general project receiving terminal and main gas transmission pipeline engineering.

(4) Provide foundation for production and operation.

It is necessary to provide the basis for the systematic management of occupational health for the construction unit after the completion of the project.

9.3.3.2 Evaluation Program

Figure 9–11 is the pre-evaluation flow chart of occupational disease hazard, for reference. The preparation work includes study of occupational health laws, regulations and standards, and other documents, technical materials relevant to construction projects, analysis of the preliminary engineering scheme, and preparing pre-evaluation plan. The core work concludes investigation of occupational health (on-site investigation, analog investigation), and analysis of construction project engineering and occupational hazards factors. The key is the construction project occupational disease hazards pre-evaluation. Finally, the pre-evaluation report is submitted after examination by experts.

9.3.3.3 Evaluation Basis and Scope

(1) Evaluation basis.

① Relevant national and local occupational health laws, regulations, rules, standards.

② Occupational disease hazards pre-evaluation power of attorney of LNG receiving terminal and gas transmission pipeline sub-projects.

③ Occupational disease hazards pre-evaluation contract of LNG receiving terminal and gas transmission pipeline sub-projects.

④ Occupational disease hazards pre-evaluation program of LNG receiving terminal and gas transmission pipeline sub-projects.

⑤ The general feasibility study report of receiving terminal and gas transmission pipeline sub-projects.

(2) Evaluation scope.

① The evaluation of the sub-project scope includes LNG jetty, LNG receiving terminal and gas transmission pipeline. The process and related activities of LNG unloading, storage, gasification, and pressure transmission are evaluated.

② Evaluation of occupational hazards factors that may occur only after the project is completed and put into operation, which does not include the occupational hazards factors in the process of project construction.

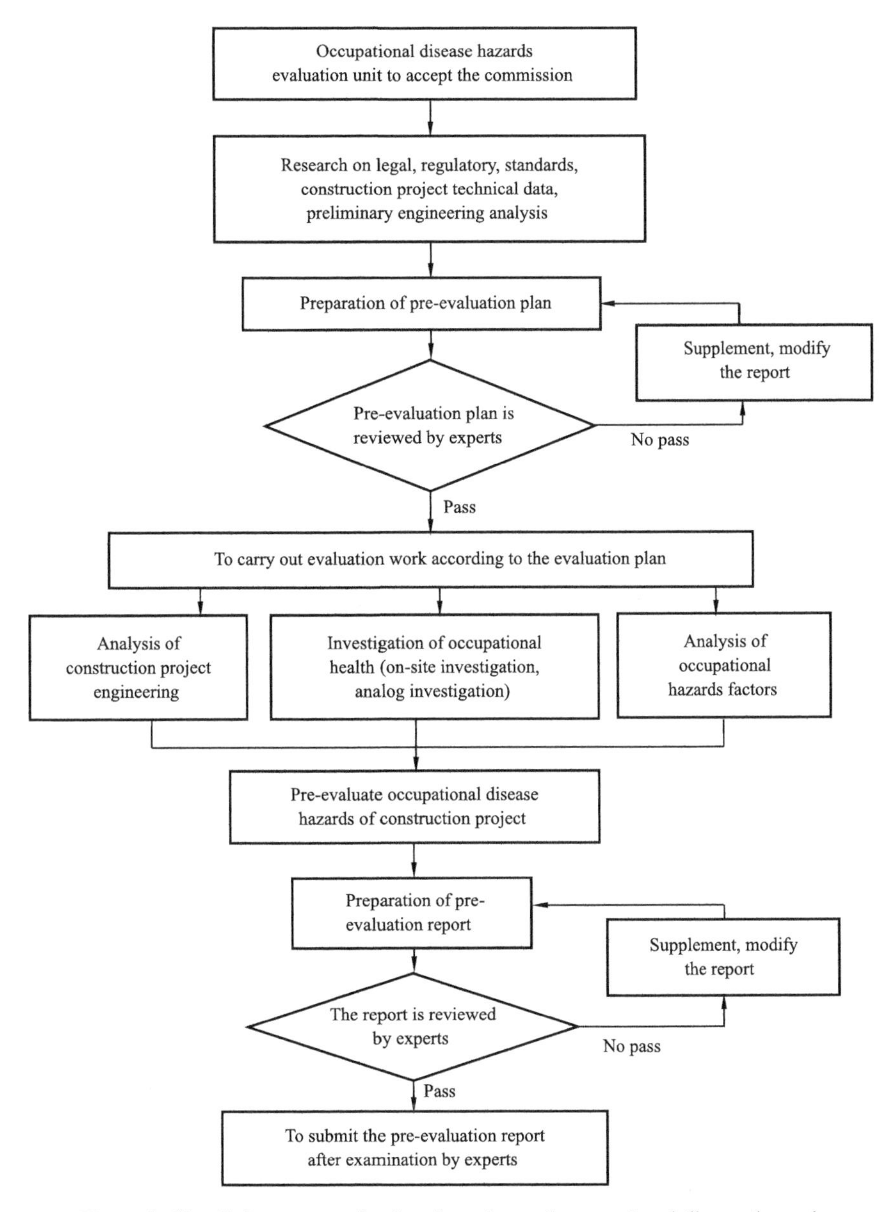

Figure 9–11 Safety pre–evaluation flow chart of occupational disease hazards

③ Evaluation contents include site selection and layout, production process and equipment layout, requirements of buildings hygiene, occupational disease hazards factors identification and evaluation, occupational health protective facilities, personal occupational protective equipment, emergency rescue measure, basic health requirements of auxiliary room, occupational health management, and occupational health budget estimation.

9.3.3.4 Contents of Occupational Disease Hazards Pre–evaluation Report

Occupational disease hazards pre-evaluation is a necessary work of project companies, and the evaluation work is embodied in the report. The report chapters are as following (Table 9–9).

Table 9-9 LNG Project Pre-evaluation Report on Occupational Hazards

Chapter	Content	
1	**General** 1.1 Project origin 1.2 Evaluation basis 1.2.1 Laws and regulations 1.2.2 Foundation 1.2.3 Project basis 1.2.4 Evaluation specification and standard 1.3 Evaluation purpose 1.3.1 According to the characteristics of the occupational health of the project, provide the occupational disease hazards protection measures	1.3.2 Implement occupational health related laws, regulations, standards and standards 1.3.3 Provide the basis for the occupational health system management of the production unit after the completion of the project 1.4 Evaluation scope and content 1.4.1 Evaluation scope 1.4.2 Evaluation content 1.5 Evaluation method 1.6 Evaluation program
2	**Engineering analysis** 2.1 General situation of construction project 2.1.1 Jetty and receiving terminal 2.1.2 Gas transmission pipeline 2.2 Natural environment of the project location 2.2.1 Geography location 2.2.2 Weather condition	2.3 Plane layout and equipment layout 2.3.1 Plane layout 2.3.2 Equipment layout 2.4 Labor force and post setting 2.5 Production process 2.5.1 Receiving terminal process flow 2.5.2 Main process flow of gas transmission station
3	**Harmful factors identification and analysis of occupational disease** 3.1 Harmful factors identification of occupational disease 3.1.1 Harmful (chemical and physical) factors produced in production process 3.1.2 Harmful factors in operation process 3.1.3 Harmful factors in production environment 3.2 The physical and chemical properties of the main occupational disease and the harm to human health	3.2.1 Harmful factors in production process 3.2.2 Unreasonable ergonomic design for human health hazards 3.3 Harmful extent analysis of occupational disease hazards 3.3.1 Analysis method 3.3.2 Selection of analog enterprise 3.3.3 Test results of analog enterprise occupational disease harm 3.3.4 Harm extent prediction in production process 3.3.5 Classification results prediction of labor condition
4	**Protective measures of occupational disease adopted in the feasibility study** 4.1 Occupational health management 4.2 Protective measures of occupational disease 4.2.1 Poison protective facilities 4.2.2 Noise control measure 4.2.3 Low temperature protective measure	4.3 Personal occupational disease protective facilities 4.4 Auxiliary chamber setting 4.5 Emergency rescue measure 4.6 Occupational health supervision 4.7 Occupational health special budget 4.8 Greening
5	**Occupational health pre-evaluation** 5.1 Site selection evaluation 5.1.1 Site selection evaluation of receiving terminal	5.1.2 Main gas transmission pipeline route selection evaluation 5.2 General plane layout evaluation 5.2.1 Jetty

Continued

Chapter	Content	
5	5.2.2 LNG receiving terminal 5.2.3 Main gas transmission pipeline and station 5.3 Production process and equipment layout evaluation 5.3.1 Production process evaluation 5.3.2 Equipment layout evaluation 5.4 Building hygienical evaluation 5.5 Occupational disease harmful factors evaluation 5.6 Occupational disease protective equipment evaluation	5.7 Personal occupational disease protective facilities evaluation 5.8 Auxiliary chamber setting evaluation 5.9 Emergency rescue measure evaluation 5.10 Occupational health supervision evaluation 5.11 Occupational health management evaluation 5.12 Occupational health special budget estimation 5.13 Determination of occupational disease hazards category
6	**Conclusions**	
7	**Recommendations**	

9.3.4 Project Environmental Impact Evaluation

LNG project environment impact evaluation work should be done by the engineering construction environment evaluation agency which has relevant qualification.

9.3.4.1 Evaluation Purposes

According to the specific situation of the project, combined with the surrounding environment of the jetty, receiving terminal and gas transmission pipeline, the environmental impact assessment work is intended to achieve the following objectives.

(1) To provide the environmental protection decision-making basis for the competent department and production department.

The fundamental purpose of LNG project environmental impact evaluation is to prevent the pollution of the environment, provide the basis for decision-making and approval of the competent department of environmental protection of construction projects, and provide technical support for the engineering design and production management of the owners.

(2) To make clear impact engineering to environment.

Through the project engineering analysis and investigation, be familiar with the process and engineering characteristics. Investigate and analyze the production and pollution links, and the situation of the surrounding environment, determine the environmental impact factors of the construction project, and determine the environmental sensitive points and the environmental protection targets of the construction project environmental assessment area.

(3) To predict the environmental effect of construction operation period.

According to the environment and engineering features, it needs to forecast the extent and scope of the impact of construction projects on ecological environment, atmospheric environment, water environment, noise environment and environmental risk, and analyze the change of the surrounding environment quality caused by the construction and operation period of the project.

(4) To propose environmental protection measures.

According to the requirements of the state on the construction projects in clean production, emission standards, energy and resources saving, discuss the project's advance, rationality of layout scheme and construction scheme. Through the analysis of the reliability of the technology of environmental protection facilities, put forward the measures and suggestions to further reduce pollution and protect ecological environment.

(5) To demonstrate environment and social benefits of the project.

Through the total amount control and environmental economic profit and loss analysis, it is proved that the clean energy (LNG) can be used to replace the conventional energy and produce environmental benefits and social benefits.

(6) To comply with laws, regulations and requirements.

According to the requirements of national relevant laws and regulations, project environment effect evaluation is a necessary condition of project approved, and is the indispensable contents in feasibility study report. It also provides the basis for the development of environmental impact assessment in the later period. At the same time, it needs to submit to the Supervision and Management Department of the State Environmental Protection Administration for record.

9.3.4.2 Evaluation Basis and Principles

(1) Evaluation basis.

① Relevant national and local laws, regulations and relevant HSE standards and specifications.

② Environment effect evaluation power of attorney of LNG receiving terminal and gas transmission pipeline sub-projects.

③ Environment effect evaluation work contract of LNG receiving terminal and gas transmission pipeline sub-projects.

④ The feasibility study report of LNG receiving terminal and gas transmission pipeline sub-projects.

(2) Principles of environment effect evaluation.

The general principle is to conscientiously implement "The Environment Protection Law of the People's Republic of China" "The Law of Environmental Impact Assessment" "Construction Project Environmental Protection Management Regulations" and other national, industry and local laws and regulations of environmental protection regulations and requirements.

① To comply with industrial policy and meet functional region planning.

It needs to earnestly study the relevant state industrial policy, based on the project construction "comply with industrial policy", carefully analyze the extent and scope of the impact of the project on the surrounding environment, so that the construction of the project "meet the functional area planning" requirements.

② To comply with the local overall planning and promote the regional economic development.

New project should be consistent with the overall planning of the region. With the LNG project as the source, promote the rise of gas power, urban gas, industrial fuel, chemical industry and other industries, and promote the development of local economy.

③ To implement "clean production" and "reaching the discharge standard".

Evaluation work should carry out the "clean production" principle. In accordance with the "clean production" requirements, implement the whole process of pollution control, thus effectively reduce the amount of pollutants and emissions. Evaluation work can take the "reaching the discharge standard" and "total quantity control" principles.

④ To take the point band line and focus on outstanding.

The characteristic of LNG project receiving terminal is a point, extending pipeline as line. Environment impact evaluation work will adopt the measure of a point extending to line, and all lines feedback. The principle is sensitive environmental protection target lot should be focused on analysis, and evaluation work should be focused on the practicality and pertinence.

⑤ To fully possess materials and assure the evaluation quality.

The evaluation work should select and use the existing monitoring data, avoid the unnecessary and duplicative work, collect the existing data of evaluation region, and assure the evaluation conclusion is real, objective, fair and clear. The natural, social environment and environment quality should be studied and used carefully. In order to save evaluation expense and accelerate the progress of evaluation, the effectiveness and timeliness should be analyzed, according to the principles of assuring evaluation product quality.

9.3.4.3 Contents of Environment Impact Assessment Report

Environmental impact assessment work is embodied in the report. According to the existing LNG project environmental evaluation report, it can be summarized as following (Table 9-10).

Table 9-10 LNG Project Environmental Impact Assessment Report

Chapter	Content	
1	**General** 1.1 Project origin 1.2 Compilation basis 1.2.1 National laws and regulations 1.2.2 Local environmental protection regulations 1.2.3 Evaluation technical guideline 1.2.4 Engineering materials 1.2.5 Others 1.3 Evaluation purposes and principles 1.3.1 Evaluation purposes 1.3.2 Evaluation principles 1.4 Evaluation work levels 1.4.1 Marine environmental assessment levels 1.4.2 Working levels of ecological environment influence	1.4.3 Evaluation levels of atmosphere environment influence 1.4.4 Evaluation levels of noise environment influence 1.5 Evaluation scope 1.5.1 Evaluation scope of marine environment 1.5.2 Atmosphere environment evaluation scope of jetty, receiving terminal and gas transmission pipeline 1.5.3 Evaluation scope of noise environment 1.5.4 Evaluation scope of onshore ecology 1.6 Evaluation period 1.6.1 Atmosphere 1.6.2 Surface water

Continued

Chapter	Content	
1	1.6.3 Investigation of onshore ecology environment 1.6.4 Noise 1.7 Environment protection targets 1.8 Evaluation standards 1.8.1 Evaluation standard of atmosphere	1.8.2 Evaluation standard of surface water 1.8.3 Evaluation standard of noise 1.8.4 Others 1.9 Evaluation focus 1.10 Evaluation of thematic settings
2	**Overview of LNG receiving terminal and gas transmission pipeline project** 2.1 Basic situation 2.2 Construction site 2.3 Project introduction 2.3.1 Composition and scale of jetty engineering 2.3.2 Composition and scale of receiving station engineering	2.3.3 Composition and scale of the trunk line engineering 2.4 Composition and characteristics of fuel gas product 2.5 Management, capacity and running hours in one year 2.5.1 Management and capacity 2.5.2 Running hours in one year 2.6 Project construction plan
3	**Engineer analysis** 3.1 Plane layout analysis 3.1.1 Jetty engineering 3.1.2 Receiving terminal engineering 3.1.3 Gas transmission pipeline engineering 3.2 Process description and pollution analysis in operation period of project 3.2.1 Jetty engineering 3.2.2 Receiving terminal engineering 3.2.3 Gas transmission pipeline engineering 3.3 Construction process, engineering quantity and environment influence factors analysis 3.3.1 Jetty engineering 3.3.2 Receiving terminal engineering 3.3.3 Gas transmission pipeline engineering 3.3.4 Qualitative analysis of construction environment impact factors	3.4 Quantitative analysis of "three wastes" emission 3.4.1 Operation period 3.4.2 Construction period 3.5 Environmental protection measures and investment 3.5.1 Jetty engineering 3.5.2 Receiving terminal engineering 3.5.3 Gas transmission pipeline engineering 3.5.4 Ecological protection and water and soil conservation 3.5.5 Others 3.6 Summary 3.6.1 Rationality analysis of general layout 3.6.2 Analysis of main environment effect factors 3.6.3 Environmental protection measures and investment
4	**LNG project site selection and environmental feasibility analysis** 4.1 Jetty and receiving terminal 4.1.1 Analysis of planning compliance 4.1.2 Environmental sensitivity analysis	4.1.3 Rationality analysis of sea area 4.2 Pipeline engineering 4.2.1 Analysis of planning compliance 4.2.2 Environmental sensitivity analysis 4.3 Conclusion
5	**General situation of construction project surrounding region environment** 5.1 Natural environment 5.1.1 Geography location 5.1.2 Geography features	5.1.3 Topography 5.1.4 Meteorology 5.1.5 Hydrology 5.1.6 Soil 5.1.7 Vegetation

Continued

Chapter	Content	
5	5.1.8 Soil and water conservation 5.1.9 Earthquake 5.2 Social environment 5.2.1 Population 5.2.2 Total output value of industrial and agricultural 5.2.3 Cultural heritage, landscape and special conservation areas 5.2.4 Port area 5.3 Environment quality status 5.3.1 Environmental air quality status 5.3.2 Surface water environment quality status 5.3.3 Current status of marine sediments 5.3.4 Current status of marine plankton and benthic ecology in the tide 5.3.5 Biological status in intertidal zone 5.3.6 Sound environment quality status	5.4 Project location of regional environmental planning 5.4.1 City overall planning 5.4.2 City Environmental Planning 5.4.3 Marine functional zoning and environmental function zoning 5.4.4 The functional categories of the receiving terminal, the sub-stations and the terminal station 5.4.5 Function categories of gas pipeline region 5.5 Regional pollution source investigation 5.5.1 Emission status of waste water pollution source 5.5.2 Emission status of waste gas pollution source 5.6 Conclusions
6	**Clean production analysis** 6.1 General 6.2 Product analysis of cleaner production 6.3 Advanced analysis of production process, equipment and facilities 6.3.1 Advanced analysis of production process 6.3.2 Advanced analysis of equipment and facilities 6.4 Energy utilization 6.4.1 LNG cold energy utilization 6.4.2 Saving energy technology and equipment 6.5 Analysis of water saving	6.6 Clean production analysis in construction period 6.6.1 Channel dredging 6.6.2 Receiving terminal and sub-transmission station construction 6.6.3 Pipeline engineering construction 6.7 Environment Management Analysis 6.7.1 Environment management agency establishment 6.7.2 Environment management system 6.7.3 Environment supervision 6.8 Conclusions
7	**Marine environmental impact evaluation** 7.1 Investigation and evaluation of water quality situation in sea area 7.1.1 Investigation of water quality situation in sea area 7.1.2 Investigation results and evaluation of water quality situation in sea area 7.2.1 Survey station locations and monitoring time 7.2.2 Monitoring project, analysis method and evaluation criteria 7.2.3 Survey results and evaluation 7.3 Plankton ecological investigation and evaluation	7.3.1 Chlorophyll and primary productivity 7.3.2 Phytoplankton 7.3.3 Zooplankton 7.3.4 Fish eggs and larvae 7.4 Benthic ecological investigation and evaluation 7.4.1 Material and method 7.4.2 Species composition and distribution 7.4.3 Composition and distribution of biomass and density 7.4.4 Composition of large benthic biological community and distribution of main species 7.4.5 Species diversity (H), evenness (J), abundance (d) and dominance (D)

Continued

Chapter	Content	
7	7.4.6 Relationship between large benthic biological distribution and environment factors 7.4.7 Analysis and evaluation of benthic biological ecology situation 7.4.8 Material collection and evaluation of benthic biological ecology 7.5 Intertidal biology investigation and evaluation 7.5.1 Material and method 7.5.2 Investigation results of soft intertidal zone 7.5.3 Lithofacies of intertidal zone 7.6 Marine environment influence evaluation 7.6.1 Analysis of main pollution sources and influence sources of engineering construction 7.6.2 Marine environment influence evaluation in construction period 7.6.3 Marine environment influence prediction and evaluation of operation period	7.7 Water intake and drainage outlet scheme and comparison 7.7.1 Marine environment influence of cold seawater discharge 7.7.2 Analysis on the influence of the dynamic conditions of the discharge water on the water intake temperature drop 7.7.3 Analysis of investment in each drainage outlet project 7.7.4 Comprehensive analysis and comparison of each intake and outlet scheme 7.8 Evaluation conclusions of marine theme 7.8.1 Evaluation of seawater and sediments quality situation 7.8.2 Evaluation of marine ecology situation 7.8.3 Marine environment influence evaluation in construction period 7.8.4 Marine environment influence evaluation of operation period 7.8.5 Water intake and drainage outlet location schemes demonstration and comparison
8	**Environmental air quality status evaluation and prediction** 8.1 Environmental air quality status survey and evaluation 8.1.1 Monitoring time 8.1.2 Monitoring items and frequency 8.1.3 Analysis method 8.1.4 Results of monitoring and statistics 8.1.5 Evaluation of environment quality status 8.1.6 Investigation results and evaluation of environment air quality status 8.2 Environment atmosphere quality situation	evaluation and impact prediction in operation period 8.2.1 Pollution meteorology material statistics 8.2.2 Atmosphere environment impact prediction 8.2.3 Atmosphere environment impact evaluation 8.3 Forecast of atmospheric environment of the sub-stations and terminal station 8.4 Analysis of atmospheric environment impact during construction period 8.5 Conclusion
9	**Environment noise status and impact evaluation** 9.1 Survey and evaluation of environmental noise 9.1.1 Noise status monitoring 9.1.2 Monitoring results and evaluation 9.2 Environment noise status impact evaluation and prediction in operation period 9.2.1 Main noise source in operation period 9.2.2 Environment noise impact prediction 9.2.3 Environment noise impact evaluation and conclusion	9.3 Environment noise impact analysis in construction period 9.3.1 Main noise source in construction period 9.3.2 Noise impact analysis in construction period 9.4 Conclusions 9.4.1 Noise environment status 9.4.2 Environment noise status impact evaluation and prediction in construction period

Continued

Chapter	Content	
9	9.4.3 Environment noise status impact prediction and analysis in construction period	9.4.4 Noise control measures
10	**Analysis of land ecological environmental impact** 10.1 Land ecological environment situation investigation and evaluation 10.1.1 Investigation of vegetation situation on land 10.1.2 Investigation of wildlife 10.1.3 Nature reservation zone 10.1.4 Evaluation of the current situation of land ecological environment 10.2 Analysis of land ecological environmental impact 10.2.1 Land ecological environment impact	in operation period 10.2.2 Land ecological environment impact in construction period 10.2.3 Influence on rare and endangered plants and animals 10.2.4 Influence on nature reservation zone, scenic spots and forest parks 10.3 Land ecological environment protection measures 10.4 Investigation of soil distribution 10.5 Conclusion and recommendations
11	**Soil and water conservation program** 11.1 Purpose 11.2 Soil and water conservation situation 11.2.1 Topography 11.2.2 Soil utilization situation 11.2.3 Soil and water erosion prevention zoning 11.2.4 Erosion characteristics and conservation situation 11.3 Prediction of the number of abandoned solid waste and soil erosion in construction 11.3.1 Responsibility scope of prevention and cure 11.3.2 The mount of spoil 11.3.3 Erosion area 11.3.4 Erosion prediction 11.3.5 Erosion hazard analysis 11.4 Soil and water conservation measures	11.4.1 Scheme principle and prevention purpose 11.4.2 Engineering measures 11.4.3 Plant measures 11.4.4 Other measures 11.5 Implementation measures of scheme 11.5.1 Leadership measures 11.5.2 Technical assurance measures 11.5.3 Coordination and arrangement of general measures 11.5.4 Implementation and arrangement schedule 11.6 Investment estimation and benefit analysis of water and soil conservation 11.6.1 Fund source and arrangement 11.6.2 Benefit analysis 11.6.3 Control degree 11.6.4 Vegetation coverage rate
12	**Environment hazard evaluation** 12.1 Project characteristics and hazard evaluation method 12.1.1 Project characteristics 12.1.2 Medium transmission 12.1.3 Object and content of environment hazard evaluation 12.1.4 Methods 12.2 Environment hazard analysis of gas transmission pipeline	12.2.1 Environment hazard identification of gas transmission pipeline 12.2.2 Investigation of accident data in long distance pipeline 12.2.3 Cause analysis of pipeline accident 12.2.4 Identification of accident hazard factors and analysis of accident consequence 12.2.5 Pipeline accident ratio and environment hazard

Continued

Chapter	Content	
12	12.2.6 Accident prevention and emergent measures 12.3 Environment hazard analysis of receiving terminal 12.3.1 Receiving terminal profile 12.3.2 Major hazard source identification 12.3.3 Potential hazard factors and identification 12.3.4 LNG tank accident frequency 12.3.5 Identification of accident hazard source and analysis of accident consequence	12.3.6 Accident prevention measures 12.3.7 Emergent measures 12.4 Hazard analysis of marine environment 12.4.1 Cause analysis of ship accident 12.4.2 Analysis of ship accident of LNG leakage impact on environment 12.4.3 Prevention and emergent protection measures of ship hazard accident in operation period 12.5 Conclusions
13	**Total pollutant discharge control** 13.1 Total control factors 13.2 Control project total accounting analysis 13.3 Total control approach	13.3.1 Approach analysis of total amount control 13.3.2 The feasibility analysis of approach 13.4 Conclusions
14	**Public participation** 14.1 Investigation scope and principles 14.2 Investigation method and objects 14.2.1 Investigation methods 14.2.2 Investigation objects 14.3 Investigation contents	14.4 public opinion analysis and sorting 14.4.1 Statistical analysis methods 14.4.2 Selection problem statistics 14.4.3 Questions and answers 14.5 Solution suggestion to public comments 14.6 Conclusions
15	**Environmental management and environmental monitoring program** 15.1 Environment management plan 15.1.1 Environment management plan in construction period 15.1.2 Environment management plan in operation period	15.2 Environment monitoring plan 15.2.1 Monitoring mechanism 15.2.2 Monitoring plan 15.2.3 Specification management of waste water discharge
16	**Environmental economic profit and loss analysis** 16.1 Economic benefit of the project 16.2 Social benefit of the project 16.2.1 Social and economic status of project area 16.2.2 Direct contribution of the project to local economy 16.2.3 Indirect contribution of the project to social and economic development 16.3 Profit and loss analysis of engineering environment economy	16.3.1 Profit and loss analysis of internal engineering environment economy 16.3.2 Profit and loss analysis of external engineering environment economy 16.3.3 Comprehensive benefit of engineering system environment 16.3.4 Technical index of engineering environment economy 16.4 Environment economy evaluation 16.4.1 Evaluation criteria 16.4.2 Economy evaluation 16.5 Conclusion

Continued

Chapter	Content	
17	**Analysis of environment protection and control measures** 17.1 Operation period 17.1.1 Waste gas 17.1.2 Waste water 17.1.3 Waste solid 17.1.4 Equipment noise control 17.1.5 Protection of pipeline facilities 17.2 Construction period 17.2.1 Land	17.2.2 Sea area 17.3 Analysis of environment sensitive goal and protection measures 17.3.1 Coefficient analysis of safety distance and pipeline design 17.3.2 Analysis of environment sensitive goal and protection measures 17.4 Environment protection measures 17.4.1 Land 17.4.2 Sea area
18	**Conclusions and recommendations** 18.1 Evaluation conclusions 18.1.1 Engineering analysis 18.1.2 Surrounding environment status of construction project 18.1.3 Prediction and evaluation of environment impact	18.1.4 Protection measure analysis of main environment 18.2 Recommendations 18.2.1 Marine engineering 18.2.2 Land engineering 18.3 Overall conclusions

9.4 HSE Management in Design Stage

9.4.1 Port and Jetty

9.4.1.1 Channel and Anchorage

(1) Channel.

The calculation of all the design rich water depth in the channel for ship to enter/ leave jetty should be determined by the relevant provisions of current industry standard "JTS 165-2013 Design Code of General Layout for Sea Port". The channel can be designed with single direction, the effective width of the channel shall also be determined by the relevant provisions of the JTS 165–2013, and the width should not be less than five times of the length of the ship design. The calculation datum of the channel design depth should be starting from the local theoretically lowest tide level.

(2) Anchorage.

The special anchorage should be established for LNG vessels. The safety net distance among anchorage, LNG jetty, channel, and other anchorages should be longer than 1000m. Anchorage scale should be determined accordance with relevant provisions of industry standard JTS 165-2013.

9.4.1.2 Jetty

(1) Safety requirements.

① The safety distance among jetty, working boat dock, storage tank and surrounding buildings, and the jetty fire protection design should be comply with the relevant requirements of "-JTS165—5—2016 : Code for Design of Liquefied Natural Gas Port and Jetty".

② The design of power supply, water supply, fire-fighting, control, communication and medical assistance and other design should be coordinated with terminal design.

③ Jetty workers should join the liquefied natural gas operations, safety operation and emergency operation training, and certificates should be obtained. Safety operation instructions, equipment operation manuals and management documents should be prepared, and the relevant personnel should be familiar with the management file contents.

④ The jetty and trestle shall be provided with manual alarm button with the necessary protection. Berthing auxiliary system, cable tension monitoring system and environment monitoring system should be set up. The quickly taking off cable device should be set up.

⑤ The electric facilities in control and electrical room of jetty should adopt explosion proof electrical facilities, and automatic fire extinguishing system should be set.

(2) Supporting facilities.

① A fixed number of permanent observation points should be set at the jetty, and the settlement, horizontal displacement and inclination of the jetty during construction and operation period are regularly observed.

② The jetty surface should set safety facilities of leakage collection.

③ The jetty should have a complete fire facilities, which shall meet the "JTS 165—5—2016 : Code for design of liquefied natural gas port and Jetty", the provisions of article 9.2.

④ Lightning protection and grounding, loading and unloading area lighting should be in accordance with the provisions of JT556 and JT/T557.

⑤ The area of the terminal platform should be considered for the maintenance vehicle operating space.

9.4.1.3 Trestle and Unloading System

(1) Trestle.

Non-combustible and flame-retardant materials should be used in trestle. Trestle width design, in full consideration of the width of the pipe belt, should be set aside sufficient width of the maintenance to vehicle access, and trestle width should not be less than 15m. LNG pipelines on the jetty and trestle should take measures to avoid damage to vehicles.

(2) Unloading system.

The loading and unloading pipe should be set up shut-off valve and drain valve. Emptying interface, shut-off valve should be set up fire protection device. Unloading arm should have cut off and disconnect function, and mobile overrun warning device, unloading system should consider the LNG loading process.

9.4.1.4 LNG Carriers Berthing and Unberthing

(1) General requirements.

① LNG carriers berthing and unberthing should comply with the regulations of wind speed, wave height, visibility, flow rate and other operation conditions in JTS165—5—2016.

② When LNG carrier is berthing, at least three tugs power not less than 2200W should operate together, and the tug should be the whole transition back type (Z type).

③ When LNG carrier is berthing, ship bow should be directed to the quickly and easily leaving direction.

④ When LNG carrier is unberthing, at least two tugs power not less than 2200W should operate together.

(2) Special requirements.

① After LNG vessel berthes, the checklist should be exchanged between ship and shore, and the unloading operation shall be carried out after the parties have confirmed it.

② Before LNG carrier unloads, relevant procedures should be transacted, according to the safety regulation of jetty facilities.

③ Before LNG ship unloads, it should be in compliance with the requirements of JTJ237, to finish the ship-shore work of cross connection.

④ The suitable communication interface should be equipped in receiving terminal, including cable, fiber optic cable and so on. Emergency shutdown system shall be tested prior to unloading.

⑤ When LNG carrier is berthing jetty, alert water area and fire forbidding zone should be set. In addition to the pilotage and escort boats, the distance from LNG ship to the channel sideline is not less than 100 m, or in accordance with the relevant standards.

9.4.2 Receiving Terminal

9.4.2.1 General Requirements of Safety Design

(1) Engineering geological conditions.

The site of receiving terminal should select the engineering geological conditions which meet the safety operation. Without the specialized demonstration, it is strictly prohibited to set earthquake-resistant in complex geological structure, the presence of late fault activities and disadvantage area.

(2) Safety distance.

The safe distance of the receiving terminal should be in accordance with the provisions of the gas diffusion concentration in GB/T 20368—2012, so that the possibility of the fire spreading to the construction red line and causing significant harm is minimal.

(3) Environment conditions.

The receiving terminal site should be selected in the location of easy excluding rain water, but not in the location of flooding and water threatening. When the site is selected in mountain area, the flash flood, mudslide and unventilated areas should be avoided. In the receiving terminal site selection, the following influences should be considered, including the large hazardous facilities, the important military facilities, key cultural relics protection area, the aircraft landing, radar navigation, astronomical observation facilities and so on.

9.4.2.2 Land Formation

(1) General requirements.

① The site of the receiving terminal should be carried out geological exploration and research, and the feasible construction scheme should be chosen.

② Foundation treatment, backfilling material, bearing strength calculation in receiving terminal should comply with the relevant regulations of GB 50007—2002, SY/T 0329—2004 and SH 3076—2013. Permanent settlement sign should be set up at the receiving terminal land formation.

(2) Special requirements.

① When constructing artificial site by filling ditch in mountains and hilly regions, we should pay attention to avoid flood flowing through the valley, to prevent collapse, loss of backfill, and to ensure the stability of excavation and filling area.

② Land formation design should fully consider the balance. Foundation treatment method or the formation of land reclamation backfill should choose reasonably and feasibly, to eliminate the uneven settlement of the site.

9.4.2.3 Plane Layout

(1) Overall consideration.

The plane layout should be coordinated with the vertical arrangement in terminal. Receiving terminal layout should be based on the characteristics of its production process, fire risk, combined with the terrain, geology, wind direction and other conditions, according to the function of the relatively concentrated layout. Storage tanks, vaporizers, condensers, pressure pumps and other important production facilities should not be arranged in the adverse geological conditions, such as earthquake rupture location. Storage tanks, loading area, process device may emit combustible gas, thus should be arranged in upwind side of annual minimum frequency wind direction, not the personnel centralized place, open fire or spark discharge site. Plane layout should be considered to facilitate construction, According to the construction conditions. Pay attention to the construction site arrangement, according to construction conditions. Terminal greening should comply with the provisions of GB 50183—2015.

(2) Pipeline in terminal.

Terminal pipeline should be arranged reasonably in the conditions of safe and easy maintenance. Various pipelines should achieve a reasonable, smooth flow, and shorten logistic distance, and ensure the coordination among pipelines and construction (structures) in the plane and vertical layout.

(3) Torch.

The elevated flare and vent tube should be arranged on the higher ground, and located in the upwind side of the annual maximum frequency wind direction in storage tank and production facilities zone. If it is arranged on downwind side of the maximum frequency wind direction, the risk of leakage of vapors clouds diffusion to the torch should be considered.

(4) Power line.

The total substation of using overhead power line should be arranged in the edge of terminal. The substation of 35kV or higher should be arranged in the terminal edge in favor of high-voltage overhead lines, and closed to the load center.

(5) Terminal road.

The layout of the road in the terminal should be combined with the vertical design and pipeline layout, and should have a smooth and convenient connection with the outside road of the terminal. Ring fire-fighting road should be set in tanks and process zone in terminal, and if restricted by terrain, it should be equipped with the end of the turnaround to the fire-fighting lane. The area should be matched by the fire vehicle models, but not less than 15m×15m. The main road should adopt two lanes in terminal, and the width of ring fire-fighting vehicle road should not less than 6m, and the radius of turning road should not less than 15 m. If the road is higher than ground 2.5m, and there are process facilities, tanks and pipelines within the scope of 15 m from the road edge, the supporting pier, dwarf wall and other protective equipment should be set in the edge of the road.

(6) Others.

① The thermal power facilities and boil room should be arranged in the edge of terminal.

② Fire-fighting garages and other garages should not be built together, and located in the region where people and traffic do not affect the rapid reaching to the fire place when implementing fire-fighting task.

③ The entrances should not be less than two, and located in different directions.

④ The net distance between walls of tank should not be less than 1/4 diameter sum of adjacent tanks, and not be less than 30m.

⑤ If the heat source of vaporizer is combustible medium, the distance among vaporizers or other heat source should be at least 15m. The adjacent vaporizers or major heat source should not be set as fire source, in the case of multiple sets of vaporizer.

⑥ The distance between integral heating vaporizers and construction red line should not be less than 30m, and the distance among them and LNG or flammable liquid collection facilities, loading and unloading joints, control room, office, workshops and other facilities should not be less than 15m.

⑦ The net distance between vaporizers should not be less than 1.5m.

⑧ The distance among LNG loading and unloading joints, storage tank, control room, office, workshops and other important facilities should not be less than 15m.

⑨ The collector tank and reliable drainage system should be set in receiving process region. Tank car loading and jetty operation region, the volume of collector tank and location should comply with the regulation of GB 50183.

⑩ If the tank car loading zone is constructed in receiving terminal, it should be arranged in the edge or outside of terminal, and can be an independent zone.

9.4.2.4 Vertical Layout

(1) General requirements.

The vertical design should be coordinated with the surrounding elevation, road and flood prevention conditions. The vertical design should meet the following requirements: rational use of terrain, to provide suitable construction sites and elevation for each single building (structure); rainwater can be quickly removed, and site is not affected by flood and waterlog disaster; meet the terminal road design requirements and provide good conditions for external road links; combined with the engineering geological and hydrogeological conditions, reasonably determine the elevation of the terminal; ensure the stability of side slope and avoid seawater intrusion.

(2) Special requirements.

The vertical design should base on the construction scale, the terrain and geological conditions of the terminal, and the choice of the vertical slope type or the ladder type. In the area where the terrain slope is large, the long axis of the building (structure) is arranged along the contour line. The receiving terminal shall be provided with a height of not less than 2.5m of the fence or wall, and the power distributing station in the terminal shall be set up to the height of the 1.5m's fence or wall.

(3) Staging construction consideration.

If the receiving terminal is constructed with different stages, vertical design should be considered unified, to ensure coordination and connection of the short and long term engineering.

9.4.2.5 Buildings / Structures

Construction of the buildings or structures for liquefied natural gas and combustible gas, should be in accordance with the relevant provisions of GB 50183 and GB 50016. Buildings or structures ventilation shall comply with the relevant provisions of GB/T 20368—2006.

The control room should adopt the positive pressure protection. Towards the side of the production, device should not set the window. Compressor plant should be open or semi open type workshop. Shift dormitory should not be built at the receiving terminal within the district boundaries.

9.4.2.6 Storage Tanks

(1) General requirements.

The storage tank is the most important device in receiving terminal. LNG storage tank should meet the following requirements: store, fill and transport LNG safely; BOG can be discharged safely; prevent air and water from entering the tank; minimize the loss of cooling capacity; avoid soil freezing caused by arch; can resist the possible external forces to the storage tank damage; operate safely between the maximum and minimum design pressure; satisfy several shutdowns, maintenances and fillings in the design life; prevent water seepage from the exterior of the tank exposed to the atmosphere; in the inner tank leak cases, both metal outer tanks and prestressed concrete full containment tanks should satisfy the low temperature and other storage requirements of LNG.

(2) Storage tank insulation.

Cold insulation material should avoid any contamination that may cause corrosion of the contact. Cold insulation materials filling in between inside and outside of the tank should consider the problems of heat expansion and cold contraction.When the cold insulation material between tank walls is expended perlite the material that can absorb the diameter deformation of large thank can be used to avoid the sinking of perlite caused by deformation.

(3) Storage tank foundation.

The earthquake resistant design of the tank foundation should be in accordance with the relevant provisions of the report of the engineering seismic safety evaluation. In order to mitigate the impact of the earthquake, the basic design can be considered to use a shock pad, which should be able to replace the large tank without stopping production. It needs to ensure that the uneven settlement is less than the allowable value of the tank cushion cap and the underlying soil can't be frozen. If it is not to avoid freezing, heating facilities should be installed. Heating facilities shall be able to repair and replace when tank is without stopping the operation; at the same time, the use one and backup one should be considered. Tank cushion cap can be used at ground, overhead and semi-underground. When the overhead type cushion cap is used, the height of the ground and the cushion cap should be ensured that the natural convection of the cushion cap is not lower than the ambient temperature 5℃.

(4) Pressure and vacuum protection.

① Instruments should be installed at suitable place to detect the pressure change. Pressure detection should be set up, such as continuous pressure measurements, independent of the high pressure and low pressure (negative pressure) detection. When the low-low pressure is detected, vaporizer compressor and tank pump should be shut down automatically. If necessary, the vacuum protection valve should be opened automatically, and the natural gas is introduced into the tank.

② Pressure difference detection should be set between the cold preservation layer and the inner tank, or in the cold-keeping layer to install separate pressure detection. The pressure and the safety valve to release the pressure between the tanks should be considered sufficient margin, to avoid unnecessary venting. Tank should be assembled at least two pressure safety valves which can be directly release pressure to the atmosphere. If the safety assessment in dicates that direct vent is not safe, it can be put into the flare system. Safety valve should be considered at least one spare.

③ Safety valve emissions should consider the maximum operating pressure of the tank, fire heat input or heat input caused by gasification, feed caused by the exhaust, feeding the flash and atmospheric pressure changes, tank circulating pump, control valve out fault and tank liquid rollover occur at the same time, several combinations of gas flow.

④ Tanks negative pressure protection is achieved by tank inner pump and BOG compressors automatic shutdown and two vacuum protection systems. The tank vacuum valve should be at least one spare. The vacuum valve flow should consider changes in atmospheric pressure, the normal operation of the pump and the BOG compressor. When calculating the flow of vacuum valve, the possibility of the above occurring simultaneously should be considered, and the flow is required to be 1.1 times than the flow needed.

(5) Bursting disc.

If the safety valve or torch / vent system is calculated without considering the liquid roll, no matter what other measures have been taken, the facilities of explosion proof film or similar bursting disc shall be installed. The bursting disc design should meet the following requirements: can be replaced under the normal production conditions; discs can't fall into the tank after bursting; the bursting discs can't damage any other components of the tank. The BOG compressor should be shut down automatically.

(6) Instrument system.

The instruments for the tank should make sure that normal production, operation and re-operation safety. Instrument system should include liquid level, pressure, temperature. The system should match the following configuration requirements at least: with the safety, operation of the relevant requirements of the tank to maintain the re-trial run should be sufficient redundancy; with the safety function (pressure, LNG liquid level, etc.), the detection instrument should be independent from the control instrument configuration. The measured results should be transmitted directly to the control room, and the alarm signal should be transmitted directly to the designated operators.

(7) Storage tank pipeline.

The cooling pipe of the storage tank shall be provided with a special large tank cooling line, which is used for the initial cooling of the large tank, and the end of the cooling pipe can be used in the form of a nozzle or annular spray pipe. If the feed line of the storage tank is a spray type design, the feed line can also be used as cooling line.

Two feed lines should be set, one is at the top and one is at the bottom of tank.

(8) Pumps in the tank.

LNG should be pump out from tank in using submersible pump. Specific requirements are follows: each pump should be equipped with a separate valve; pump outlet and downstream valve should be installed between the downstream valves. Measures should be taken to avoid water hammer.

Based on the characteristics of the pump, a small flow of cycle pipeline should be set up. The design should consider the exhaust line off the pump cylinder produced natural gas.

Pump shaft should be equipped with vibration monitoring.

(9) Storage tank structure monitoring.

The storage tank shall be equipped with a structural test facility to monitor the structure of the storage tank (including the bottom of the tank), and the monitoring facilities shall be able to display normal values, alarm and emergency alarm. The design of the monitoring facility shall ensure that there is an abnormal situation and sufficient time to take corrective action.

① Temperature detection. Temperature monitoring should consider to set up three sets of temperature monitoring: one is installed inside the tank wall and bottom of tank, which is used to monitor the cooling and heating process when stop operation; one is installed outside of the insulation layer of the tank wall and the bottom of the tank, which is used to detect leakage and due to the reduction of the insulation effect; one is installed outside of the concrete tank wall of full containment tank and the supporting point of the concrete cap, which is used to monitor the temperature gradient.

All sensor parameters shall be recorded in the control room, and there should be a warning to confirm any leakage. The sensor is fully covered, and it can be ensured to detect any leakage and temperature gradients.

If the bottom of the tank is provided with a heating system, the power consumption of the heating system should be recorded continuously.

② Settlement detection. The equipment of settlement detecting should be set up.

③ Internal tank leakage monitoring. The internal and external tanks should be provided with a nitrogen system, and set temperature detection or combustible gas detection to monitor the leakage of the inner tank. For all tanks with no connection to the thermal insulation layer, a circulating nitrogen system should be set up in the insulation layer, and the sealing degree of the main vessel is monitored by detecting the hydrocarbon component in the nitrogen purge.

9.4.2.7 Technology of Storage Tank Area

(1) BOG recovery and treatment.

The receiving terminal should install the BOG recovery unit to collect the BOG. Measures should be taken to prevent air from entering the vapor recovery system. In the design of piping, process vessels or other equipment that may occur in vacuum, the equipment shall be considered to be able to withstand the corresponding negative pressure; or measures should be taken to prevent the damage caused by the vacuum to the facility. If it is used by introduction of gas to prevent the vacuum, the introduction of gas can't form combustible mixture within the system.

① BOG collection system. The design of system should guarantee that low temperature gas can not be directly released into the atmosphere in the process of normal operation. Meanwhile, the system should at least collect the flashed off BOG from storage tanks and all containers which contain LNG and the BOG emissions from gas safety relief system of LNG equipment and pipeline.

The material of the BOG system should have low temperature properties. The insulation layer thickness of BOG system pipeline shall be consistent with the insulation layer thickness of low-pressure LNG pipeline which has the same diameter, BOG to the torch/vent system pipeline besides. The maximum working pressure of the BOG system should be the maximum pressure that can be reached when the gas safety relief system is opened. The drain valve shall be installed in the low level of all the main line or torch line (upstream of the torch collection tank), and is connected with the drain collection system.

② Gas return system for LNG ship. The gas return system of LNG ship shall be set. If the return gas pressure of the LNG ship's gas return system is insufficient, pressurization device can be used to fill up gas to the LNG ship.

(2) Gas compressor.

Materials of the gas compressor shall adapt to the temperature and pressure under all scope of its work. Gas compressor should be manual or automatic stopping, and corresponding valve shall be set up at the same time to make the compressor to be isolated maintenance. Check valve shall be set on the outlet pipeline of compressor. Compression equipment of combustible gas shall set emptying device, and the vent gas shall be collected and then discharged in the security zone.

(3) LNG pressurization system.

Materials of the LNG pressurization system shall adapt to the temperature and pressure under all scope of its work, the matching between the various materials shall be paid attention to at the same time.

During maintenance, each pump shall be ensured the independence of isolation, drainage and nitrogen purging. If the pumps are in parallel, check valve shall be installed between the pump and downstream isolation valve, preventive measures shall be taken to avoid water hammer at the same time. Barrel pump should ensure adequate vapor venting capability. Exhaust vent and (or) safety valve shall be set on each pump. Vibration monitoring devices should be installed on the pump shaft. Pump barrel shall set a valve for liquor drainage; the installation of drain valve shall be easy to operate.

(4) Gasification.

Gasification is usually carried out using the vaporizer.

① Vaporizer material should be suitable for treating LNG. Heated vaporizer materials have to match the heating fluid (can't rust or corrosion) or set protective layer in the part of contact with the heating fluid.

② Block valve. On the parallel vaporizers, each vaporizer shall have block valves at the inlet and outlet. The design temperature of each vaporizer outlet valve and pipe fitting that in the upstream of the outlet valve and safety valve, shall be set the temperature of the LNG. Automatically turn-off device shall be set to prevent LNG entering into the transmission system or gas temperature higher or lower than the design temperature of the transmission system. The automatic shut-off equipment should be independent with other control system. When the parallel vaporizer stops running, the inlet shall adopt double valve isolation. Safety measures should be taken to deal with the accumulation of LNG or natural gas between the two valves, except for ambient vaporizer whose inlet pipeline is less than or equal to 50mm.

③ Heating Vaporizer. The heat source cut-off device shall be equipped on the heating vaporizer. The device may be local and remote control. Remote control points from the vaporizer should be at least 15m. A shut-off valve shall be installed on the heating vaporizer of LNG inlet pipeline, and the valve from the vaporizer is at least 15m. If the vaporizer is installed in the building, the shut-off valve should be away from the building at least 15m. Shut-off valve should be local control or remote control, while the valve should be avoided due to external ice and unable to operate. If long-distance heating vaporizer is using flammable fluid heat medium, shut-off valves should be installed on the cold and heat pipes of the heat transfer fluid system, and these valves' control point from the vaporizer should be at least 15m.

④ Protective layer. If the vaporizer is using protective layer (paint, metal spraying, electroplating, hot plating zinc powder, etc.), the protective layer should be in a stable state under the condition of temperature of LNG and maximum temperature of heating fluid. Regularly maintenance spray should be carried out on protective layer.

⑤ Natural gas loop. At the outlet of the vaporizer, the pipeline material should be selected according to the lowest temperature that may appear. Main considerations are temperature set point of the self-closing isolation valve, thermal transients before temperature stability and temperature reduction caused by low pressure expansion of the gas.

⑥ Safety relief valve. The relief valve shall be installed on each vaporizer, and the relief valves discharge ability should be satisfy the heating vaporizer or process vaporizer. The relief valve's discharge capacity of heating vaporizer should meet the following requirement: under the pressure not higher than 10% of the vaporizer's maximum allowable working pressure, the discharge flow rate shall be 110% of the rated gasification natural gas flow rate. The relief valve's discharge capacity of environmental vaporizer should meet the following requirement: under the situation of pressure not higher than 10% of the vaporizer's maximum allowable working pressure, the discharge flow rate shall be 150% of the rated gasification natural gas flow rate. If heat source of the Integrated heating vaporizer or long-distance heating vaporizer is installed in buildings, measures should be taken to prevent the accumulation of harmful products of combustion.

(5) Torch and vent systems.

The torch and vent system should consider the maximum flow capacity possible. The arrangement of the torch shall be determined according to the wind direction, and the risk of the combustible gas and the flame will be reduced to a minimum.

9.4.2.8 Natural Gas Transmission

(1) Pressure adjustment and measurement.

In the first gas station, the design of pressure adjustment and the measurement process should meet the gas process design requirements, such as pressure and temperature, flow rate and variable conditions, and should meet the needs of the opening, shutdown and maintenance.

(2) Pigging system.

Pigging technology should adopt non-stop gas sealed pigging process. Pigging by the indicator should be installed on the section of entrance and exit of the station, and the indication signal is transmitted to the station. Pigging device transceiver cylinder structure should meet the pigging device or detector requirements. Quick opening blind plate on the pigging device transceiver cylinder should not face the important building's spacing which is less than or equal to 60m. Cleaning operation of the sewage disposal should be collected and processed, not free discharge.

(3) Odorization.

According to customers' demand to add smell, the minimum amount of the odor agent should be in accordance with the 20% concentration of the explosion limit of the natural gas leakage, which can be detected by the normal sense of smell. The addition of an odor agent is commonly used which has an obvious smell of thiol, sulfur ether or other sulfur compounds.

9.4.2.9 Process Piping

Process piping should be laid on the ground, buried and trench shall not be used. The terminal pipeline installation design should be adopted to reduce vibration and thermal stress of the measures. Casing protection shall be adopted when the pipeline crossing the roadway. The main and auxiliary process system should be as open as possible to install, and the clearance height is not less than 5m.

(1) Materials.

Manufacturing materials can withstand both the normal operating temperature and the extreme temperature under a state of emergency. Through the insulating or other means to delay the pipeline failure caused by extreme temperature, until the operators to take measures, in case of sufferring from overflow high temperature, the pipeline can be isolated.

(2) Insulation.

Insulation materials should be selected in line with the requirements of appropriate standards. The pipe insulation materials used in the area where fire may occur shall have fire resistance, and in hot or cold environments without losing thermal insulation properties. There shall be used a low chloride content insulation materials. Avoid using porous insulation materials which may cause adsorption of methane gas, unless the insulation material itself is not permeable vapor, or moisture barrier is set up.

(3) Installation.

① Pipe connections. Pipe connections with nominal diameter greater than 50mm shall be welded or flanges. Thread interface and the flange interface shall be used as little as possible. Flange or other transitional connection technology by experimental verification should be used between different metals. Gaskets shall be used for fireproofing materials.

② Valves. The extended valve and stem valve shall use packing to seal, and installation position will not due to valve freezing caused by leakage or false operation. If the extended stem installed in low temperature pipes to deviate from the plumb line more than 45°, to ensure its correct action, shut-off valve shall be set on the container and storage tank interface. The local control and remote control of the valve should be installed in the storage of LNG or low temperature gas system. Valves and valve actuators should be able to operate in freezing conditions. 200mm and above emergency shut-off valve should also be equipped with power operation and manual mechanism. The storage tank isolation valve on the LNG pipeline which connects to the storage tank should be able to isolate LNG storage tank and ensure its safe parking.

③ Welding. Welders who participate in welding should hold corresponding project qualifi-

cation certificate issued by the welder examination committee. The scope of the welder welding should be consistent with the qualification certificates, and should be certified.

In the welding impact test of the material, the qualified welding procedure should be chosen, so that the low temperature performance damage of the pipeline material is minimal. In the welding of the thin pipe, the selection of welding procedures and technology should be the minimum risk of burning. It should not be used for gas welding.

(4) Inspecting and testing.

Weld inspection shall be carried out according to the provisions of JB 4730. The pressure test temperature of carbon and low alloy steel pipeline shall be higher than its brittle transition temperature.

When use water as test medium, the strength test pressure shall be 1.5 times of design pressure; when use compressed air as the test medium, the strength test pressure shall be 1.25 times of design pressure. When test tightness, the test pressure shall be equal to the design pressure. If choose water to carry out experiment, the system should be completely dry after the test. The acquisition and collection of pressure testing data should be complete, and the main datas are pressure test program, pressure test reports, pressure, temperature data, pressure test case records.

(5) Replacement.

The replacement media should be an inert gas. In order to facilitate the replacement of all process and combustible gas pipeline, the vent tube and piping cleaning interface shall be set up. The mixed gas is discharged to the torch or vent tube, and the residual is injeeted into the liquid recovery system in the process of replacement.

(6) Safety valves.

Pressure safety devices should be installed, to minimize the possibility of failure of the pipeline or accessories. Under normal circumstances, do not use the thermal insulation material for heat insulation of the safety valve, and thermal expansion safety valve shall be installed to prevent the pipeline over pressure of the liquid or cold vapor which isolated by shut-off valves. Pressure setting of the thermal expansion safety valve should be equal to or lower than the design pressure of the protected pipeline. When install safety valve, it shall be paid attention to the vent towards the least risk direction of personnel and other equipment, and reduce the possibility of combustible gas cloud contacting with fire.

(7) Corrosion protection.

For metal parts that may be subjected to corrosion, shall take anti-corrosion measures. In the process of storage, construction, manufacturing, test and use, the austenitic stainless steel and aluminum alloy shall be protected from collision; the corrosive tape and other packaging materials should not be used for the pipeline or pipeline components. If the heat preservation material will cause aluminum or stainless steel corrosion, corrosion inhibitors or waterproof layer should be used. All underground metal parts should be adopted coating and cathodic protection which comply with the relevant standard requirements. Special attention should be paid to the protection of the cathode coating of metal components near sea.

9.4.2.10 Instrument and Control Systems

Receiving terminal control system should include the following different independent systems: process control system, safety instrument system, access control system, closed circuit television surveillance system and radio call system.

(1) Control systems.

① Process control system. The process control system shall enable the operator to complete the drive, parking and normal operation of the device. Its functions include providing device remote control of the operating interface, providing visual / sound alarm, process parameters tracking records (real-time and trend) of the display, the device's routine control and monitoring, automatic sequence, timing and logic control functions, storage and printing process history trend / alarm / system alarm / operator action.

The process control system should have the same reliability as the safety level of the receiving terminal, and the Fail Safe Design should be used. When the safety instrument system starts up the emergency stop, the process control system should be able to receive the corresponding information, so as to adjust the process operation condition, in order to meet the new requirements.

② Safety instrument system. Safety instrument system should be based on the design of the programmable logic controller (PLC), and be independent of the process control system. The Fail Safe type should be adopted.

The system can automatically or manually trigger the corresponding chain protection, perform the isolation of individual equipment or units, to prevent the risk or accident of the occurrence of process disturbances or other reasons which cause the safety, environment, equipment, or potential damage to the environment.

Safety instrument system of field data acquisition unit shall be independent from the process control system. Its features include monitoring protection equipment, monitoring protection ancillary facilities, triggering emergency shut-off through the corresponding button and logic, and providing event sequence records.

The reliability of the safety instrument system shall consistent with the safety level of receiving terminal; the emergency shut-off should be able to perform to make the device in a state of safety.

For key input signal in the safety instrument system, there needs a pass through voting system to ensure the effectiveness of the alarm or the turn off signal. The safety related signal needs hard wire connection.

Some signal input in safety instrumented system should have the function of bypass, for the maintenance and on-line testing of equipment without affecting the normal safety production, and the time limit bypass for the process system is to drive the interlocking signal.

③ Access control system. The entire receiving terminal shall be zoning control, each entrance and exit point should be controlled by checkpoint with different permission, and checkpoint's opening should be authorized by the specified access control. Access control should be able to determine the access level and the personnel statistics through checkpoints into the terminal.

④ Closed-circuit television surveillance system. The receiving terminal shall be equipped with closed-circuit television surveillance system, to make the operator and security personnel monitor the situation of the receiving terminal area and near boundary area within the scope in real time.

⑤ Radio call system. The receiving terminal shall be equipped with radio call system for radio call and alarm in a state of emergency. The radio call system should be fully independent with other control systems.

(2) Emergency shutdown system.

LNG receiving terminal needs to set the emergency shutdown system (ESD). When the system performs an emergency stop, the source of the LNG and other flammable and explosive medium should be cut off or isolated, and shutdown equipment that will result in intensification and expand of the accident on account of continue running, or take necessary vent. In addition to meet the requirements of this regulation, equipment also meets the requirements of the emergency shutdown system (such as safety level, fire protection).

According to risk assessment, once detected, a causal map is established to ensure that the emergency stop can be normal as defined by the causal logic.

The control room and field should be equipped with emergency stop manually trigger device, and it can perform the corresponding logic shut-off after the trigger. Device should have protective measures to prevent false operation. Manually trigger device should be located in the area where it can reach in time when accident happens, at least 15m away from the protection equipment, and its design function shall be clearly marked.

Emergency stop should have self-locking function. To ensure the safety in production, and the system can return to normal state after it is approved by artificial confirmation and reset.

(3) Liquid level monitoring.

① LNG storage tanks. LNG storage tank shall be equipped with two sets of independent liquid level gauge, which can be able to monitor the level of the tank in real time. Each level should be able to provide high and low level alarm. High level alarm values should be set so that the operator has sufficient time to suspend the LNG feeding to avoid liquid level exceeding the maximum allowable filling height. The low level of the alarm should be sufficient to allow the operator to have sufficient time to stop the tank inside the pump. The design, selection and installation of liquid level meter should consider the difference of the density of LNG in the tank, and should meet the requirements of on-line maintenance.

In addition to the above two methods to monitor the level of liquid, liquefied natural gas storage tanks should be equipped with a separate level of high detection. Signal should be used in hard wire connection mode and have an access to the safety instrumented system, and in the liquid level to set up to trigger the interlocking cut off liquefied natural gas feed. A voting method should be adopted to ensure the validity of the signal.

② Flammable and explosive process liquid container. Flammable and explosive liquid storage tanks within the receiving terminal shall be equipped with level gauge, to meet the requirements of real-time monitoring. If the tank could be excessive filling, independent liquid level detection and cut-off device should also be installed.

(4) Density monitoring.

For safety reasons, LNG storage tanks should be equipped with a densimeter, which should be selected to meet the requirements of online maintenance.

(5) Pressure monitoring.

In addition to be equipped with pressure testing instrument of respectively used for process control and safety interlocks, the LNG storage tank should also be equipped with absolute pressure testing instrument. These three should be independent of each other, and installed above the position of highest liquid level in the storage tank. On vacuum jacketed equipment, there should be equipped with corresponding instrument or interface in order to check the absolute pressure in annular space. For the detection of low temperature medium pressure and differential pressure, design and installation of lead pressure pipe should ensure the medium to full gasification. LNG or flammable medium submersible pump, the cable filling nitrogen sealed cavity should provide pressure detection and introduce safety interlock.

(6) Temperature monitoring.

The low temperature storage tanks and containers on site assembly shall be equipped with a temperature monitoring device to control the cooling temperature, or as an auxiliary means for checking and calibrating the liquid level meter. The foundation of cryogenic vessels and equipment, which may be adversely affected by land icing or frost, should be equipped with temperature monitoring system.

(7) Flow detection.

Flow detection should be set at the main entrance or exit of process medium, according to process requirements.

9.4.2.11 Electrical Equipment

(1) Requirements for electrical equipment in hazardous areas.

All electrical equipment, instrumentation equipment and apparatus in the hazardous area shall comply with the provisions of the 1 to 7 chapter of GB 3836. Electrical equipment and wiring shall comply with the relevant provisions of the 7.6 section of GB/T 20368.

(2) Power requirements.

LNG terminal load level should be consistent with the requirements of current national standard GB 50052. The particularly important load should be set emergency power to ensure the personnel safety, environmental safety and device safety after the main power failure.

(3) Junction boxes and local control panel.

The junction boxes and local control panel should meet the requirements of electrical equipment in hazardous areas. The protection level of outdoor equipment should meet the requirements of local environmental and climatic conditions.

(4) Grounding and shielding.

All indoor and outdoor electrical equipment of the uncharged metal shell and process requirements of the grounding of non-electric equipment should be reliable grounding.

① Anti-static grounding protection. Production storage and transportation process in the receiving terminal will produce static electricity, which accumulated in pipes, vessels, tanks and processing equipment that should be static grounding. When the grounding system is connected to other utilities grounding system, the grounding resistance should meet the minimum requirements.

When a flammable tanker is loading or unloading, the terminal should be equipped with anti-static grounding protection device and automatic suspend in the ground handling operations that do not meet the requirements of the situation.

② Stray or impressed current. If stray currents or common impressed current (such as cathodic protection) exist in the loading and unloading systems, protective measures should be taken to prevent sparks.

③ Ground loops. Ground loops should be consistent with the existing domestic norms or standards. Ground loops shall be used as the connecting line of the green / yellow insulated cable. Grounding materials should use copper or galvanized steel, buried in the production area around, and grounding resistance should be less than 4Ω.

④ Lightning protection. All buildings and structures should comply with GB 50057 and local provisions, and set the lightning protection system. Tanks that not contact with the earth well should be set lightning protection and grounding to protect personnel, equipment and infrastructure.

⑤ Lighting. The total plant should be installed safety and emergency lighting to ensure the safety of personnel evacuation in emergency situations. Security lighting should be self-rechargeable, and the battery standby time is at least 30 minutes. Storage tanks and other elevated structures should be equipped with warning lights in accordance with aviation and maritime safety rules.

9.4.2.12 Fire Protection and Safety

(1) Safety assessment.

According to the safety pre-evaluation report, the receiving terminal safety assessment including the following contents: type, number and location of the devices; the possible consequences caused by leaked or overflowing combustible gas and liquefied natural gas, and the need to control open flames; device type, number and location that possible happen non-process and electrical fires; methods to protect equipment and structures in case of fire; fire water system; other fire facilities; equipment and process of emergency shutdown systems, and tank or equipment needed pressure relief in case of fire; sensor type and location of emergency shutdown system or subsystem; emergency measures taken and outside support in case of fire accident; required protective equipment, training and qualifications for terminal personnel; when a fire happens, whether the function of LNG terminal equipment and materials support structure are affected within the stipulated time; all control devices and cables must take fire-control measure to ensure that operations can continue in case of fire. Fire stations may be considered to build together with other companies and so on.

(2) Detection system.

For cases with high risk, two sets of testing equipment should be set up. The gas detection system should give out audible alarm when measured gases and vapors are not more than 25% of the lower explosive limit. Fire and gas leakage detection device layout principles shall comply with the provisions of GB 50183—2004.

(3) Extinguishing system.

Fire protection water system should be equipped with different power source of fire pump. The terminal fire water system configuration principle should be in conformity with the provisions of GB 50183—2004, and the configuration principle of the jetty fire water supply system shall comply with the relevant provisions of JTS 165—5—2016. When the tanks and equipment which need cooling in fire should be installed with sprinkler systems around, foam extinguishing system, dry powder extinguishing system, emergency shutdown system should be complied with the provisions of GB 50183—2004.

(4) Coating.

Coating should be considered the fire-resistant time of equipment and structures in the pool fire and jet fire two cases. Liquefied natural gas unit, metal equipment, pipes and metal structure surface must be coated with protective coating. The surface of the concrete structure should be adopted fire paint in accordance with the requirements. The pipe rack inside and torch tower should be treated with fireproofing and anti-corrosion.

(5) Personnel safety and escape.

The terminal personnel emergency and protection should comply with the provisions of GB 50183—2004. In the appropriate location of the terminal, the channel indicator and weather vanes should be set up to help escape.

9.4.2.13 Utility Systems and Other Auxiliary Works

(1) Seawater System.

① Materials. Seawater system materials should be selected according to the fluid properties and geographical environment: austenitic stainless steel cannot be used for seawater circuit; chlorination system should avoid using PVC. It should particularly pay attention to the consistency of the materials to avoid galvanic corrosion.

② Sea water pump. Pumps and other metal components should have effective cathodic protection.

③ Chlorination devices. Chlorination and other measures should be taken to limit the water pipeline microorganisms or mollusks breed. It is better to use hypochlorite detection system to prevent excessive chlorination.

(2) Instrument air.

Instrument air system should be configured with at least two air compressors as well as spare. When the main power source fails, the instrument air should be able to ensure the normal supply for some time, in order to meet the requirements of maintaining the terminal safety. Instrument air dew point requirements should be met.

(3) Nitrogen system.

The receiving terminal should take into account the nitrogen need of continuous and intermittent.

9.4.2.14 Tanker Loading System

If we consider the tanker loading, loading facilities should meet the following requirements, including layout, piping system, instrument control, electrical and fire protection.

(1) Process piping systems.

Cut-off valves should be installed at the end of each branch and system boundaries. For automatic cut-off valve, the closure time should be determined by analyzing the water hammer to prevent the failure of the pipe and equipment. If it is more than the allowable stress, it should put down the stress level by extending the closing time of the valve or other safety measures.

The liquid and gas mains shall be equipped with emergency shut-off valve, at least 7.6 m away from the loading area, but not far than 30m. These valves should be easy to use in emergency situations.

Piping system should be installed with a check valve to prevent backflow as required, and should be close to back flow interface as near as possible. System should replace air or other gas with a safe means, and should be set vent tube and piping cleaning interface, in order to facilitate the replacement of all processes and combustible gas pipelines.

One or two loading lines are usually arranged on a loading platform.

(2) Pressure and flow control.

LNG loading mains can be set with pressure control valves. Walve post pressure control should be complied with the requirements of tanker. Vapor return mains also should be set with pressure control valve.

Flow meter, flow control valve or weighbridge can be set in each loading line, and can use centralized control system or local controller to control loading flow.

(3) Tanker loading area.

The tankers into the loading area should be special tankers which are produced and approved by national laws and regulations. Tanker loading areas should have sufficient size, the vehicles can be as low as possible to move or turn, and it is independent with the receiving terminal. The access control system should be set up. Tanker import scheduling and filling operations control should be integrated in a unified management system to ensure the safety of tanker loading.

(4) Loading arm and hose.

Cryogenic rotating joint of loading arm shall pass cryogenic tests. The loading arm type should consider the connections of tanker's tail and side. To provide appropriate support to loading arm, the balance should be considered in the weight of the icing on the loading arm. The flange or quick connector can be used in tanker and loading arm connection, and can be installed with emergency facility. The use of loading hose should be consistent with the risk assessment. The hose design should be consistent with the corresponding standard, and hose should be tested once a year at least. And the appearance should be checked before each use to find whether it is damaged or defective.

(5) Grounding and communications.

LNG tanker filling should provide anti-static grounding protection facilities, and grounding can be test and stay the hardwired connection with control system. For example, the filling process stops automatically if grounding is failure. Handling location should be equipped with communication facilities for operators contacting with distant workers who control the loading and unloading. Communication can be used for explosion-proof telephone, radio system, radio or signal lamp.

(6) Collection tank.

Collecting tank can be set to collect LNG by nitrogen purging before loading arm breaks, and also should be set the embankment around for anti-leakage spread.

(7) Collector tank.

Loading zones' possible leaking LNG should be discharged into a collector tank which is equipped with foam generator. The location of foam generator should be considered with the prevailing wind direction.

(8) Metering.

To determine whether the loading capacity meets the requirements of custody transfer, it should use weighbridge to measure weight.

9.4.3 Long Distance Transportation Pipeline

9.4.3.1 Influence on Pipe Network by Natural Conditions

For the pipeline project, the impact of natural conditions mainly reflects in lightning, soil, rainfall, geological disasters, earthquake.

(1) Thunder and lightning.

Lightning damage to the pipeline mainly reflects in following aspects.

① Pipes overhead part and the ground part (e.g. station piping, process facilities.) are excellent lightning receiver comparing to the entire buried pipelines. Under the condition of the air existing thunderclouds, it is possible to form an induced charge center, thereby suffer the threat of a direct lightning.

② When thunderclouds take shape in the sky, a large area of the ground will form an electrostatic field, and surface of the buried pipelines induce the opposite charge the same as the ground. When the charge accumulates to a certain degree which meets the discharge conditions, there will be a strong discharge process. If such discharge cannot leak into the earth rapidly by leak point of insulating layer, then the pipe insulation or poor parts will generate high voltage and cause the secondary discharge.

③ Metal pipe itself is a good conductor, which is easy to become a large lightning discharge channel, and lightning stroke.

④ The pipe not only can induce positive thunder (i.e. positive charge accumulating on the pipe), also can induce negative thunder (i.e. negative charge accumulating on the pipe). The positive and negative thunders have different impact on pipeline, especially for the operation of cathode protective equipment.

(2) Rainfall.

South China such as Guangdong province, Fujian province have abundant rainfall. Overburden layer along the pipeline will be loose and slope soil will be washed away due to the annual flood and storm, which will lead the risk of the pipe baring on ground. The flood triggered by heavy rain is easy to wash away the top and bottom overburden of the pipe, making the pipes exposed and hanging, even twisted and broken down.

(3) Landslide.

Adverse geological region which the pipe network goes through should be identified, monitored and controlled. After storm, water and soil loss will lead to landslide and pipes exposion, which will bring a certain threat to the safety of this network operation.

(4) Collapse.

The river bank and reservoir slope which are composed of clay sediment or argillaceous expansive rock, often occur strong bank collapse due to river lateral erosion, undermining and reservoir filling, and endanger the safety of the pipeline.

9.4.3.2 Influence on Pipe Network by the Social Environmental and Construction Conditions

(1) Social environment influence.

Today, the pipeline transportation technology continues to develop, accidents caused by design, equipment, facilities and similar factors are reduced, but the third party damages such as population density, settlements and human activity make the accident more frequent. It mainly reflectes in drilling pipe to steal gas, illegally constructing buildings on the pipeline or in pipeline corridor, expanding the city and endangering pipelines, repairing roads on the pipeline or in pipeline corridor, dredging river for sand and stone , farming, industrial and commercial activities, and construction activities. The characteristic of transmission medium determines that the gas pipeline in the event of leakage, fire and explosion will occur, causing heavy casualties and property losses.

(2) Construction conditions influence.

LNG pipeline is long, passing through the region densely covered with water network, developed traffic. Along major pipeline, it is required crossing rivers route, highways, railways, first-class and secondary highways, township roads and small rivers, leading to difficult pipeline construction.

Therefore, the influence of related external conditions to the pipeline project mainly reflects in the level of construction team and the quality level of construction. Low level of construction team or poor quality of construction will easily lead to the early invalidation and scrap of buried pipeline.

9.4.3.3 Pipe Network Design

(1) Line trend design principles.

Design principles are: strictly conduct relevant national and industry standards, norms, procedures; on the basis of the topography, engineering geology and geographic location of gas supply along the line, try to keep away from important military facilities, flammable warehouses, security zone of state key protection units, the area of perennial economic crops, important infrastructure facilities of farmland, populated areas, fishpond zone, landslides, collapse, and debris flow and other adverse geological zones; combining with the development and plan of farmland area, water conservancy, industrial and mining enterprises, railway and highway, avoid conflicts with the pipeline route. The design principles meet the requirements of "GB 50251—2015: Code for Design of Gas Transmission Pipeline Engineering" on route selection.

(2) General design.

In order to avoid the damage to the pipeline in the long run, the pipeline is generally laid underground. When the natural gas pipeline network is buried, there is a longitudinal friction between the pipe network and the soil. The soil depth is bigger, then the friction is bigger. Therefore, the buried depth of the pipeline is considered as the depth of the agricultural tillage. According to the GB 50251—2015, pipelines in different regions are laid in different ways.

(3) Line corner design.

The line horizontal rotation and vertical rotation are priority to the use of elastic installation. Limited by terrain and other factors, when the elastic lay is difficult, it can use cold-formed bend with radius of curvature $R = 40D$. When cold bend is also restricted, it can use hot-bending bend, generally radius of curvature $R \geqslant 5D$.

(4) Pipeline crossing design.

① Crossing the river. Gas transmission trunk line and branch line may cross large, medium sized rivers, even the straits. In order to ensure the safety of the pipeline and not affect the environment, large and medium sized rivers are suitable for directional drilling through. The small river crossing through the region of the pipeline is generally suitable for dig through. If the geological conditions and the site conditions permit, the horizontal hole drilling rig (plus casing) through or small directional drilling also can be used. The top depth of the pipe is not less than the maximum scour line 0.5m. If the scour depth is unknown, and the depth is not less than 1.5m, it can set concrete ballast blocks according to the circumstances. If strait formation changes great, for the large differences of lithology, and not continuously distributing, it can adopt shield to go through.

② Crossing the railway, highway. Pipeline crossing the railway, expressway, one or two grade highway, top concrete casing shall be used. When pipeline diameter is 500mm or less than 500mm, the outside diameter of concrete casing is 1.2m. When pipeline diameter is more than 500mm, the outside diameter of concrete casing is 1.5m. Insulation support is provided in the casing, and the casing end is sealed by the insulation material of the asphalt tagel. Pipeline crossing three or less grade of highway, cross hole drilling shall be used. Insulation support is arranged in the casing. At the end of the casing, it is sealed with high shrinkage heat shrink sleeve seal. When pipeline passes a lower level of road, method can be used without casing excavation through the way, but the design coefficient of the gas pipeline shall improve a grade.

③ Crossing other pipelines and electricity, communication cables. When the gas pipeline and other pipelines intersect, the vertical distance is not less than 0.3m. When the net distance is less than 0.3m, the two pipelines are provided with a solid insulation, and the two sides of the pipelines at the point of intersection of each pipe section extending over 10m should use special reinforced insulation. When underground electric power and communication cables intersect, and gas pipelines buried under the cables, the pipeline top and cable space is not less than 0.5m. The pipeline section and cables which two sides of the cross points extend over 10m should adopt special reinforced insulation levels.

9.4.3.4 Pipeline Anticorrosion

(1) Corrosion environment in the area of pipe network.

When the gas transmission line crosses the river network, fish ponds, farmland, houses, roads and underground structures, the soil along the pipeline will be varying degrees of corrosion.

(2) Influences of high voltage transmission line on pipeline.

There are many gas pipelines and high voltage transmission lines in parallel and in a public corridor, and the high voltage transmission line level of the individual location is more than 220kV. The main influences of high voltage transmission lines on the pipeline are as follows.

① Capacitive coupling. Because the pipeline itself has an external anti-corrosion insulation, before and after the pipeline assembly, welding and being buried under ground, the inductive coupling capacitance exist. Before burying pipeline, if the pipeline is longer, the high-voltage electrostatic induction will cause harm to construction workers on the ground.

② Induction effect. When the pipes and power lines are long distance parallel or oblique approach, the magnetic field generated around the transmission line will generate two times of AC voltage on the buried pipeline. Too high induced voltage will bring hazard to pipe production and management personnel.

③ Resistance effect. Resistance effect is also called failure effect. When faults occur in high-voltage transmission line, kilovolt high-voltage fault current between transmission tower and earth electrode may flow into the pipeline, which threatens the operator of the pipeline near and distant.

④ Breakdown of pipeline anticorrosive coatings. The high voltage induction voltage, although the existence of its time is very short, only about 0.5s, it is a threat to the personal safety, and also can break down pipe anticorrosive coating, even the formation of electric arc burning through the pipeline.

⑤ Causing pipeline corrosion. In parallel with power lines, pipelines will generate interference voltage, thus generate DC and cause pipeline corrosion. To eliminate the hazard, it should take temporary static grounding, electrical isolation, and zinc anode grounding protection systems during different periods.

(3) Pipeline corrosion protection.

① The anticorrosive coating selection. The requirements for the corrosion protection specification, long distance buried pipeline coating in general should have the following characteristics: good insulation, good stability, good cathode peeling strength, sufficient mechanical strength, good adhesion with the pipe, resistant to plant roots penetrating, widespread paint source, reliable quality, low cost, continuous mechanized production, to meet the needs of construction and easy to repair in the field.

According to the above requirements, Guangdong Dapeng Project uses three layer polyethylene coating with high reliability and safety. Special heat shrinkable sleeve is used in directional drilling joint coating; double epoxy coating is used in hot bend, outside wrapped with polyethylene tape around for protection; three layer structure heat shrink sleeve for the corrosion protection is used in pipe welded antiseptic joint. In the latter part of the project, special coating is also used for corrosion protection joint.

② Cathodic protection. Cathodic protection is a method through cathode polarization of the metal surface to inhibit metal corrosion. Cathodic protection usually has two ways: impressed current method and sacrificial anode method. In general section, impressed current cathodic protection is used; sacrificial anode protection is used in section where used metal casing through road, railway, large and medium rivers as well as soil resistivity is less than 20Ω · m. In the parallel section of pipeline and high voltage transmission line, the protection of the current cathodic protection is adopted and in the meanwhile, the protection of the AC interference current drainage is adopted. If impressed current system has not yet put into operation within the next six months after the pipeline is buried, banding sacrificial anode will be taken as the temporary protection. After applying the cathodic protection measures, the tube ground potential at any part of the pipe network is at least -0.85 V (relative to the copper / copper sulfate reference electrode).

9.4.3.5 Line pipe, Station and Block Valve Chamber

(1) Project line pipe.

Natural gas delivered by pipeline is from the LNG receiving terminal, which belongs to grade Ⅰ of the Chinese standard (GB 17820), which has no corrosion to pipes. Engineering line can be used to transport non acidic medium to design, but should consider the regional levels. Just like Guangdong Dapeng Project, some pipeline goes through the mountain section, undulating terrain, high mountains and steep slopes, and should use X65 grade steel for pipe network. When the diameter is ϕ559mm above, the longitudinal double submerged arc welding of UOE steel shall be used. When the diameter is ϕ273mm, the straight electric resistance welding (ERW) steel shall be used. If the transport of natural gas is from different sources in the future, it should be based on the temperament characteristics of mixed gases, and select the appropriate pipe.

(2) Gas transmission station setting.

① Site selection of gas transmission station. The principles are: close to the pipeline network, convenient for gas pipeline network to entry and exit; meet the requirements of the regional arrangement of fireproof distance; meet the requirements of construction site.

② Layout in distribution stations. Each distribution station is generally divided into the production area, auxiliary production area and life management area, which should be arranged independently and meet safety requirements.

③ Building structure. Engineering buildings should be chosen according to the standard requirements of the fire resistance rating, and the main building should be in accordance with the local seismic fortification intensity.

(3) Block valve chamber set.

"GB 50251: Code for Design of Gas Transmission Pipeline Engineering" states that block valve chamber should be set at a place with convenient transportation, open terrain, and high terrain; the maximum interval between the block valve chamber should comply with the following requirements: pipeline mainly in first level areas is not more than 32km; pipeline mainly in second level areas is not more than 24km; pipeline mainly in third level areas is not more than 16km; pipeline mainly in fourth level areas is not more than 8km.

According to the level of areas along the gas pipeline, combined with the distribution of station and large rivers, block valve uses gas-liquid linkage straight-through type of ball valve, using drop rate to close automatically and manual reset mode operation by go-devil, and remote control.

(4) Deal with artificial damage.

Strengthen ties with the local government, strengthen pipeline safety knowledge and protection, and educate the surrounding residents. Promote the surrounding residents to comply with "Oil and Gas Pipelines Protection Ordinance" strictly, and ensure the safe operation of the pipeline. Carry out safety education and publicity work in communities and schools along the pipe network, and organize fire safety and pipeline safety activities together with the local government.

9.4.3.6 Automatic Control System of Pipe Network

(1) Design safety principles.

In line with the GB 50251—2015, the monitoring system for long distance pipeline network has the following requirements.

① The system must have high reliability, stability and flexibility in order to ensure the safe and reliable operation of production. The system can self-diagnose and form a report regularly to monitor the working status of the system, so as to maintain the system.

② Hardware, network and software of control centers should have 50% expansion capacity at least, and should allow the scalability of future databases, memory, disk capacity, communication channels and so on. Station control system, RTU should have 25% expansion capacity at least.

③ Important parts of the system should be set up redundant settings. When the fault occurs, the fault can be automatically switched, and the data is automatically backed up to provide a reliable guarantee for the operation and management, in order to ensure the system normal, reliable and smooth.

④ The main channel of data transmission between dispatching control center and each station control system is used by optical cable communication, and telecommunications public network is taken as an alternate channel. The two can be switched automatically.

⑤ The system should have a strong man-machine dialogue function.

⑥ The software of the SCADA system should be modular, easy to add, delete, and positioning in the SCADA system configuration. Hardware in performance should be modular too, allowing future expansion in capacity and function.

⑦ In the case of external power shutdown, UPS can guarantee the normal operation of the main equipment 4h of the dispatching center, and ensure the normal operation 2h of the station control system and instrument.

⑧ Equipment selection should consider the performance stability, high reliability, high performance price ratio, to meet the required accuracy requirements, on-site environmental and process requirements, in line with the principle of environmental protection requirements.

⑨ In explosion hazardous location, electrical instrumentation and electrical equipment are designed in explosion-proof type generally, electrical equipment and electrical connections generally base on explosive dangerous zone 1 to select and design. The selected electrical equipment must have a recognized authority issued anti-explosion certification which complies with the relevant standards.

(2) Automatic control system design.

① Station control system. Main functions include: temperature of each station in and out, pressure measurement are detected and uploaded to the control room; each station's gas component is detected and uploaded to the control room for the flow calculation; the pressure before and after the natural gas separator is on-site displayed at different stations; the temperature of each station is displayed before the heater, the temperature and pressure after the heater are detected, and the temperature signal is uploaded to the control room; the display of the automatic control system of natural gas measurement, flow rate and pressure of the equipment of the sub stations to the city gate station; the display of the automatic control system of natural gas metering and pressure / flow at the end station of each power plant; and combustible gas detection and alarm shall be equipped in each station.

② Flowmeter system. Flowmeter system is used by high precision ($\leqslant \pm 0.5\%$) of the flowmeter—gas ultrasonic flowmeter and gas turbine flowmeter, and is calibrated periodically in accordance with relevant regulations.

③ Flow regulation and control systems. Metering pipelines are used by flow / pressure automatic selection and regulation system from distribution stations or end station to the city gas gate station and power plants. When the downstream pressure of control valve exceeds the set value, the control valve can be switched automatically for pressure control to make the pressure out of the station not exceed the set value.

④ Pressure control of metering pipeline. To prevent the downstream pressure of control valve exceeding the limit, it can take two steps. First, increase the design pressure of control valve downstream (distribution stations to the city gate station, the end station to the power plant) appropriately. Second, install two sets of safety shut-off valve in series in upstream of each measurement loop control valve. Under normal conditions, the safety shut-off valve is in a full open state. When the downstream pressure exceeds the limit and causes safety shut-off valve closed, it can be reset (open) manually, and also can be automatically reset remotely (open).

(3) The items that should be noted in the design of automatic control system.

① In the humid and rainy coastal areas, sometimes air humidity can be more than 70%. Therefore, the instruments, apparatuses and automatic control equipment of the control center and each station must meet the requirements of air humidity, and the control room should set up drying ventilation system, in order to avoid the electronic circuit board in the long term in the humid environment, such as corrosion, short circuit and other undesirable phenomena. Meanwhile, lightning protection measures also should be taken to prevent the effects of lightning on the transmitted signal and damage of control equipment, communication lines.

② Instruments, apparatuses, and control equipment selection should note the requirements of openness (interchangeability requirements), that is, when the selected device needs to be replaced, not necessarily only use the product of the original manufacturer, it also can use other manufacturers' products for instead. Therefore, the choice of equipment should meet the general standard (preferably international standard).

③ To ensure the safe operation of the SCADA system, hardware and software firewalls should be set up to prevent the virus infection infecting the operation system from enterprise network or other areas.

④ The abnormality of automatic control system of the power supply will directly lead to the nonfunction of the system. So the total station control centers should be equipped with first order load power supply and UPS.

9.5 HSE Management in Construction Stage

9.5.1 Brief Introduction of Safety Accidents in the Process of LNG Project Construction

Judging from the domestic and foreign LNG project construction process, in the storage tank, pipeline, receiving terminal construction, the contractor's employees have had the casualty accidents, some of them belong to serious casualty accidents, and LNG project company should pay highly attention to them. Here we recount of the accidents whose purpose is to let other project company to learn a lesson and in order to reduce the accident rate of HSE in the construction of LNG project.

9.5.1.1 Casualty Accident Caused by the Pipeline Explosion during the Pressure Test

(1) Accident description.

February 6, 2009, 11:30, in a LNG receiving terminal, 36" natural gas pipeline explosion occurred during the period of the pressure test, causing 1 person dead, 3 persons injured, 5 persons minor injuries, and the deceased was a construction contractor staff.

Pressure test of natural gas pipeline length is 533 m, diameter is 914.4mm, and wall thickness is 46mm. Working pressure is 9.2MPa, design pressure is 13.6MPa, and gas test pressure is 15.64MPa. On the fifth day, the pressure of the test had risen to 12.3MPa, and then the test pressure pipe exposed. The explosion occurred in the reserved for project phase II project of 36" pipe end flange. The flange was fractured from one end of the root of pipe, and flew out about 100 meters. The pressure relief pipe's swinging caused damage to the vaporizer and associated facilities.

(2) Accident investigation.

The situation observed by the accident investigation team on the fracture is: fracture is located in the root of the flange, the fracture surface is almost neat, 3~4cm from the weld. Subsequently, a performance test on the fractured flange body sampling was carried out, and investigations found that the axial mechanical properties of flange cone zone material did not meet the relevant technical requirements, especially the Charpy impact energy value, tangential and axial were not in accordance with the relevant technical requirements, even far away greatly. After normalizing and tempering heat treatment, re-sampling for detection again, plasticity and toughness indexes had not yet reached the requirements, and material embrittlement was serious.

Judging from the macro and micro analysis and metallographic analysis on the fracture surface, the material embrittlement of flange was obvious, and there also existed vulcanized inclusion segregation phenomenon. No welding defects were in macro and micro test welding area, and the fracture of flange occurred outside of the weld joints and heat affected zone. Seen from the macroscopic detection, a groove was left by the small end corner of flange cone surface, which formed an obvious stress concentration effect in service.

From the macro and micro analysis, it can be concluded that, the accident flange crack is brittle fracture by the appearance of multiple sources. The main reason of cracking is that the mechanical properties of materials (especially the toughness index) is lower than the standard requirement, and due to the effect of stress concentration. The groove at corner of flange cone surface small end becomes the inducing factor of cracking.

When cracking, the gas pressure was 12.3MPa, lower than the design pressure of 13.6MPa, even lower than the pneumatic test pressure of 15.64MPa. Ruling out the possibility that flange fracture was caused by weld defects, the conclusion is that the cracking was caused by the low stress brittle failure (that is the pressure was lower than the material yield strength which resulted in the fracture).

(3) Lessons.

① The quality of all equipment, materials, and accessories in high-pressure systems should be controlled starting from raw material.

② Before high pressure system carries out pressure test, it should calculate the total energy that may release in the accident state, and determine the safe distance and protective measures.

③ The risk analysis for the process units is not enough, which should be paid great attention especially in gas pressure test.

9.5.1.2 Personnel Suffocation Casualty Accident In the Process of Tank Filling Perlite

(1) Accident description.

March 7, 2006, at 3 pm around, when a receiving terminal was filling with perlite in the No.1 outer tank and inner tank in between, an Indian perlite supplier personnel on site supervision work in the process of filling of perlite, accidentally fell from the height of the platform 6～7m, into the loose perlite 4m in inner tank. After the accident happened, the on-site doctor and rescue team immediately came to the scene, and the local fire department also came to the scene to participate in the rescue. They used long time oxygen equipment for rescue work. Until March 8, 9 : 20, the person was found. Through further perlite blowing moved, at 12 : 00, the accident person's body was moved out of the tank. Due to the body continuing to sink, until March 9 in the morning around 1 : 00, the person was moved to the top of the tank, and was confirmed suffocated to death.

(2) Accident analysis.

Possible reason: the staff came from subcontractor, when he worked within the annular space of the no.1 storage tank, accidentally fell from high and dropped into an uncompacted perlite; his body was equipped with seat belts, but not tied before.

(3) Lessons.

① Seat belts must be buckled up when working aloft.

② The subcontractor must clear the dangers of working aloft and strengthen the training of employees.

9.5.1.3 Other Casualty Accidents.

(1) Steel mesh slipped down in the construction of storage tank.

According to the report of Xinhuanet, on June 16,2009,7:50, in the scene of a LNG receiving terminal project located in south of China, the vault block hoisting work was being implemented in the construction of No.1 storage tank. The steel mesh reinforcements happened to fall, causing a major casualty accident that 8 people were killed, 14 people were injured, and 3 people seriously injured. The cause of the accident preliminary was found out. It was operator's error when workers operated the tower crane, resulting in the tower crane boom to touch the template which has been installed on the vault. The template rapidly slid down from 30m high and knocked to the construction workers who were under it, causif heavy casualy accident. According to another sources, after the accident, the loss of the No.1 tank was expected to billions of dollars.

(2) The bulldozer turned over in the terminal construction.

February 3,2004, at 9:50 around, in an earthwork construction site of LNG receiving terminal, a bulldozer driver was working on the earthwork platform which was 15m high. Since there is no assessment on the field conditions, when the bulldozer left from the east ramp, the ramp collapsed and led the bulldozer turned over, and the driver fell off before bulldozer turned over and crushed driver to death soon afterwards. The conclusion was that the driver's illegal driving on the unsafe ramp led to the accident.

(3) Personnel casualties in pipeline operation.

At about 12 O'clock on October 17,2005, a contractor of LNG project pipeline engineering was doing pipeline operation, violating the safety specification of excavation in the project company: inadequate grading, the digging earth stacking on the edge of the pipe trench, which caused inconvenience of going in and out of the passageway. An employee accidentally fell into the pipe trench, causing serious neck injury to death.

9.5.2 HSE Management for Contractors

From the above accident introduction and analysis, in the LNG project construction, it is prone to frequently produce safety accidents in installation and testing process with high risk operations, so as we should focus on strengthening the HSE management of contractors.

9.5.2.1 Contractors Selection.

LNG project is actually a collaborative work that shareholders invest, the project company takes the lead to operate independently, and a number of contractors are involved to take part in the design, construction and consultation. Therefore, the contractor's HSE management is a very important work to the project company management, and now bringing the contractor's HSE management into the management of the project company is a common practice of China's LNG project companies. To do this job well, manager needs to do the following aspects.

(1) High level of HSE management is one of the subject conditions in the contractor selec-

tion.

① Questionnaire survey. The contractor's pre-qualification work starts from questionnaire survey. Questionnaires is issued by HSE department of the project company, filled out by intention bidders, through comparison of several bidders; the score is given as one of comprehensive score aspects.

② Determination of prospective contractors. After a survey into the pre-qualification stage, carry out further in-depth investigation of HSE, to understand the HSE philosophy, policy, the previous HSE performance, HSE management system and institutions, and take them as one of the factors of evaluation.

③ Contractor bid winner. HSE content of the tender by the project company HSE departments to clarify, including the project HSE management regulations and the management procedures of the document review. The highest level of HSE management is selected as successful bidder.

(2) Supervision for the contractor HSE management in the process of construction.

① Before start review. The main review of some HSE documents includes the construction of HSE manuals, risk analysis procedures, emergency plans.

② Admission training. Mainly by the project company HSE department, carries on training to the contractor to instill the company's HSE philosophy, policy, HSE management procedures, management regulations.

③ According to the requirements of the project company, the contractors modify their own HSE management procedures and documents, including the site HSE management and cooperation with HSE personnel of the project company, and issue entrance permits to those contractors who reach the requirements of the project company.

④ Contractor's HSE management procedures and regulations up to standard can't represent the real implementation capacity. The contractor's execution in the construction of the HSE is the most important, so the project company HSE departments shall not relax on the contractor's supervision and inspection, also including HSE meeting, on-site management procedures and means, accident event handling procedures, emergency treatment, the rectification measures and HSE management continuous improvement.

9.5.2.2 HSE Management Organization during Construction

According to the EPC and the sub project division of the project, establish the HSE organization. The following is an LNG project of the receiving terminal and pipeline engineering as example, and the HSE management mechanism is introduced, for readers' reference.

(1) The receiving terminal HSE management.

Construction safety management of the receiving terminal falls mainly on the general contractor's head. In the construction period, owner has a construction safety manager who is responsible for receiving terminal construction, the other set up a senior safety engineer and three security engineers. Owner's safety managers and general contractors consist for an integrated safety management team, to carry on-site construction safety management together. Subcontractors have their own safety managers responsible for their own construction project management. From the perspective of the entire process of project construction, this management mode is effective as a whole, to ensure the information communication and exchange between owner's and general contractor, so that the owner can go deep into the specific management of the construction site, reduce the friction between the two sides and strengthen the control efforts of the subcontractors.

But the following two aspects should be paid attention to in this kind of management mode.

The first, the status of the independent supervision of owner should be highlighted. When conflict occurs between owner and general contractor on the HSE management, it should be in accordance with the HSE management terms in contract. Clear owner's HSE final jurisdiction, to ensure the safety and smooth of the project.

The second, when the receiving terminal uses a foreign general contractor, who is responsible for the safety, while they bring a lot of advanced management experience and system, but they are insufficient understanding of the importance of domestic safety management in the coordination of the work with the subcontractors, which causes some miscommunication and cultural differences, leading to non-uniform thinking.

For foreign HSE management staff, the best way is to adopt the good and advanced HSE practice and implement the system in operation, rather than by foreigners directly commanding subcontractor's front-line operations personnel.

(2) Pipeline HSE management.

From the existing situation, pipeline construction safety management falls mainly on the owner's head, and the pipeline was divided into several sections according to the length to set up the bid management team. Such as an LNG project, pipeline has set up a total of six bid management teams and a shield management team. Each management group is equipped with a construction manager, a field engineer, a quality assurance engineer, a site safety engineer, a site construction engineer, total five people. At the same time, each section of owner management team is located in the section of the job site, along with the construction unit to move with the construction activities. The site safety engineer is dispatched by the owner's HSE department. In business, the site safety engineer's work is arranged by each section of the construction manager, and report to the construction manager; vertically managed by HSE department, providing oversight and guidance to ensure the company's safety management system, got compliance and implementation in professional. There also sets a senior safety engineer above each tender of the safety engineer, responsible for overall supervision and coordination of on-site safety management. As the construction ends smoothly, this construction management mode is proved to be very effective. Not only the safety management can make the pipeline construction site closely cooperated with the construction department, but also the safety not be neglected by the reason of schedule. The HSE department can provide independent supervision and management outside the construction department.

9.5.2.3 Multi-contractor Construction HSE Management Interfaces

(1) Multi-contractors involved in construction brings difficulty for HSE management.

From the view of China's existing LNG projects, by a number of contractors involved in construction, LNG project model can be divided into several sub-projects according to the specific project management. Such as Guangdong Dapeng LNG project receiving terminal, a total of 13 contractors participated in the construction with two major parts of jetty and pipeline. Fujian LNG project was divided into four major parts such as receiving terminal, storage tanks, jetty and pipelines, and a total of 15 contractors participated in the construction. Shanghai LNG project receiving terminal was divided into four major parts such as jetty, terminal, storage tanks and pipelines, and a total of 9 contractors participated in the construction.

From above we can know, because so many contractors are involved in the design, procurement and construction, the difficulty for HSE management is increased.

(2) The project company shall play a leading role in HSE management.

① The owner's responsibility can't be replaced. Just because of multiple contractors involving in the project, although some main engineering project adoptes the form of EPC contractor, the owner must play a leading role in HSE management, coordinate the relationship among the general contractor and the subcontractors, the various contractors in the HSE management, act as a supervision and coordination and go-between role, and shall never think that the EPC general contractor can relax the owner's HSE management.

② To urge the general contractor to take responsibilities for the owner's HSE management. Almost all the built LNG projects have taken sub-project EPC package with different scale and content. In addition to the audit of its HSE management procedures and requirements in the bidding stage, important one is to see the general contractor's management capabilities during HSE construction, whether HSE Management staff is in place and structured. It should also urge the general contractor to divide HSE management interfaces for the subcontractors, and do not leave blank.

③ To pay more attention to subcontractors' safety performance assessment. It should take the performance appraisal to the safety performance of each subcontractor, and form an incentive mechanism that pays attention to safety award and illegal operation punishment in the subcontractors. Such as Fujian LNG project company, the subcontractor gives material reward to employees who acquire outstanding achievements in the safety performance assessment, which achieves good results.

9.5.3 Construction Period HSE Management Focus

9.5.3.1 Health and Safety Risk Identification

When the LNG project enteres the construction stage, the project company should carry out health and safety risk level and safety evaluation, provide the basis for the formulation of the construction period of health and safety objectives and management programs, and ensure the health and safety risk of the project company to get effective process control.

(1) The contents of health and safety risk identification.

① Construction environment. In the process of LNG project construction, it includes geography, meteorological condition, technological process, device layout, in the workplace.

② The contractor's construction facilities. It includes construction preparation, construction equipment, devices, facilities.

③ Contractor's HSE awareness. It includes all involved in the design and construction process of conventional and unconventional activities, construction and operating personnel HSE awareness.

(2) Common hazard types.

Common hazards include mechanical energy hazards, electricity harm, thermal damage, chemical hazards, radiation energy hazards, biological hazards, personnel and engineering factors hazards.

(3) Common accident types.

Common accidents include high-altitude fall, the object hit, lifting injuries, mechanical injuries, fire damage, electric shock injuries, vehicle damage.

(4) Regional division.

According to the receiving terminal, pipeline and sub-transmission station facilities and equipment to focus on regional and functional areas. The plane location map is divided in operation area, for preparation of each block with risk level and size identification.

(5) Quantitative assessment of health and safety risk.

In order to manage the health safety factors of LNG project, it is necessary to design the quantitative evaluation of the form and the relevant description. Generally, the two methods of expert scoring and analogy are adopted, and the following describes the common method of expert scoring (Table 9-11).

Table 9-11　Health and Safety Risk Rating Scale

Evaluation items	Judgment	Score	Evaluation score
A: Occurrence of health or safety accident probability	Entirely possible be expected	10	
	Quite likely	6	
	Likely, but not often	3	
	Less likely to	1	
	Highly unlikely	0.5	
B: Frequency in a health hazard and dangerous environment	Continuous exposure	10	
	Exposure of daily working hours	6	
	Once a week or accidental exposure	3	
	Exposure once a month	1	
	Several times a year	0.5	

Continued

Evaluation items	Judgment	Score	Evaluation score
C: Health and safety impact degree	Several deaths or cause major damage to property	40	
	One death or cause major damage to property	15	
	personnel serious injuries or cause some damage to property	7	
	personnel slight wound or cause minor damage to property	3	
	Compelling, is not conducive to the basic HSE safety requirements	1	
L: Quantitative assessment of health and safety risk $L = A * B * C$			

(6) Regional division of dangerous levels.

According to the regional division and the health and safety risk quantitative evaluation table, identification on the basis of the above damage, accident factors , size and frequency of the appearing risk are given by the experts scoring. According to the formula of Table 9–11 to calculate the L value, the larger the value of L, the greater the risk is, thus differentiate the levels of risk, such as the red area, the yellow area, the orange area, the blue area and so on, representing that the danger level from strong to weak.

9.5.3.2 HSE Daily Management

(1) Doorkeeper, personnel and equipment access management.

Electronic credit card system shall be used, to strictly manage the access of personnel. Only through the safety admission education, the staff can receive a personal photo IC card, to ensure the effective implementation of safety education. At the same time, the individual labor protection equipment, such as safety shoes and safety helmet, shall be inspected in entrance. Personnel who do not have worn personal protective equipment correctly shall not be allowed to enter the construction site.

(2) Construction vehicles, equipment entrance inspection and labeling.

All construction vehicles and equipment must be checked by specialized personnel before entering the construction site. The inspection mainly includes all kinds of effective certificates inspection and appearance check. Issue qualified label when they are qualified, and post on the vehicles or equipment. On the qualified label, there is only one identification number for each vehicle or equipment and effective in the quarter, and it needs to review and change the label when cross-quarter. Label of each quarter is strict distinction. Construction vehicles and equipment need to be check by the doorkeeper at the entrance; only with a valid label qualified is allowed to enter the construction site. In order to control the unqualified vehicle and construction equipment not to enter the scene, it is necessary to lay a solid foundation for the safe operation of vehicles and equipment.

(3) Scaffold listing.

The construction involves a large number of the work at height and needs to build scaffold.

On receiving terminal site, it shall be made of steel tubular scaffold or other scaffold which has been proved to be qualified. When the scaffold erection is over, it must be approved by professionals. After the inspection, hang up green label at the entrance; for the unqualified, under construction or rectification, hang up a red label and ban for operation; for not being able to set up the scaffolding, hang up yellow label, which means staff can go up to work, but must wear a safety belt, in order to ensure the safety of the work at height and scaffolding.

(4) Operation safety analysis.

For any larger or more complex operations, the operation safety analysis must be finished first. Operation safety analysis splits the operation into individual steps which need to carry through, to determine the risk factors and the corresponding safety prevention control measures of each step as well as the emergency measures, to ensure the safety of these dangerous operations.

(5) Zero tolerance.

For serious violations of the safety system and behaviors which may cause serious consequences, such as the work at height operation without the safety belt, the site will implement the zero tolerance system. Violation of personnel will be immediately expelled from the construction site, especially in the construction rush period. It is described the perlite injection process in this chapter, with people falling and suffocation deaths, which is due to high altitude operation not wearing seat belts and brings serious consequences.

(6) Safety training.

The project company and the general contractor shall arrange special safety training engineer to give various kinds of special safety training to all personnel of subcontractors who are approach in, especially the safety rules and regulations of training. It is suggested to find one or two training engineers with rich on-site experience to shoulder this task.

(7) Pre-shift meeting.

Every day before the start of construction activities, team leader shall explain the job contents and steps for the construction personnel according to the construction work situation of the day at the construction site, as well as the relevant risk factors and prevention control measures. Practice has proved that this is the best way to inform the construction personnel of the damage that concrete the specific and safety precaution measures. The project company shall ensure it by means of constant checks and audits.

(8) Construction work permit system.

For the following types of work, such as thermal, excavation, confined space, ray, water, blasting, pressure test, large hoisting, manned hanging basket, it needs to perform construction work permit management system. The implementation of the system needs to be set out with the office of the work permit, the need for a variety of operating license application scope and the validity of a suitable provision, which is helpful for improving the safety control of the site construction work, worth promoting.

9.5.3.3 HSE Emergency Management

(1) Purpose.

Clarify the responsibility and communication channels of main departments and relevant personnel under the state of emergency. Guide the contractor's emergency preparedness and response, then put forward emergency plans to the accidents that may take place, to ensure the emergency rescue work to be completed efficiently and orderly by project company under the state of emergency, and to minimize the loss caused by disasters or epidemics.

(2) Emergency classification.

Gas leakage, fire explosion, personnel heavy casualties, major equipment accidents, radioactive / hazardous substances leakage, outbreak of infectious disease, natural disasters (earthquakes, typhoons), violence incident.

(3) Emergency measures.

① The emergency organization. When the above emergency events occur, the project company shall report to all shareholders in the first reaction time, and set up the corresponding emergency events committee and office. According to the nature of the event, the organizational structure emphasizes on different parts.

② Emergency exercise. The LNG project company's HSE department and EPC general contractor jointly prepare all kinds of emergency plans in advance, and time to time exercise for a variety of emergency situations in the construction site, as for exercise like typhoon emergency drill for southern of China.

③ Emergency resources. Establish the relevant emergency technical information database, query for the effective disposal plan and practice to the existed emergency event, coordinate and call the relevant personnel and materials, and contact emergency technical experts.

④ Plan start. The start command is issued by director of the project company emergency events committee. Ensure the emergency meeting report time for the first time, emergency response office notices and carry out putting the personnel who comes from the various emergency functional departments in place, and organize on-site command and disposal.

⑤ Emergency response records. Emergency response office shall record the full process of emergency events. The information must be quick, efficient and accurate. The information includes telephone, fax, reports, video or audio recording, timely upload, issued. Pay attention to the secrecy of emergency events at the same time.

⑥ To organize rescue. The corresponding emergency plans shall be started immediately. Pay attention to people-oriented, first to the personnel rescue and relief, including the internal organization of the evacuation, escape, the establishment of emergency gathering place, allocation of rescue resources, guarantee of communication and relief channels smoothly, counting the personnel. For equipment and environmental accidents, the area should be cordoned off as soon as possible, and ensure evacuees, power outages, gas cut-off, to minimize the loss caused by the diffusion.

⑦ External communication channels. For casualties, infectious diseases, radioactive/ harmful material leakage, it shall set up a network organization which on-site medical staff, local counterpart hospital are included in, and establish communication channels of counterpart. For major equipment accident, establish communication channels with the equipment vendors; for fire explosion, get in touch with the local fire department as soon as possible; for natural disasters, keep in touch with the local seismological bureau and the typhoon emergency center; for violence, contact with the local public security departments. Speak with one voice on information released to media.

⑧ Emergency rehabilitation work. Include statistical of personnel casualties and equipment losses, site assessment, data archiving of the accident analysis and corrective actions to prevent similar accidents to happen again, and finally report to the shareholders for record.

9.6 HSE Management in Commissioning Production

9.6.1 HSE Risk Analysis in Commissioning Production

When the LNG project enters the commissioning production stage, it should be supposed to enter the feeding operations and carry out a special HSE inspection. At this stage, HSE management is focused on discovering and solving problems, which includes the receiving terminal, jetty and pipeline safety risk assessment, the environmental impact assessment, and key facilities inspection, such as long-distance pipelines, distribution stations, storage tanks, unloading arms, station pipe manifold gas tightness and pressure test, laying the foundation for the overall project acceptance.

9.6.1.1 Safety Risk Evaluation

The HSE evaluation of LNG receiving terminal, jetty and pipeline is carried out on the basis of the previous two stages, but the evaluation of this stage is more important. The reason is through the construction change, the surrounding environment, the overall production facilities, equipment, process area completely finalized, the evaluation work is more in line with the actual and close to the production conditions.

(1) LNG terminal risk evaluation.

In order to fully assess the commissioning production risks of LNG terminal, according to LNG project management practices, all the involved environmental conditions and safety production conditions should be assessed, and the assessment units and contents are shown in Table 9–12.

(2) Jetty risk evaluation.

The risk of jetty operation mainly considers the safety of engineering construction and implementation of scheme, the safety evaluation under conditions of LNG leakage and explosion. Evaluation units and contents are shown in Table 9–13.

Table 9-12 Division of Evaluation Units for Receiving Terminal

Evaluation units		Evaluation contents
Evaluation of the safety environment of the receiving terminal		The impact of receiving terminal on the surrounding units of production, business activities and residents living. The impact of surrounding units of production, business activities and residents living on the receiving terminal
Evaluation of safety production conditions of receiving terminal	General layout safety evaluation	Plane layout, vertical layout
	Storage tank safety evaluation	Accident causation analysis, process program and facilities safety evaluation, leakage diffusion and fire explosion quantitative modeling
	Process area safety evaluation	Accident causation analysis, process program and facilities safety evaluation, leakage diffusion and fire explosion quantitative modeling
	Safety evaluation of loading and unloading process	Accident causation analysis, process program and facilities safety evaluation, leakage diffusion and fire explosion quantitative modeling
	Safety evaluation of tanker loading area	Accident causation analysis, process program and facilities safety evaluation, leakage diffusion and fire explosion quantitative modeling
	Safety evaluation of piping systems and components	LNG / natural gas pipeline systems and components
	Safety evaluation of automatic control system	DCS systems
	Safety evaluation of electrical system	Power supply, power substation, lightning protection and anti-static
	Safety evaluation of fire extinguishing system	Fire-fighting apparatus and facilities
	Safety evaluation of public works	Drainage works, civil engineering, communication systems
	Other aspects of safety evaluation	Torch, sea water systems, maintenance rooms, laboratories

Table 9-13 Division of Evaluation Units for Jetty

Evaluation units	Evaluation sub-units
Safety evaluation of engineering construction program	Safety evaluation of jetty site selection General layout safety evaluation Safety evaluation of jetty loading and unloading process and equipment Safety evaluation of fire extinguishing system Safety evaluation of related auxiliary projects Risk analysis evaluation of ship berthing and departing operations Impact assessment of natural conditions Other factors safety evaluation

Continued

Evaluation units	Evaluation sub-units
Leakage diffusion hazard evaluation	Leakage rate, leakage amount, evaporation rate, diffusion concentration calculation and hazard evaluation
Fire explosion risk assessment review	fire explosion hazard index, pool fire, flash fire

(3) Pipeline risk evaluation.

Pipeline risk evaluation firstly should be distinguished between land and subsea pipelines, because evaluation contents of the two pipelines are different. Land pipeline not only should consider the natural environment, but also take more concerns to the life and property of the people along the route; while subsea pipeline should consider whether it has affected the local fishery production and waterways. Evaluation units and contents are shown in Table 9-14.

Table 9-14 Division of Evaluation Units for Pipeline

Evaluation unit	Evaluation sub-units	Evaluation Contents
Land pipeline	Construction program safety evaluation	Route selection, pipe laying, pipe anticorrosion, pipe used, station and block valve chamber setting, automatic control system, auxiliary production facilities, construction
	Safety evaluation of natural condition, social environment influence	Impact evaluation of natural conditions, such as flood, landslide and other natural conditions. Impact evaluation of social environment, such as densely populated areas, schools, cultural relics, military facilities
	Evaluation of gas pipeline accident simulation	Leakage diffusion, fire explosion accident simulation and risk analysis
	Evaluation of engineering station accident simulation	Leakage diffusion and fire explosion accident simulation
Subsea pipeline	Seabed topography	Evaluation from the surrounding environment, the future planning, natural disasters and other aspects of the impact on design
	Evaluation of subsea pipeline accident simulation	Evaluation of the harmful factors to subsea pipeline
	Human activities	Evaluation of the hazardous factors in the operation of submarine pipeline from the aspects of the surrounding water and fishery activities

9.6.1.2 Identification and Evaluation of Environmental Factors

Environmental factors identification contents should consider three kinds of state of normal, abnormal and emergency, and also take into account atmospheric emissions, water discharge, waste gas management, land pollution, noise, utilization of raw materials and natural resources and other social problems.

(1) Division of evaluation area.

According to the jetty, receiving terminal, pipeline three functional areas, combined with the specific LNG project natural environment and social environment, carry out environmental evaluation.

(2) Risk level evaluation of environmental factors.

The expert scoring method can be used to identify and evaluate the environmental factors, and to evaluate the risk level of environmental factors in different regions (Table 9–15).

Table 9–15 Risk Level Evaluation Form of Environmental Factors

Evaluation items	Appear situation judgment	Score	Evaluation Score
A: Occurrence frequency or probability of environmental factors risk	Completely possible	10	
	Easy to happen	6	
	Possible, but not always	3	
	Small possibility	1	
	Highly unlikely	0.5	
B: Environmental impact	Serious contamination and spread a wide range, damage the ecological environment seriously, not easy to treat	10	
	Contamination and spread the nearby area, cause some damage to the ecological environment, difficult to treat	6	
	Contaminate work area, difficult to treat	3	
	Contaminate work area, easy to treat	1	
	Light pollution	0.5	
C: Impact levels on the project company's finance, reputation, and business	Serious loss of reputation, business and property, social media exposure, government punishment, property loss of more than 100,000 RMB	40	
	Reputation, business, property loss, internal media exposure, government agencies propose rectification, property loss more than 50,000 RMB	15	
	Reputation, business, and property are damaged, internal criticism, government agencies propose rectification and compensation, property damage more than 10,000 RMB	7	
	Reputation, business, property damage is not large, the company's internal rectification, customer suggestions for improvements, property loss less than 10,000 RMB	3	
	Reputation, business, and property not incur loss, only internal company makes rectification requirements	1	
L: Risk level evaluation $L=A*B*C$			

According to the formula in Table 9–15, the L value is calculated, the greater the L value, the greater the risk of the environment. It can be used to divide the different levels of environmental risk.

9.6.1.3 Evaluation Report

(1) Safety problem.

In addition to qualitative analysis, there is a quantitative analysis of the potential risk and harmful frequency of the area, the connecting and interface, equipment and parts, speak with the data. It also needs to put forward security measures, give preventive measures, formulate countermeasures, and put forward security recommendations.

(2) Environmental problems.

The noise pollution and natural resource utilization and other social problems of the environmental evaluation are proposed, and the corresponding measures are put forward.

9.6.2 HSE Special Acceptance

9.6.2.1 General Requirements

(1) LNG receiving terminal.

After the LNG receiving terminal gets mechanical completion, the project company should contact the local fire department in a timely manner to carry out special inspection for the receiving terminal (including the port) fire extinguishing system. At the same time, project company should also timely contact with the local meteorological bureau to conduct a special inspection for the receiving terminal (including port) lightning protection system. Before LNG receiving terminal gets final acceptance, the project company should promptly contact with the local occupational health, labor safety and health and environmental protection departments to implement the special monitoring and evaluation.

(2) Jetty.

Jetty and LNG terminal at the same time carry out special inspection of fire protection system. Before jetty engineering gets preliminary acceptance, the project company should contact with the local occupational health, labor safety and health and environmental protection departments to implement the special monitoring and evaluation. At the same time, the construction side should promptly contact with the local approved bodies to review the "Safety Inspection Evaluation Report".

(3) Pipeline.

Before the pipeline starts construction, the construction planning, construction permits, land permits need to be approved. For design and construction of pressure vessels, fire protection, and lightning protection system, it needs the administrative departments in charge to participate, after mechanical completion and before pre-commissioning to carry out the acceptance. The construction side should contact with the local occupational health, labor safety and environmental protection departments for the implementation of special monitoring and evaluation.

9.6.2.2 Special Inspection

(1) Fire control acceptance inspection.

According to the requirements of "Provisions on the Administration of Fire Supervision and Audit of Construction Projects", after the completion of the construction project fire facilities, project company should be entrusted with the qualification of building fire facilities testing units to carry out technical tests, to obtain building fire facilities and technical test report. After fire control inspection is qualified, the project company should put forward the application of fire control acceptance check to public security fire supervisory body with building fire facility technical test reports, fill out the "Application Form for Construction Fire Inspection", and organize fire inspection and acceptance. The public security and fire supervision agency issue a construction project fire acceptance opinion.

① The main contents of constructional engineering fire control acceptance inspection.

The overall layout includes: fireproof distance, fire lane and water sources, which may affect fire safety; fire risk categories and fire resistance rating of buildings; building fire prevention and smoke zoning and building structure; safe evacuation and fire elevator; fire-fighting water supply and automatic fire extinguishing system; fire prevention design of smoke control, smoke exhaust and ventilation and air conditioning systems; fire-fighting power supply and power distribution; fire emergency lighting, emergency broadcast and evacuation signs; automatic fire alarm system and fire control room; fire prevention design of interior decoration of buildings; building fire extinguisher configuration; explosion proof design of A and B class factory buildings with explosive danger; other contents of the fire design in the national engineering construction standards.

② Acceptance focus.

1) Check the contents of the completion drawings, data and the contents of the "Fire Inspection and Acceptance Declaration Form" of the construction project and the audit opinion of fire agencies whether are consistent with the project.

2) Check fire protection problems proposed in the "Constructional Engineering Fire Protection Design Audit Opinion" whether to be rectified in the project.

3) Check the quality and performance of the construction and installation of all kinds of fire-fighting facilities and equipment.

4) Make spot test of fire facilities function and linkage situations.

(2) Occupational health acceptance inspection.

In trial production stage, according to the requirements of "Measures for the Classification and Management of Occupational Disease Hazards in Construction Projects", the project company should entrust with the corresponding qualification of occupational health technical service institutions for occupational disease prevention facilities operating conditions and workplace occupational disease risk factors for monitoring and occupational disease hazard control effect evaluation, and obtain reports. When the occupational hazards control evaluation is qualified, the project company should put forward the application of occupational health protection facilities completion acceptance to health administrative department. After acceptance, project company should obtain submissions issued by health administrative department. After the completion of the main construction projects, when in the trial production, the supporting occupational protective equipment must be put into trial operation with the main project at the same time.

The evaluation report of occupational hazards control effect in construction project should include following contents: the construction project overview; the basis, scope and content of occupational hazards control effect evaluation; trial operation; occupational hazard factors and harm extent in construction projects; operation and effect of occupational protective equipment; evaluation and conclusion.

When the construction project is completed and accepted, it shall apply to the health administrative department for acceptance of occupational disease prevention facilities, and submit the report of the construction project completion acceptance report and the occupational hazard control effect evaluation.

(3) Labor safety and hygiene inspection.

In trial production stage, according to the requirements of "Provisions of the Labor Safety and Health Inspection and Supervision System for Construction Projects", the project company should entrust with the qualification of the unit to carry out the work condition testing, the degree of hazard classification and related equipment safety and health testing and inspection, to obtain the testing and inspection data; to write the "Special Report on Labor Safety and Health", whose contents include labor safety and health, equipment operation, the effect of measures, testing and inspection data, the existing problems and measures to be taken. The labor safety and health improvement issues in the pre-acceptance should be made on schedule, and timely reported to labor administrative department, then submit the application for safety and health inspection to the safety production supervision and management department for approval. After the labor safety and health facilities and technical measures are approved by the department, the project company shall handle the examination and approval procedures for the labor safety and health of the construction project in a timely manner, and finally handle "Occupational Safety and Health Project Acceptance and Approval Form".

(4) Environmental protection acceptance.

The project company should, after the completion of the construction project, apply to the environmental protection administrative department for environmental protection acceptance according to the requirements of "Measures for the Administration of Environmental Protection Final Acceptance of Construction Projects". Based on the environmental protection acceptance monitoring or survey results, and through on-site inspection and other means, evaluate whether the construction project achieve environmental protection requirements of the activities.

The scope of environmental protection final acceptance of construction project includes: the environmental protection facilities, including engineering, equipment, devices, monitoring tools, and various ecological protection facilities which are built or equipped for the prevention of pollution and protection of the environment; environmental impact report (table) or environmental impact registration form, and other environmental protection measures required by relevant provisions of the project design document.

Before the trial production of the construction project, the project company should submit the trial production application to administrative department of environmental protection, then conduct the trial production after approval. In the trial production stage, the construction unit should entrust environmental monitoring station, environmental radioactivity monitoring station or environmental impact assessment unit with the corresponding qualifications for monitoring and evaluation of operation effect of environmental protection facilities, and obtain "Application for Environmental Protection Acceptance of Construction Project Final Completion (Table)" or the "Environmental Protection Acceptance Survey Report (Table)". After the monitoring of environmental protection is qualified, the project company shall submit "The Application for Environmental Protection and Acceptance of Construction Project" to the competent department, and attached with "Environmental Protection Acceptance Monitoring Report (Table)" or "Environmental Protection Inspection Report (Table)" for the environmental acceptance. After acceptance, conduct approval procedure.

Environmental protection acceptance conditions of completed construction project are: pre-construction environmental protection review and approval process is complete; technical data and environmental protection archives is complete; environmental protection facilities and other measures have been completed or implemented based on the requirements of approved environmental impact report (table) or environmental impact registration form and design documents; environmental protection facilities has passed the test and its pollution prevention ability meets the needs of the main projects; the installation quality of environmental protection facilities meets the professional engineering inspection practices procedures and inspection assessment issued by national authorities. Conditions for the normal operation of environmental protection facilities include: qualified personnel, sound operating procedures and corresponding regulations, enough raw materials and power supply, meeting other requirements of deliverable; pollutant emissions conform to the environmental impact report (table) or environmental impact registration form and requirements of the standard proposed in the design document and approved total pollutant emission control index; various ecological protection measures are implemented in accordance with the requirements of the environmental impact report (table) regulations; the environment that has been damaged can be recovered in the process; the measures has been taken for rehabilitation in accordance with the regulations; environmental monitoring items, points, institutions and staffing, are in line with the environmental impact report (table) and the requirements of the relevant provisions; the environmental impact report (table) is proposed for the environmental sensitive point of environmental impact assessment indicators for verification, clean production, environmental protection measures in the construction period to carry out engineering environmental supervision situation, according to the requirements have been completed. The environmental impact report book (table) requires the project company to take measures to reduce emissions of other facilities, or requires construction project site where the local government or the relevant departments to take "Regional Reduction" measures to meet the total pollutant emission control requirements. The corresponding measures have been implemented.

(5) Lightning protection design acceptance.

The application of completion acceptance of lightning protection system should submit the following materials: the application form of completion acceptance of lightning protection system; lightning protection design approval; lighting protection engineering construction units and personnel qualification certificates.

The contents of completion acceptance of lightning protection system include: legitimacy of the application materials and authenticity of the content; whether lightning protection system meet the operating requirements and relevant national technical standards prescribed by the State Council meteorological authority; whether in accordance with the audit approval of the construction plans.

(6) Acceptance inspection of boiler and pressure vessel and special equipment.

Boilers, pressure piping and special equipment are required to apply for special equipment use registration, which should meet the following basic requirements: special equipment users have a legal identity (natural person shows the identity card, while legal person and other organizations show registration certificate); special equipment should be designed, manufactured, installed, supervised and test by units which have obtained permission, and safety performance should conform to the safety requirements of technical specifications; the user has management personnel, technical personnel and special equipment operators who have compatible certificate; it has maintenance capability, or establishes a contractual relationship with professional maintenance units, thus can maintain equipment timely; the user has created a sound quality management system, various management systems, emergency measures, and can operate effectively; the user has set safety equipment technical file; the users can ensure the special equipment use, conforming to the basic requirements of special equipment safety specifications.

(7) Acceptance of soil and water conservation.

Pipeline project involves land acquisition and conservation. The project company should implement capital, supervision, management and other assurance measures in accordance with approved programs, make engineering design, invitation for bids, construction, and organization under the approved program, strengthen the management of contractors, and practically implement soil and water conservation "Three Simultaneous" system. Procurement of stone, sand, gypsum and other production materials need to choose specified stock ground, clear the responsibility for prevention of water and soil loss, and put on records to the local water administration department. Regularly inform the implementation of the program of water and soil conservation to the provincial water administrative department, and accept the supervision and inspection of the head of the water administration department. Entrust appropriate monitoring institution to undertake monitoring task of water and soil loss, regularly monitor and submit reports to the relevant water administration department.

The project company should strengthen soil and water conservation supervision works to ensure the quality of construction. Preparation unit should send the approved soil and water conservation program report to the project site at all levels of competent authorities within 30 days, then send the service return receipt to Ministry of Water Resources Division Soil and Water Conservation Department. Construction side should follow the provisions of "Measures for the Administration of the Acceptance of Water and Soil Conservation Facilities in the Development and Construction Projects", cooperating with water administrative department to timely apply and organize the water and soil conservation facilities acceptance.

9.7 HSE Audit and Continuous Improvement

HSE audit is the premise of continuous improvement. The following LNG project in various stages of the HSE audit contents are described, to put forward the idea that HSE focuses on continuous improvement and objectives.

9.7.1 HSE Audit Forms and Systems

9.7.1.1 Project HSE Audit Forms

Routine HSE management procedures and specified regular internal audit can be carried out once or twice a year. For a major event of a major accident, the audit can't be carried out irregularly. From the perspective of foreign oil companies' HSE audit form, it can be divided into three categories: internal audit, external audit and superior audit. The following are introduced.

(1) Internal audit.

Internal audit refers to the LNG project company internal staff who audits its own system, or specially audit for a system. The project company select experienced personnel from the LNG HSE department to form audit team, to conduct audit through questionnaires, questions, discussions, on-site inspection.

(2) External audit.

External audit refers to the professional advisory personnel outside the LNG project company. The audit team members tend to be qualified personnel, such as the system of external auditors. This kind of audit is usually conducted for getting the system certificate or maintaining the validity of the certificate, such as ISO14001, ISO18001. Audit methods are often in the scope of the provisions of the standard documents and audit staff's audit experience and skills.

(3) Superior audit.

Superior audit refers to audit by HSE management department of shareholders, and in most cases, the audit team members are often composed of HSE managers of the shareholders and external experts. According to the project's requirements and the members' own experience, audit methods are usually including interviews, review of relevant documents and records, the scene view. Interview range is from LNG project company general manager, senior management to front-line staff.

9.7.1.2 Project HSE Audit System

LNG project company should formulate HSE audit system, including the composition, forms, procedures, and corrective measures of HSE audit. The following describes a superior company HSE audit procedures.

In the process of construction of receiving terminal and pipe network project, based on BP company's project HSE censorship, a LNG project company carries out periodic review of the HSE work in seven stages. At each stage, the review panel are composed of the shareholders or external experts invited by project company. HSE audit team generally consists of various professional personnel, and most are experts in engineering design, construction, operation, HSE and other aspects. Review work is generally arranged at the company location. The first day is for the mobilization and site visits; on the second day, usually talk with the key management personnel of the project company and review grouping data and record; on the third day, continue to review and prepare preliminary results of the review; on the fourth day, review results feedback and ask for opinions. After the review, the review team leader will finalize the review report and recommendation draft, and set the final draft after asking opinions for each member of the team, then submit to the manager level of the project company. Final report is generally completed within one week after the end of the review. After receiving the receipt of manager level, the project company will organize staff to implement the recommendations in the report.

9.7.2 HSE Stage Audit

9.7.2.1 HSE Audit in the Early Stage of Research

(1) General concept.

HSE audit in the early stage of research includes pre-feasibility study and feasibility study two phases, and the specific time is decided by the project company HSE department according to the situation. The general framework of the HSE system has formed basically, and the purpose of the review is to identify the project company HSE systematic frame. The strategy is suitable for the project feasibility study phase of the HSE review.

(2) The main contents of the review.

① Throughout the life cycle of the project, all the relevant HSE concerns can be used in the new technology, whether the project construction site has been determined.

② To confirm whether it has verified and complied with the requirements of all applicable laws and regulations, and shareholders.

③ To confirm whether it has established a suitable HSE plan.

④ To confirm whether it has established appropriate HSE risk management strategies.

⑤ To confirm the maneuverability of research strategy, and whether the strategy is suitable for the project feasibility study phase of the HSE review.

9.7.2.2 HSE Audit in the Design Stage

(1) The main audit contents in the preliminary design stage.

① Whether it has carried out special HSE research. Whether it has implemented the recommendations of the research report.

② To confirm whether the HSE project design has met or exceeded legal requirements. Whether it has identified and complied with engineering specifications, standards.

③ To confirm whether the engineering design quality assurance program has been established.

④ To confirm whether engineering change management program has been established.

⑤ To confirm whether the document management system has been established.

⑥ To confirm whether the resources and training strategy have been established.

Preliminary design phase of the HSE review is generally close to the completion of the preliminary design, when the main design results have been made clear.

(2) The main audit contents in the detailed design stage.

① To confirm whether various HSE special studies, the hazard and operability studies have been completed, and whether the recommendations of the study report have been implemented.

② To confirm whether the change management process has obtained enforcement and compliance, and whether the harm that caused by the change has been reviewed to ensure the integrity of HSE.

③ To confirm whether the expert review work has been carried out and whether examined proposals have been implemented.

④ To confirm whether engineering design controls and inspection have been implemented.

⑤ To confirm the HSE management system, including whether the HSE plans have been implemented effectively.

⑥ To confirm whether the detailed design results have considered the HSE issues adequately, and whether detailed design documents and drawings have been ready for construction.

During the detailed design stage, HSE review is carried out generally before the detailed design is close to completion, Piping and Instruments Diagram (PID) and logic control causal map have been released, and the ground important construction activities start.

9.7.2.3 HSE Audit in the Construction Stage

(1) The main audit contents in the construction stage.

① To confirm whether the quality control of construction projects can ensure the quality safety of building facilities.

② To confirm whether HSE change management system has been established and is effectively implemented in the construction project process.

③ To confirm whether construction workers have been trained, and whether the competency assessment management systems of construction managers and constructors have been established and implemented.

④ To confirm HSE management system, including whether the construction HSE plan has been established and effectively implemented, and whether HSE performance during construction has been guaranteed.

(2) HSE review focus.

HSE audit is generally carried out during the construction phase, especially when the important installation work starts. HSE audit in the construction phase must be conducted at the construction site.

9.7.2.4 HSE Audit in Acceptance and Commissioning Stage

(1) HSE audit in acceptance stage.

Acceptance stage test, without feeding for commissioning, the main HSE audit contents include the following aspects.

① To confirm acceptance work has been completed satisfactorily and facilities already have the debugging condition.

② To confirm commissioning and operation personnel have been adequately trained. Tools and equipment are ready , have the ability to debug or run, and all the necessary debug and run procedures have been ready and can be used at any time.

③ To confirm that the project company is ready for commissioning.

④ To confirm emergency response arrangements have been formulated and ready.

⑤ To confirm the design deviation has been fully assessed to ensure not affect HSE performance.

Large construction projects such as LNG project, generally need to carry out two test before the HSE review.

Large-scale construction projects such as LNG project, generally require two times of HSE review before commissioning. The first time is generally two months before feeding, which is in order to ensure enough time to work for corrective training and document preparation. The second time is generally a few days before feeding, to ensure that all the necessary preparations are ready. If local laws require safety audit before commissioning, the project company should be strictly enforced.

(2) HSE audit in the commissioning phase.

The main contents of HSE audit during the commissioning phase are as follows.

① To confirm whether the HSE performance of operational equipment and facility meet the design requirements.

② To confirm whether the HSE lessons in the detailed design phase, the construction phase and the early operational phase of the project are already fully collected and shared within the company.

③ HSE audit in the commissioning stage is generally conducted within 6～12 months after the feeding for commissioning.

9.7.3 HSE Continuous Improvement

9.7.3.1 Preparatory Work

The purpose is through the relevant HSE internal self-examination, external inspection, problems and pitfalls are found by the superior audit, on the bases of full information to pave the way for the rectification.

(1) Deep investigation.

Through HSE audit during each stage above, it needs to collect the problems and risks which do not comply with HSE regulations and procedures, including accident report, HSE weekly, HSE monthly, HSE annual and HSE accident inspection and rectification reports.

(2) Analysis.

To analyze carefully on the basis of this, mainly from four aspects.

① Human factors, such as HSE awareness, psychological factors, physiological factors, responsibility and so on.

② Equipment factors, such as equipment quality standards requirements, failure occurs during use and troubleshooting, and other potential safety risks.

③ Organizational factors, such as project organization structure and staffing, HSE management procedures, regulations, continuous improvement philosophy and so on.

④ Laws and regulations, standard factors, such as the revision, adding of national and local laws and regulations, changing and upgrading various standards and so on. Such element changes cannot be controlled, so only implements unconditionally.

9.7.3.2 Rectification

(1) Improvement of HSE management system.

On the basis of the audit, the project company conducts HSE management evaluation, finds out the institutional settings, management procedures, the provisions of the gap with the benchmark company, takes corrective and preventive measures timely, and eliminates factors that do not meet the project company HSE regulations and procedures in order to achieve continuous improvement.

(2) Implementation units.

This work is generally done by the engineering department and the HSE department of the project company, the contractors and the relevant units. Rectification is a process, here including the institutions to adjust settings, staffing, the original management procedures to amend and improve the implementation of improved operation rules, equipment replacement and rectification plan, submission of post-evaluation and the rectification report and so on.

In fact, HSE audit and continuous improvement described in this section are just for convenience, and HSE audit and continuous improvement go throughout the entire LNG project cycle.

9.7.3.3 Rectification Problems Focus

(1) Leadership responsibilities.

As a senior leader of LNG project company, especially what the "Top leaders" concern and commit to the HSE, often plas a decisive role in project HSE performance. From several domestic and international companies' survey, in the evaluation elements of corporate HSE management system, all take the "leadership" as a prime factor to evaluate, because of the leadership and decision-making power of invest resources. If the project company HSE management level is improved, which is often reflected in the awareness improvement of HSE guidelines, ideas, policies, and objectives of leadership, the adjustment of HSE organization, and the increase of personnel and capital investment.

(2) Contractor management.

From the survey conducted by the foreign oil companies, the personnel deaths of contractors' company is 2.78 times of that of the oil company. From China's LNG project construction, almost all fatal accidents occur in the contractors, therefore, to strengthen the HSE management of contractors is top priority. From the perspective of the project company HSE management, it should pay attention to the completion of contractors HSE management structure, staffing, HSE procedures and requirements, at the same time to supervise and improve the execution of the contractor in the actual work.

(3) Risk assessment and measures.

Specific to a particular job, HSE management risk assessment is the primary work, and only recognize the HSE risks in advance will it be reliable. Proposing preventive measures against the risk is the prerequisite to reduce accidents. For risk assessment, it needs to learn the mature previous experience of the assessment, and also needs to improve the evaluation and proposes countermeasures to make this work improve constantly.

(4) Crisis and emergency management.

Accidents may happen unexpectedly any time. In the commissioning production process of LNG project construction, natural disasters may occur unexpectedly, so it must do a series of preparation work in advance to prevent crises and risks bringing the loss of life and property to the project, and all kinds of contingency plans are the basis. Juding from the Chinese LNG project running, it also appeares major casualties, checking whether emergency plan meets the actual needs through practice. Improving the effectiveness of the plan is the key, which needs to contact with local government departments closely, and publicize to local authorities in advance that LNG project dangerous may occur during operations, so that when a crisis occurs, the local government can support. In the absence of an emergency situation, the exercises of contingency plan are necessary.

(5) The comprehensive evaluation of HSE in the project acceptance.

After the completion of the project, the project company has to conduct a comprehensive evaluation and acceptance, including HSE, which will help the performance evaluation of the entire project, also help investment holding company accumulate HSE management experience used in the new project, and the project company needs to conduct HSE inspection, registration and archiving.

Chapter 10 LNG Project Cost Management

10.1 The Concepts of LNG Project Cost Management

10.1.1 The Definition of LNG Project Cost Management

10.1.1.1 Cost Definition

The so-called cost is to promote and operate the LNG project, the need to invest in a variety of forms of human, material and financial resources, here in the form of currency to calculate its sum. According to the cost input relevance, input stage and executable degree, the costs can be divided into following categories.

(1) Classifications by cost input relevance.

① Direct costs. Direct costs include labor, raw materials, equipment and other expenses for work packages or specific tasks. These expenses represent direct cash flow expenses, as the project progresses, the direct costs are usually separated from the management cost, and the sum of low levels of the projects only included in the direct cost.

② Indirect costs. It is also called overhead expense. It is the cost that cannot be connected with specific deliverable, but it is the cost which plays a role in the whole project. The overhead expense, include the consulting fee, executive salaries, training and business travel expense. Someone regards the expense that connected with project deliverable or work packages called direct management cost. Overhead expense is usually allocated as a percentage of total direct costs.

③ Total costs. It is the sum of the direct costs and the indirect costs of all deliverables. If the project is undertaken by contractor, it should be considered to add a percentage as contractor's profit. The LNG project company (owner) has to take direct costs as the cost control of performance evaluation. The reason is the project company can only control the direct costs.

(2) Classifications by cost input stage.

① The stage before the feasibility research report approved. For the project is expected to progress, it is required to list the costs incurred in the early stages of research, such as market research, LNG resource negotiations and contract signing (sometimes the event occurs in the latter), pre-feasibility studies and feasibility studies, environmental assessment costs, commission special research costs, consulting fees, agency fees, technical annex and business negotiation costs, investigation costs, conference fees, expert hire fee, etc., the sum of monetary form of the human, financial and material. At this stage, the work is to be initiated by the LNG project sponsor, early project team and the project office to complete together, but this part of the cost is often difficult to control, generally is not listed as a LNG project company cost management.

② The stage after the project approved. Generally, the project company has been set up at this time. The budget is prepared according to project schedule, approved preliminary design estimates and signed contract. Specifically, it includes project FEED design, preliminary design, detailed design, construction drawing design, land acquisition, field formation, etc., the sum of monetary form of the human, financial and material resource input before construction. The above work can be managed by bidding. The detailed plan budget is listed according to the LNG project work plan arrangement and research outline.

③ The construction stage. Project construction costs, include labor, materials, machinery fee, outsourcing of turnkey engineering costs and other direct fees and management fees, etc. Usually, the project construction costs account for more than 80% of the total costs, also is the focus of costs management and control.

④ The acceptance stage. The project examination and acceptance, including single debugging, the overall debugging, safety and environment evaluation, production and operation organizations transfer, etc., all above costs input.

(3) Classification by cost of executable degree.

In this book, the cost system as follows:

① Cost estimate. After the project feasibility study, project investment of the content is clear, the calculation of the amount of investment is called cost estimate for the scale of investment and financing options for the project investment decision-making reference department, the industry and policy is strong, but the execution is weak.

② Budgetary estimate. On the basis of project feasibility study to carry out preliminary design of the project, including detailed estimates of the former cost, and it has distinctive regional and market characteristics, more practically, this cost calculation result is called budgetary estimate. The budget estimate shall include all the expenses required from the start of construction to complete the inspection and acceptance before delivery of the project. The project budgetary estimate approved by the board of directors is the total investment of LNG project control benchmark.

③ Cost budget. On the basis of project preliminary design, to carry out the project construction drawing design, according to the approved construction design drawings, construction drawings quota and unit valuation table, the various rates standard, then to draw up the project cost, which is called the project cost budget, only to prepare the unit works budget and comprehensive budget, it does not include the cost of the project in the early stages, it is the basis for annual budget in the construction period.

④ Annual budget (including the year adjustment of the budget). It is defined on the basis of the budget of construction plan, the investment is taken a year as unit is called the annual budget. Annual budget is the basis for the preparation of the annual capital (liquidity) demand plan with strong execution degree.

⑤ Cost settlement. Also called the completion of the settlement, in the completion of the project acceptance stage, the cost settlement is generally prepared by the contractor, according to the changes in construction process, the construction drawing budget changed, and the actual amount is compared, compared with the relevant quota standards after re-accounting, also taking into account due to omissions and missing items in the construction drawing of all the project cost. It is the foundation for the project owner to audit.

⑥ Final accounts. Final acceptance is generally carried out in the LNG project overall test run for a period of time (usually half a year), generally, final accounts is formulated by the project legal person unit or commissioned unit, which is the total cost of the whole process of the project summary calculation actually occurred, in the preparation of final accounts of summary and list of assets, it is necessary to pay attention to comprehensive, real and objective reflect the project cost to evaluate the actual effect of investment projects. It is the final approval point of the project cost.

⑦ Cost accuracy. Because of the basis of the above information, and the role is different, its accuracy by a "progressive elaboration" of the process, the later one is more close to the actual investment than the previous one. Compared with final accounts, generally cost estimate bias 10%~30%, budgetary estimate and budget deviation bias in 5%. Under normal circumstances, the settlement is part of the final accounts, is the basis for final accounts. Final accounts can't exceed the budget, the budget can't exceed the budgetary estimate, the budgetary estimate can't exceed cost estimate.

The budgetary estimate, budget and final accounts are called "Three Calculation".

10.1.1.2 Definition of LNG Project Cost Management

LNG project cost management is through cost planning, resource planning, cost estimation, cost budgetary estimate (budget) and cost control process, so as to strengthen management, improve the cost management system and the level of cost accounting, reduce construction and operating costs, to complete the LNG project in the scope of the approved budget, in order to achieve profit target.

10.1.1.3 Definition of Total Life Cycle Cost Management

From the LNG project pre-research, project construction and project operations extended to the entire project production period, the generalized cost management is called "whole life cycle cost management", and it is relative with the whole life of the project management concept in Chapter 1—From LNG Project Company, the first concern is the cost of resources needed to complete the project tasks (costs), however, China LNG project operation mode is after the completion of the project, and the operating company is the original project company evolved, the project stage cost investment have direct impacts on the entire production operation level of profitability. In the project life cycle, therefore, consider operation period of cost reduction, putting forward the concept of whole life cycle cost management means great significance.

10.1.2 The Significance and Principles of LNG Project Cost Management

10.1.2.1 The Significance of LNG Project Cost Management

(1) Cost management is the pre-requisite of project profitability.

The ultimate goal of project is profit. In a sense, the lower the project investment cost is, the greater the possible profit in the operating period. Cost management provides an effective way to track, monitor and control all kinds of expenses during the whole process of the project, analyzes the reasons and put forward corrective measures to ensure the cost control in reasonable scope, in the case of less people, financial, material input, to achieve the maximum economic benefit, in order to achieve the objectives of the project's profitability.

(2) Cost management is an important aspect of project management.

From the above division of the LNG project stages, the corresponding cost management focus is not the same. In the early stage of the project, LNG resource of negotiation is the key, the lower the price of LNG resources, the greater profit space for projects. In the project design stage, the qualified design unit selected, closer to the user selected location and relate small amount of engineering construction, the investment will be reduced. In the project construction stage, it is the key to select experienced contractors, the use of bulk purchase and time difference acquisition, at the same time, to strengthen the project schedule control and cost control. In project acceptance stage, the contractors, the supervision units, the owner and the local government coordination, responsible for the project acceptance is its core. Only by grasping the cost management of each link, can ensure the overall project investment minimum.

(3) Cost management is the effective measure to promote the project management.

Cost management is not chest thumping, but the need for a scientific methods and means, it includes a strict management system and procedures, statistical methods and information feedback. The general manager of project company is the first responsible person, but it also needs to establish the cost control department and a sound cost control responsibility system, only to establish a sound system of cost expenditure statistics, so as to analyze the reasonable cost of expenses in a timely manner. These methods and measures promote the other project management, such as schedule, quality, communication management and coordination, so as to promote the overall levels of project management.

10.1.2.2 The Principles of LNG Project Cost Management

(1) The principle of comprehensive management.

LNG project cost management means that the whole process, the full range, all links, the stages of management. Cost savings can not only depend on a short period or a single incident, but also the implement of the entire project life cycle, and even consider the entire operating period. From the definition of the project cost, the cost is the sum of each work package and each activity resource input, which requires us from every expense very careful in reckoning, only every

minimum expenditure to ensure the lowest overall cost.

(2) The principle of seeking truth from facts.

LNG project involves a wide range of technical complexity, some materials and equipments need to import, maybe in some design and construction management to rely on foreign forces, therefore, in terms of cost management and control, under the full investigation, should be imported equipment material must be imported, and foreign technology must be used. According to the reasonable cost price standard for the domestic and international to prepare the cost budget, the cost control should reflect on the reality and rationality.

(3) The principle of obeying the procedures.

Cost management and control is the core of project management. To do this work, we must formulate strict, reasonable and convenient for operation in advance the cost of examination and approval procedures and processes, such as the implementation of cost approval unit, cost plan, cost audit unit and expense plan execution, etc. Combining with the market forecast, work arrangement, the project schedule to prepare the cost plan, to carry out the responsibility at all levels, make the cost management as rules-based, illegal-rectified.

(4) The principle of stressing the key events.

Cost management is a complicated process, but in the actual practice, it should be paid attention to the key work of cost management, because seizing the key points is to seize the main contradiction, such as the initial LNG resource negotiations, the price is the focus of cost management work. The lower the LNG price, the greater the profit margin for the project. Forthermore, if the receiving terminal is close to customer market, the project amount and pipeline investment can be reduced, thus greatly reducing the cost of the whole project.

(5) Flexible principle.

Cost reduction methods have many ways; we must have a flexible application. When LNG resource buyer's market is formed, it should take advantage of the market competition, to obtain preferential LNG price; When LNG resource seller's market is formed, it can be combined with more investor's LNG projects at the same time, before or a after the construction, taking bulk purchases to obtain LNG resource, materials and equipment at the lower prices, or trying to use domestic material replace imported in order to reduce the price; or reducing the cost of finance by loan interest through negotiations with the banks, or reducing the some risks by insurance, and so on.

(6) Upper and lower binding principle.

Cost savings is not a matter for general manager of the project company alone, nor is it just a matter for not the control department and the financial sector, but a matter for every cost responsible department and every employee, it shall be to encourage all employees to save, and to implement in their own work, also to encourage the supervision company and the cost of the contractor control behavior, in order to form a joint force amory all project participants to contribute to cost control.

(7) Exception principle.

Every stage of the project operation, not everything can be as we wish, when we cannot meet the above principles of reducing costs, but also must promote the key moment of the project, taking a long-term view, accept some loses and meanwhile in the important cost links and high cost items to get preferential proportion, so as to achieve the goal of low cost on the whole.

10.1.3 LNG Project Cost Management Process

10.1.3.1 Three Components of the Cost Management Process

The general flow chart of LNG project cost management is shown in Figure 10–1. We can see from the chart, the first part is the basis of LNG project cost management, and the contents are composed of project scope and work breakdown (including work package resource estimation), schedule, cost quota, market quotation and project contracts. The middle part is the core of cost management, cost management system from the beginning to the resource planning, and lay the foundation for the cost estimate, budgetary estimate, budget preparation, each layer of work is the premise of the next layer of work, in the preparation process of the cost plan, and can be fed back to various levels, the middle part of the working depth and confirm has a direct impact on the effectiveness of cost management. Cost control is throughout the entire project life cycle activities.

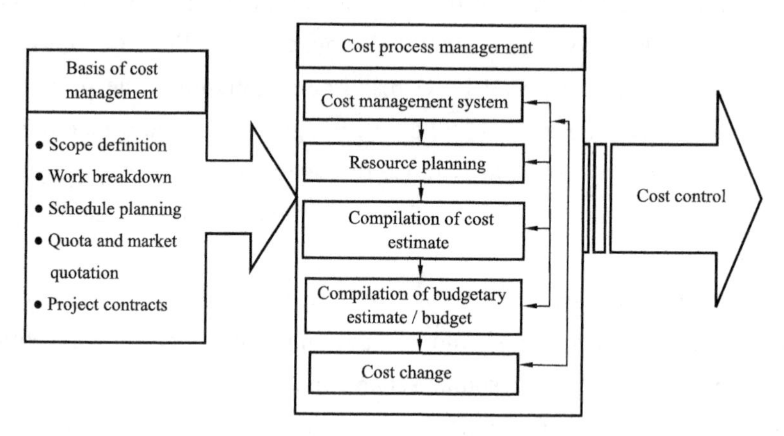

Figure 10–1 LNG project cost management flowchart

10.1.3.2 The Relationship Among the 3 Components of the Process

(1) The relationship between the basis of cost management and cost process management.

From the cost management basis, the scope of the project determines the size of the workload, work breakdown to further clarify the work schedule to reflect the project to go through the length of time, from another reflection of the cost of investment, project quota and market conditions provide a benchmark for calculating the cost, in addition to the project contract the project workload, also associated with the cost price; cost management system to ensure the implementation of cost management from organization and system, only when the cost management basis is clear, in order to ensure the smooth process of cost management, to provide reliable data for resource planning, project investment plans only resources to ensure the accurate cost estimate in

place so as to provide conditions for budgetary estimate and budget accuracy.

(2) Relationship between cost process management and cost control.

As shown in Figure 10–1, the cost management process is its core, including the cost management system, organization system and institution system which are reflected in financial management system. The resource plan is jointly formulated by the project company, the design company and the various engineering contractors, which reflects the resource input of LNG project. The cost estimate is based on the feasibility study, which is based on the money invested in the resources of each task, through the accumulation of the amount of money into the currency. The budgetary estimate is the sum of the cost of the investment resources and constraints or assumptions under consideration, which is completed on the basis of preliminary design, completed by the design unit. The budget is completed in the construction drawing design stage, provides cost quantitative data for individual sub-projects, unit works and work package budget, and the project cost change is based on changes in the scope of work, work breakdown, schedule changes, market and make changes, and the work of the middle part is also called the component part of the static cost management. Only the above work is in place, in order to expand the effective cost control, it runs through the different stages of the whole project cycle.

(3) The relationship between and cost control the former two.

Cost control is a dynamic management content, in addition to its basis of cost management and cost management process as the premise, also according to the change of internal and external environment factors, timely adjustment and coordination, to enable the project to operate under effective cost control. If the cost process control is smooth and in place, we can find insufficient cost management basis in turn, can help the project cost management department to consider what on the basis of in-depth investigation and supplement for cost management review process provides entry point is the imperfect management system? Or the cost of estimating missing items? Or the cost of the budgetary estimate, the budget did not consider the trend of price changes, etc. Similar to the project schedule management, cost management only above 3 part is tightly linked, consistent with a combination of static and dynamic, also on the cost of change when necessary and timely adjustment, in order to ensure the completion of the project cost estimates in the rational.

10.2 LNG Project Cost Management System and General Practice

10.2.1 LNG Project Cost Management Organization System

10.2.1.1 Cost Management Organization

(1) Cost management before the establishment of project company.

Cost management is the highlight of the LNG project management, to seize cost management is to seize the core of the project management, the project sponsor or project team have to invest a lot of energy and manpower to strengthen cost management. LNG project cost management is a part of project plan management. In the early days of the project team, the cost management is assumed or entrusted by project investor's budget or financial department.

(2) Cost management of project company.

Like the continuous improvement of the overall organization of the LNG project, the establishment and staffing of the cost management institutions also have a continuous improvement process. According to the existing LNG project company cost management practice, the organizational arrangement for the project cost management are not the same, but there is a common characteristic under the leadership of project company's board of directors by the budget management committee management to carry out cost control, the control department of the project company is the daily organization of the cost management. But some project company puts the cost management functions under the financial department, the project company departments in charge of their respective costs according to the authority, to manage and control their costs (Figure 10-2).

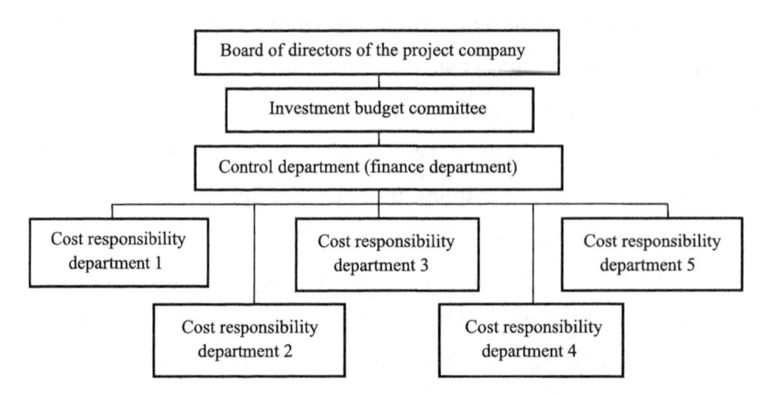

Figure 10-2 LNG project company cost management organization

10.2.1.2 Cost Management Organization and Responsibilities

(1) Board of directors.

The Board of Directors is the highest authority in the project cost management, and its rights and obligations are determined by the company's articles of association. The cost management responsibilities are as follows:

① To review the policies, regulations and system of project cost management, etc..

② To review the principles and procedures for the preparation of project expenses.

③ To approve project cost planning, feasibility study cost estimate, preliminary design budgetary estimate and annual budget.

④ To review the major cost adjustment plan.

(2) The budget management committee.

The budget management committee is a special agency for the implementation of cost management, in the form of budget meeting to review the project budget, the committee is composed of director (generally by the general manager of the project company), deputy director (by deputy general manager or chief financial officer of the company) and the members (by responsible person of the each department). Its responsibilities are as follows:

① To review project estimate, budgetary estimate and annual budgets.

② To adjust budgetary estimate decomposition, coordinate the differences in the budget preparation and execution.

③ To decompose and issue the annual budget approved by board of directors;

④ To review the budget implementation report, to organize the implementation of the budget analysis meeting;

⑤ According to the needs, to review the budget of the adjustment and superaddition, and to submit to the board of directors for approval;

⑥ To assess the implementation of the budget responsible departments.

(3) The cost control department.

The cost control department is responsible for the project cost management of routine affairs. Its duties are as follows:

① To formulate the policies, regulations and systems of budgetary estimate and budget management;

② To issue the principles and procedures for estimate budgetary and make budget, guide and coordinate with the compilation of departmental budgets;

③ To review and balance the draft budget submitted by each department, compile the draft budget of the project, report it to the budget management committee for approval;

④ To check and supervise each responsible department for budget implementation.

⑤ To sum up the budget execution situation report and budget variance analysis report, submit to the budget management committee;

⑥ When necessary, to propose the opinion and scheme of budget adjustments and superaddition, submit to the budget management committee;

⑦ To assist the budget management committee, coordinate and deal with the problems appeared in the process of budget implementation;

⑧ To handle the other budget management issues assigned and authorized by budget management committee.

(4) The cost responsible departments.

Project company departments are responsible for the implementation of the cost of the department. Its responsibilities are as follows:

① To draw up the budget draft of the responsible department;

② To organize and executive the issued annual budget of responsible department;

③ To prepare the budget performance reports and variance analysis reports of responsible department according to the prescribed time limit, submit them to cost control department;

④ If necessary, to submit the budget adjustments and superaddition opinions and proposal;

⑤ To handle the other budget management issues assigned and authorized by budget management committee.

10.2.2 General Practice of LNG Project Cost Management

10.2.2.1 Summary

LNG construction project cost effective control is an important part of the construction management, throughout the whole process of the project construction. The cost management system is a part of financial management system.

(1) To pay close attention to engineering cost.

The engineering cost is the priority among priorities of project cost control, how to determine and control the project cost reasonably, control the construction of investment within the approved budget, correct the deviation at any time is the core of cost management, in order to ensure the realization of the project investment objectives, which is the main purpose and task of the project management of project company.

(2) To build the stage cost basis.

In the construction of the project, from the beginning of the preparatory work for the project, we can not do it without the cost management, cost calculation, payment here including cost frame count, cost estimate, budgetary estimate, final accounts, each of the previous cost is the basis for the latter.

(3) To finalize the final completion of the project.

Final accounts is the starting point for the final confirmation of the project, only the above cost calculation is based on solid, we should pay close attention to the cost management of change, the workload change, verify the resource inputs and the corresponding costs, strictly supervise and manage the contractor, in order to avoid conflicts and disputes, as soon as possible to obtain the final accounts approved by all parties.

10.2.2.2 Estimate, Budgetary Estimate, Budget Detail

The project estimate, budgetary estimate and budget details including engineering cost and expenses, liquidity and management. The details are as follows.

(1) Engineering cost.

Project cost estimate, budgetary estimate and budget are divided into the receiving terminal engineering estimate, budgetary estimate and budget (including tanks, vaporizer equipment), jetty engineering estimate, budgetary estimate and budget, pipeline engineering estimate, budgetary estimate and budget, supporting engineering estimate, budgetary estimate and budget and so on. Where the cost of civil engineering, equipments, materials and equipment purchase costs, installation costs can form a fixed asset, construction and other costs may form the formation of fixed assets, intangible assets and deferred assets.

(2) The circulating fund.

Generalized circulating fund refers to the current assets of all project company, including cash, inventory (material, work in process and finished goods), accounts receivable, securities, advance charge, other items, and also known as the working capital of project operation. The circulating fund in a narrow sense=current assets−current liabilities, also known as net circulating fund. The more net circulating fund, the stronger paying short-term debt, so the credit status is also higher, financing in the capital market is easy, the cost is relatively low. The Project Company through the preparation of the annual budget for the project cycle, it provides a basis for the circulating fund.

(3) Management cost breakdown.

Project company management costs and engineering costs are separated. Project company management costs should be drew up, includes cost estimate, budgetary estimate and budget in detail, such as the project team and the project company of various functional departments of staff salaries, fixed assets (vehicles, office equipment, personal computers, etc.) office rent, IT, communication facilities and services, office stationery, car rental fees, transportation fees, communication fees, travel expenses, conference fees, business consulting fees, advertising fees, training fees, staff recruitment fees etc.

10.2.2.3 Estimate, Budgetary Estimate, Budget Preparation and Management

(1) Basis of preparation.

According to the relevant national laws and regulations and policies; project joint venture agreement and the articles of association of the company, the project construction management system, project management strategy, company expenses and financial and accounting system, bidding system; project scope, schedule, work break down, project construction cost quota standards and market conditions, project construction contracts, agreements and on the basis of others.

(2) The responsible units for the compilation of cost estimate, budgetary estimate and budget.

① Project cost estimate is compiled by the LNG project feasibility study report preparation unit, led by the project company.

② Project budgetary estimate is compiled by the functional departments of the project company and the preliminary designed units together. Led by the planning department to coordinate the various functions of the departments, to prepare the management fees, and finally by the designed units to summarize the project budgetary estimate.

③ Project budget is compiled by construction drawing designed units according to the project composition, based on the project budgetary estimate, the unit works is planned to invest resources and budget quota and index are used to calculate the cost. Similarly, led by the planning department to coordinate the various functions of the departments, to prepare the management fees, and finally by the construction drawing designed units to summarize the project budget.

(3) The responsible units for the compilation of the annual budget.

① The project company of all the departments within the scope of budget management, according to the project budget, contents and requirements, to prepare for the framework plan of the department under its responsibilities and the budget draft of the next year, after approved by the company leadership in charge of the department, submit to the planning department.

② After reviewing and balancing the next annual framework plan, the next annual budget submitted by the various departments, the planning department sums and prepares the next year budget of the project company and submits it to the budget management committee for review.

③ The budget management committee shall, at the end of the year, organize and review the company's next year framework plan, the next year budget draft, and submit to the company's board of directors for consideration and approval.

10.2.2.4 Budgetary Estimate, Budget Approval Management and Budget Execution Process Control

(1) Budgetary estimate, budget examination approval management.

The shareholders of company organizes expert to review the project budgetary estimate, after approved by the board of directors of the company, and as a benchmark for project investment; the annual budget approval authority is belong to the board of directors, the director of the budget management committee should convene a joint review of the shareholder's business competent department, and then submits to the board of directors for approval. After the approval of the project budget, the project company control the implementation of the annual budget.

(2) Budget implementation process control.

① Control principle.

The budget execution process control should be based on the basic principle of "With money into the dish, payment by budget". Over-budget items or extra budget items, are not allowed to be used and incurred until they are approved by the company's financial management authority.

② Decomposition of responsibility.

a. The contract planning department is responsible for project budget and other budget (in addition to financing costs, interest of construction period, engineering insurance, health safety and environmental protection fees), the engineering department to assist, if the project has set up a project teams, which is directly responsible for the project teams.

b. The finance department is responsible for financing costs, construction period interest and engineering insurance cost three party budget control.

c. The QHSE (or HSE) department is responsible for health, safety and environmental protection fee budget control.

d. The Office or the human resources department is responsible for project company management fees budget control.

e. The production preparation department is responsible for the deferred assets budget control.

engineering department (or project teams), administrative office and HR department may

transfer the budget control responsibilities under the work to other functional departments under the work required, but the transfer of such obligations must be in written report to the company with the approval of the budget management committee and copy to the control department for the record.

(3) Cost approval authority.

Project cost approval authority by the LNG project investment side of the budget management regulations and the practice of a great impact, the specific cost of the approval authority is non-unification. Generally, LNG project investment is huge, the amount of investment funds up to billions of Yuan (RMB), so, at all levels of approval to use the amount of funds amounted to one million (CNY). Under the jurisdiction of a project company LNG budget management authority introduced as follows (which: $X_1>X_2$, $X_3>X_4$):

① Before use of the budget allowance of the project, it is necessary to report to the board of directors for approval, the amount is less than X_1 (CNY) needs to be approved by general manager of the project company, the amount is X_1 or more (including X_1) needs to be approved by the chairman of the board of directors.

② Civil engineering, equipment installation engineering in the total approved budget, such as the case of project change and design change need to break the annual budget, the amount is less than X_2 (CNY), the general manager of project company can be approved in advance, and report to the board for the record; the amount is more than X_2 but less than X_1, the chairman may approve in advance, and report to the board for the record; the amount is more than X_1 should be reviewed and approved by the board of directors.

③ The project company management fees is in the total amount of the approved budget, In case of special needs of project construction, it must be broken through the annual budget, the amount is less than X_4 (CNY), the general manager of project company can be approved in advance, and report to the board for the record; the amount is more than X_4 but less than X_3, the chairman may approve in advance, and report to the board for the record; the amount is more than X_3 shall be reviewed and approved by the board of directors.

(4) Matters that should be paid attention to in the process of budget execution.

① The budget within the project budget is controlled implementation by the responsible person of the department, the planning contract and the finance department are responsible for supervision. Before the use of budget, except for the amount of funds used within the department responsible person for the approval authority or fixed and periodic occurred items within the jurisdiction of project funds, such as wages, bonuses, allowances and subsidies, or department self-control of funds, the responsible departments should first put forward to the use of the application, fill in the application form, examination and approval according to the prescribed procedures. Approved the use of the original application form is stayed in the finance department, and a copy is stayed in the budget department. When applying for the first payment, the application payment department shall submit a copy of the application form together with other supporting materials to the finance department.

② Each individual budget should not breakthrough, and the budget items should not be diverted. In case of special circumstances need to be a breakthrough, firstly, it is needs to make an application, state the reasons, examined and adopted by the budget management committee, and it can be used after approved by the company's board of directors.

③ Implementation of the annual budget inspection system. During the middle of the year, according to the actual implementation of the first half of the project budget, it is necessary to re-predict annual budget for the year, to accurately estimate the total budget expenditure, and prepare for the compilation of the annual budget for second year.

④ Superaddition and adjustment of the annual budget. Annual budget in the implementation process, such as the case of the following factors, which can apply for superaddition or adjustment. These factors including: national price, tax policy adjustment, natural disasters, such as force majeure, project schedule adjustment, project plan changes and other factors approved by the board of directors. Responsible departments must state the reasons for the preparation of the need to adjust or supplement the project budget, submit to the planning and contract department for review, then the budget management committee for consideration and report to the board of directors for approval after the implementation. The board of directors is not in session, under special circumstances need adjustment or supplementary budget of the project, the item company reported to the board of directors, directors can sign in the form of a resolution of the board of directors.

10.2.2.5 Budget Implementation Analysis and Evaluation

(1) Analysis meeting system.

It needs to use the analysis meeting system to review the implementation of the budget, and analysis meeting will be organized once every quarter to analyze the implementation of the budget, check and find out the reasons for the differences, and to propose appropriate measures to improve according to the impact factors. Analysis meeting will be directly organized by the budget management committee or authorized organization, the budget management committee members and the responsible person participate in the meeting, special and emergency situations can be held at any time to analyze the budget, in order to solve the urgent problems in the budget implementation process.

(2) Analysis and assessment method.

① The responsible departments should follow up and check the implementation of the budget in a timely manner, each department prepares their own budget implementation and the annual rolling forecast report, submits to the planning and contract department for a number of days before the analysis meeting. The planning and contract department is responsible for the comprehensive analysis of the implementation of the budget responsible department report, and formulates the project company budget implementation report, submits to the budget management committee for consideration.

② The budget implementation report needs to deal with the budget deviation analysis, explain the reason, distinguish the responsibility belongs, put forward the corrective action and the opinion.

③ The budget management committee needs to assess the budget report and budget implementation of the budget responsible departments in accordance with the budget management assessment methods, incorporate into the company's integrated appraisal management.

10.3 LNG Project Cost Estimate Management

10.3.1 General Concept of LNG Project Cost Estimate Management

10.3.1.1 The Definition of LNG Project Cost Estimate

LNG project cost estimate refers to the work according to the project feasibility study, it is necessary to conduct in-depth research on project resources investment, carry on the earnest arrangement on the input resources plan, by expert judgment, similar project quota data, statistical analysis and mathematical model, the current market various items' resources price information, basic implementation of the various activities of the project cost, estimate the total cost of the project covering from preparation to completion and acceptance.

10.3.1.2 LNG Project Cost Estimate Preparation Purpose

In the integrity of the cost management, cost estimate is one of the early basic works in cost management. Generally, this job in the project company have been sets up, the engineering design, construction scheme, equipment and materials procurement management strategy have been formulated, domestic and abroad market research has been completed, the feasibility study report has been compiled, this is the best time to make a cost estimate. The significance of the cost estimation are as follows.

(1) Market oriented to control project cost.

LNG project is operated under a certain market environment, therefore, the cost estimate needs to be competitive in the market, it is necessary to establish and improve the market oriented, efficiency oriented management model, highlights the importance of cost management in project management, so as to control costs and expenses effectively, improve the competitiveness of the project.

(2) Benchmark for judging the price of a bid.

In LNG project operation, project bidding is the main way for project engineering design and construction, the project company control department or finance department, only have the project cost estimate, in order to have a benchmark of the price in the next step of the business tender, do not appear tender business bid price quantity level error, if not do this step, we can not control the project cost from the source.

(3) To provide basic data for project cost budgetary estimate (budget) calculation.

From the project cost management process, the cost estimation is the premise and basis for cost budgetary estimate calculation, only do a solid job of cost estimate, in order to make budgetary estimate work smoothly, so that the cost management is in the correct operation of the track.

10.3.1.3 LNG Project Cost Estimation Itself characteristics

(1) The specific activities of project inherent uncertainty.

"Is the use of investment initiators can provide LNG resources?" "Or on the project to find LNG resources are from abroad?" "What is the specific workload size of receiving terminal construction site?" "How to implement the pipeline routing?" "Are large user projects approved by the government?" "What is the number of alternative energy projects?" and so on. Because at this time in the feasibility study phase of the project, there are some factors still uncertained.

(2) Uncertainty about the project size and the amount of resources consumed.

It includes the equipment and materials purchased abroad or domestically the project company's own human resources are uncertain, it refers to some specific activities of the scale in the implementation process of the project and the amount of consumption of resources may be relatively large or may also be less.

(3) Uncertainty about the price of the project's resources.

What is the price of LNG resources? What is the EPC total contract price? What is the price of the main imported equipments and materials? What is the new technology or the purchase price of a patent? This refers to the realization of the project in the process of some specific activities and consumption of resources consumed by the price of the abnormal fluctuations and changes.

(4) Uncertainty to control the contingency cost.

The uncertainty of project risks is absolute, while certainty is relative. In order to realize the overall risk control of the project and management, the most fundamental task is to identify the various risks that exist in a project and determine the risk cost; secondly to control the occurrence and development of risk events, directly or indirectly, control risk costs. Since in the earlier project, it is difficult to carry out the forecast of reserve funds, including the cost of the risk and the non-foreseeable fee.

10.3.1.4 Project Cost Account Settings

(1) Completely EPC contract mode.

① Direct costs: EPC contracts agreed price, mainly including: the receiving terminal, jetty, gas transmission line engineering construction, equipment procurement, installation costs, supporting engineering direct investment, contracting unit management costs and contracting unit profits.

② Indirect costs: the project management costs, mainly including: the project team and the project company of various functional departments of personnel salaries, fixed assets acquisition fee for management use, office rent, IT facilities and services, office stationery, car rental, transportation costs, communication costs, travel expenses, conference fees, business consulting fees, advertising expenses, hospitality, training and recruitment fees.

(2) Independent design and construction mode.

① Sub-project cost. According to the project, the project is divided into the receiving terminal sub-project (including storage tanks, gasification facilities), jetty sub-project and pipeline sub-project costs and ancillary works costs(the cost of a variety of labor), materials costs(material resource consumption and occupancy costs), consultancy fees (a variety of consulting services and expert fees), equipment costs (depreciation, rental fees, etc.), other costs (such as insurance, statutory profit of subcontractors, etc.) and unforeseen costs (for the prevention of project change management reserves).

② Management fees of the project company or the project team, the same as above.

③ Other expenses, including the pre-project costs.

(3) Mixed mode.

When LNG project has both the sub-project of EPC contractor and the independent design and construction, the cost of two subjects are integrated. It should be noted that, no matter which kind of modes, indirect administrative costs of the project are necessary to happen.

10.3.2 The Basis and Method of LNG Project Cost Estimate

10.3.2.1 Cost Estimate Basis and Accuracy Requirements

(1) The main basis.

① The cost structure, the estimation index, calculation method and other relevant calculation project cost documents issued by the completed LNG project;

② LNG engineering construction other cost calculation methods and cost standards released by specialized agency, as well as the price index released by the government sector;

③ The construction content and quantity of each item in the project of LNG project feasibility study report;

④ Description of project investment resources (see the Chapter 7 of this book, 7.2.4 section of the work package and resource management). It is summarized in the form of currency and provides the basis for the cost management of each stage.

(2) The accuracy requirements.

① The contents and the cost of the project is complete, the calculation is reasonable, do not repeat calculation, do not raise or lower the estimated standard, ensure no missing items.

② When there is a standard or condition difference between the index and the specific engineering, the selection and adjustment should be carried out.

③ The accuracy of the investment estimation should meet the requirements of the preliminary designed estimates.

10.3.2.2 The Methods of LNG Project Cost Estimation

(1) Analogous estimation method.

This method is also called the top-down method, which is to estimate the cost of a new project by comparing the actual cost of the similar projects, estimation of project cost accuracy by using this method is related to the estimate personnel or consulting expert who adopted project types and scopes of substantive similarity, for the use of foreign similar project information, it should be considered the exchange rate factor and the difference between the international market and the Chinese market, etc.

(2) Parameter Estimation Method.

It is necessary to use historical statistics, the LNG project characteristic parameters to create a mathematical model to estimate project costs. For example, the use of building unit area cost or unit cubic cost, pipeline unit length cost, design of LNG gasification capacity cost, the total project cost is obtained by multiplying the amount of the project plan and the cost of the unit, also need to correct by index rising prices. The accuracy of this method depend on the matching degree between the amount of historical data and the price of the model.

(3) Work package statistical method.

The work package statistical method is also called bottom-up method, this method is the first to give a LNG project work package of resources cost statistics, then it is necessary to estimate the cost for labor, materials and work, finally add up to obtain the total cost of the project.

(4) The seller bidding price balance method.

For some of the equipments and materials can be used at home and at abroad, through domestic and international bidding, comparing prices, we can get a reasonable price or contract price; for some of the equipment and materials must be imported from abroad, it is necessary to adopt the international competitive bidding method for overseas acquisition. All of those need to balance the domestic market and foreign market analysis, in the buyer's market or the seller's market conditions, respectively, to take different price strategies. In short, the total cost of equipment materials or a project contract can be obtained through bidding price analysis.

(5) Standard quota method.

In order to strengthen the cost management, it needs to collect the LNG industry civil engineering, equipment installation engineering base quota and market labor service, material price information. At the same time, to play the LNG project sponsor their own cost database, the database is based on the tracking of LNG project cost and individual equipment, materials, labor costs, financing interest, insurance, management fees and a series of data at home and abroad, to establish the labor quota database, equipment using the quota database, material consumption quota database, the management of the cost of the database, the workload list database, but for the new project, it is necessary to consider the resources invested time, assumptions and constraints, the total investment calculation of the specific LNG project can be used after the data correction.

(6) Software tools method.

The use of computer software to estimate project costs. If cost estimating software, computer worksheets, simulations and statistical tools, these tools can overcome the human estimating vulnerabilities, while speeding up the computing speed, it is the cost estimates future direction of

development.

In fact, in the process of preparing the cost estimate, several kinds of combination tools proved to be effective are selected and flexibly used.

10.3.3 LNG Project Cost Estimate Results

10.3.3.1 LNG Project Cost Estimate Report and Estimate Tables

(1) The cost estimate report.

The cost estimate report as an independent report, can also be used as part of the feasibility study report, can refer to the next section of the budget report format and content compilation. We won't introduce much detail here.

(2) Cost estimate tables.

The project cost estimate sheet is the possible cost estimate of the resources required to complete the project (can refer to the next section of cost budgetary estimate form). The table content according to the cost management department to grasp how much and the accuracy of the information, the content can be detailed or simple, the project scope of all work and sub-projects resources need to be put into the project cannot be missed, including but not limited to labor, equipments, materials, supplies, and the price index, exchange rate, unforeseen expenses etc.

10.3.3.2 Related Support Details Files

(1) Estimate detailed description.

The contents of the cost estimate form should be accompanied by a detailed description to support the estimation of the cost.

(2) Other supporting documents.

The contents including the project scope of work description, detailed description of the preparation of the estimation, assumptions, constraints and estimation error descriptions.

10.4 LNG Project Cost Budgetary Estimate Management

10.4.1 General Concepts of Project Cost budgetary Estimate Management

10.4.1.1 Definition of Preliminary Design Cost Budgetary Estimates

In the preliminary design or the expansion of the preliminary design stage, the designed units in accordance with the requirements of the owner, calculate the construction of the proposed project from the beginning of the project to the use of the whole process of the construction costs of the document, which is an important part of the designed documents.

10.4.1.2 Definition of LNG Project Cost Budgetary Estimate Management

The LNG project cost budgetary estimate is based on cost estimate, through the project preliminary design process, the investment cost is studied in detail, taking the budgetary estimate cost quota and index of similar projects, combined with the current market prices and forecasts to calculate the specific tasks involved in the project cost of the corresponding task, which covers all the expences required for LNG project from the start to the completion and acceptance before delivery of the management of the work. The results are the project cost budgetary estimate report and coat budgetary estimate tables. Approved by the board of directors, the project cost budgetary estimate is the LNG project total investment control benchmark.

10.4.1.3 Principles for the Compilation of LNG Project Cost Budgetary Estimate

(1) The project owner is responsible for the overall work.

Project owners through bidding select qualified preliminary design unit, project owner is responsible for the overall work, preliminary designed units clarify the unit cost and price by using estimates for the quota and index, makes the specific budgetary estimate and lays the foundation for the preparation of the budget.

(2) The use of LNG engineering quota.

The preliminary designed units are the implementation of the main responsibility of the unit quota design with construction contractor unit participation, the project owner organize and coordinate project work of quota design, taking the static cost estimate investment in feasibility report as the project as project design "Limit", the control object of the quota design is the static investment part.

(3) The project scope of feasibility study is the foundation.

Project engineering quantity should be based on the design engineering quantity and material standard in approved feasibility study stage, the main control contents of the project include: structure form of building engineering, design standard, volume, area, length and total three materials (steel, wood, cement (concrete)) and so on, installation of all kinds of accessories, equipment, spare parts, quantity, quality, weight etc.

(4) The establishment of a large amount of raw materials for floating price mechanism.

In view of the LNG project for long period, large contract amount, contract signing time and the future of a long period of time, the bulk of raw materials' prices may be changed, at that time, the locked price may damage the interests of the one side in the future; to be fair, the use of a suitable floating pricing mechanism for both sides of the contract is basically equal to opportunity, fair and reasonable, and it is conducive to the two sides to resolve the system risk, to complete the common goal of the contract, and achieve a win-win situation.

(5) The flexible choice of "Three New" cost increase.

Here the "Three New" refers to the new technology, new equipment and new process. the "Three New" may be not considered in the feasibility study stage, but it can reduce production cost, also meet safe, reliable, economical and applicable principles, in the preliminary design stage to modify the scheme and needs to increase investment, through consultation with the

owners of the project.

10.4.1.4 The Basis of LNG Project Cost Budgetary Estimate

(1) The data provided by the project owner unit to the design unit.

① Feasibility study report and investment cost estimate;

② The regional construction unit valuation table (or quota), the regional material budget price and the indirect cost quota;

③ The construction site of the land requisition, demolition, demolition, rental fees and compensation fees;

④ Equipment bidding contract price or actual delivery prices;

⑤ Traffic and transport prices;

⑥ Area wage standards, the approved material budget price and machine-team costs;

⑦ The budgetary estimate quota or budgetary estimate index, project overhead quota, other related fee standard issued by State or provincial and municipal authorities;

⑧ National or provincial provisions of other engineering costs;

⑨ Estimates of similar projects and technical and economic indicators;

⑩ The project company management fees.

(2) Construction units to provide information to design unit.

① Approved by the department in charge of the construction organization design and construction scheme and measures for special projects;

② Construction land leasing fees and demolition, relocation, compensation;

③ Agreed upon by the construction unit to provide other information and data.

10.4.1.5 Classification and Role of LNG Project Cost Budgetary Estimate

(1) The total budgetary estimate.

Total budgetary estimate is to determine the entire construction project from the project approval to the completion of the whole process of the cost of documents required by the sub-projects comprehensive budgetary estimate, other costs of construction and budgetary allowance.

(2) Sub project composite budgetary estimate.

Sub project composite budgetary estimate is to determine the construction cost of a sub-project documents, which is also the preparation of each sub-project within the professional unit works of the sub-project budgetary estimate summary compiled.

(3) The unit works budgetary estimate.

The unit works budgetary estimate is a document to determine the cost of each unit works of the sub-project construction, and is also the basis for the compilation of the comprehensive budgetary estimate of the sub-project. The basic compilation unit of the design budgetary estimate is a sub-project. After the completion of the preparation of the unit works budgetary estimate, it is necessary to summarize into a sub-project comprehensive budgetary estimate, further consolidated comprehensive budgetary estimate to obtain the total budgetary estimate for the construction project.

(4) The role of cost budgetary estimate.

① The cost budgetary estimate is the basis for determining and controlling the cost of basic capital construction and the plan of basic capital construction. After the approval of the relevant departments of the total budgetary estimate is the maximum amount of the total cost of the project construction.

② The cost budgetary estimate will provide the basis for the design of technical and economic analysis, multi program evaluation and optimization of project financing.

③ The cost budgetary estimate provides the objectives of cost control for the next stage of the construction drawing design.

④ Sub-project comprehensive budgetary estimate and project total project budgetary estimate are the basis for determination of the lump sum cost index and the approved or revised cost budgetary estimate is foundation for signing, the lump sum contract and controlling the lump sum amount.

⑤ The budgetary estimate is an important basis for the project owner unit to carry out the project accounting, the comparison of construction project "Three Calculation", and evaluation of the project cost and the economic effect.

10.4.2 LNG Project Cost Budgetary Estimate Preparation

10.4.2.1 Prophase Cost Calculation

Prophase cost is usually the work cost need to be paid before the project company set up (Table 10–1). Sometimes, the cost is classified as part of project expenses. Because this kind of charge is before the establishment of the project company, it is generally not as the object of project cost control, the cost is calculated according to the actual amount.

Table 10–1 LNG Project Prophase Cost Budgetary Estimate

Compilation date		Amount of money			
No.	Content	Folding (Subtotal) RMB	Among them RMB	Among them USD	Calculation basis or description 10^4 RMB, 10^4 USD
1	**Land acquisition fee**				
1.1	Land				
1.2	Sea area				
1.3	Work funds				
2	**Pre Project management fee**				
2.1	Leading group and preparatory office expenses				

Continued

Compilation date		Amount of money			
No.	Content	Folding (Subtotal) RMB	Among them RMB	Among them USD	Calculation basis or description 10^4 RMB, 10^4 USD
2.2	Owner committee fee				
2.3	Executive office expenses				
3	**Pre-feasibility study cost**				
4	**Feasibility study cost (Receiving terminal, jetty and pipeline)**				
4.1	Special research fee				
4.2	FEED design				
4.3	Preliminary design				
4.4	Detailed design				
4.3	**Expert and consultant fees**				
4.3.1	LNG Resources and transportation				
4.3.2	Engineering investigation				
4.3.3	Market				
4.3.4	Finance and financing				
4.4	**Special commission fee**				
4.4.1	Hydrological and meteorological observation				
4.4.2	Engineering seismic analysis				
4.4.3	Environmental protection assessment				
4.4.4	Safety assessment and review				
4.4.5	Biding document preparation				
4.3.6	Financing scheme				
5	**Pre project cost**				
6	**Project personnel training fee**				
7	**Research test fee**				
8	**Engineering supervision cost**				
9	**Project construction technical service fee**				
10	**Interest in construction period**				
11	**Others**				
12	**Contingency sum**				

10.4.2.2 Construction Cost Calculation

Construction cost is the cost that required for the construction of permanent buildings and structures.

(1) Unit construction investment calculation method.

Unit construction cost multiplied by the total quantity of construction projects equal to the total cost of the project. Generally, it needs to use the unit volume investment, such as the main plant, control building, multiplied by the total volume of the corresponding construction project, storage tank foundation volume, BOG vaporizer base volume, multiplied by the corresponding unit building volume investment, etc.

(2) Physical quantity investment calculation method.

The unit physical quantities investment multiplied by the total quantity of physical engineering is equal to the total construction cost. The unit cubic meter of earth and stone investment multiplied by the total amount of earth and stone is equal to the total investment of earth and stone. The total pipeline investment is obtained by multiplying the pipeline investment per meter by the total pipeline length.

(3) Budgetary estimate index investment calculation method.

For without the above indicators, and the cost of construction project takes relatively large investment in total project cost, can be used to budgetary estimate index calculation method, the calculation method should have detailed engineering materials, building materials price index, the cost of the project, investment of time and workload.

(4) The cost budgetary estimate table.

By using the above calculation method, to calculate the cost budgetary estimate of each building (Table 10–2).

Table 10–2 LNG Project Construction Cost Budgetary Estimate

No.	Name of building (structure)	Unit	Engineering quantity	Unit price (10^4 RMB)	Total cost (10^4 RMB)
1					
2					
…					

10.4.2.3 Cost Calculation of Equipment and Materials

(1) The domestic equipment and materials purchase cost calculation method.

① Equipment purchase costs = each of the equipment factory price multiplied by the total number of equipment and miscellaneous expenses (transportation + handling + warehouse).

② Purchasing cost of tools and equipment, generally it is taken accounted for a certain proportion of equipment costs or other methods to calculate the cost.

③ Material acquisition cost = per unit of measurement of material factory price multiplied by the total volume+ transport miscellaneous expenses (transport + loading and unloading + warehouse storage).

④ It also including field production of non —standard equipment costs, the production of furniture and the cost of the purchase of the corresponding transport costs.

⑤ Domestic equipments and materials purchase budgetary estimate table.

(2) The foreign equipment purchase cost calculation method.

① Imported equipment purchase costs are composed by imported equipment, imported ancillary costs and domestic transport and miscellaneous expenses. The price of imported equipment according to the delivery location and delivery mode, divided in to FOB and CIF. Imported ancillary costs including: foreign freight, foreign transportation insurance, import tariff and import value-added tax, foreign trade fees, bank financial fees and customs fees.

② On the basis of the CIF calculation formula.

Import equipment CIF = FOB price + foreign transportation fee + foreign transportation insurance.

Among them: foreign transport costs = FOB price×transport rates (or foreign transport costs) = unit transport price×transport volume, foreign transport insurance = (FOB price + foreign transport charges) ×foreign insurance rates.

③ Several other ancillary costs of imported equipment are usually calculated by the following formula:

a. Import tariffs = CIF value of imported equipment × RMB exchange rate ×Import tariff rate.

b. Import VAT = (CIF value of imported equipment × RMB exchange rate + Import tariff + consumption tax) × Value added tax rate.

c. Foreign trade handling fee =CIF value of imported equipment × RMB exchange rate × Foreign trade commission rate.

d. Bank financial expenses = The price of imported equipment × bank RMB exchange rate × Bank financial rate.

e. Customs house fees = CIF value of imported equipment × RMB exchange rate ×Customs clearance rate.

④ Foreign equipment and materials purchase budgetary estimate table.

Based on the above equipment, tools and materials acquisition cost formula to prepare the budget estimate table (Table 10–3).

10.4.2.4 Installation Engineering Cost Calculation

(1) The installation of engineering cost composition.

The equipments to be installed should calculate the cost of the installation works, usually in accordance with the engineering standards issued by the specialized agencies of the industry, fees and standards. Including all kinds of mechanical and electrical equipment assembly and installation costs.

① The working platform, the ladder and the installation project cost which is connected with the equipment, and the platform is attached to the pipeline laying project cost of the equipment being installed;

Table 10-3 Imported Equipment Purchase Budgetary Estimate

No.	equipment name	Number	FOB price	Foreign freight	Foreign transportation insurance	CIF price	import tariff	consumption duty	Value added tax	Foreign trade handling fee	Bank finance charge	Customs supervision fee	Domestic freight	Total purchase cost of equipment
1														
2														
…														
Note: If it is difficult to calculate the ancillary costs of imported equipment by a unit set, can be calculated according to the total offshore price of imported equipment														

② Installation equipment insulation, heat preservation, anti-corrosion and other engineering costs;

③ Single trial operation and joint load free trial operation cost, etc.

(2) The calculation formula.

Specific calculations maybe in accordance with the rate of assembly, installation costs per ton of equipment or per unit of the cost of installation of physical engineering amount of settlement, as follows:

① By the rate: Installation cost = Equipment price×installation rates;

② By the tonnage: Installation cost = equipment tonnage×per ton installation rate;

③ By engineering physical quantity: Installation cost = the installation of the physical amount×the installation cost index;

④ Installation engineering cost budgetary estimate.

According to the above calculation formula, to calculate the equipments installation cost, and prepare for budgetary estimate (Table 10-4).

Table 10-4 LNG Project Equipment Installation Cost Budgetary Estimate

No.	Installation engineering name	Unit	Number	Index (rate)	Installation costs (10^4RMB)
1	Equipment in receiving terminal				
	A				
	B				
2	Equipment in Jetty				
	A				
	B				
3	Equipment in Pipeline				
	A				
	B				
……	……				
	Total				

10.4.2.5 Other Cost Calculation of Engineering Construction

Other cost calculation results of the project are as follows (Table 10–5), according to the actual price of each subject, at least in accordance with the contract price or rate calculation.

Table 10–5 Engineering Construction Other Expenses Budgetary Estimate Form

No.	Cost name	Calculation basis	Cost rate or standard	Total price
1	**Project construction management fee**			
1.1	Construction project legal person management fee			
1.2	Equipment service charge			
1.3	Spare parts purchase fee			
1.4	Engineering insurance premium			
2	**Project construction technical service fee**			
2.1	Engineering supervision cost			
2.2	Equipment supervision fee			
2.3	Construction project post evaluation cost			
3	**Production preparation fee**			
3.1	Manage vehicle purchase cost			
3.2	Tool, equipment, office, production and living furniture purchase cost			
3.3	Production staff training and advance entry fee			
3.4	Subsystem and the whole set of start-up commissioning fee			
4	**Others**			
4.1	Subsidy for construction safety measures			
4.2	Engineering quality supervision and inspection fee			
4.3	Budget quota preparation management fee			
5	**Imported equipment and materials related costs**			
5.1	Foreign drawing data translation fee			
5.2	External technical staff reception fee			
5.3	Transportation equipment purchase fee			
5.4	Personnel to go abroad fee			
……	……			
	Total			

10.4.2.6 Other costs

(1) Project construction contingency expenses.

Due to contingency expenses that may occur in the implementation of the project, it needs to be calculated in advance for reservation, in order to prevent the occurrence of the actual situation. The contingency expenses can be taken into account of construction, equipments and equipment purchase, installation and other costs of the project construction as the base for the calculation, multiplied by the contingency expenses rate.

(2) Price increase reserve calculation.

Because the LNG project construction period is long, equipment, materials, labor and other price increases may occur during the construction period, therefore, it needs to consider capital reserves in advance. The reserve for price increase cost can be taken into account of building construction costs, equipments and tools for the purchase, installation cost as the base for the calculation. The calculation formula is:

$$PC = \sum_{t=1}^{n} I_t \left[(1+f)^t - 1 \right] \quad (10\text{–}1)$$

where PC——Price increase reserve;

I_t——In year t, the sum of construction engineering costs, purchase cost of equipment and tools, and installation costs;

f——Construction period price rise index;

n——Construction period.

In the construction period, the price rise index, if the government departments have the provisions in accordance with the provisions of the implementation, if there is no provision, adopted by preliminary designed units researchers forecast.

(3) Construction, equipment installation EPC project quotation.

If the project company using EPC tender to determine the contractor for construction and installation of equipments, the quotation of the contract shall be considered in the construction and installation cost, which is composed of three parts: direct cost, indirect cost and legal profit.

① Construction and installation costs.

a. Direct cost. The calculation formula is:

Direct costs = Engineering quantity×construction or installation quota (bidding)unit price + area material price difference + other direct cost.

b. Indirect costs. The calculation formula is:

Indirect costs = Direct costs (or labor costs) ×indirect rates (including construction management and other indirect costs).

c. Statutory profit. The calculation formula is:

Statutory profit = (direct costs + indirect costs) ×statutory rate of profit

② Equipment acquisition costs.

It is composed of equipment cost and equipment transportation cost.

③ Other independent costs.

It is the cost in accordance with the provisions and requirements of the "Other Engineering and Cost" quota.

10.4.2.7 LNG Project Cost Budgetary Estimate Report and Forms

(1) The report and contents.

① The report can be referenced budgetary estimate report format and contents to write, including the general situation of the engineering and the main technical and economic indicators, the total budgetary estimate table, professional summary budgetary estimate table, sub-project budgetary estimates, other accessories and attached lists.

② Project cost budgetary estimate tables need to meet the improvement of quality of budgetary estimates and economic analysis, the accumulation of economic indicators and other requirements, It must be strictly in accordance with the provisions of the form, that is, according to the content of the project division, arrangement order, cost nature division and other requirements, compile summary item by item.

(2) The support detail files.

The cost budgetary estimate tables need to give detailed description to support the cost items, including the plan scope of work description, budgeting link details, assumptions, constraints and the estimation error description.

10.5 LNG Project Cost Budget Management

10.5.1 The General Concept of LNG Project Cost Budget Management

10.5.1.1 The Definition of LNG Project Cost Budget Management

LNG project cost budget is on the basis of cost budgetary estimate, through the construction design drawing, to further implement the workload and resources, by budget quota index and contract price as the principle of valuation, to carry out the construction project cost management. The results are the project cost budget report and attached lists.

10.5.1.2 The Premise of Preparation for LNG Project Cost Budget

(1) Engineering contractors (including EPC) have been selected.

Through project contractor's bidding, in particular the EPC bidding activities to determine the constructors of the LNG project, the project company has a further understanding for the contract price and turnkey project equipment, material prices, through comparison with the cost budgetary estimate and price clarification, to lay the foundation for the preparation of the budget.

(2) Domestic and foreign equipments and materials procurement inquiry work is nearly completed.

For some non-EPC approach projects, such as pipeline and supporting engineering, on the basis of cost budgetary estimate, the project company can make a wide range of inquiry for equipments and materials at home and at abroad, in order to make the comparative shopping, to understand the current market situation, to prepare for the preparation of the budget.

(3) The project company's management strategy has been determined.

Before the start of the project, the project company should be according to its technical and management strength, determine the project management strategy, including which sub-projects to use the EPC contractor, which sub-projects to take self-management and procurement, but the contractor to undertake the construction project, which sub-projects need to hire technical consultant. Only when these strategies are determined, we can define the scope of the investment and management of the human resources for the project company, on the basis of this, we can calculate the management cost of the project company.

10.5.1.3 The Basic Principles of LNG Project Cost Budget Management

(1) The combination of cost estimation and scientific forecast.

Cost budget is completed on the basis of cost budgetary estimate, the tools and methods of cost budget are the same as the cost budgetary estimate, but due to the control department in the budget period to grasp the cost information more and more comprehensive, at this time, scientific forecasting will become possible. On the basis of obtaining the price information of each bid, the project company must analyze—to find out the price quotation and the change law of domestic and foreign equipments.

(2) The comprehensive balance of the foreign contract and project management.

At present Chinese LNG project is a set of receiving terminal, storage tank, jetty, pipeline and ancillary works. It is necessary to borrow the domestic and foreign technologies and management experience, such as jetty and pipeline which can borrow domestic engineering design and construction team; storage tanks, gasification process can be carried out at home and at abroad for the tender, please have the qualifications of the company to undertake engineering design and construction. Therefore, budget management should work closely with the contractor, worked out reasonable construction drawing budget. At the same time, the project company control department should according to the scientific forecast result, carries on the entire project the comprehensive balance, get expert approved LNG project budget, and report to the budget committee and the board of directors for approval.

(3) Overall cost budget control and breakdown of responsibility.

LNG project company should firmly grasp the right of budget control, the control department is working expenses operating entity, we must clear the duties and rights of the department first, in addition with control department to supervise and control the design and construction of each sub project, especially to organize the each functional department of project company to prepare a reasonable management budget, control the implementation of the management budget. At the same time, the control department goes deep into the cost management process of each contractor by means of management, the cost management procedures and practices, reasons and timely

analysis of contractor expenses deviation, so as to judge the rationality of the change of the cost of the contractor, decompose the cost management responsibilities to the various contractors, to form the control department to take the lead, functional departments separately check, the contractor's costs self-management and control of joint forces, together with the contribute to cost budget management.

10.5.1.4 Analysis of Factors Affecting Project Cost Budget

(1) The project size.

LNG project cost budget size should be related to the project size, generally speaking, the higher the capacity to receive LNG tonnage and gasification, the longer the gas transmission pipeline, the greater the investment, on the contrary, the lower the investment. However, these are not simply additive relationships, in a certain investment environment, the project should be achieved economies of scale first, on this basis, the combination of that can withstand natural gas price of the user market, to determine the size of the LNG project, compare with the size of the same type of LNG project investment budget.

(2) Localization degree of equipment and materials.

Generally speaking, the more the imported equipments and materials, the greater the investment budget. In project based on the analysis of the resources, we should know the size of the degree of localization, such as Guangdong Dapeng LNG project domestic rate reached to 65%, to a certain extent, reduce the investment, but to carry out a detailed analysis, the same period as some domestic materials and equipment prices maybe higher than foreign countries, we have to consider importing from abroad, can not treat it as the same.

(3) The urgency of the project.

LNG project urgency is closely linked with the national and local demand, if the state and local urgent need, we should speed up the process of the project, so that we might not reach its capacity requirements on technology and management, and the project must be accelerated, we can only use more design and construction of manpower at home and at abroad, and even hire more experts for consultation, in this way, the budget of the investment will increase.

(4) The degree of mastery of the project details by the project company.

The project company staff consists of two aspects of technology and management, if the project manager and members participated in similar LNG project technology and management, took a lot of work in the project, in the new project will be able to play their technical and management advantages, can reduce the outsource and reduce the investment budget. On the contrary, it can only use external forces, such as contractors, supervisors, consultants and consulting firms to increase the budget.

10.5.1.5 The Main Tasks of Project Budget Management

(1) Preparation of project tasks and work package expenses budget.

The tasks or work package cost budget is the basis of comprehensive budget, construction drawing design unit is the preparation of the cost budget department, but the project company control department is the responsible department of budget, the two sides should cooperate closely, compared with the existing historical data analysis, and the market of the future market price forecasting, also analyze the EPC bidding, equipments and material information and then through the expert judgment and scientific accounts, work out the cost budget list of each task and unit works. The format of the specific budget list can be considered in accordance with the project company's expense management habits and the setting of pinancial expense accounts, so that it can be used as the basis for expense statistics.

(2) Preparation of the comprehensive budget.

With the above LNG project budget for each task and work package, It is easy to work out the LNG project comprehensive budget, but need to pay special attention to the task or work package budget, overall project coordination and balance, to avoid repeated workload statistics of different contractors, also to prevent the leakage of the scope of the project, to reflect the integrity of LNG project scope of work content and supporting expenses. This work may be related to each of the work package and task budget with multiple feedback. In the end, the project budgetary estimate can be compared with the total project budget, and form the comprehensive budget.

(3) The arrangement of project comprehensive budget input time.

After the project comprehensive budget book completed, an important task is to sort out the time of the input resources, the time here has several layers of meaning, the first meaning is whether the time budget to avoid the non-working day weather window; the second meaning is the working face whether meet the needs of different contractors working at the same time, the third meaning is whether to meet the cost budget and payment request; the forth meaning is put into the other resources are in place in time. In the case of the time required to meet the above requirements, the cost of the budget is meaningful, on the basis of this, the schedule of cost budget input is compiled according to the time sequence. Therefore, this work is often in the cost and time matching conditions, after repeated mutual feedback and correction, in order to achieve the coordination and the overall balance.

(4) The preparation of project cost budget plan and detailed instructions.

Based on the above three work, to prepare the project cost budget report, together with all the tasks and the work package budget table, the time sequence cost budget input plan. In the report, it needs to the key unit works and its budget time input to give a detailed arrangements description, in particular the budget of the assumptions and constraints.

10.5.2 LNG Project Cost Budget Management Methods

10.5.2.1 The Project Budget Plan Program Composition

(1) The basis for the compilation of the construction drawing budget.

① Approved design budgetary estimate;

② Corresponding budget quota or district unit valuation table;

③ Approved construction drawings;

④ The technical complexity of the project;

⑤ Material budget price, etc.

(2) The domestic and international marketing survey.

① Equipment and materials. For the LNG project, firstly, to distinguish a certain period of time, conduct extensive market research on the proposed acquisition of foreign and domestic procurement of equipments and materials lists, including demand, cost, alternative materials, equipments and technical advantages and price differences, so as to understand the impact of market prices on the project cost.

② Contractor selection. The LNG project has which sub-projects to external contracting, is to select domestic contractors or foreign contractors, or the combination of both contractors, to understand their qualifications and contracted engineering quality, contract price level etc..

③ Supervision company selection. Whether the national requirements of implementing supervision system management of large engineering project, or LNG project itself, it is necessary to employ the supervision company for the supervision and inspection of the quality of the project, the project company should make fully use of the mechanism, in the domestic market for the supervision company's performance and quality survey, prepare for the selection of supervision companies.

(3) Data collection.

In addition to collect the data of the above equipments and materials, it is also to extensively collect the domestic and foreign similar project total cost information, sub-project cost information, the national cost quota index and the enterprise's cost indicators.

(4) The selection of budget calculation method and model.

The above cost budgetary estimate methods are also applicable to the budget, but according to the collected data, especially for the time and accuracy requirements, we should analyze the various methods for matching and choose a method, select three or four subsidiary; also according to the source of data, we also choose and compile the mathematical model which can forecast the specific cost calculation.

(5) Cost budget calculation.

The mathematical model and the information and data are selected to predict and calculate the cost budget. Including the cost of the work package resources input, the sub-project resources input and the cost of the entire LNG project resources input, the "S" curve of cumulative budget cost can be drawn under the time axis and the cost axis.

(6) The analysis of calculating results.

Whether the prediction of historical data is reasonable, one is to pass the recognition of experts, but also through the actual situation of the market test, the analysis work is very important.

10.5.2.2 Compilation of Budget Report and Budget Plan for Construction Project

(1) The project budget report contents.

The project budget is a programmatic document of cost control, which directly affect the quality of cost management work late, If the cost budget report is comprehensive with market research in-depth, basic information is implemented with mathematical model with strong pertinence, the prediction method is appropriate with analysis work in place, the classification table is detailed, it will provide reliable basic data for the budget committee and the board of directors, to promote the role of budget decision actively. The following describe the main contents of the project budget report:

① The background and the scope of LNG project;

② The analysis and statistics of project management, project cost budget and other expenses budget;

③ Project design and construction strategies;

④ Analysis of each engineering resource investment;

⑤ Analysis of resource investment assumptions and restrict conditions;

⑥ Analysis, statistics and cost prediction of each work package resources;

⑦ Project total investment cost forecast, analogy, analysis and summary table;

⑧ Prediction, analysis and ranking of the relationship between project resource input time and cost;

⑨ Risk analysis and countermeasures of resource input;

⑩ Conclusions and suggestions.

(2) Compilation of budget plan.

① The engineering construction and equipment installation investment budget table (Table 10-6), including the main statistics of receiving terminal, jetty, gas transmission pipeline, and other construction project and equipment installation costs.

Table 10-6 Construction and Installation Project Investment Budget

Compilation date:				Unit: 10^4 Yuan, RMB		
No.	Items	Budgetary Estimate	Cumulative completed amount of budget	The annual budget	Current period budget	Plan to finish the work load
One	**Receiving Terminal**					
1	Jetty area					
2	Storage tank area					
3	Process area					
4	Public area					
5	Field and civil engineering					

Continued

Compilation date:				Unit: 10^4 Yuan, RMB		
No.	Items	Budgetary Estimate	Cumulative completed amount of budget	The annual budget	Current period budget	Plan to finish the work load
6	Other engineering					
Two	**Jetty**					
1	Harbor turn area dredging					
2	Hydraulic engineering					
2.1	LNG dock					
2.2	LNG berth pier					
2.3	The workboat dock					
2.4	The workboat pier					
3	Power supply and lighting engineering					
4	Water supply and drainage engineering					
5	Other engineering					
Three	**Gas transmission pipeline**					
1	Line project					
1.1	Trunk line					
1.2	Branch					
2	Field engineering					
2.1	Terminal					
2.2	Head office and controlling center					
2.3	Distribution station 1					
2.4	Distribution station 2					
2.5	End station					
3	Communication engineering					
4	Other engineering					
Total						

② Equipment investment budget (Table 10-7), including the main statistics of the sub-items by the equipment classification of investment costs.

Table 10-7 Equipment Investment Budget

Date of establishment:											Unit: 10^4RMB, and 10^4USD
No.	Equipment number	Equipment name	Budgetary Estimate		Completed investment		Annual budget		Current period budget		Me-mo
			Amount RMB	Among US$	Amount RMB	Among US$	Amount RMB	Among US$	Amount RMB	Among US$	
1		According to equipment classification									
2											
…											

③ Management expense budget table (Table 10-8), including the main statistical departments of project company staff wages and benefits, insurance, travel, office, reception, transportation and other expenses.

Table 10-8 Administrative Expenses Budget

Date of establishment:			Unit: Million, RMB and Million, USD	
No.	Items	Amount: RMB	Among: US$	Calculation Basis or Instructions
1	Total			
2	Salary, bonus			
3	Employee services and benefits			
4	Labor union expenditure			
5	Staff education funds			
6	Unemployment insurance premium			
7	Labor insurance premium			
8	Hosing fund			
9	External staff fee			Including salary, bonus, welfare, social insurance expense and housing fund, etc.
10	depreciation			
11	Office expenses			
12	Travel expenses			Including travel expenses, travel allowance, living allowance in different places
13	Traffic expense			Including car rental, road maintenance, bridge crossing, traffic fee and other

Continued

Date of establishment:			Unit: Million, RMB and Million, USD	
No.	Items	Amount: RMB	Among: US$	Calculation Basis or Instructions
14	Insurance expense			
15	Rental expense			Including office and residence
16	Repair charge			
17	Legal fare			
18	Environmental protection and sewage charges			
19	Greening fee			
20	Material consumption fees			
21	Low-value consumption goods			Including office furniture and office equipment
22	Audit fee			
23	Technological research fees			
24	Business entertainment fees			
25	Foreign guests entertainment fees			
26	Taxes			
29	Water and electricity fees			Including water fee, electricity fee and heating fee
30	Correspondence fees			Including telephone bill, mobile phone bill and other bill of office and living
31	Conference expense			
32	Overseas travel fee			Including crafting fee, traffic fee, and life board fee , etc.
33	Advertising expense			
34	Guard, fire cost			
35	Labor protection fees			Including labor supplies, summer cooling costs, health care costs
36	Book and data fees			
37	Other			Including the party, Youth League and other organization fund, enterprise culture fund
38	Unexpected expenses			Above 10% of the total
	Total			

④ Office fixed assets budget (Table 10–9), including the office equipment for all departments as the fixed assets for the cost of statistics.

Table 10–9 Office Fixed Assets Budget

Date of establishment:				Unit: Million, RMB and Million, US dollars			
No.	Fixed assets name	Specifications and models	Budgetary Estimate	Annual budget	Current budget	Use department	Memo
1							
2							
…							

10.6 LNG Project Cost Control

10.6.1 General Concept of LNG Project Cost Control

10.6.1.1 The Definition of LNG Project Cost Control

LNG project cost control, simply to say that the actual investment of the project is no more than the actual cost of the project budgetary estimate. It is based on a prior set of project budgetary estimate or changes to the baseline plan, through the use of a variety of measures and methods for the management of project cost, to make the project actual cost control in the budgetary estimate management. Project cost control involving a variety of factors that can cause the change of project cost control, the cost control in the project implementation process and the actual cost of the project change control three aspects.

10.6.1.2 The Basis of Project Cost Control

(1) Project cost management plan.

Project cost management plans are different at different stages, the feasibility study stage of cost estimate, the preliminary design stage of the cost budgetary estimate and construction drawing stage of the total project budget, annual budget, budget adjusted in the middle of year, cost investment plan by time, which used as the benchmark to measure and control the overall project cost. For LNG this large multi sub-projects' portfolio, have multiple cost baselines and input resource bases, which are the original basis of cost control. Generally speaking, project cost control objectives are not to exceed the project budgetary estimate.

(2) The project daily report.

In the process of project implementation, in accordance with the provisions on the administration of construction of the project company, generally it is necessary to take engineering daily report, weekly report, monthly report and annual report to record the project implementation in the process of staff and mechanical input, material consumption, equipment installation, engineering progress, schedule analysis, quality and HSE records, special consulting

with problems and solutions in construction. Undoubtedly, these datas reflect the implementation of the project's cost directly and indirectly.

(3) Project cost management performance report.

In the project implementation process, the project budget performance report, specifically to analyze the project cost and resource performance, including the completed and unfinished deliverables, completed and not completed the cost of delivery of the results of the analysis, the completion of the delivery results still need to be put into the cost of total project budget forecast, etc.

(4) Approved project cost change.

For the approved project cost changes, it is necessary to give the reasons for the change of the cost, background analysis, key roles of cost change to project, the new cost of investment plans, changes brought about other changes in project management, etc..

(5) Contract change clauses.

The project contract change documents, equipment and materials supply in the contract change text, the design contract change documents are part of the contract documents, contract changes conclude project contents, quantity, specifications, quality, duration and terms of payment.

10.6.2 LNG Project Engineering Cost Control Method

10.6.2.1 General Cost Control Measures

(1) Standard design.

① Engineering standard design refers to using general standard drawings as much as possible in engineering design to promote industrialization level, speed up the progress of the project, save materials and reduce construction costs.

② At present, China has constructed 10 LNG projects, and the storage tank adopts full containment type ($16\times10^4 m^3$), the storage tank construction with repeated construction, similar production ability and properties, very suitable for the use of standard design.

③ The use of standard design can accelerate the designed speed, shorten the designed cycle and cost saving design; can make the construction process setting, improve labor productivity and save material, reduce construction costs; accelerate the construction preparation and precast concrete, accelerate the construction speed, and reduce the cost of construction and installation work.

(2) The quota design.

① The quota design is based on the approved cost estimates to control preliminary design, based on the approved preliminary design budgetary estimate to control construction drawing design, the cost control of the last stage is the target of the cost control in the next stage of the design, the cost control target of this stage is decomposed into each professional, and then it is decomposed into each unit works and portioned works.

② It is needed that every professional in the guarantee to meet the use of the function of the premise, according to the distribution of cost limit control design, strictly control of technical design and construction design of unreasonable changes in order to ensure that the total cost of the limit is not exceeded.

③ Quota design is running throughout the project feasibility study, preliminary investigation, preliminary design, detailed survey, technical design, construction drawing design, etc.

④ In each of the professional, every design should take the limit design as the key work content, and the implementation of the quota design is an important guarantee for the realization of the quota target.

10.6.2.2 LNG Project Cost Control Measures in Different Stages

(1) Feasibility stage.

① In accordance with the provisions of the management of the project company, it needs to take the bidding process, select the research institute in accordance with the feasibility study;

② To provide the project background information to the research institute as far as possible, cooperate with the research institute to carry out the feasibility study;

③ To promote the decomposition of the cost of the project, refined to the specific sub-projects;

④ To require research institutes to optimize the feasibility plans, efforts to control or reduce construction costs;

⑤ To strictly control in accordance with the requirements for feasibility study of the depth and scope of work, reduce the leakage and missing items;

⑥ To give fully play to the subjective initiative of the feasibility study personnel, encourage the innovative programs and awareness in the feasibility study report;

⑦ To timely organize experts to review the feasibility study of single special report and the overall report, propose the amendments and implement on the cost estimation.

(2) Preliminary design and construction design drawing stage.

① Based on the quota index.

No matter it is the preliminary design cost estimate or the construction plan design expense budget, it must take the corresponding stage cost budgetary estimate quota and the expense budget quota index as the foundation, based on the scope of the project construction, repeat the implementation of specific project, no basis no estimating.

② Contract price as the criterion.

When the bidding work has been carried out for equipments, materials and spare parts, it is necessary to take the contract price as the criterion, it is even more close to the actual than the corresponding quota indicators, such as the cost of the budgetary estimate, the cost budget, cost control error also can be reduced.

③ Market price as the premise.

When some ancillary equipmens, materials, spare parts are too late for bidding activities at that time. It could be subject the current market condition and actual procurement prices, thus obtaining complete cost budgetary estimate and budget.

(3) Bidding and acquisition stage.

① In the process of bidding or purchasing evaluation, it needs to adhere to the qualification and technical evaluation in line with the requirements of the item company supplier, according to the principle of reasonable low price Award.

② According to the company's bidding or acquisition strategy, to prepare the tender and the base bid price documents, Implement the bid or purchase quota.

③ During the process of technical and commercial clarification, it is necessary to divide all fees in the process of contract execution, avoid ambiguity costs brought about changes or additions to the late, for both the expense dispute to bury next hidden trouble.

④ To actively prepare contract terms, ensure that the owner in the contract price, object, period, quality and other key elements of the rights and the interests.

⑤ To define the owner's obligation scope during the contract performance period, and to reduce the cost of performing the contract.

⑥ Engineering quantity list and its preparation.

a. The engineering quantity list is a detailed list of portioned or itemized works, measures, other items and corresponding quantities to be planned for LNG project. The engineering quantity list is a part of the bidding document.

b. Preparation of the engineering quantity list shall be by the person who has the ability to prepare the bidding documents, or with the appropriate qualifications of engineering cost consulting and tendering agency and other intermediary agencies to meet the basic engineering design drawings and construction specifications.

⑦ Tender offer.

a. Tender offer is a bidder who wants to obtain the qualification of project construction contract, in accordance with the engineering quantity list in tender documents to conduct engineering evaluation, to propose bidding strategy and determine the tender price, in order to win the bid and through the implementation of the project to achieve the legitimate economic benefits of activities.

b. The engineering quantity list use comprehensive unit price valuation, the comprehensive unit price refer to the labor cost, material cost, mechanical use fee, management fee and profit of a specified unit of measure in the project quantity list and risk factors.

⑧ Bid evaluation.

a. The bid evaluation means that the bid evaluation criteria and measures specified in the tender documents should be evaluated and examined in accordance with the criteria and measures, to determine the activities of the bid winning units. The basis of bid evaluation is bidding documents and bidding price.

b. Bid evaluation criteria

a) Comprehensive evaluation method, according to the "Bidding and Tendering Law" forty-first provisions, the bid of the winning bidder is most closely matched with the "Maximum to

Meet the Requirements of the Comprehensive Evaluation Criteria in the Tender Documents".

b) Low bid method, in the other bidding factors are in line with the requirements of the tender documents, after review, to determine the lowest bidder winning bid.

(4) The construction stage.

① According to the principles of the company bidding methods, to choose the best construction contractor, carefully review the contractor's qualification, performance, construction organization design and construction scheme, to ensure to complete the task of building construction in a reasonable period, conduct the quota construction contract according to the cost break-down.

② To strengthen project monitoring, give fully play to the supervisory role of supervision company, according to procedures strictly control engineering change, for the designed change, material change, work change, engineering outside of contract should be analyzed, put forward to reasonable cost treatment method.

③ The task of cost control in construction stage is to implement according to the designed requirements, so that to control the actual expenditure in the construction budget, to reduce designed changes, efforts to reduce costs, to conduct the settlement and final accounts after the project completion.

④ The basic principle of cost control in construction stage is put the cost plan as the cost control goal, in the construction process, to regularly compare the actual cost with the target value, by comparing the deviation of the two, put forward to the improvement measures to ensure the realization of the cost control objectives.

⑤ The main methods of cost control in the construction stage are as follows:

a. Fund use plan.

a) The premise of cost control is to set up the goals of cost control, the compilation of the plan for the use of funds and the cost of the target decomposition can be decomposed by the cost structure, by the sub-projects and the time.

b) By the cost of the decomposition of the use of funds plan, refers to the construction of civil engineering, equipment purchase costs, installation costs, other costs of construction to decompose and compile.

c) By the sub project decomposition of the funds use plan, due to the budgetary estimate and budget are mostly based on individual sub-projects and unit works of the project, the total cost of the project will be decomposed into individual sub-project and unit works.

d) By the time schedule to decompose the funds used plan.

b. Engineering measurement and payment

a) Engineering measurement refers to the provisions on the calculation of engineering quantity according to the designed documents and contract, the project supervision units check the declared amount of the engineering have been completed by contractor. Engineering quantity measured by supervision units are the certificate of payment to the contractor.

b) According to the contract, the main settlement of the project funds has the following kinds: monthly settlement, once the completion of the settlement, subsection settlement, other settlement methods agreed by the two parties.

c) Progress payments of the project are generally carried out according to the actual project completion of the month settlement (Payment procedures are shown in Figure 10-3), leaving a certain proportion of the final payment, after the completion of the project to handle the completion of the settlement and the liquidation.

d) After the project completion acceptance report is approved by the owner, the contractor should submit the completed settlement report and settlement data to the owner, both sides to carry out the project completion settlement. Supervision unit audits the final settlement report submitted by the contractor. After consultation with the owner and the contractor with consensus, issue the final settlement document and the final payment certificate. The engineering warranty is generally not less than 3% of the cost of the construction contract, which will be returned to the contractor after the expiration of the quality guarantee period.

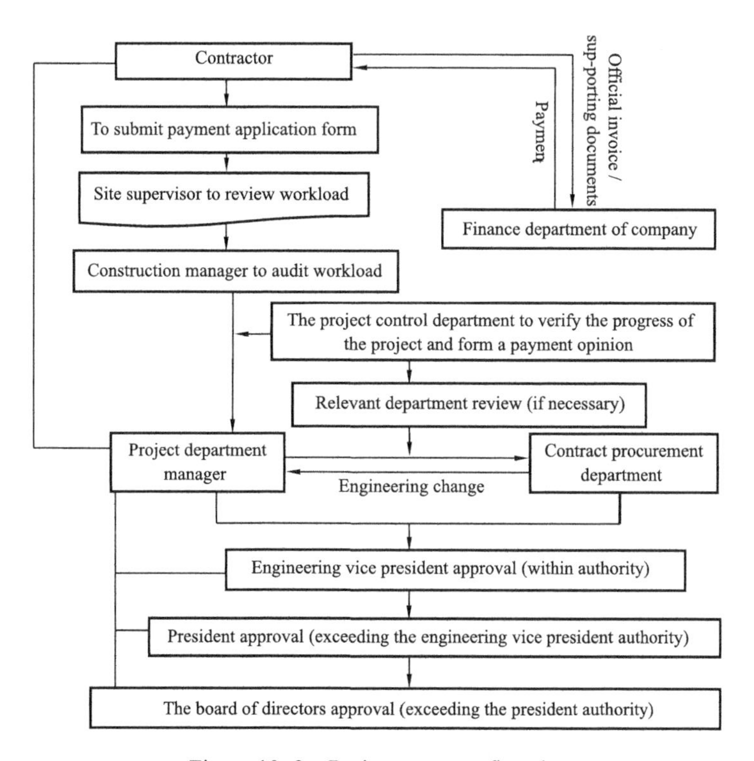

Figure 10-3 Project payment flowchart

(5) The financial settlement and final accounts stage.

① In strict accordance with the provisions of the procedures and contract, according to the designed documents and drawings, approved construction organization design and on-site visa to carry out the contractor's works of the settlement and final accounts;

② According to the company's financial system, to review payment support documents and invoices strictly;

③ To control the contractor's payment amount and schedule strictly, and supervise the progress payment paid;

④ To strengthen the engineering insurance and tax planning work, in a timely manner to deal with the claims of accidents, review and payment of taxes and fees;

⑤ To strengthen the special managements of margin and performance bond;

⑥ Based on the results of the project financial settlement and final accounts, the comparison and analysis, to provide financial support for the performance evaluation.

⑦ Summary table of the final accounts (Table 10–10).

Table 10–10 Summary of the Final Accounts of the LNG Project

Unit: 10^4RMB

Item	Approved budgetary estimate	Project final accounts	Savings (Budgetary estimate - Final accounts)	Saving rate (%)
1. Project cost				
1.1 Civil Engineering				
1.2 Equipment purchase				
1.3 Installation engineering				
2. Other expenses				
3. intangible assets				
4. Basic reserve				
5. Loan interest				
6. Basic working capital				
Total				

(6) Construction claims.

① Construction site conditions, climatic conditions, changes in the construction schedule and the terms of the contract, technical specifications, construction drawings and other factors will make the construction of the inevitable claims. The claim managements are important parts of the cost control in the construction phase.

② The construction claims are divided into two types: the claims made by the contractor to the owner and the claims made by the owner to the contractor, therefore, the owners should understand in response to the claims of the situation, so as not to be passive.

③ The main circumstances of the claim by the contractor to the owner:

a. Adverse natural conditions and man-made obstacles caused by claims;

b. Claims made by the owner or by the owner in charge of the engineering change;

c. Cost claim for extension of project;

d. Accelerated construction claims;

e. The owner's claim for wrongful termination of the project;

f. The delay payment claims;

g. Other, such as the contract, the provisions of the contractor may rise in price, the relevant laws and regulations to the owners' claims.

10.6.2.3 The Compilation of LNG Project Budget Execution Report and Budget Execution Table

(1) LNG Project budget execution report.

LNG project budget execution report is an important document for project budget monitoring, according to the provisions of the management of the project company, it needs to prepare the project budget performance report regularly or irregularly. The report is based on the actual operation of the project budget implementation, reflecting the cost management work, according to the statistics and analysis of the implementation of the pre-budget summary, the late budget implementation trends forecast, key analysis of major budget changes, the next step is to improve the measures, the purpose is to strive to budget control within the budget. The following introducing the main contents of the project budget implementation report, sometimes including the determination of the weight of every sub-project and the performance evaluation of the evaluation content.

① The summary of the LNG project stage budget performance;

② Stage project management and other cost analysis and statistics;

③ Analysis of the integrated weight coefficient of each sub project and its investment cost;

④ Analysis of resource investment in stage sub-projects;

⑤ Analysis on the assumption and restriction of resource input;

⑥ Prediction, analysis and statistics of the work package resources of the sub-projects;

⑦ Stage project total investment cost forecast, analogy, analysis and summary table;

⑧ Stage project resources input of time and cost prediction, analysis and ranking;

⑨ Stage major budget change analysis and countermeasures;

⑩ The performance evaluation analysis and the next stage of the project budget forecast.

(2) Preparation of budget performance report and tables.

① Construction project investment budget execution table (Table 10-11). Corresponding to the construction project investment budget, it is necessary to count the difference between the current budget and the actual amount budget of the receiving terminal, jetty and gas transmission trunk line, analyze the difference between the total budget and the cumulative amount of the actual occurrence.

② The project installation investment budget execution table (Table 10-12). Corresponding to the installation project investment budget, it needs to count the difference between the total budget and the cumulative amount of the receiving terminal, jetty, gas transmission line and other installation process, to analyze the difference between the total budget and the cumulative amount of the actual occurrence.

Table 10-11 Construction Investment Budget Implementation Status Table

Compilation date:														Unit: 10^4RMB
No.	Items	Current budget		Actual budget for the current period		Current difference	Cumulative budget			Cumulative actual budget		Cumulative difference		Difference analysis
		Amount of money	among USD	Amount of money	among USD	Amount of money	%	Amount of money	among USD	Amount of money	among USD	Amount of money	%	
1	Land acquisition cost													
1.1	Lands													
1.2	……													
	Total													

Table 10-12 Installation Investment Budget Implementation Status Table

Compilation date:														Unit: 10^4RMB
No.	Items	Current budget		Actual budget for the current period		Current difference	Cumulative budget			Cumulative actual budget		Cumulative difference		Difference analysis
		Amount of money	among USD	Amount of money	among USD	Amount of money	%	Amount of money	among USD	Amount of money	among USD	Amount of money	%	
1	Receiving terminal													
1.1	Trestle													
1.2	Storage tank													
	……													
	Total													

③ Management expense budget execution table (Table 10–13). Corresponding to the administrative expense budget, it is needs to count the difference between the total budget and the cumulative amount of statistical items for every department of the company staff wages and benefits, insurance, office, travel, reception and transportation costs, the analysis of the difference between the total budget and the cumulative amount of the actual occurrence.

④ Equipment investment budget execution table (Table 10–14). Corresponding to the equipment investment budget table, including the difference between the current budget and the actual amount of the current period, the difference between the total budget and the cumulative amount of the actual occurrence.

⑤ Office fixed assets budget execution status table (Table 10–15).Corresponding to the office fixed assets budget table, including the difference between the current budget and the current actual amount, the analysis of the difference between the total budget and the cumulative amount of the actual occurrence.

10.6.3 Performance Analysis of Project Cost Control

10.6.3.1 Performance Evaluation Method

(1) Definition of performance evaluation.

The performance evaluation analysis is an important method for the project cost control, the basic idea of this method is to compare the budgeted cost of work completed, the planned budget and the actual cost of the work have been completed, to find out the deviation and the deviation magnitude, in order to analyze the project cost change trends and make scientific prediction and judgment.

(2) The relevant definitions.

① Planned Value (PV) refers to the budget costs for the planned time points before the planned completion of the task.

② Earned Value (EV) refers to the budgeted cost for the actual completion of the planned tasks within the given time period.

Is the budgeted cost of the actual completion of the planned tasks within the given time period.

③ Actual Cost (AC) is the total actual cost of completion of the planned tasks within the given time period. It is necessary to note that the AC must be relative to PV and EV in terms of definition and scope of content (such as only the direct number of days, including direct costs, or indirect costs, or all costs, etc.).

(3) Two difference variables.

① Cost Variance (CV), calculating formula:

$$CV = EV - AC \tag{10–2}$$

When the project is completed, the cost deviation is equal to the difference between the budget and the actual cost. It is clear that the CV value is bigger than 0 for the project cost savings, less than 0 for the project cost overrun.

Table 10-13 Management Expense Budget Implementation Status Table

Compilation date:														Unit: 10^4RMB
No.	Items	Current budget		Actual budget for the current period		Current difference	Cumulative budget			Cumulative actual budget		Cumulative difference		Difference analysis
		Amount of money	among USD	Amount of money	among USD	Amount of money	%	Amount of money	among USD	Amount of money	among USD	Amount of money	%	
1.	Salary and bonus													
2.	Employee welfare													
3.	……													
	Total													

Table 10-14 Equipment Investment Budget Execution Status Table

Compilation date:														Unit: 10^4RMB
No.	Items	Current budget		Actual budget for the current period		Current difference	Cumulative budget			Cumulative actual budget		Cumulative difference		Difference analysis
		Amount of money	among USD	Amount of money	among USD	Amount of money	%	Amount of money	among USD	Amount of money	among USD	Amount of money	%	
1.	According to equipment classification and name													
2.														
3.	……													
	Total													

Table 10-15 Office Fixed Assets Investment Budget Execution Status Table

Compilation date:														Unit: 10^4RMB
No.	Items	Current budget		Actual budget for the current period		Current difference	Cumulative budget			Cumulative actual budget		Cumulative difference		Difference analysis
		Amount of money	among USD	Amount of money	among USD	Amount of money	%	Amount of money	among USD	Amount of money	among USD	Amount of money	%	
1.	Office													
1.1	Computer													
3.	……													
	Total													

② Schedule Variance (SV), calculating formula:

$$SV = EV - PV \quad (10\text{-}3)$$

When the project is completed, all the planned value has been realized, so the difference between the two is 0. It is necessary to pay attention to is that instead of using the time unit to measure the schedule deviation, but using the monetary units to measure the schedule deviation. It is clear that the SV value is bigger than 0 for the project ahead of schedule, less than 0 for the project behind schedule.

(4) The two index variables.

① Cost Performance Index (CPI), calculating formula:

$$CPI = EV / AC \quad (10\text{-}4)$$

If CPI value is less than 1, it indicates that the cost is beyond the budget, if bigger than 1, said the cost is less than the budget.

② Schedule Performance Index (SPI), calculating formula:

$$SPI = EV / PV \quad (10\text{-}5)$$

If the SPI value is less than 1, it indicates that the project is behind schedule, and if it is bigger than 1, it indicates that the project is ahead of schedule.

10.6.3.2 Cost Forecast Method

The project cost forecast is based on the information and knowledge that have been mastered in the course of the project operation, to estimate and forecast the future situation. But to analyze the factors that cause the deviation is to continue to play a role or by chance, in the future whether these factors continue to exist, or no longer happen. According to the deviation factor analysis, the different calculating methods are given.

(1) The relevant definitions.

① Estimate To Complete (ETC): expected costs for the remaining work to be done.

② Estimate At Completion (EAC): the estimated total cost when the task is to be completed within the prescribed scope of work.

(2) The calculation of ETC.

① Calculation of atypical deviation based on ETC, if the current deviation is looked at is atypical, and future predictions will not have a similar bias, can take the following formula:

$$ETC = BAC - EV^{C} \quad (10\text{-}6)$$

Where: BAC for the completion of the budget, EV^{C} for the current cumulative realized value.

② Calculation of typical deviation based on ETC, if the current deviation is looked at can represent the future deviation, can take the following formula:

$$ETC = \left(BAC - EV^{C}\right) / CPI^{C} \quad (10\text{-}7)$$

Where: CPI^{C} for cumulative cost performance index.

(3) The calculation of EAC.

① EAC calculation, on late deviation influencing factors change. If the deviation of the past influence factor's consciousness is not in place, or the late deviation influence factors changed, the following formula can be used:

$$EAC = AC^{C} - ETC^{N} \qquad (10\text{–}8)$$

Where: AC^C for the current actual costs, ETC^N for the project company to provide the new ETC value.

② EAC calculation, on similar deviation of influencing factors not changed.

EAC calculation of similar deviation in the later period, if the current deviation is atypical, and the prediction will no longer occur in the future, the EAC may take the following formula:

$$EAC = AC^{C} - BAC - EV^{C} \qquad (10\text{–}9)$$

③ CPI calculation using cumulative EAC. The following formula can be adopted to predict the current deviation, which can represent the typical deviation of the future deviation:

$$EAC = AC^{C} + \left(BAC - EV^{C}\right) / CPI^{C} \qquad (10\text{–}10)$$

10.6.3.3 LNG Sub Project Cost Schedule Comprehensive Weighted Assessment Method

LNG project is a combination of super large project with multiple sub-projects, generally includes LNG storage tanks, receiving terminal, jetty, pipeline and several supporting sub-projects. How to bring sub-project cost, schedule management and control into a project management is a systematic project evaluation problem. The following describes the Fujian LNG project in practice management in an attempt to provide readers with reference.

(1) The parameters selection.

① Direct project cost (C_i) of every sub-project. It Includes engineering equipment and materials and construction costs, to a certain extent, reflects the degree of difficulty of the project, but there are exceptions, such as large-scale equipment and installation, investment is large, but maybe easier than complex construction projects.

② Each sub project construction period (T_i). Construction period will be one of the parameters, to a certain extent, reflects the degree of difficulty of the project, also has the exceptions, such as some pipeline projects, the real time and energy consuming is not the construction but land acquisition.

③ Coefficient of brainstorming (E_i). Based on the engineering cost and duration parameters of the sub project, the experts use Delphi method to take other difficult factors into comprehensive consideration, finally score the degree of difficulty of the sub project. Certainly, the scores including the existence of human experience and judgment factors.

(2) Coefficient of choice (Table 10–16).

① $D_i\%$ is the percentage of the sum of each sub project costs, computation formula:

$$D_i\% = 100\, C_i \sum_{i=1}^{m} C_i \qquad (10\text{–}11)$$

② $U_i\%$ is the percentage of the sum of the construction duration, computation formula:

$$U_i\% = 100\, T_i \sum_{i=1}^{m} T_i \qquad (10\text{–}12)$$

③ Weight factor (F_i). Weight factor is defined as the product of the percentage of the sub-project cost and the percentage of the time duration.

$$F_i = D_i\% * U_i\% \qquad (10\text{–}13)$$

④ G_i % is the percentage of weight factor F_i for each sub-project, computation formula:

$$G_i\% = 100\, F_i \sum_{i=1}^{m} F_i \qquad (10\text{–}14)$$

⑤ Weight coefficient (Y_i). The weight coefficient is defined as the product of brain storm coefficient and weight factor.

$$Y_i - E_i * U_i\% \qquad (10\text{–}15)$$

⑥ Weighted coefficient W_i (%). It needs to define the ratio of the weight coefficient to the total weight coefficient.

$$W_i = 100\%\, Y_i \sum_{i=1}^{m} Y_i \qquad (10\text{–}16)$$

Table 10–16 is numerical values calculated by the above formula, the last line is the weight of each sub project coefficient, the size of the weight coefficient reflects the weight and importance on the sub project engineering in LNG project. From the comprehensive weight coefficient, the weight coefficient of LNG tank, receiving terminal, pipeline engineering is 32%, 27%, 26%, respectively, the three sub-projects play decisive roles in the whole LNG project, is the highlight of the project implementation process management. We can also use the same method, for every sub-project to carry out its own comprehensive cost schedule weight coefficient analysis.

Table 10–16 Fujian LNG Sub–projects Comprehensive Weight Coefficient

	Storage Tank		Receiving Terminal		Jetty		Pipeline		Auxiliary eng.	
Direct expense (million) C_i (%)	C1	D1%	C2	D2%	C3	D3%	C4	D4%	C5	D5%
	892.91	15.6	2085.38	36.5	444.58	7.8	2148.72	37.6	140.00	2.5
Coefficient of brain-storming E_i	280		100		250		110		250	
Duration (days) T_i	T1	U1%	T2	U2%	T3	U3%	T4	U4%	T5	U5%
	1096	23.0	1096	23.0	944	19.8	944	19.8	692	14.5

Continued

	Storage Tank		Receiving Terminal		Jetty		Pipeline		Auxiliary eng.	
Weight factor F_i ($D_i\%*U_i\%$)	F1	G1%	F2	G2%	F3	G3%	F4	G4%	F5	G5%
	359.1	16.8	838.6	39.3	154.0	7.2	744.2	34.9	35.5	1.7
Weight coefficient Y_i ($E_i*G_i\%$)	Y1	W1%	Y2	W2%	Y3	W3%	Y4	W4%	Y5	W5%
	4717.0	32.1	3934.4	26.7	1806.1	12.3	3840.9	26.1	416.9	2.8
The weighted coefficient of weights $W_i\%$	32		27		12		26		3	

10.6.3.4 Cost Schedule Diagram Analysis

Cost schedule diagram is one of the main methods to evaluate the effect of schedule control costs. The diagram puts the curve of the planned value (base line), realized value and actual cost into a picture, so we can analyze the relationship of these three, and also draw the actual cost forecast (Figure 10–4), in order to control the cost by the cost control department and project manager. Figure 10–3 is the implementation of a LNG project simulated, by the end of the PV line gives the total cost of 8 billion RMB, at the time of the control point 50, but also the planned value is 50%, that is in accordance with the plan to spend 4 billion RMB, but the actual cost of 3.6 billion RMB, but schedule (monetary unit, rather than time units) have reached 60, that is equivalent to 75% of the task has been completed. Through the three curves can be seen, the current cost of this LNG project and the schedule deviation are positive, the implementation is good, according to the existing situation forecast, the entire cost maybe 6.7 billion RMB, which meaning budget savings.

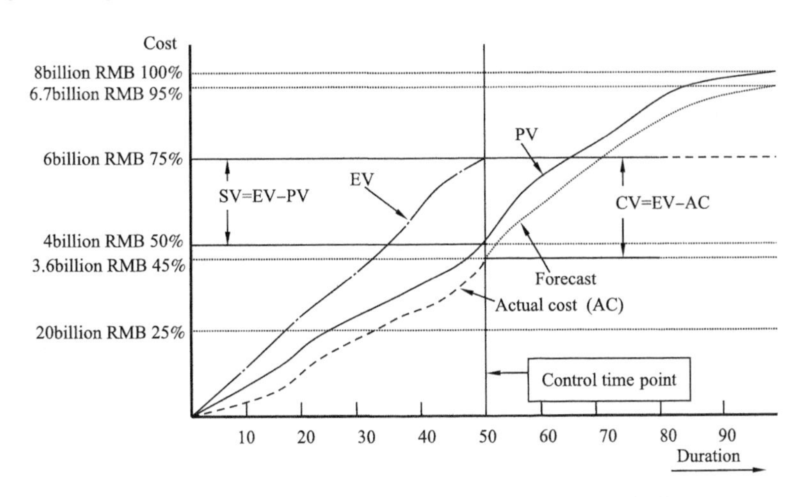

Figure 10–4 LNG project cost schedule diagram analysis

10.6.3.5 Other Methods

Cost control in addition to the above methods, there are policy correlation methods, target task correlation methods, data statistics analysis methods, survey methods, expert experience methods, case analysis methods and results of back-stepping methods, etc.. According to the project condition, cost management personnel can flexibly apply those or their combination to the project.

10.6.4 LNG Project Cost Control Management Results

10.6.4.1 To improve the Project Management System

(1) Organizational system.

From the budget management organizational system, the project company sets up the budget control committee, the control department and the finance department for daily management. In the actual management process, the committee's work functions and management procedures maybe adjusted, especially the budget management position setting and function of the control and finance department will be updated, with the full implementation of sub-project, more personal should be added in time. In the late period of the project, we should pay attention to explore three kinds of sub projects: jetty, receiving terminal and pipeline in different management mode (EPC and owner management) of sub project quantity and budget management personnel's matching relationship, give the quantitative relationship, but also to explore the owners, contractors and supervision company in the project management mode of communication mechanism and process.

(2) Institutional system.

The cost management system should be revised and supplemented by the project operation to verify and revise the cost resource planning process, cost estimation process, cost budget process and annual (monthly) budget plan. By backtracking methods, to formulate a more reasonable resource plan for the project scale, which can provide reference for the investment project later, and with the different stages of the project operation, to select the more appropriate time for preparation of the cost estimation and cost budget with budget results accuracy. For the preparation of the cost plan, it needs to complete and improve all kinds of inspection forms, so that the cost control based on more solid data statistics.

10.6.4.2 Benchmark of Preparation for the Budget

(1) Engineering quantity standard.

Includeing jetty, receiving terminal and pipeline engineering quantity statistics, summed up the different natures of the project, such as channel dredging engineering quantity and mechanical equipment type, input human matching relationship; in storage tank construction, relationship between working surface size and construction machinery, personnel density; the relationship between work package assignment rationality and sub project, etc.

(2) Equipment and materials.

The equipment and materials in LNG projects with the same annual receiving capacity are roughly the same, therefore, the statistics for the three sub-projects, jetty, receiving terminal

and pipeline equipment and materials required, will be a guiding role for the new project, for example, the $16 \times 10^4 m^3$ of the tank which require amount of 9% nickel steel, steel reinforcing bar, special cement and so on, it is valuable reference to the purchase of storage tank materials of similar scale.

(3) Human Resources.

Different types of projects, there will be a reasonable personnel matching with front-line workers and management personnel for contractors, supervisor personnel and owner's management personnel, such as pipeline construction tenders, each section of the personnel with appropriate number will provide valuable information for owner's management.

(4) Time.

Through the operation of the project, the construction of the different nature of the project will be explored a set of reasonable time arrangement. Firstly, the duration of the arrangements, such as jetty sub-project, there are two kinds of construction: off-shore engineering and onshore engineering, according to seasonal climate, construction time should be arranged in reasonable period, in order to ensure the two engineering of mutual non-interference and mutual cohesion. Secondly, the time window arrangement, such as the South China rainy in spring and summer, according to the engineering category to choose the best time for construction, especially in the typhoon season, we should make good prevention and avoidance.

(5) Cost baseline.

Through cost control process management, we can get a new round of cost estimation, including jetty like wharf and trestle, the storage tanks, vaporizers, pipes cut-off valve chamber, equipment, materials, direct costs and indirect management costs in gas station construction process.

10.6.4.3 Cost Control Effective Methods

(1) The best combination methods.

General cost control work often require a combination of multiple methods to be used in order to verify each other, such as the evaluation of the storage tank project cost control, can take the performance evaluation analysis method and cost schedule analysis, considering both cost and schedule and only consider the cost will produce an illusion.

(2) Stage cost control.

The construction project is completed step by step, the cost will spend by cumulative workload of several stages, so the cost control should be carried out at the expense of the stage plan, the cost and the work plan is linked, In the decision-making stage, a comprehensive analysis is needed, in order to determine the total cost, then control the design, procurement, construction, acceptance and other phase plan, corresponding to the cost plan, but not random overrun.

(3) To create cost database.

To a series of data such as single equipment, materials, labor costs, financing interest, insurance and management costs obtained from the operation of Chinese LNG projects. It creates

China's own LNG project cost database, which will be a valuable asset for the project company, after new project's resource input time, assumptions and constraints and data correction, the new LNG project cost database can be obtained.

10.6.4.4 Experience and Lessons

After the completion of the project, generally the summary of the project will be carried out, to form the completion report, the experience and lessons will be the main contents.

(1) Experiences.

The completion of the project, including project leadership effort and sweat, but also reflect all personnel involved in the project experience and wisdom, the experiences are mainly reflected in the technical details, construction standards; in the project management process control, the system arrangement, its experience will provide the valuable reference for the new project.

(2) Lessons.

The project company should conscientiously sum up lessons before the project is put into normal production after completion and acceptance .The experience is certainly a reference for new project, but sometimes the lessons are more honorable than the experience, because it can warn the late comers to go out of the new path of innovation from the lessons of the formers.